Vesta and Ceres: Insights from the Dawn Mission for the Origin of the Solar System

The NASA Dawn mission, launched in 2007, aimed to visit two of the most massive protoplanets of the main asteroid belt: Vesta and Ceres. The aim was to further our understanding of the earliest days of the Solar System, and compare the two bodies to better understand their formation and evolution. This book summarizes state-of-the-art results from the mission, and discusses the implications for our understanding not only of the asteroid belt but of the entire Solar System. It comprises of three parts: Part I provides an overview of the Main Belt asteroids and provides an introduction to the Dawn mission; Part II presents key findings from the mission; and Part III discusses how these findings provide insights into the formation and evolution of the Solar System. This is a definitive reference for academic researchers and professionals of planetary science, asteroid science, and space exploration.

SIMONE MARCHI is a staff scientist at the Southwest Research Institute in Boulder, Colorado. He has been involved in several space missions and is the deputy project scientist for the NASA Lucy mission, co-investigator for the NASA Psyche mission, co-investigator for instruments on board ESA BepiColombo and JUICE missions. He was co-investigator for the Dawn mission and associate scientist for instruments on board ESA Rosetta. He has won multiple awards including the Paolo Farinella Prize (2017), NASA's Susan Mahan Neibur Early Career Award (2014), and several Group Achievement Awards from NASA and ESA.

CAROL A. RAYMOND is a principal scientist at Caltech's Jet Propulsion Lab. She has been involved in mission and instrument leadership for NASA and program science support at JPL. She was the deputy principal investigator on the NASA Dawn Discovery Mission and principal investigator of the Europa Magnetometer System (ICEMAG) for the NASA Europa Clipper mission. Her research focuses on geophysics and interiors of small bodies and moons, and magnetic fields in the Solar System. She has received multiple awards including three NASA Exceptional Public Achievement Medals (2013, 2016, 2019), the Shoemaker Award from the American Geophysical Union (2018), and the National Aeronautic Association Collier Award to the Dawn Flight Team (2015).

CHRISTOPHER T. RUSSELL is a distinguished professor in the Earth, Planetary and Space Sciences department at the University of California, Los Angeles. He was the principal investigator of NASA's Dawn ion-propelled mission to Vesta and Ceres. He has been awarded the American Geophysical Union Macelwane Award and Fleming medal, COSPAR's Space Science Award, NASA's Exceptional Scientific Achievement Medal (2012) and Distinguished Public Service Medal (2017), and the National Aeronautic Association Collier Award to the Dawn Flight Team (2015). He has built and operated instruments on many space missions.

Cambridge Planetary Science

Series Editors:

Fran Bagenal, David Jewitt, Carl Murray, Jim Bell, Ralph Lorenz,
Francis Nimmo, Sara Russell

Books in the Series:

† Reissued as a paperback

VESTA AND CERES

Insights from the Dawn Mission for the Origin of the Solar System

Edited by

SIMONE MARCHI

Southwest Research Institute, Boulder, Colorado

CAROL A. RAYMOND

California Institute of Technology

CHRISTOPHER T. RUSSELL

University of California, Los Angeles

CAMBRIDGE
UNIVERSITY PRESS

CAMBRIDGE
UNIVERSITY PRESS

University Printing House, Cambridge CB2 8BS, United Kingdom

One Liberty Plaza, 20th Floor, New York, NY 10006, USA

477 Williamstown Road, Port Melbourne, VIC 3207, Australia

314–321, 3rd Floor, Plot 3, Splendor Forum, Jasola District Centre, New Delhi – 110025, India

103 Penang Road, #05–06/07, Visioncrest Commercial, Singapore 238467

Cambridge University Press is part of the University of Cambridge.

It furthers the University's mission by disseminating knowledge in the pursuit of education, learning, and research at the highest international levels of excellence.

www.cambridge.org
Information on this title: www.cambridge.org/9781108479738
DOI: 10.1017/9781108856324

First published 2022

Printed in the United Kingdom by TJ Books Limited, Padstow Cornwall

A catalogue record for this publication is available from the British Library.

ISBN 978-1-108-47973-8 Hardback

CONTENTS

The plate section is to be found between pages 118 and 119

CONTRIBUTORS

ELEONORA AMMANNITO
Italian Space Agency, Italy

KATHERINE R. BERMINGHAM
Rutgers University, USA

RICHARD P. BINZEL
Massachusetts Institute of Technology, USA

PHILIP BLAND
Curtin University, Australia

WILLIAM F. BOTTKE
Southwest Research Institute, USA

DORIS BREUER
German Aerospace Center, Germany

DEBRA L. BUCZKOWSKI
Johns Hopkins Applied Physics Laboratory, USA

JULIE CASTILLO-ROGEZ
Jet Propulsion Laboratory, California Institute of Technology, USA

JEAN-PHILIPPE COMBE
Bear Fight Institute, USA

MARIA CRISTINA DE SANCTIS
National Institute for Astrophysics, Italy

MARCO DELBO
Côte d'Azur University, France

BETHANY EHLMANN
California Institute of Technology, USA

ANTON I. ERMAKOV
University of California, USA

KONSTANTIN GERBIG
Max Planck Institute for Astronomy, Germany, and
Yale University, USA

SUNAO HASEGAWA
Japan Aerospace Exploration Agency, Japan

RALF JAUMANN
Free University of Berlin, Germany

MARTIN JUTZI
University of Bern, Switzerland

HUBERT KLAHR
Max Planck Institute for Astronomy, Germany

THOMAS S. KRUIJER
Lawrence Livermore National Laboratory, USA, Leibniz
Institute for Evolution and Biodiversity Science, Germany,
and Free University of Berlin, Germany

SIMONE MARCHI
Southwest Research Institute, USA

HARRY Y. MCSWEEN JR.
University of Tennessee, USA

ANDREAS NATHUES
Max Planck Institute for Solar System Research, Germany

DAVID NESVORNÝ
Southwest Research Institute, USA

THOMAS PRETTYMAN
Planetary Science Institute, USA

ANDREA RAPONI
National Institute for Astrophysics, Italy

MARC D. RAYMAN
Jet Propulsion Laboratory, California Institute of Technology,
USA

CAROL A. RAYMOND
Jet Propulsion Laboratory, California Institute of Technology,
USA

SEAN N. RAYMOND
University of Bordeaux, France

CHRISTOPHER T. RUSSELL
University of California, USA

JENNIFER E. C. SCULLY
Jet Propulsion Laboratory, California Institute of Technology, USA

MICHAEL J. TOPLIS
University of Toulouse, France

FUMIHIKO USUI
Kobe University and Japan Aerospace Exploration Agency, Japan

PIERRE VERNAZZA
Aix-Marseille University and Astrophysics Laboratory of Marseille, France

DAVID A. WILLIAMS
Arizona State University, USA

NAOYUKI YAMASHITA
Planetary Science Institute, USA

PREFACE

The NASA Dawn spacecraft took off from Cape Canaveral in September 2007 atop a Delta II rocket starting an ambitious journey to Vesta and Ceres, the two most massive worlds in the largest reservoir of asteroids in the Solar System, the Main Belt. The mission's name was chosen to testify to a bold intent: execute a journey in space and time toward the dawn of the Solar System.

Prior to the Dawn launch, Earth-bound observations of Vesta and Ceres revealed intriguing features – from Vesta's rugged shape to Ceres' tenuous water exosphere – but these objects remained fuzzy speckles of light, even through the lenses of the most powerful telescopes. What we knew, however, was enough to justify a dedicated space mission to explore Vesta and Ceres. Yet our pre-existing knowledge turned out to be only the tip of the iceberg of what Dawn discovered.

Since 2007, the science of Main Belt asteroids has considerably evolved. Remote telescopic observations have increased in spatial resolution, leading to detailed views of large Main Belt asteroids, including Vesta and Ceres. These new developments, along with an historical recount of how Dawn came to be, are discussed in Part I of the book.

With Dawn's exploration of Vesta (2011–2012) and Ceres (2015–2018), these two worlds came into focus. Breathtaking details emerged of how large collisions sculpted Vesta, liberating massive amounts of material in the inner Main Belt and providing the source of an important family of meteorites recovered on Earth. Ceres' complex geology, which may rival that of the Earth and Mars, showed recent cryovolcanic activity. Part II of this book is dedicated to these highlights and many more discoveries of the Dawn mission. The outstanding collection of papers presented in this part, led by members of the Dawn Science Team, follows a strong tradition of books showcasing the findings of space missions. But the present book is more than that.

By the time Dawn completed its mission in 2018, our understanding of the formation of the Solar System had greatly evolved thanks to new theoretical models and to a new trove of meteorite geochemical data, and Dawn observations of Vesta and Ceres provide new, vital constraints to synergistically interpret models and data. The broader implications of the Dawn legacy are discussed by prominent scientists from various disciplines in Part III of this book.

The idea for this book was born in the aftermath of the international workshop "The Main Belt: A Gateway to the Formation and Early Evolution of the Solar System" held in Sardinia, Italy, 2019 June 4–7. Many of the chapters' authors attended the meeting and embraced lively discussions about Dawn, Vesta, and Ceres, with the Mediterranean Sea in the background, less than 300 miles away from Palermo Observatory, where Ceres was discovered in 1801.

The editors hope this book will serve as a solid reference for the younger generations as well as for more seasoned researchers to successfully pursue future exploration of the Main Belt. We certainly have learned a lot thanks to Dawn, and yet we know that we have barely scratched the surface of what Main Belt asteroids can tell us about the dawn of our Solar System.

Part I

REMOTE OBSERVATIONS AND EXPLORATION OF MAIN BELT ASTEROIDS

Remote Observations of the Main Belt

PIERRE VERNAZZA, FUMIHIKO USUI, AND SUNAO HASEGAWA

1.1 INTRODUCTION

It might be no exaggeration to say that the history of asteroid science has always been driven by the history of research on the largest Main Belt asteroids (with diameters greater than ~100 km), with Ceres and Vesta being the most studied bodies. Because of their brightness and scientific interest due to their large size (between that of the most common asteroids and planets), the largest asteroids have always been privileged targets for every new generation telescope and/or instrument. Today, the asteroid belt contains ~230 such bodies (see Table 1.1 for a complete list of D ≳ 200 km bodies and Table 1.2 for a complete list of D ≳ 100 km bodies). Spectrophotometric observations have been carried out for all of these bodies in the visible and/or near-infrared range. Among the 25 spectral types defined within the Bus-DeMeo asteroid taxonomy based on principal components analysis of combined visible and near-infrared spectral data spanning wavelengths from 0.45 to 2.45 μm for nearly 400 asteroids (DeMeo et al., 2009), only O-, Q-, and Xn-type asteroids are absent among D > 100 km bodies. C-complex asteroids (B, C, Cb, Cg, Cgh, Ch), S-complex asteroids (Q, S, Sa, Sq, Sr, Sv), P/D type asteroids (low albedo X, T, and D-types), and the remaining types (1 A-type, 4 K-type, 2 L-type, 1 R-type, 1 V-type, 1 high albedo X-type, 2 Xe-type, 9 Xc-type, 15 Xk-type) represent, respectively, 61%, 10%, 13%, and 16% of all D > 100 km asteroids.

Early photometric observations of these bodies were key in establishing the existence of a compositional gradient in the asteroid belt (Gradie & Tedesco, 1982), with S-types being located on average closer to the Sun than C-types and P-/D-types being the farthest from the Sun. On the basis of these observations, scenarios regarding the formation and dynamical evolution of the asteroid belt and that of the Solar System in general have been formulated (e.g., Gomes et al., 2005; Morbidelli et al., 2005; Tsiganis et al., 2005; Bottke et al., 2006; Levison et al., 2009; Walsh et al., 2011; Raymond & Izidoro, 2017). These dynamical models suggest that today's asteroid belt may not only host objects that formed in situ, typically between 2.2 and 3.3 AU, but also bodies that were formed in the terrestrial planet region (Xe- and possibly S-types), in the giant planet region (Ch/Cgh- and B/C-types) as well as beyond Neptune (P/D-types). In a broad stroke, the idea that the asteroid belt is a condensed version of the primordial Solar System has

progressively emerged. Notably, these observations along with those of more distant small bodies (giant planet trojans, trans-Neptunian objects) were instrumental in imposing giant planet migrations as a main step in the dynamical evolution of our Solar System. On the basis of these datasets, the idea of a static Solar System history has dramatically shifted to one of dynamic change and mixing (DeMeo & Carry, 2014). See Part III of this book for more details on this topic.

The study of the largest Main Belt asteroids is not only important because of the clues it delivers regarding the formation and evolution of the belt itself but also because many of these bodies are likely "primordial" remnants of the early Solar System (Morbidelli et al., 2009), that is their internal structure has likely remained intact since their formation (they can be seen as the smallest protoplanets). Many of these bodies thus offer, similarly to Ceres and Vesta detailed in the present book, invaluable constraints regarding the processes of protoplanet formation over a wide range of heliocentric distances (assuming that the aforementioned migration theories are correct).

In the present chapter, we review the current knowledge regarding large (D ≳ 100 km) Main Belt asteroids derived from Earth-based spectroscopic and imaging observations with an emphasis on D > 200 km bodies including Ceres and Vesta. Our motivation is to provide a meaningful context for the two largest Main Belt asteroids visited by the Dawn mission (see Chapter 2) and to guide future in-situ investigations to the largest asteroids – that's why small (D < 100 km) asteroids, which are essentially the leftover fragments of catastrophic collisions, are not discussed here.

1.2 SPECTROSCOPIC OBSERVATIONS OF LARGE MAIN BELT ASTEROIDS

Detailed reviews concerning the compositional interpretation of asteroid taxonomic types and their distribution across the Main Belt can be found in Burbine (2014, 2016), DeMeo et al. (2015), Reddy et al. (2015), Vernazza et al. (2015b), Vernazza and Beck (2017), and Greenwood et al. (2020) and will not be repeated with the same level of detail in this chapter. Rather, we put the emphasis on the currently proposed connections between the various compositional classes present among the largest Main Belt asteroids and the two main classes of extra-terrestrial materials, namely meteorites and interplanetary dust particles (hereafter IDPs). The two largest asteroids, Vesta and Ceres, "heroes" of the present book, illustrate well the Main Belt paradox: some asteroids appear well sampled by meteorites (Vesta) whereas others don't (Ceres). IDPs may be more appropriate analogues

FU was supported by JSPS KAKENHI: grant nos. JP19H00725, JP20H00188, and JP20K04055. SH was supported by JSPS KAKENHI (grant nos. JP15K05277, JP17K05636, JP18K03723, JP19H00719, and JP20K04055) and by the Hypervelocity Impact Facility (former facility name: the Space Plasma Laboratory), ISAS, JAXA.

Table 1.1 *Volume equivalent diameter (Deq), geometric albedo, spectral type following the Bus-DeMeo taxonomy, semi-major axis (a), eccentricity (e), and inclination (i) for the largest (D > 200 km) Main Belt asteroids listed according to decreasing values of their size.*

The albedo and diameter values represent the averages of the values reported in Tedesco et al. (2002), Usui et al. (2011), and Masiero et al. (2011, 2014). Spectral types were retrieved from Bus and Binzel (2002), Lazzaro et al. (2004), and DeMeo et al. (2009)

Object	Deq (km)	Geom. alb.	Spectral type	a (AU)	e	i (deg)
1 Ceres	939.4	0.087	C	2.77	0.08	10.59
4 Vesta	525.4	0.404	V	2.36	0.09	7.14
2 Pallas	513.0	0.140	B	2.77	0.23	34.83
10 Hygiea	434.0	0.059	C	3.14	0.11	3.83
704 Interamnia	332.0	0.044	Cb	3.06	0.16	17.31
52 Europa	314.0	0.049	C	3.09	0.11	7.48
511 Davida	303.2	0.064	C	3.16	0.19	15.94
65 Cybele	286.3	0.049	Xk	3.42	0.11	3.56
87 Sylvia	280.0	0.043	X	3.48	0.09	10.88
31 Euphrosyne	268.0	0.047	Cb	3.16	0.22	26.28
15 Eunomia	267.3	0.196	S	2.64	0.19	11.75
107 Camilla	254.0	0.040	X	3.49	0.07	10.00
3 Juno	249.0	0.210	Sq	2.67	0.26	12.99
451 Patientia	242.9	0.069	C	3.06	0.07	15.24
324 Bamberga	232.9	0.061	C	2.68	0.34	11.10
16 Psyche	222.0	0.156	Xk	2.92	0.13	3.10
48 Doris	220.8	0.064	Ch	3.11	0.07	6.55
88 Thisbe	212.0	0.052	C	2.77	0.16	5.21
423 Diotima	211.2	0.053	C	3.07	0.04	11.24
19 Fortuna	211.0	0.048	Ch	2.44	0.16	1.57
13 Egeria	205.0	0.080	Ch	2.58	0.09	16.54
7 Iris	204.0	0.244	S	2.39	0.23	5.52
29 Amphitrite	204.0	0.185	S	2.55	0.07	6.08

for these bodies. We first start by summarizing the results of Earth-based spectroscopic campaigns devoted to constrain the surface composition of Ceres and Vesta and then continue by summarizing current knowledge regarding the surface composition of D > 100 km asteroids.

1.2.1 Focus on Earth-based Spectroscopic Observations of Ceres and Vesta and Comparison to Dawn Measurements

1.2.1.1 (1) Ceres

In the 1970s, Ceres was identified as a carbonaceous chondrite-like asteroid based on a low albedo (∼0.05–0.06) (Veverka, 1970; Matson, 1971; Bowell & Zellner, 1973) and a relatively flat reflectance from 0.5 to 2.5 μm (Chapman et al., 1973, 1975; Johnson & Fanale, 1973; Johnson et al., 1975). A few years later, a ∼3.1 μm absorption feature was discovered in its spectrum (Lebofsky, 1978; Lebofsky et al., 1981) and was interpreted as indicative of the presence of hydrated clay minerals similar to

those present in carbonaceous chondrites at the surface of Ceres. Subsequent work proposed ammoniated saponite (King et al., 1992) and water ice (Vernazza et al., 2005) as the origin of this band. The presence of water ice in the subsurface of Ceres was predicted based on the detection of OH escaping from the north polar region (A'Hearn & Feldman, 1992). Rivkin et al. (2006b) reported the presence of carbonates and iron-rich clays at the surface of Ceres based on spectroscopic measurements in the 2–4 μm range. This compositional interpretation was refined a few years later (Milliken & Rivkin, 2009) and an assemblage consisting of a mixture of hydroxide brucite, magnesium carbonates, and serpentines was proposed to explain Ceres' spectral properties. Recent observations with the AKARI satellite have revealed the presence of an additional absorption band in the 2.5–3.5 μm range that is located at 2.73 μm (Usui et al., 2019), while also confirming the presence of a band at 3.06–3.08 μm (Usui et al., 2019). Measurements performed by the VIR instrument onboard the Dawn mission have shown that the ∼3.06 μm band assigned to ammoniated phyllosilicates by King et al. (1992) is the most

Table 1.2 *Volume equivalent diameter (Deq), geometric albedo, spectral type following the Bus-DeMeo taxonomy, semi-major axis (a), eccentricity (e), and inclination (i) for the largest (D > 100 km) Main Belt asteroids.*

The albedo and diameter values represent the averages of the values reported in Tedesco et al. (2002), Masiero et al. (2011, 2014), and Usui et al. (2011). Spectral types were retrieved from Bus and Binzel (2002), Lazzaro et al. (2004), and DeMeo et al. (2009). The spectral type in braces was determined using the Bus and Binzel (2002) taxonomy

Object	Deq (km)	Geom. Alb.	Spectral type	a (AU)	e	i (deg)
1 Ceres	939.4	0.087	C	2.77	0.08	10.59
2 Pallas	513.0	0.140	B	2.77	0.23	34.83
3 Juno	249.0	0.210	Sq	2.67	0.26	12.99
4 Vesta	525.4	0.404	V	2.36	0.09	7.14
5 Astraea	114.0	0.236	S	2.57	0.19	5.37
6 Hebe	196.0	0.220	S	2.43	0.20	14.75
7 Iris	204.0	0.244	S	2.39	0.23	5.52
8 Flora	140.0	0.226	Sw	2.20	0.16	5.89
9 Metis	168.0	0.189	K	2.39	0.12	5.58
10 Hygiea	434.0	0.059	C	3.14	0.11	3.83
11 Parthenope	154.5	0.186	Sq	2.45	0.10	4.63
12 Victoria	129.7	0.134	L	2.33	0.22	8.37
13 Egeria	205.0	0.080	Ch	2.58	0.09	16.54
14 Irene	149.7	0.239	S	2.59	0.17	9.12
15 Eunomia	267.3	0.196	S	2.64	0.19	11.75
16 Psyche	222.0	0.156	Xk	2.92	0.13	3.10
18 Melpomene	146.0	0.190	S	2.30	0.22	10.13
19 Fortuna	211.0	0.048	Ch	2.44	0.16	1.57
20 Massalia	140.8	0.229	S	2.41	0.14	0.71
21 Lutetia	98.0	0.184	Xc	2.44	0.16	3.06
22 Kalliope	161.0	0.171	X	2.91	0.10	13.72
23 Thalia	105.1	0.270	S	2.63	0.24	10.11
24 Themis	189.6	0.074	C	3.14	0.13	0.75
27 Euterpe	113.9	0.218	S	2.35	0.17	1.58
28 Bellona	122.7	0.187	S	2.78	0.15	9.43
29 Amphitrite	204.0	0.185	S	2.55	0.07	6.08
31 Euphrosyne	268.0	0.047	Cb	3.16	0.22	26.28
34 Circe	117.1	0.051	Ch	2.69	0.10	5.50
35 Leukothea	104.1	0.066	(C)	2.99	0.23	7.94
36 Atalante	111.5	0.060	Ch	2.75	0.30	18.43
37 Fides	110.0	0.182	S	2.64	0.17	3.07
38 Leda	116.1	0.063	Cgh	2.74	0.15	6.97
39 Laetitia	164.0	0.238	Sqw	2.77	0.11	10.38
40 Harmonia	116.7	0.209	S	2.27	0.05	4.26
41 Daphne	187.0	0.055	Ch	2.76	0.28	15.79
42 Isis	106.7	0.152	K	2.44	0.22	8.53
45 Eugenia	186.0	0.051	C	2.72	0.08	6.60
46 Hestia	126.0	0.051	Xc	2.53	0.17	2.34
47 Aglaja	142.0	0.064	C	2.88	0.13	4.98

Table 1.2 (*cont.*)

Object	Deq (km)	Geom. Alb.	Spectral type	a (AU)	e	i (deg)
48 Doris	220.8	0.064	Ch	3.11	0.07	6.55
49 Pales	161.3	0.052	Ch	3.09	0.23	3.17
50 Virginia	94.4	0.041	Ch	2.65	0.28	2.83
51 Nemausa	144.0	0.078	Cgh	2.37	0.07	9.98
52 Europa	314.0	0.049	C	3.09	0.11	7.48
53 Kalypso	110.7	0.044	Ch	2.62	0.21	5.17
54 Alexandra	143.0	0.066	Cgh	2.71	0.20	11.80
56 Melete	120.0	0.060	Xk	2.60	0.24	8.07
57 Mnemosyne	114.5	0.209	S	3.15	0.12	15.22
59 Elpis	168.0	0.043	C	2.71	0.12	8.63
62 Erato	97.0	0.063	B	3.13	0.17	2.23
65 Cybele	286.3	0.049	Xk	3.43	0.11	3.56
68 Leto	127.2	0.215	S	2.78	0.19	7.97
69 Hesperia	134.8	0.150	Xk	2.98	0.17	8.59
70 Panopaea	137.1	0.050	Cgh	2.61	0.18	11.59
74 Galatea	122.6	0.042	(C)	2.78	0.24	4.08
76 Freia	173.1	0.042	C	3.41	0.16	2.12
78 Diana	145.5	0.051	Ch	2.62	0.21	8.70
81 Terpsichore	124.0	0.042	C	2.86	0.21	7.80
85 Io	165.0	0.054	C	2.66	0.19	11.96
86 Semele	119.2	0.049	Cgh	3.11	0.21	4.82
87 Sylvia	280.0	0.043	X	3.48	0.09	10.88
88 Thisbe	212.0	0.052	C	2.77	0.16	5.21
89 Julia	140.0	0.172	S	2.55	0.18	16.14
90 Antiope	108.0	0.087	C	3.16	0.16	2.21
91 Aegina	109.0	0.042	Ch	2.59	0.11	2.11
92 Undina	121.0	0.280	Xk	3.19	0.10	9.93
93 Minerva	159.0	0.048	C	2.76	0.14	8.56
94 Aurora	199.0	0.041	C	3.16	0.09	7.97
95 Arethusa	144.5	0.062	Ch	3.07	0.15	13.00
96 Aegle	172.9	0.051	T	3.05	0.14	15.97
98 Ianthe	109.8	0.043	(Ch)	2.69	0.19	15.58
104 Klymene	127.8	0.054	Ch	3.15	0.16	2.79
105 Artemis	121.0	0.045	Ch	2.37	0.18	21.44
106 Dione	174.7	0.067	Cgh	3.18	0.17	4.60
107 Camilla	254.0	0.040	X	3.49	0.07	10.00
111 Ate	144.9	0.053	Ch	2.59	0.10	4.93
114 Kassandra	99.0	0.090	K	2.68	0.14	4.93
117 Lomia	158.6	0.048	(X)	2.99	0.03	14.90
120 Lachesis	170.2	0.050	C	3.12	0.06	6.95
121 Hermione	187.0	0.061	Ch	3.45	0.13	7.60
127 Johanna	121.7	0.053	Ch	2.76	0.07	8.24

Table 1.2 (*cont.*)

Object	Deq (km)	Geom. Alb.	Spectral type	a (AU)	e	i (deg)
128 Nemesis	185.4	0.053	C	2.75	0.12	6.25
129 Antigone	126.0	0.147	Xk	2.88	0.21	12.23
130 Elektra	199.0	0.064	Ch	3.12	0.21	22.86
134 Sophrosyne	113.3	0.045	(Ch)	2.56	0.12	11.60
137 Meliboea	147.8	0.049	(C)	3.12	0.22	13.41
139 Juewa	167.2	0.047	(X)	2.78	0.18	10.91
140 Siwa	114.0	0.063	Xc	2.73	0.22	3.19
141 Lumen	136.4	0.050	Ch	2.67	0.21	11.89
144 Vibilia	141.0	0.051	Ch	2.66	0.24	4.81
145 Adeona	150.1	0.044	Ch	2.67	0.15	12.64
146 Lucina	134.5	0.052	Ch	2.72	0.07	13.10
147 Protogeneia	121.8	0.062	C	3.14	0.03	1.93
150 Nuwa	145.5	0.043	C	2.98	0.13	2.19
153 Hilda	186.6	0.055	X	3.97	0.14	7.83
154 Bertha	187.0	0.047	(C)	3.20	0.08	20.98
156 Xanthippe	113.6	0.048	Ch	2.73	0.23	9.78
159 Aemilia	130.0	0.059	Ch	3.10	0.11	6.13
162 Laurentia	98.6	0.055	(Ch)	3.02	0.18	6.10
164 Eva	107.7	0.039	(X)	2.63	0.34	24.47
165 Loreley	173.0	0.045	(Cb)	3.12	0.08	11.22
168 Sibylla	148.8	0.054	(Ch)	3.37	0.07	4.64
171 Ophelia	110.6	0.071	(Cb)	3.13	0.13	2.55
173 Ino	152.6	0.069	Xk	2.74	0.21	14.21
175 Andromache	110.8	0.071	Cg	3.18	0.23	3.22
176 Iduna	124.3	0.080	(Ch)	3.19	0.17	22.59
181 Eucharis	118.2	0.095	Xk	3.13	0.20	18.89
185 Eunike	168.2	0.057	C	2.74	0.13	23.22
187 Lamberta	131.8	0.059	Ch	2.73	0.24	10.59
190 Ismene	197.3	0.043	X	4.00	0.17	6.16
191 Kolga	102.3	0.040	Cb	2.89	0.09	11.51
194 Prokne	172.6	0.050	Ch	2.62	0.24	18.49
196 Philomela	148.8	0.195	(S)	3.11	0.02	7.26
200 Dynamene	133.4	0.050	(Ch)	2.74	0.13	6.90
203 Pompeja	111.3	0.045	(C-complex)	2.74	0.06	3.18
206 Hersilia	99.3	0.062	(C)	2.74	0.04	3.78
209 Dido	138.2	0.049	(Xc)	3.14	0.06	7.17
210 Isabella	85.8	0.048	Cb	2.72	0.12	5.26
211 Isolda	149.2	0.056	Ch	3.04	0.16	3.89
212 Medea	150.0	0.039	X	3.12	0.11	4.26
216 Kleopatra	121.0	0.145	Xe	2.80	0.25	13.10
221 Eos	105.6	0.139	K	3.01	0.10	10.88
225 Henrietta	119.3	0.042	(B)	3.39	0.26	20.87

Table 1.2 (*cont.*)

Object	Deq (km)	Geom. Alb.	Spectral type	a (AU)	e	i (deg)
227 Philosophia	97.6	0.056	(X)	3.17	0.19	9.11
229 Adelinda	104.2	0.037	X	3.42	0.14	2.08
230 Athamantis	112.0	0.163	S	2.38	0.06	9.44
233 Asterope	106.0	0.157	Xk	2.66	0.10	7.69
238 Hypatia	151.4	0.042	(Ch)	2.91	0.09	12.39
241 Germania	178.7	0.052	C	3.05	0.10	5.51
247 Eukrate	147.0	0.051	(Xc)	2.74	0.24	24.99
250 Bettina	111.0	0.137	Xk	3.15	0.13	12.81
259 Aletheia	185.3	0.041	(X)	3.13	0.13	10.82
266 Aline	112.1	0.048	Ch	2.80	0.16	13.40
268 Adorea	148.9	0.040	X	3.09	0.14	2.44
275 Sapientia	110.9	0.042	C	2.77	0.16	4.76
276 Adelheid	123.9	0.046	X	3.12	0.07	21.62
279 Thule	124.9	0.043	D	4.28	0.01	2.34
283 Emma	142.0	0.032	C	3.05	0.15	7.99
286 Iclea	103.9	0.043	(Ch)	3.19	0.03	17.90
303 Josephina	103.4	0.052	(Ch)	3.12	0.06	6.87
308 Polyxo	147.3	0.045	T	2.75	0.04	4.36
324 Bamberga	232.9	0.061	C	2.68	0.34	11.10
328 Gudrun	120.1	0.045	C/Cb	3.11	0.11	16.12
334 Chicago	182.7	0.050	C	3.90	0.02	4.64
344 Desiderata	133.9	0.058	Xk	2.60	0.32	18.35
345 Tercidina	101.5	0.058	Ch	2.33	0.06	9.75
349 Dembowska	174.4	0.280	R	2.93	0.09	8.25
350 Ornamenta	114.3	0.063	(Ch)	3.11	0.16	24.91
354 Eleonora	159.9	0.187	A	2.80	0.12	18.40
356 Liguria	134.2	0.051	Ch	2.76	0.24	8.22
357 Ninina	104.6	0.053	(B)	3.15	0.07	15.08
360 Carlova	135.0	0.039	(C)	3.00	0.18	11.70
361 Bononia	150.5	0.041	D	3.96	0.21	12.62
365 Corduba	101.6	0.037	(C)	2.80	0.16	12.78
372 Palma	192.6	0.064	(B)	3.15	0.26	23.83
373 Melusina	98.7	0.042	(Ch)	3.12	0.14	15.43
375 Ursula	193.6	0.049	C	3.12	0.11	15.94
381 Myrrha	127.6	0.055	(Cb)	3.22	0.09	12.53
386 Siegena	167.0	0.053	(C)	2.90	0.17	20.26
387 Aquitania	97.0	0.171	L	2.74	0.24	18.13
388 Charybdis	122.2	0.044	(C)	3.01	0.06	6.44
393 Lampetia	121.9	0.064	(Xc)	2.78	0.33	14.88
404 Arsinoe	98.0	0.047	(Ch)	2.59	0.20	14.11
405 Thia	124.4	0.048	Ch	2.58	0.24	11.95
409 Aspasia	164.0	0.060	Xc	2.58	0.07	11.26

Table 1.2 (*cont.*)

Object	Deq (km)	Geom. Alb.	Spectral type	a (AU)	e	i (deg)
410 Chloris	116.6	0.058	(Ch)	2.73	0.24	10.96
412 Elisabetha	100.6	0.043	(C)	2.76	0.04	13.78
419 Aurelia	122.5	0.043	C	2.60	0.25	3.93
420 Bertholda	148.7	0.038	D	3.41	0.03	6.69
423 Diotima	211.2	0.053	C	3.07	0.04	11.24
426 Hippo	123.0	0.051	B	2.89	0.10	19.48
444 Gyptis	165.6	0.047	C	2.77	0.18	10.28
445 Edna	97.6	0.037	(Ch)	3.20	0.19	21.38
451 Patientia	242.9	0.069	C	3.06	0.07	15.24
455 Bruchsalia	98.5	0.052	(Xk)	2.66	0.29	12.02
466 Tisiphone	111.0	0.072	(Ch)	3.35	0.09	19.11
469 Argentina	127.4	0.039	(Xk)	3.18	0.16	11.59
471 Papagena	132.0	0.232	Sq	2.89	0.23	14.98
476 Hedwig	124.5	0.045	Xk	2.65	0.07	10.94
481 Emita	112.5	0.047	(Ch)	2.74	0.16	9.84
488 Kreusa	161.4	0.052	(Ch)	3.17	0.16	11.52
489 Comacina	138.8	0.044	(X)	3.15	0.04	13.00
490 Veritas	114.5	0.064	Ch	3.17	0.10	9.28
491 Carina	99.1	0.063	(C)	3.19	0.09	18.87
505 Cava	102.8	0.060	Xk	2.68	0.25	9.84
506 Marion	108.0	0.045	(X)	3.04	0.15	17.00
508 Princetonia	134.9	0.050	(X)	3.16	0.01	13.36
511 Davida	303.2	0.064	C	3.16	0.19	15.94
514 Armida	105.6	0.040	(Xe)	3.05	0.04	3.88
517 Edith	96.3	0.037	C	3.16	0.18	3.19
521 Brixia	118.9	0.060	Ch	2.74	0.28	10.60
522 Helga	97.2	0.044	(X)	3.63	0.08	4.44
532 Herculina	191.0	0.211	S	2.77	0.18	16.31
536 Merapi	159.5	0.042	Xk	3.50	0.09	19.43
545 Messalina	112.2	0.042	(Cb)	3.21	0.17	11.12
554 Peraga	102.9	0.039	Ch	2.38	0.15	2.94
566 Stereoskopia	167.0	0.042	(X)	3.38	0.11	4.90
570 Kythera	102.4	0.051	D	3.42	0.12	1.79
595 Polyxena	110.6	0.091	(T)	3.21	0.06	17.82
596 Scheila	114.3	0.038	T	2.93	0.16	14.66
602 Marianna	128.9	0.051	(Ch)	3.09	0.25	15.08
618 Elfriede	133.6	0.050	(C)	3.19	0.07	17.04
635 Vundtia	99.0	0.045	(B)	3.14	0.08	11.03
654 Zelinda	128.3	0.042	(Ch)	2.30	0.23	18.13
683 Lanzia	103.0	0.108	(C)	3.12	0.06	18.51
690 Wratislavia	157.9	0.044	(B)	3.14	0.18	11.27
694 Ekard	105.3	0.037	(Ch)	2.67	0.32	15.84

Table 1.2 (*cont.*)

Object	Deq (km)	Geom. Alb.	Spectral type	a (AU)	e	i (deg)
702 Alauda	195.0	0.056	C	3.19	0.02	20.61
704 Interamnia	332.0	0.044	Cb	3.06	0.16	17.31
705 Erminia	137.4	0.042	(C)	2.92	0.05	25.04
712 Boliviana	127.4	0.049	(X)	2.58	0.19	12.76
713 Luscinia	101.8	0.044	(C)	3.39	0.17	10.36
733 Mocia	102.6	0.041	(X)	3.40	0.06	20.26
739 Mandeville	111.2	0.054	Xc	2.74	0.14	20.66
747 Winchester	178.0	0.048	(C)	3.00	0.34	18.16
748 Simeisa	106.6	0.039	(T)	3.96	0.19	2.26
751 Faina	111.9	0.050	(Ch)	2.55	0.15	15.61
762 Pulcova	149.0	0.055	C	3.16	0.10	13.09
769 Tatjana	106.3	0.044	(C-complex)	3.17	0.19	7.37
772 Tanete	130.6	0.050	C	3.00	0.09	28.86
776 Berbericia	155.6	0.063	Cgh	2.93	0.16	18.25
780 Armenia	99.7	0.045	(C)	3.11	0.10	19.09
786 Bredichina	101.3	0.060	(X/Xc)	3.17	0.16	14.55
788 Hohensteina	119.2	0.060	(Ch)	3.12	0.13	14.34
790 Pretoria	159.1	0.045	(X)	3.41	0.15	20.53
804 Hispania	158.8	0.052	(C)	2.84	0.14	15.36
814 Tauris	111.0	0.045	(C)	3.15	0.31	21.83
895 Helio	134.2	0.049	(B)	3.20	0.15	26.09
909 Ulla	114.8	0.036	X	3.54	0.09	18.79
1015 Christa	100.6	0.043	(Xc)	3.21	0.08	9.46
1021 Flammario	101.0	0.045	C	2.74	0.28	15.87
1093 Freda	112.2	0.043	(Cb)	3.13	0.27	25.21
1269 Rollandia	108.2	0.045	(D)	3.91	0.10	2.76

plausible interpretation (De Sanctis et al., 2015), while the assemblage with brucite was not confirmed. The VIR instrument has further revealed the presence of water ice (Combe et al., 2016), carbonates (De Sanctis et al., 2016), organics (De Sanctis et al., 2017), and chloride salts (De Sanctis et al., 2020) at the surface of Ceres. See Chapters 7 and 8 for more detail.

Notably, a "genetic" link between Ceres and carbonaceous chondrites available in our collections has progressively been questioned with time (Milliken & Rivkin, 2009; Rivkin et al., 2011). First, the ~3.06 μm band present in Ceres spectrum differs from what is seen in carbonaceous chondrite spectra. Second, the band depth of the 2.73 μm band is shallower than that of CM-like (Ch-/Cgh-type) asteroids (Figure 1.1; Usui et al., 2019). Third, Ceres' spectrum possesses a broad absorption band centered on ~1.2–1.3 μm that is not seen in spectra of aqueously altered carbonaceous chondrite (Figure 1.1). Vernazza et al. (2015a) and Marsset et al. (2016) have tentatively attributed this absorption band to amorphous silicates (mainly olivine), whereas Yang and Jewitt (2010) proposed magnetite instead. Observations in the mid-infrared wavelength range have further reinforced

the existence of compositional differences between Ceres and carbonaceous chondrite meteorites (Vernazza et al., 2017, Figure 1.2). They have further revealed the presence of anhydrous silicates at the surface of Ceres in addition to phyllosilicates and carbonates.

1.2.1.2 (4) Vesta

It has been well known since the 1970s that Vesta possesses a high albedo (0.3–0.4) (e.g., Allen, 1970; Cruikshank & Morrison, 1973; Tedesco et al., 2002; Ryan & Woodward, 2010; Usui et al., 2011; Hasegawa et al., 2014) and that its surface is basaltic in composition (McCord et al., 1970; Larson & Fink, 1975; McFadden et al., 1977). Vesta's spectrum displays two diagnostic absorption bands centered at ~0.9 μm and ~2 μm which imply the presence of pyroxene at its surface. A comparison of these observations with laboratory measurements of meteorites has revealed a spectral similarity between Vesta and that of the howardite–eucrite–diogenite (HED) achondritic meteorites (e.g., McCord et al., 1970; Feierberg & Drake, 1980; Feierberg et al., 1980;

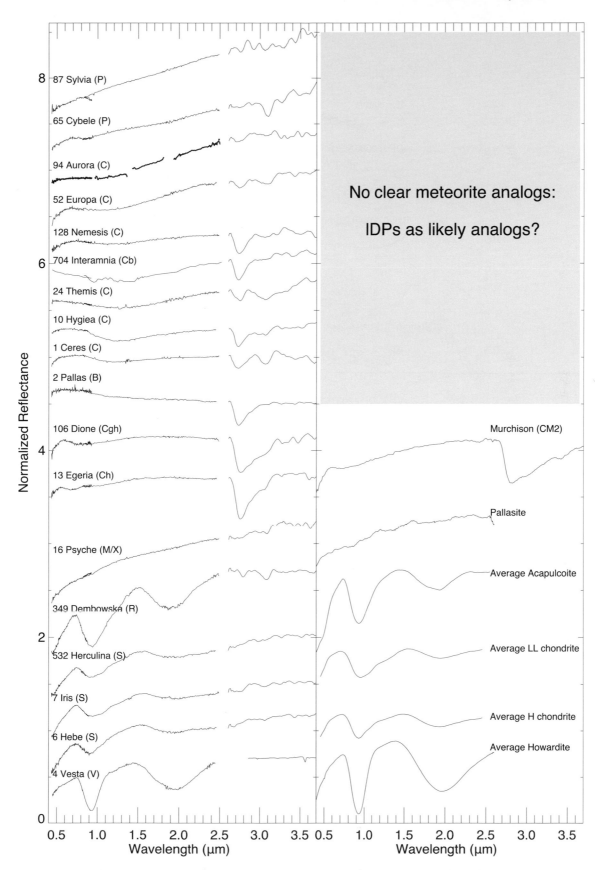

Figure 1.1 Sample of reflectance spectra of D > 100 km asteroids in visible to near-infrared wavelengths obtained with ground-based telescopes (Bus & Binzel, 2002; Hasegawa et al., 2003, 2017; Hardersen et al., 2004; Lazzaro et al., 2004; Rivkin et al., 2006a; DeMeo et al., 2009; Vernazza et al., 2014; Binzel et al., 2019; some spectra were also retrieved from the smass.mit.edu database) and the AKARI satellite (Usui et al., 2019) representative of the compositional diversity among these bodies. The asteroid spectral types are indicated in parentheses. Spectra of analogue meteorites are also shown next to each asteroid type (data are retrieved from the RELAB spectral database). The gray region denotes the asteroid types for which no clear meteoritic analogues exist.

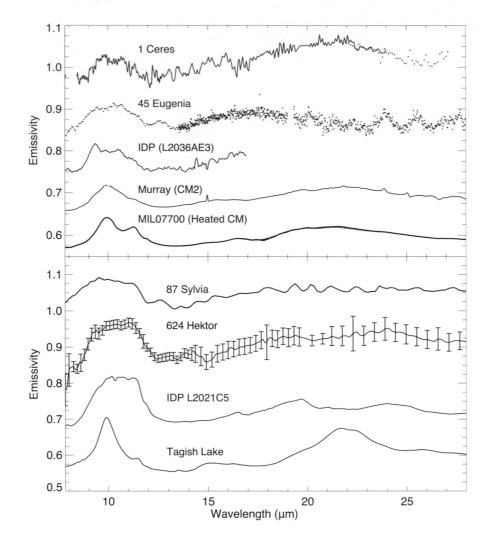

Figure 1.2 Emissivity spectra of C- (top) and P/D-type asteroids (bottom) compared to meteorite and IDP spectra. The data were retrieved from Emery et al. (2006), Brunetto et al. (2011), Marchis et al. (2012), Merouane et al. (2014), and Vernazza et al. (2017). Here, we illustrate the typical mismatch between carbonaceous chondrites and most C-, P-, and D-type asteroids (Vernazza et al., 2015a). IDPs instead appear as more convincing analogues for these objects.

Hiroi et al., 1994), opening the possibility that Vesta could be the parent body of this meteorite group (e.g., Drake, 2001). Spectroscopic observations in the 3-μm region have delivered contrasting results. Hasegawa et al. (2003) were the first to report the presence of a shallow 3-μm absorption that they interpreted either as contamination by CM-like impactors or solar wind implantation. These results were not confirmed by Vernazza et al. (2005), whereas Rivkin et al. (2006a) could not formally rule out a shallow (∼1%) absorption band. Finally, the presence of olivine has been reported at the surface of Vesta on the basis of spectral and color data (Binzel et al., 1997; Gaffey, 1997; Dotto et al., 2000; Heras et al., 2000).

The heterogeneity of Vesta's surface composition has been studied since the late 1970s (e.g., Blanco & Catalano, 1979; Degewij et al., 1979; Binzel et al., 1997; Gaffey, 1997; Vernazza et al., 2005; Rivkin et al., 2006a; Carry et al., 2010c), revealing a non-uniform surface composition which was later confirmed by the data obtained by the Dawn spacecraft's visible and infrared spectrometer (De Sanctis et al., 2012). The Dawn measurements have further confirmed that the mineralogy of Vesta is consistent

with HED meteorites while also revealing the presence of olivine-rich areas in unexpected locations far from the Rheasilvia impact basin (Ammannito et al., 2013). For a detailed discussion see Chapter 3.

The primordial Vesta material is no longer found on Vesta alone but also among its family members. Early dynamical studies have suggested the existence of a Vesta dynamical family (Williams, 1979; Zappalá et al., 1990). Follow-up spectroscopic observations at visible wavelengths have confirmed the existence of such a family, revealing 20 small (diameters <10 kilometers) main-belt asteroids with optical reflectance spectral features similar to those of Vesta and eucrite and diogenite meteorites (Binzel & Xu, 1993). Since then, more than 15,000 objects have been identified as likely Vesta family members (Nesvorny, 2015), and several of these candidates have been confirmed via spectroscopic observations (e.g., Hardersen et al., 2014, 2015; Fulvio et al., 2018). We recall here that the term "Vestoid" is usually employed to designate a member of the Vesta family in the literature. Whereas the Vestoids are all located in Vesta's vicinity, between 2.1 and 2.5 AU, some V-types (see Tholen & Barucci, 1989; Bus & Binzel, 2002; DeMeo et al., 2009 for a definition of asteroid taxonomies) have also been discovered in the near-Earth space (Cruikshank et al., 1991; Binzel et al., 2004, 2019; Thomas et al., 2014) as well as in the middle and the outer belt (e.g., Lazzaro et al., 2000; Roig & Gil-Hutton, 2006; Moskovitz et al., 2008; Roig et al., 2008; Hardersen et al., 2018). Whereas V-type NEAs might have originated in the Vesta family and escaped the inner belt via the ν_6 and 3:1 resonances, it is unlikely to be the case for most middle and outer Main Belt V-types (e.g., Fulvio et al., 2018; Hardersen et al., 2018). (1459) Magnya, in particular, located at 3.15 AU, is unlikely to be from Vesta (Lazzaro et al., 2000; Hardersen et al., 2004). These results imply that the basaltic population present in the asteroid belt is not limited to Vesta and its family. Rather, it encompasses a larger number of bodies.

1.2.2 Extra-Terrestrial Analogues of D > 100 km Main Belt Asteroids

1.2.2.1 Asteroid Types with Plausible Meteoritic Analogues

Meteorites, which are mostly fragments of Main Belt asteroids, have played a key role in our understanding of the surface composition of asteroids and their distribution across the Main Belt. Their spectral properties have been measured in laboratories over an extended wavelength range (from the visible to the mid-infrared; e.g. Gaffey, 1976; Cloutis et al., 2010, 2011, 2012, 2013; Beck et al., 2014) and used for direct comparison with those acquired for asteroids via telescopic observations (see Burbine, 2014, 2016; DeMeo et al., 2015; Reddy et al., 2015; Vernazza & Beck, 2017, and references therein). Because most minerals present in meteorites possess diagnostic features either in the near- or mid-infrared and because quality measurements in these ranges became available for a large number of asteroids in the late 1990s, we can say that the Golden Age of asteroid compositional studies really started toward the end of the second millennium.

About 60% of the spectral types defined by DeMeo et al. (2009) (A, Cgh, Ch, K, Q, R, S, Sa, Sq, Sr, Sv, V, and some X, Xc, and Xk types) can be linked to (or entirely defined by) a given meteorite class. Hereafter, we make a brief summary of these associations as well as a review of the spectroscopic observations of D > 100 km bodies belonging to these classes:

- A-type asteroids mostly comprise the parent bodies of differentiated meteorites such as brachinites and pallasites, and possibly those of the undifferentiated R chondrite meteorites (Sunshine et al., 2007). (354) Eleonora is the only D > 100 km A-type asteroid and it contains more than 80% of the mass of all A-type bodies (DeMeo et al., 2019). Its composition is compatible with that of differentiated meteorites (Sunshine et al., 2007).
- Ch- and Cgh-type asteroids comprise the parent bodies of CM chondrites (Vilas & Gaffey, 1989; Vernazza et al., 2016, and references therein). The Main Belt contains 61 CM-like bodies with diameters greater than 100 km. The analysis of the visible and near-infrared spectral properties of 34 of these bodies were reported by Vernazza et al. (2016). They showed that the spectral variation observed among these bodies is essentially due to variations in the average regolith grain size. In addition, they showed that the spectral properties of the vast majority (unheated) of CM chondrites resemble both the surfaces and the interiors of CM-like bodies, implying a "low" temperature (<300°C) thermal evolution of the CM parent body(ies). It

follows that an impact origin is the likely explanation for the existence of heated CM chondrites.
- K-type asteroids comprise the parent bodies of CV, CO, CR, and CK meteorites (e.g. Bell, 1988; Burbine et al., 2001; Clark et al., 2009). The four D > 100 km K-type asteroids are (9) Mertis, (42) Isis, (114) Kassandra, and (221) Eos, with Eos being the largest remnant of a large outer Main Belt family (e.g., Broz et al., 2013).
- The R-type class counts only one object so far, (349) Dembowska. Of interest to the present chapter, this object is a D > 100 km body and its meteoritic analogues may comprise both the lodranite and acapulcoite achondritic meteorites. While (349) Dembowska's visible and near infrared spectrum appear very similar to those of H-like S-type asteroids (Vernazza et al., 2015a), the depth of its 1- and 2-μm bands is much larger than that observed in S-type spectra (see Figure 1.1). As noticed by Vernazza et al. (2015a), both lodranites and acapulcoites have spectral properties and ol/(ol + low Ca-px) ratios that are very similar to those of ordinary chondrites and H chondrites in particular. Yet, given their lower iron content, the depth of their 1- and 2-μm bands are larger than those of H chondrites (see Figure 1.1), making them plausible analogues for a body such as (349) Dembowska.
- S-complex asteroids (Q, S, Sa, Sq, Sr, Sv) comprise the parent bodies of the most common type of meteorites, namely ordinary chondrites (~80% of the falls; see Vernazza et al., 2015b for a detailed review on this topic), and also possibly those of some differentiated meteorites such as lodranites and acapulcoites (Vernazza et al., 2015b). Vernazza et al. (2014) analyzed the visible and near-infrared spectral properties of nearly 100 S-type asteroids among which 23 bodies have diameters greater than 100 km (out of 24 in total). They found that the surface composition of these bodies is compatible with that of H, L, and LL ordinary chondrites, and that H-like bodies are located on average further from the sun than LL-like ones. This is somewhat counterintuitive given that H chondrites are more reduced than LL chondrites, suggesting a formation closer to the sun.
- (4) Vesta is the only D > 100 km V-type, and its connection with HED meteorites has been described in Section 1.2.1.
- X-complex asteroids (X, Xc, and Xk types) are representative of a great compositional diversity that largely exceeds (in terms of the number of compositional analogues) the number of taxonomic types. Their geometric albedos range from values below 0.05 to about 0.20. Notably, while some low albedo (<0.1) X-complex bodies likely comprise the parent bodies of rare types of meteorites such as Tagish Lake or CI chondrites, the remaining low albedo X-complex bodies appear unsampled by available meteorites in our collections (see Section 1.2.2.2). So far, there isn't a single D > 100 low albedo X-complex asteroid that can be unambiguously linked to either the Tagish Lake meteorite or CI chondrites. Concerning the high albedo (>0.1) X-complex bodies, they likely comprise the parent bodies of (1) iron meteorites (e.g. Cloutis et al., 1990; Shepard et al., 2015, and references therein), (2) enstatite chondrites (ECs) and aubrites (e.g., Vernazza et al., 2009, 2011; Ockert-Bell et al., 2010; Shepard et al., 2015), and (3) CB and CH chondrites (e.g., Hardersen et al., 2011; Shepard et al., 2015). The compositional interpretation of the visible and near-infrared spectral properties and/or the radar data for many D > 100 km

X-complex asteroids can be found in Ockert-Bell et al. (2010), Hardersen et al. (2011), Vernazza et al. (2009, 2011), and Shepard et al. (2015).

1.2.2.2 Asteroid Types with No Clear Meteoritic Analogues

About 40% of the spectral types defined by DeMeo et al. (2009) (B, C, Cb, Cg, D, L, O, T, Xe, Xn, and most low albedo X, Xc, and Xk types) cannot be unambiguously linked to (or entirely defined by) a given meteorite class (e.g., Sunshine et al., 2008; Vernazza et al., 2015a, 2017).

In a few cases, this may be due to the rarity of the spectral type with L, O, Xe, and Xn types representing less than ~1% of the mass of the Main Belt (DeMeo & Carry, 2013). As a matter of fact, our meteorite collections do certainly not sample every single Main Belt asteroid, and it may thus not be a surprise that some rare asteroid types are absent from our meteorite collections.

In most cases, however, the spectral mismatch between asteroid and meteorite spectra must be telling us something important about the nature of these unsampled yet abundant asteroid types (B, C, Cb, Cg, D, T, and low albedo X-, Xc-, and Xk-type asteroids represent at least 50% of the mass of the asteroid belt; DeMeo & Carry, 2013). Notably, these asteroid types (particularly the B, C, Cb, and Cg types) comprise – similarly to S-types – a high number of large asteroid families such as the Hygiea, Themis, Euphrosyne, Nemesis, and Polana-Eulalia (Pinilla-Alonso et al., 2016) families (see Broz et al., 2013 and Nesvorny, 2015 for further information on asteroid families). Considering that asteroid families are a major source of meteorites (this is well supported by the connection between the Vesta family and the HED meteorites), we should be receiving plenty of fragments from these bodies. Yet, this seems not to be the case, at least not under the form of consolidated meteorites. Note that metamorphosed CI/CM chondrites have been proposed in the past as analogues of B-, C-, Cb-, and Cg-type surfaces (Hiroi et al., 1993). However, such a possibility is presently untenable for the majority of these asteroids for three reasons, namely the paucity of these meteorites among falls (~0.2% of meteorite falls) compared to the abundance of B-, C-, Cb,- and Cg-types and of families with similar spectral type, the difference in density between these meteorites and these asteroids (see Section 1.3.2), and the difference in spectral properties in the 3-μm region and in the mid-infrared region (e.g., Figure 1.2) between the two groups.

Vernazza et al. (2015a) proposed that these asteroid types might (at least their surfaces/outer shell) consist largely of friable materials unlikely to survive atmospheric entry as macroscopic bodies. It may thus not be surprising that these asteroid types are not well represented by the cohesive meteorites in our collections. Vernazza et al. (2015a) proposed that interplanetary dust particles (IDPs) as well as volatiles may, similarly to comets, be more appropriate analogues for these bodies. Available density measurements for these asteroid types ($\rho < 2$ g/cm^3 for D > 200 km bodies and in particular $\rho < 1.4$ g/cm^3 for the two largest low albedo X-type asteroids with D > 250 km; see Section 1.3.2) support such a hypothesis as they indicate that these bodies cannot be made of silicates only, and must comprise a significant fraction of ice(s). The discovery of several active objects among these asteroid types provides additional support for their comet-like nature (e.g., Hsieh & Jewitt, 2006; Jewitt, 2012; Jewitt et al., 2015). In addition, water

ice may not only be present in the interior of these bodies but also at their surfaces (e.g., Campins et al., 2010; Rivkin & Emery, 2010; Licandro et al., 2011; Hargrove et al., 2012, 2015; Takir & Emery, 2012).

An analogy for these asteroid types with IDPs had already been proposed by Bradley et al. (1996) based on IDP spectra collected in the visible domain. In recent years, this association has been strengthened thanks to the availability of quality spectroscopic measurements for these asteroid types in the mid-infrared range (e.g., Barucci et al., 2002; Emery et al., 2006; Licandro et al., 2012; Marchis et al., 2012; Vernazza et al., 2013). Thanks to these datasets, it has been shown that the surfaces of some of these asteroid types (mainly low albedo X-, T-, and D-types and some C-types) are covered by a mixture of anhydrous amorphous and crystalline silicates (Vernazza et al., 2015a, 2017; Marsset et al., 2016), namely a composition that is similar to that of chondritic porous IDPs. Given that silicate grains in the interstellar medium (ISM) appear to be dominantly amorphous, these observations suggest a significant heritage from the ISM for these bodies. Furthermore, D-, low albedo X-, and T-type objects appear enriched in crystalline olivine with respect to pyroxene, whereas B-, C-, Cb-, and Cg-type asteroids tend to have about as much crystalline pyroxene as olivine (Vernazza et al., 2015a), suggesting two main primordial reservoirs of primitive small bodies as well as a compositional gradient in the primordial outer protoplanetary disk (10–40 AU).

Spectroscopic observations of these bodies in the so-called 3 μm region (which covers approximatively the 2.5–3.5 μm wavelength range; Rivkin et al., 2002, 2015) have allowed refining their surface compositions and thermal histories. In this wavelength region, all water-related materials (phyllosilicates including ammonium-bearing ones, water ice, brucite, to name a few) as well as absorbed water molecules in regolith particles (e.g., in lunar rocks or soils; Clark 2009) exhibit one or several absorption features. In the case of hydrated minerals, the absorption band is located at ~2.7 to ~2.8 μm (depending on the phyllosilicate mineralogy/hydration state), whereas in the case of water ice it is located at ~3.05 μm. Hereafter, we summarize the findings of the two main spectroscopic surveys of D > 100 km asteroids in the 3-μm region:

- Takir and Emery (2012) and Takir et al. (2015) reported the observations of several tens of D > 100 km main-belt asteroids (mainly C-, low albedo X-, and T-types) with the NASA Infrared Telescope Facility (IRTF) located on the summit of Mauna Kea, Hawaii, and classified them into four spectral groups based on the absorption band centers and shapes ("sharp," "rounded," "Ceres-like," and "Europa-like"). The "sharp" group exhibits a characteristically sharp 3 μm feature attributed to hydrated minerals, whereas the "rounded" group exhibits a rounded 3 μm band attributed to H_2O ice. The "Ceres-like" group and "Europa-like" groups have narrow 3 μm band features centered at ~3.05 μm and ~3.15 μm, respectively. Whereas the "sharp" 3-μm feature appears similar to that observed in laboratory spectra of CM and CI chondrites and is mostly seen in the spectra of Ch- and Cgh-types, the other features are unlike those seen in meteorites (Rivkin et al., 2019). Whereas members of the "sharp," "Ceres-like," and "Europa-like" groups are concentrated in the 2.5–3.3 AU region, members of the "rounded" group are concentrated in the 3.4–4 AU region. Unlike the "sharp" group, the "rounded" group did not experience aqueous alteration (Takir & Emery, 2012).

- The AKARI satellite made low-resolution spectroscopic observations in the 2.5–5 μm region with enough sensitivity to be able to characterize the surface composition of many D > 100 km Main Belt asteroids (Usui et al., 2019). AKARI observed 66 asteroids, including 23 C-complex asteroids, 22 X-complex asteroids, and 3 D-type asteroids. Most C-complex asteroids (17 out of 22; ~77%), especially all CM-like bodies (Ch- and Cgh-types) and all B- and Cb-type asteroids, are found to possess a ~2.7 μm feature associated with the presence of hydrated minerals at the surfaces of these objects, in agreement with Rivkin's (2012) results. Some C-complex asteroids such as (94) Aurora, however, do not show any obvious feature in the 3 μm band, suggesting an anhydrous surface composition. Most low-albedo X-complex (P-type) asteroids but only one D-type asteroid possesses an absorption feature at ~2.7 μm, similarly suggesting the presence of hydrated minerals at their surfaces. For these objects, however, the depth of the ~2.7 μm is shallower than that of C-type bodies, opening the possibility of a lower abundance of phyllosilicates at the surfaces of these bodies. It is interesting to notice that both the IRTF and AKARI surveys point to a compositional gradient among C-, P- (T- and low albedo X-), and D-type bodies, with C-complex bodies having the greatest abundance of phyllosilicates at their surfaces, followed by P-type bodies and finally D-type asteroids whose surfaces are mostly anhydrous. In addition, aqueously altered bodies are essentially concentrated inward of ~3.3 AU, whereas dry ice-rich bodies are essentially concentrated outwards of ~3.3 AU.

The presence of anhydrous silicates at the surface of some of these bodies implies they formed >5 Myrs after calcium–aluminum-rich inclusions (Neveu & Vernazza, 2019; Figure 1.3). Their anhydrous surface composition would otherwise have been lost due to melting and ice-rock differentiation driven by heating from the short-lived radionuclide ^{26}Al. It follows that IDP-like asteroids with anhydrous surfaces formed much later than the meteorite parent bodies, including CM-like bodies (Neveu & Vernazza, 2019; Figure 1.3).

In summary, IDPs (and volatiles) may reflect asteroid compositions at least as well as meteorites. A definitive confirmation of the link between IDPs and these asteroid classes will however require the acquisition of a consistent set of visible, near-infrared, and mid-infrared spectra (0.4–25 μm) for all classes of IDPs.

1.3 IMAGING OBSERVATIONS OF MAIN BELT ASTEROIDS

Whereas our understanding of the surface composition of asteroids and its distribution across the asteroid belt has improved enormously over the last decades (see Section 1.2) the same cannot be said regarding their internal structure, which is best characterized by the density. To constrain the density, one needs to fully reconstruct the 3D shape of a body to estimate its volume and to determine its mass from its gravitational interaction with other asteroids, or preferably, whenever possible, with its own satellite(s).

The lack of accurate density measurements for D > 100 km asteroids is due to the fact that disk-resolved observations of these bodies (which are needed to reconstruct their 3D shape) have until recently only been obtained with sufficient spatial resolution for a few bodies, either by dedicated interplanetary missions (e.g., Galileo, NEAR Shoemaker, Rosetta, Dawn, OSIRIS-REx, Hayabusa 1 & 2) or by remote imaging with the Hubble Space Telescope (HST), and adaptive-optics-equipped ground-based telescopes (e.g., VLT, Keck) in the case of the largest bodies (Vesta and Ceres).

The drastic increase in angular resolution provided by the new-generation adaptive-optics SPHERE/ZIMPOL instrument at the VLT with respect to the HST (about a factor of 3) implies that the largest Main Belt asteroids (with diameters greater than 100 km; angular size typically greater than 100 mas) become resolvable worlds and are thus no longer "extended point sources." With the SPHERE instrument, craters with diameters greater than approximately 30 km can now be recognized at the surfaces of

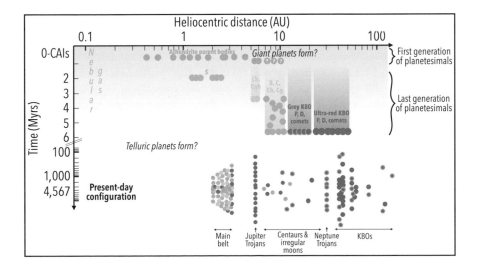

Figure 1.3 Postulated sequence of events tracing the time, place, and duration of formation of small bodies (top) to present-day observed characteristics (bottom; vertical spread reproducing roughly the distribution of orbital inclinations). The accretion duration is shown as gradient boxes ending at the fully formed bodies. Numerical simulations suggest that volatile-rich IDP-like bodies (blue dots; B, C, Cb, Cg, P, D, comets, grey and ultra-red KBOs) accreted their outer layers after 5–6 Myrs (adapted from Neveu & Vernazza, 2019).
A black and white version of this figure will appear in some formats. For the color version, refer to the plate section.

Main Belt asteroids, and the shapes of the largest asteroids can be accurately reconstructed (e.g., Marsset et al., 2017).

In the present section, we first summarize the imaging observations of (1) Ceres and (4) Vesta performed from Earth prior to the arrival of the Dawn mission and further discuss how they compare to the Dawn observations. Second, we provide an overview of what is currently know regarding the shape, topography, and density of the largest Main Belt asteroids.

1.3.1 Focus on Earth-based Imaging Observations of Ceres and Vesta and Comparison to Dawn Measurements

1.3.1.1 (1) Ceres

Prior to the arrival of the Dawn mission in 2015, the highest resolution images of Ceres had been acquired with both the HST (Thomas et al., 2005) and the Keck II telescope on Mauna Kea (Carry et al., 2008), leading to a resolution at the surface of Ceres of the order of ∼30 (HST) to ∼50 (Keck) km.

Thomas et al. (2005) used the limb profiles of 217 HST images to constrain the shape and spin of Ceres. They found that the shape of Ceres can be well described by an oblate spheroid of semi-axes a = b = 487.3 ± 1.8 km and c = 454.7 ± 1.6 km (1σ), and a volume-equivalent radius of 476.2 ± 1.7 km. In addition, both the limb profiles and the tracking of Ceres' main bright spot (emplaced in what is known today as Occator crater) allowed constraining the spin to 291° ± 5° (right ascension) and 59° ± 5° (declination). The shape was further used to place constraints on Ceres' internal structure. First, it suggested that Ceres is a relaxed object in hydrostatic equilibrium. Second, using the determined density of 2.077 ± 0.036 g/cm^3 along with the a and c semi-axes, Thomas et al. (2005) concluded that Ceres is most likely a differentiated object. These HST images were further used by Li et al. (2006) to constrain Ceres' geometric and single-scattering albedos (respectively, 0.087 ± 0.003 and 0.070 ± 0.002 at 535 nm) and to produce the very first spatially resolved surface albedo maps of Ceres. The maps reveal a small albedo contrast (typically < 4% with respect to the average for most of the surface) with the bright spot being ∼8% brighter than Ceres' average albedo.

Carry et al. (2008) performed a similar analysis using Keck/NIRC2 adaptive optics J/H/K imaging observations. Whereas they also found that Ceres' shape can be well described by that of an oblate spheroid, they derived different semi-axes values with respect to the HST results (Thomas et al., 2005), namely a = b = 479.7 ± 2.3 km and c = 444.4 ± 2.1 km (1σ), and a volume-equivalent radius of 467.6 ± 2.2 km, implying a ∼5% higher density, namely 2.206 ± 0.043 g/cm^3. Concerning the internal structure, the Carry et al. (2008) study agreed with the findings by Thomas et al. (2005) that Ceres must be differentiated. Note that observations by Dawn imply that Ceres' interior is only partially differentiated (Park et al., 2016; Chapter 12). Finally, Carry et al. (2008) also constrained the spin axis to 288° ± 5° (right ascension) and 66° ± 5° (declination) and produced albedo maps in the different bands (J, H, K), revealing an albedo variegation similar to the one reported by Li et al. (2006).

Recently, Ceres has been imaged by the SPHERE instrument at the VLT revealing so far the most spectacular view of this object from Earth at a resolution of 4.4 km/pixel (Vernazza et al., 2020; Figure 1.4). A reflectance map based on the best-quality image

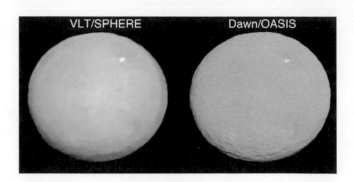

Figure 1.4 Comparison of the VLT/SPHERE deconvolved images of Ceres (left) with a synthetic projection of the Dawn 3D shape model produced with OASIS and with albedo information (right). The deconvolved image (left) shows a clear–dark–clear border, which is a deconvolution artefact.

(Vernazza et al., 2020) reveals a much higher albedo contrast (around ∼20%) than previously reported (Li et al., 2006; Carry et al., 2008).

When confronting Earth-based observations of Ceres with those of the NASA Dawn mission, it appears that the true dimensions of Ceres (a = 483.1 ± 0.2 km, b = 481.0 ± 0.2 km, and c = 445.9 ± 0.2 km, and a volume-equivalent radius of 469.7 ± 0.2 km; Russell et al., 2016; Park et al., 2019) as well as its density (2.162 ± 0.003 g/cm^3; Russell et al., 2016) fall in between those determined early on by Thomas et al. (2005) and Carry et al. (2008) – although closer to those of Carry et al. (see Table 1.3). In addition, Ceres' exact pole coordinates (a right ascension of 291.43° ± 0.01° and a declination of 66.76° ± 0.02°; Park et al., 2019) are very close to those derived earlier by Carry et al. (2008).

While the superior spatial resolution of the Dawn images has not drastically changed our knowledge of Ceres' dimensions and rotation axis, the same cannot be said regarding the surface topography and the variegation of the albedo across its surface. Schröder et al. (2017) have shown that Ceres' brightest spot corresponds to several extremely bright yet small areas within Occator crater, whose maximum size is ∼10 km in diameter (Cerealia Facula) and whose average visual normal albedo is 0.6 ± 0.1, reaching locally a visual normal albedo close to unity at a scale of 35 m/pixel (see Chapter 10). The amplitude of the albedo variegation is thus much higher than that deduced from the SPHERE images considering that the average albedo is around 0.1.

Regarding the surface topography, Ceres possesses a heavily cratered surface with a paucity of large craters, suggesting that relaxation has occurred (Marchi et al., 2016). In contrast, craters can hardly be identified (at least not unambiguously) in the VLT/SPHERE images. This may be due to the morphology of D > 30 km craters (size limit above which craters can be resolved with VLT/SPHERE) as suggested by Vernazza et al. (2020). D > 30 km craters on Ceres are essentially flat floored, implying a smaller contrast between the crater floor and the crater rim than in the case of simple bowl shaped craters. Adding to that, a nonperfect correction of the atmospheric turbulence may lead to a global absence of contrast due to the topography in the deconvolved VLT/SPHERE images of Ceres. The same phenomenon is also observed in the case of other C-complex asteroids observed by VLT/SPHERE, such as (10) Hygiea (Vernazza et al., 2020) and (704) Interamnia (Hanus et al., 2020).

Table 1.3 *Vesta and Ceres' physical and geological properties from Dawn compared to those derived from Earth.*

	Vesta (Earth-based)	Vesta (Dawn)	Ceres (Earth-based)	Ceres (Dawn)
Surface composition	HED meteorites (McCord et al., 1970)	- HED meteorites (De Sanctis et al., 2012) - Localized olivine-rich areas (Ammannito et al., 2013)	- Ammoniated phyllosilicates (King et al., 1992) - Carbonates (Rivkin et al., 2006b; Vernazza et al., 2017) - Brucite (Milliken & Rivkin, 2009)	- Mg- and NH_4-bearing phyllosilicates (De Sanctis et al., 2015; Ammannito et al., 2016) - Carbonates (De Sanctis et al., 2015, 2016) - Aliphatic organics (De Sanctis et al., 2017) - Localized water ice (Combe et al., 2016)
Meteorite analogue?	Yes (HEDs)	Yes (HEDs)	No	No
Volume equivalent radius	264.6 ± 5 km (Thomas et al., 1997a)	262.7 ± 0.1 km (Russell et al., 2012)	467.6 ± 2.2 km (Carry et al., 2008)	469.7 ± 0.1 km (Russell et al., 2016)
Dimensions	(289 × 280 × 229) ± 5 km (Thomas et al., 1997a)	(286.3 × 278.6 × 223.2) ± 0.1 km (Russell et al., 2012)	(479.7 × 479.7 × 444.4) ± 2.5 km (Carry et al., 2008)	(483.1 × 481 × 445.9) ± 0.2 km (Russell et al., 2016)
Spin axis	RA: 308° ± 10° Dec: 48° ± 10° (Thomas et al., 1997b)	RA: 309.03° ± 0.01° Dec: 42.23° ± 0.01° (Russell et al., 2012)	RA: 288° ± 5° Dec: 66° ± 5° (Carry et al., 2008)	RA: 291.42° ± 0.01° Dec: 66.76° ± 0.02° (Russell et al., 2016)
Mass	$2.6 ± 0.3 × 10^{20}$ kg (Konopliv et al., 2011)	$2.59076 ± 0.00001 × 10^{20}$ kg (Russell et al., 2012)	$9.31 ± 0.06 × 10^{20}$ kg (Konopliv et al., 2011)	$9.384 ± 0.001 × 10^{20}$ kg (Russell et al., 2016)

1.3.1.2 (4) Vesta

A few years after the discovery of the prominent Vesta family (Binzel & Xu, 1993), HST imaging observations of Vesta revealed the presence of an impact crater ~460 km in diameter near its south pole (Thomas et al., 1997a), thus strengthening the hypothesis of a collisional origin for Vesta-like bodies. The HST images were further used to constrain Vesta's spin, shape, and density, and to produce albedo, elevation, and compositional maps of its surface (Binzel et al., 1997; Thomas et al., 1997a, 1997b).

Thomas et al. (1997a) found that the shape of Vesta can be well described by a triaxial ellipsoid of semi-axes a = 289 ± 5 km, b = 280 ± 5 km, and c = 229 ± 5 km, and a volume-equivalent radius of 264.6 ± 5 km. By combining the volume with the best mass estimates, they derived a mean density in the 3.5−3.9 g/cm³ range. In addition, the spin was constrained to 308° ± 10° (right ascension) and 48° ± 10° (declination) (Thomas et al., 1997b). The pole solution as well as the dimensions, including the volume-equivalent diameter, are very close to those derived from the Dawn imaging data (see Russell et al., 2012 and Table 1.3).

The HST images further revealed a strong albedo variegation across the surface (Binzel et al., 1997), as well as the presence of a prominent central peak within the impact basin whose height was estimated to be of the order of 13 km above the deepest part of the floor (Thomas et al., 1997a). The Dawn mission actually revealed the existence of two overlapping basins in the south polar region and a central peak whose height rivals that of Olympus Mons on Mars (e.g., Jaumann et al., 2012; Marchi et al., 2012; Russell et al.,

2012; Schenk et al., 2012). The images of the Dawn mission also allowed the production of a high-resolution map of the albedo across Vesta's surface, revealing the second greatest variation of normal albedo of any asteroid yet observed after Ceres (the normal albedo varies between ~0.15 and ~0.6; Schröder et al., 2014; see also Chapter 6).

Recently, VLT/SPHERE images have recovered the surface of Vesta with a great amount of detail (Figure 1.5; Fetick et al., 2019). Most of the main topographic features present across Vesta's surface can be readily recognized from the ground. This includes the south pole impact basin and its prominent central peak, several D ≥ 25 km sized craters, and also Matronalia Rupes, including its steep scarp and its small and big arcs. On the basis of these observations, it follows that next-generation telescopes with mirror sizes in the 30–40 m range (e.g., the Extremely Large Telescope, hereafter ELT) should in principle be able to resolve the remaining major topographic features of (4) Vesta (i.e., equatorial troughs, north–south crater dichotomy), provided that they operate at the diffraction limit in the visible range.

1.3.2 Overview of Earth-based Imaging Campaigns of Main Belt Asteroids

The large angular diameter of (1) Ceres and (4) Vesta at opposition (up to ~840 mas and ~700 mas, respectively) explains why these two bodies have been the subjects of in-depth studies using first-generation high angular resolution imaging systems such as HST

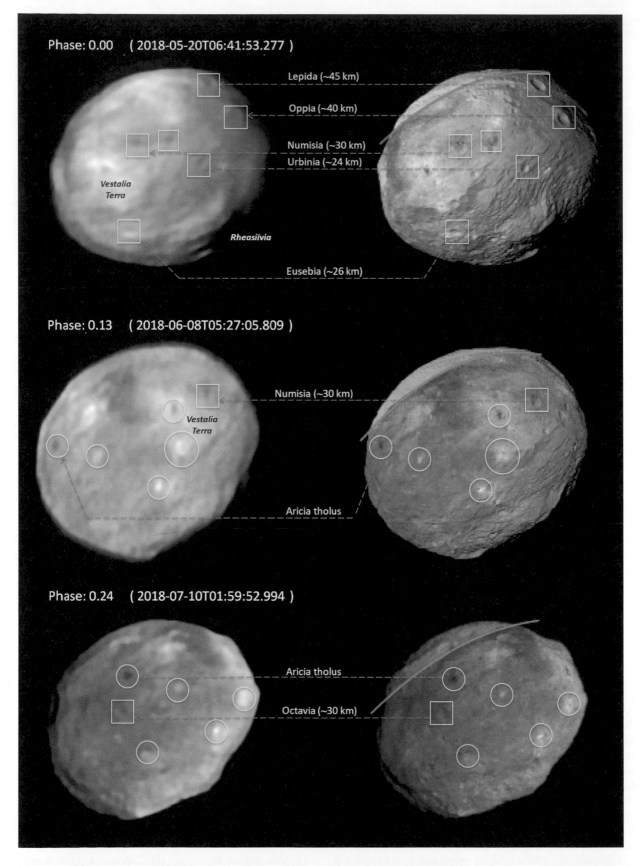

Figure 1.5 Comparison of the VLT/SPHERE deconvolved images of Vesta (left column) with synthetic projections of the Dawn 3D shape model produced with OASIS and with albedo information (right column). No albedo data is available from Dawn for latitudes above 30° N (orange line). The main structures that can be identified in both the VLT/SPHERE images and the synthetic ones are highlighted: craters are embedded in squares and albedo features in circles (from Fetick et al., 2019).

A black and white version of this figure will appear in some formats. For the color version, refer to the plate section.

(resolution of $\sim$50 mas). There are a few other Main Belt bodies that possess relatively large angular diameters at opposition ((2) Pallas: $\sim$540 mas; (324) Bamberga: $\sim$380 mas; (3) Juno, (7) Iris, and (10) Hygiea: $\sim$325 mas) and that could have been valuable targets for dedicated observing campaigns using either HST or Keck/NIRC2. This has, however, only been the case for the largest of them ((2) Pallas; Schmidt et al., 2009; Carry et al., 2010a). Overall, before the advent of the SPHERE instrument at the VLT, sparse Keck/NIRC2 imaging data had been collected for many D $>$ 100 km asteroids. Furthermore, these data were rarely acquired when the targets were at opposition, leading to non-optimal resolution for the observations of these bodies (Hanus et al., 2017b). Hereafter, we describe some important constraints that have been collected for D $>$ 100 km asteroids based on high angular resolution imaging observations.

1.3.2.1 Asteroid Densities and Internal Structures

Density is the physical property that constrains best the interior of asteroids. Unfortunately, the latter has only been measured for a handful of asteroids (mostly in the case of multiple asteroids but also via in-situ space missions). Importantly, most D $<$ 100 km asteroids are seen as collisionally evolved objects (Morbidelli et al., 2009) whose internal structure can be largely occupied by voids (called macroporosity, reaching up to 50–60% in some cases; Carry, 2012; Scheeres et al., 2015), thus limiting our capability to interpret meaningfully their bulk density in terms of composition(s). On the contrary, large bodies (D $>$ 100 km) are seen as primordial remnants of the early Solar System (Morbidelli et al., 2009); that is their internal structure has likely remained intact since their formation (they can be seen as the smallest protoplanets). For most of these objects, the macroporosity is likely minimal ($<$20%) and their bulk density is therefore an excellent tracer of their bulk composition.

Following the discovery by the Galileo spacecraft of the first Main Belt asteroid's satellite (Dactyl, around (243) Ida; Chapman et al., 1995; Belton et al., 1995), high angular resolution imaging observations of the largest Main Belt asteroids were performed during the subsequent $\sim$15 years to search for the presence of moons/companions. Note that accurate 3D shape reconstruction algorithms based on AO data such as KOALA (Carry et al., 2010b) and ADAM (Viikinkoski et al., 2015a) were not available during that time frame, leaving the moon search and the subsequent characterization of its orbit as the most compelling science objective. Indeed, direct imaging observations of multiple-asteroid systems are the most efficient approach from Earth for deriving precise asteroid masses. Specifically, the images provide constraints on the orbital parameters of the moon(s) and hence the total mass of the system. In the case of (a) small companion(s), the total mass is dominated by the primary implying that the mass of the primary can be well constrained (usually with a $<$10% uncertainty). The only other way to constrain masses of large asteroids with similar precision is via dedicated interplanetary missions, either a fly-by for the largest ones (as in the case of (21) Lutetia) or a rendezvous (e.g. Dawn mission, OSIRIS-REx, Hayabusa 1 & 2). Note that the orbital properties of the moon(s) can be used to constrain the dynamical quadrupole J_2; by comparing the latter with the J_2 derived from the 3D shape of the primary, one can search for the presence or absence of a mass concentration within the primary, hence a differentiated internal structure (e.g., Pajuelo et al., 2018).

At the time of writing the Asteroid III book, seven multiple systems were known among the largest (D $>$ 90 km) Main Belt asteroids (Merline et al., 2002), and this number increased by a factor of two at the time of writing the Asteroid IV book (Margot et al., 2015). Since then, only one multiple system has been discovered among large Main Belt asteroids ((31) Euphrosyne and its moon; see Vernazza et al., 2019). Among these 15 multiple systems, six are triple systems, the first to be discovered being the one of (87) Sylvia (Marchis et al., 2005). Two thirds of these multiple systems belong to the C-complex, while the remaining systems belong either to the P or M classes. Surprisingly, there are currently no large S-types with known companions. Whereas the observed S/C type dichotomy may be interpreted as a consequence of compositional differences (hence different material strengths) implying a different response to impacts between the water-poor silicate-rich S-types and the water-rich C-types, the existence of the metal/silicate-rich Kalliope and Kleopatra systems complicates the picture.

Hereafter, we summarize current knowledge regarding the bulk composition of the main taxonomic classes (Figure 1.6). In the case of the largest S-type asteroids, the derived densities are consistent with those of ordinary chondrites (Viikinkoski et al., 2015b; Hanus et al., 2017b, 2019; Marsset et al., 2017), implying – contrary to asteroid (4) Vesta – an absence of large-scale differentiation for these bodies, which is in agreement with the thermal history of ordinary chondrites (Huss et al., 2006; Monnereau et al., 2013). Similarly, the density of the largest CM-like bodies (Cg/Cgh types) is consistent with that of CM chondrites, suggesting a homogeneous internal structure for these bodies (Carry et al., 2019). Concerning the remaining large (D $>$ 200 km) C-complex bodies (C, Cb, Cg, and most B-types apart from Pallas), their density is in the 1.4–2.0 g/cm^3 range (Marchis et al., 2012; Hanus et al., 2020; Vernazza et al., 2020), implying a high water fraction for these bodies. Modeling results by Beauvalet and Marchis (2014) in the case of (45) Eugenia further suggest the presence of a denser core, possibly consistent with an early differentiation. A differentiated interior has also been advocated in the case of the largest C-type, (1) Ceres (e.g., Thomas et al., 2005; Park et al., 2016; Chapter 12). It is too early to conclude that all large water-rich C-types underwent differentiation, yet thermal evolution models suggest that such a trend was quasi unavoidable for D $>$ 200 km objects consisting initially of a 70% rock, 30% ice composition by mass (Neveu & Vernazza, 2019). Finally, large P-type asteroids for which precise density estimates are available indicate a density in the 1.0–1.3 g/cm^3 range that appears substantially lower than that of water-rich C-types. It is worth mentioning that, for the two largest bodies – (87) Sylvia and (107) Camilla – the presence of a small dense core is predicted (Pajuelo et al., 2018, Carry et al., 2021). It is presently unclear whether these bodies are more water-rich than C-types or if they formed with substantial porosity (70–75%), as observed in the case of comet 67P/Churyumov-Gerasimenko (Jorda et al., 2016).

1.3.2.2 Asteroid Shapes

Because of their large masses, Solar System bodies with diameters larger than $\sim$900 km possess rounded, ellipsoidal shapes, consistent with hydrostatic equilibrium, which is a part of the current IAU definition of a planet or a dwarf planet. On the other side of the

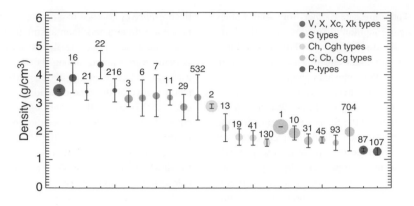

Figure 1.6 Density of some of the largest asteroids. Asteroids are grouped following their spectral classification. The relative sizes of the dots follow the relative diameters of the bodies in logarithmic scale. Error bars are 1-sigma. The science based on these density estimates can be retrieved in Paetzold et al. 2011, Russell et al. 2012, 2016, Viikinkoski et al. 2015b, Marsset et al. 2017, Hanus et al. 2017a, b, Pajuelo et al. 2018, Carry et al. 2019a, b, Ferrais et al. 2020, Hanus et al. 2020, Marsset et al. 2020, Vernazza et al. 2020, Vernazza et al. 2021, Yang et al. 2020 and Dudzinski et al. 2020.
A black and white version of this figure will appear in some formats. For the color version, refer to the plate section.

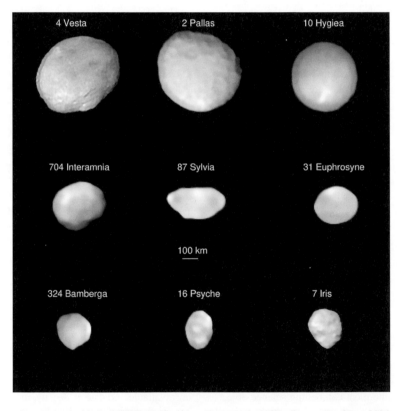

Figure 1.7 VLT/SPHERE images deconvolved with the MISTRAL algorithm (Fusco et al., 2003) of a sample of D > 200 km asteroids. The relative sizes are respected, and the scale is indicated on the plot. The objects appear according to decreasing values of their volume equivalent diameter (see Table 1.1). The science based on these images can be retrieved from Fetick et al. (2019), Hanus et al. (2019, 2020), Ferrais et al. (2020), Marsset et al. (2020), Vernazza et al. (2020), Yang et al. (2020) and Carry et al. (2021).

mass range, very small bodies (diameters < 100 km) tend to possess highly irregular shapes, with the notable exception of some D < 5 km bodies that are affected by the so-called YORP effect (Yarkovsky–O'Keefe–Radzievskii–Paddack; Rubincam, 2000; Vokrouhlický et al., 2003), and which have similar shapes to a spinning top (e.g., Ryugu, or Bennu; Nolan et al., 2013; Watanabe et al., 2019). However, it remains to be tested at what size range the shape of a typical minor body transits from a nearly rounded equilibrium shape to an irregular shape and to what extent this size

range depends on factors such as the bulk composition of the minor planet or its collisional and thermal history (Hanus et al., 2020).

Recently, an imaging survey of a substantial fraction of all D > 100 km Main Belt asteroids (sampling the main compositional classes) has been conducted using VLT/SPHERE (Vernazza et al., 2018, 2020; Viikinkoski et al., 2018; Carry et al., 2019; Fetick et al., 2019; Hanus et al., 2019, 2020; Ferrais et al., 2020; Marsset et al., 2020; Yang et al., 2020; Figure 1.7). So far, it appears that water-rich asteroids with D > 250 km have shapes close to equilibrium (Ferrais et al., 2020; Yang et al., 2020). At similar sizes, rocky bodies such as (3) Juno appear more irregular. Still, rocky or even metal-rich bodies such as Psyche with D > 200 km have shapes that are closer to that of an ellipsoid than D < 200 km bodies (Ferrais et al., 2020), suggesting that the transition from a regular equilibrium shape to a fully irregular shape is smooth. The equilibrium shapes of water-rich bodies suggest that these bodies were fluid during their history, either as the result of early thermal heating leading to aqueous alteration of their interiors or because of large impacts and subsequent reaccumulation (e.g., Vernazza et al., 2020; Yang et al., 2020).

1.3.2.3 Cratering History of D > 100 km Asteroids

Prior to the first light of the adaptive-optics SPHERE imaging system at ESO/VLT in 2014, the characterization of the surface topography of asteroids had only been performed via dedicated in situ space missions (Galileo, NEAR Shoemaker, Hayabusa 1&2, Rosetta, Dawn, OSIRIS-REx) with the notable exception of Vesta's south pole impact basin detected by HST (Thomas et al., 1997a). The SPHERE survey of D > 100 km asteroids (Vernazza et al., 2018) has drastically augmented the number of asteroid craters detected from Earth. For example, Fetick et al. (2019) detected nine craters larger than 25 km in diameter on Vesta (e.g., Figure 1.5), Hanus et al. (2019) identified eight impact craters 20–40 km in diameter on (7) Iris, while Marsset et al. (2020) identified 36 craters larger than 30 km in diameter on (2) Pallas. The large number of D > 30 km craters on Pallas appears to be consistent with its high inclination ($\sim35°$) implying a more violent collisional history than that of Vesta or Ceres. Also, VLT/SPHERE observations of asteroid (89) Julia (Vernazza et al., 2018), a D $\sim$ 140 km S-type asteroid and the parent body of a small collisional family that consists of 66 known members with D < 2.5 km, have revealed the presence of an impact crater (~75 km wide) that could be the origin of this family. These results highlight the fact that the geological history of asteroids can now be investigated from the ground for a large number of asteroids, thus nicely complementing higher resolution data obtained in situ for a few bodies.

1.4 CONCLUSION AND PERSPECTIVE

When confronting the measurements of the global physical properties (average surface composition, size, shape, spin, rotation period, density) obtained from Earth with those obtained in-situ for both Vesta and Ceres, it appears that the results are about the same within errors (Table 1.3). This underlines the reliability of Earth-based observations and of the methods employed to analyze the data, especially when they are collected from the ground requiring atmospheric correction. Notably, recent VLT/SPHERE imaging observations have demonstrated in a striking manner how the gap between interplanetary missions and ground-based observations is getting narrower (Fetick et al., 2019; Vernazza et al., 2020). With the advent of extremely large telescopes (ELT, GMT, TMT), the science objectives of future interplanetary missions to Main Belt asteroids will have to be carefully thought out so that these missions will complement – not duplicate – what will be achieved via Earth-

based telescopic observations. For instance, future ELT adaptive-optics imaging of main-belt asteroids will allow us to resolve craters down to ~5 km in size, implying that we shall be able to characterize their global geological history from the ground. This implies that geological and geomorphological studies (based on high angular resolution imaging data) might become – during the coming decade – the prime science objectives that will be conducted from Earth in the case of large D > 100 km asteroids, dethroning studies related to the nature/origin of these bodies (deduced from the analysis of their spectra). Interplanetary missions to Main Belt asteroids performing cosmochemistry experiments, landing, and eventually a sample return should be preferred at the forefront of *in-situ* exploration as these would ideally complement the investigations conducted from Earth. In particular, IDP-like asteroids as well as the rare L-, Xe-, Xn- and O-types should be preferred targets as little is known regarding these bodies due to the paucity/absence of samples for these objects in our collections.

REFERENCES

A'Hearn, M. F., & Feldman, P. D. (1992) Water vaporization on Ceres. *Icarus*, 98, 54–60.

Allen, D. A. (1970) Infrared diameter of Vesta. *Nature*, 227, 158–159.

Ammannito, E., DeSanctis, M. C., Ciarniello, M., et al. (2016) Distribution of phyllosilicates on the surface of Ceres. *Science*, 353.

Ammannito, E., DeSanctis, M. C., Palomba, E., et al. (2013) Olivine in an unexpected location on Vesta's surface. *Nature* 504, 122–125.

Barucci, M. A., Dotto, E., Brucato J., et al. (2002) 10 Hygiea: ISO infrared observations. *Icarus*, 156, 202–210.

Beauvalet, L., & Marchis, F. (2014) Multiple asteroid systems (45) Eugenia and (87) Sylvia: Sensitivity to external and internal perturbations. *Icarus*, 241, 13–25.

Beck, P., Garenne, A., Quirico, E., et al. (2014) Transmission infrared spectra (2–25 μm) of carbonaceous chondrites (CI, CM, CV-CK, CR, C2 ungrouped): Mineralogy, water, and asteroidal processes. *Icarus*, 229, 263–277.

Bell, J. F. (1988) A probable asteroidal parent body for the CV or CO chondrites (abstract). *Meteoritics*, 23, 256–257.

Belton, M. J. S., Chapman C. R., Thomas, P. C., et al. (1995) Bulk density of asteroid 243 Ida from the orbit of its satellite Dactyl. *Nature*, 374, 785–788.

Binzel, R. P., DeMeo, F. E., Turtelboom, E. V., et al. (2019) Compositional distributions and evolutionary processes for the near-Earth object population: Results from the MIT-Hawaii Near-Earth Object Spectroscopic Survey (MITHNEOS). *Icarus*, 324, 41–76.

Binzel, R. P., Gaffey, M. J., Thomas, P. C., et al. (1997) Geologic mapping of Vesta from 1994 Hubble Space Telescope images. *Icarus*, 128, 95–103.

Binzel, R. P., Rivkin, A. S., Stuart, J., et al. (2004) Observed spectral properties of near-Earth objects: Results for population distribution, source regions, and space weathering processes. *Icarus*, 170, 259–294.

Binzel, R. P., & Xu, S. (1993) Chips off of asteroid 4 Vesta: Evidence for the parent body of basaltic achondrite meteorites. *Science*, 260, 186–191.

Blanco, C., & Catalano, S. (1979) UBV photometry of Vesta. *Icarus*, 40, 359–363.

Bottke, W. F., Nesvorny, D., Grimm, R. E., Morbidelli, A., & O'Brien, D. P. (2006) Iron meteorites as remnants of planetesimals formed in the terrestrial planet region. *Nature*, 439, 821–824.

Bowell, E., & Zellner, B. (1973) Polarizations of asteroids and satellites. In T. Gehrels (ed.), *Planets, Stars and Nebulae Studied with Photopolarimetry*. Tucson: University of Arizona Press, pp. 381–403.

Bradley, J. P., Keller, L. P., Brownlee, D. E., & Thomas, K. L. (1996). Reflectance spectroscopy of interplanetary dust particles. *Meteoritics and Planetary Science*, 31, 394–402.

Broz, M., Morbidelli, A., Bottke, W. F., et al. (2013) Constraining the cometary flux through the asteroid belt during the late heavy bombardment. *Astronomy & Astrophysics*, 551, A117.

Brunetto, R., Borg, J., Dartois, E., et al. (2011) Mid-IR, Far-IR, Raman micro-spectroscopy, and FESEM-EDX study of IDP L2021C5: Clues to its origin. *Icarus*, 212, 896–910.

Burbine, T. H. (2014) *Asteroids. Planets, Asteroids, Comets and The Solar System, Volume 2 of Treatise on Geochemistry*, 2nd ed., ed. Andrew M. Davis. Amsterdam: Elsevier, pp. 365–415.

Burbine, T. H. (2016) Advances in determining asteroid chemistries and mineralogies. *Chemie der Erde – Geochemistry*, 76, 181–195.

Burbine, T. H., Binzel, R. P., Bus, S. J., & Clark, B. E. (2001). K asteroids and CO3/CV3 chondrites. *Meteoritics & Planetary Science*, 36, 245–253.

Bus, S. J., & Binzel, R. P. (2002) Phase II of the small main-belt asteroid spectroscopic survey: A feature-based taxonomy. *Icarus*, 158, 146–177.

Campins, H., Hargrove, K., Pinilla-Alonson, N., *et al.* (2010) Water ice and organics on the surface of the asteroid 24 Themis. *Nature*, 464, 1320–1321.

Carry, B. (2012) Density of asteroids. *Planetary and Space Science*, 73, 98–118.

Carry, B., Dumas, C., Fulchignoni, M., et al. (2008) Near-infrared mapping and physical properties of the dwarf-planet Ceres. *Astronomy and Astrophysics*, 478, 235–244.

Carry, B., Dumas, C., Kaasalainen, M., et al. (2010a) Physical properties of (2) Pallas. *Icarus*, 205, 460–472.

Carry, B., Kaasalainen, M., Leyrat, C., et al. (2010b) Physical properties of the ESA Rosetta target asteroid (21) Lutetia. II. Shape and flyby geometry. *Astronomy and Astrophysics*, 523, A94.

Carry, B., Vachier, F., Berthier, J., et al. (2019) Homogeneous internal structure of CM-like asteroid (41) Daphne. *Astronomy and Astrophysics*, 623, A132.

Carry, B., Vernazza, P., Vachier, F., et al. (2021) Evidence of differentiation of the most primitive small bodies. *Astronomy and Astrophysics*, 630, A129.

Carry, B., Vernazza, P., Dumas, C., & Fulchignoni, M. (2010c) First disk-resolved spectroscopy of (4) Vesta. *Icarus*, 205, 473–482.

Chapman, C. R., McCord, T. B., & Johnson, T. V. (1973) Asteroid spectral reflectivities. *The Astronomical Journal*, 78, 126–140.

Chapman, C. R., Morrison, D., Zellner, B. (1975) Surface properties of asteroids: A synthesis of polarimetry, radiometry, and spectrophotometry. *Icarus*, 24, 104–130.

Chapman, C. R., Veverka, J., Thomas, P. C., et al. (1995) Discovery and physical properties of Dactyl, a satellite of asteroid 243 Ida. *Nature*, 374, 783–785.

Clark, B. E., Ockert-Bell, M. E., Cloutis, E. A., et al. (2009) Spectroscopy of K-complex asteroids: Parent bodies of carbonaceous meteorites? *Icarus*, 202, 119–133.

Clark, R. N. (2009) Detection of adsorbed water and hydroxyl on the moon. *Science*, 326, 562–564.

Cloutis, E. A., Gaffey, M. J., Smith, D. G. W., & Lambert, R. St. J. (1990) Reflectance spectra of "featureless" materials and the surface mineralogies of M- and E-class asteroids. *Journal of Geophysical Research*, 95, 281–293.

Cloutis, E. A., Hardensen, P. S., Bish, D. L., et al. (2010) Reflectance spectra of iron meteorites: Implications for spectral identification of their parent bodies. *Meteoritics and Planetary Science*, 45, 304–332.

Cloutis, E. A., Hiroi, T., Gaffey, M. J., Alexander, C. M. O'D., & Mann, P. (2011) Spectral reflectance properties of carbonaceous chondrites: 1. CI chondrites. *Icarus*, 212, 180–209.

Cloutis, E. A., Hudon, P., Hiroi, T., & Gaffey, M. J. (2012) Spectral reflectance properties of carbonaceous chondrites 4: Aqueously altered and thermally metamorphosed meteorites. *Icarus*, 220, 586–617.

Cloutis, E. A., Izawa, M. R. M., Pompilio, L., et al. (2013). Spectral reflectance properties of HED meteorites + CM2 carbonaceous chondrites: Comparison to HED grain size and compositional variations and implications for the nature of low-albedo features on Asteroid 4 Vesta. *Icarus*, 223, 850–877.

Combe, J.-Ph., McCord, T. B., Tosi, F., et al. (2016) Detection of local H_2O exposed at the surface of Ceres. *Science*, 353, aaf3010.

Cruikshank, D. P., & Morrison, D. (1973). Radii and albedos of asteroids 1, 2, 3, 4, 6, 15, 51, 433, and 511. *Icarus*, 20, 477–481.

Cruikshank, D. P., Tholen, D. J., Hartmann, W. K., Bell, J. F., & Brown, R. H. (1991) Three basaltic earth-approaching asteroids and the source of the basaltic meteorites. *Icarus*, 89, 1–13.

De Sanctis, M. C., Ammannito, E., Capria, M. T., et al. (2012) Spectroscopic characterization of mineralogy and its diversity across Vesta. *Science*, 336, 697–700.

De Sanctis, M. C., Ammannito, E., McSween, H. Y., et al. (2017) Localized aliphatic organic material on the surface of Ceres. *Science*, 355, 719–722.

De Sanctis, M. C., Ammannito, E., Raponi, A., et al. (2015) Ammoniated phyllosilicates with a likely outer Solar System origin on (1) Ceres. *Nature*, 528, 241–244.

De Sanctis, M. C., Ammannito, E., Raponi, A., et al. (2020) Fresh emplacement of hydrated sodium chloride on Ceres from ascending salty fluids. *Nature Astronomy*, 4, 786–793.

De Sanctis, M. C., Raponi, A., Ammannito, E., et al. (2016) Bright carbonate deposits as evidence of aqueous alteration on (1) Ceres. *Nature*, 536, 54–57.

Degewij, J., Tedesco, E. F., & Zellner, B. (1979) Albedo and color contrasts on asteroid surfaces. *Icarus*, 40, 364–374.

DeMeo, F. E., Alexander, C. M. O., Walsh, K. J., Chapman, C. R., & Binzel, R. P. (2015) The compositional structure of the asteroid belt. In P. Michel, F. E. DeMeo, & W. F. Bottke (eds.), *Asteroids IV*. Tucson: University of Arizona Press, pp. 13–42.

DeMeo, F. E., Binzel, R. P., Slivan, S. M., & Bus, S. J. (2009) An extension of the Bus asteroid taxonomy into the near-infrared. *Icarus*, 202, 160–180.

DeMeo, F. E., & Carry, B. (2013) The taxonomic distribution of asteroids from multi-filter all sky photometric surveys. *Icarus*, 226, 723–741.

DeMeo, F. E., & Carry, B. (2014) Solar System evolution from compositional mapping of the asteroid belt. *Nature*, 505, 629–634.

DeMeo, F. E., Polihook, D., Carry, B., et al. (2019) Olivine-dominated A-type asteroids in the Main Belt: Distribution, abundance and relation to families. *Icarus*, 322, 13–30.

Dotto, E., Müller, T. G., Barucci, M. A., et al. (2000) ISO results on bright Main Belt asteroids: PHT-S observations. *Astronomy and Astrophysics*, 358, 1133–1141.

Drake, M. J. (2001) The eucrite/Vesta story. *Meteoritics & Planetary Science*, 36, 501–513.

Emery, J. P., Cruikshank, D. P., & van Cleve, J. (2006) Thermal emission spectroscopy (5.2–38 μm) of three Trojan asteroids with the Spitzer Space Telescope: Detection of fine-grained silicates. *Icarus*, 182, 496–512.

Feierberg, M. A., & Drake, M. J. (1980) The meteorite–asteroid connection: The infrared spectra of Eucrites, Shergottites, and Vesta. *Science*, 209, 805–807.

Feierberg, M. A., Larson, H. P., Fink, U., & Smith, H. A. (1980) Spectroscopic evidence for two achondrite parent bodies: asteroids 349 Dembowska and 4 Vesta. *Geochimica et Cosmochimica Acta*, 44, 513–524.

Ferrais, M., Vernazza, P., Jorda, L., et al. (2020) Asteroid (16) Psyche's primordial shape: A possible Jacobi ellipsoid, *Astronomy & Astrophysics*, 638, L15.

Fetick, R., Jorda, L., Vernazza, P., et al. (2019) Closing the gap between Earth-based and interplanetary mission observations: Vesta seen by VLT/SPHERE. *Astronomy & Astrophysics*, 623, A6.

Fulvio, D., Ieva, S., Perna, D., et al. (2018) Statistical analysis of the spectral properties of V-type asteroids: A review on what we known and what is still missing. *Planetary and Space Science*, 164, 37–43.

Fusco, T., Mugnier, L. M., Conan, J.-M., et al. (2003) Deconvolution of astronomical images obtained from ground-based telescopes with adaptive optics. In P. L. Wizinowich, & D. Bonaccini (eds.), *Adaptive Optical System Technologies II. Proceedings of the SPIE 4839.* Bellingham, WA: SPIE, pp. 1065–1075.

Gaffey, M. J. (1976) Spectral reflectance characteristics of the meteorite classes. *Journal of Geophysical Research*, 81, 905–920.

Gaffey, M. J. (1997) Surface lithologic heterogeneity of asteroid 4 Vesta. *Icarus*, 127, 130–157.

Gomes, R., Levison, H. F., Tsiganis, K., & Morbidelli, A. (2005) Origin of the cataclysmic late heavy bombardment period of the terrestrial planets. *Nature*, 435, 466–469.

Gradie, J., & Tedesco, E. (1982) Compositional structure of the asteroid belt. *Science*, 216, 1405–1407.

Greenwood, R. C., Burbine, T. H., & Franchi, I. A. (2020) Linking asteroids and meteorites to the primordial planetesimal population. Geochimica. *Et Cosmochimica Acta*, 27, 377–406.

Hanus, J., Marchis, F., Viikinkoski, M., et al. (2017a) Shape model of asteroid (130) Elektra from optical photometry and disk-resolved images from VLT/SPHERE and Nirc2/Keck. *Astronomy & Astrophysics*, 599, A36.

Hanus, J., Marsset, M., Vernazza, P., et al. (2019) The shape of (7) Iris as evidence of an ancient large impact ? *Astronomy & Astrophysics*, 624, A121.

Hanus, J., Vernazza, P., Viikinkoski, M., et al. (2020) (704) Interamnia: A transitional object between a dwarf planet and a typical irregular-shaped minor body. *Astronomy & Astrophysics*, 633, A65.

Hanus, J., Viikinkoski, M., Marchis, F., et al. (2017b) Volumes and bulk densities of forty asteroids from ADAM shape modeling. *Astronomy & Astrophysics*, 601, A114.

Hardersen, P., Cloutis, E., Reddy, V., et al. (2011) The M-/X-asteroid menagerie: Results of an NIR spectral survey of 45 main-belt asteroids. *Meteoritics & Planetary Science*, 46, 1910–1938.

Hardersen, P., Gaffey, M. J., & Abel, P. A. (2004) Mineralogy of Asteroid 1459 Magnya and implications for its origin. *Icarus*, 167, 170–177.

Hardersen, P., Reddy, V., Cloutis, E., et al. (2018) Basalt or not? Near-infrared spectra, surface mineralogical estimates, and meteorite analogs for 33 Vp-type asteroid. *The Astronomical Journal*, 156, 11.

Hardersen, P., Reddy, V., & Roberts, R. (2015) Vestoids, part II: The Basaltic nature and HED meteorite analogs for eight Vp-type asteroids and their associations with (4) Vesta. *The Astrophysical Journal Supplement Series*, 221, 19.

Hardersen, P., Reddy, V., Roberts, R., & Mainzer, A. (2014) More chips off of asteroid (4) Vesta: Characterization of eight Vestoids and their HED meteorite analogs. *Icarus*, 242, 269–282.

Hargrove, K. D., Emery, J. P., Campins, H., & Kelley, M. S. (2015) Asteroid (90) antiope: Another icy member of the Themis family? *Icarus*, 254, 150–156.

Hargrove, K. D., Kelley, M. S., Campins, H., et al. (2012) Asteroids (65) cybele, (107) Camilla and (121) Hermione: Infrared spectral diversity among the cybeles. *Icarus*, 221, 453–455.

Hasegawa, S., Kuroda, D., Yanagisawa, K., & Usui, F. (2017) Follow-up observations for the asteroid catalog using AKARI spectroscopic observations. *Publications of the Astronomical Society of Japan*, 69, 99.

Hasegawa, S., Miyasaka, S., Tokimasa, N., et al. (2014) The opposition effect of the asteroid 4 Vesta. *Publications of the Astronomical Society of Japan*, 66, 89.

Hasegawa, S., Murakawa, K., Ishiguro, M., et al. (2003) Evidence of hydrated and/or hydroxylated minerals on the surface of asteroid 4 Vesta. *Geophysical Research Letters*, 30, 2123.

Heras, A. M., Morris, P. W., Vandenbussche, B., & Müller, T. G. (2000) Asteroid 4 Vesta as seen with the ISO short wavelength spectrometer. In M. L. Sitko, A. L. Sprague, & O. K. Lynch (eds.), *Thermal Emission Spectroscopy and Analysis of Dust, Disks and Regoliths, Astronomical Society of the Pacific Conference Series*, 196. San Francisco, CA: ASP, pp. 205–213.

Hiroi, T., Pieters, C. M., & Takeda, H. (1994) Grain size of the surface regolith asteroid 4 Vesta estimated from its reflectance spectrum in comparison with HED meteorites. *Meteoritics*, 29, 394–396.

Hiroi, T., Pieters, C. M., Zolensky, M. E., & Lipschutz, M. E. (1993) Evidence of thermal metamorphism on the C, G, B, and F asteroids. *Science*, 261, 1016–1018.

Hsieh, H. H., & Jewitt, D. A. (2006) Population of comets in the main asteroid belt. *Science*, 312, 561–563.

Huss, G. R., Rubin, A. E., & Grossman, J. N. (2006) Thermal metamorphism in chondrites. In D. S. Lauretta, & H. Y. McSween Jr. (eds.), *Meteorites and the Early Solar System II.* Tucson: University of Arizona Press, pp. 567–586.

Jaumann, R., Williams, D. A., Buczkowski, D. L., et al. (2012) Vesta's shape and morphology. *Science*, 336, 687–690.

Jewitt, D. (2012). The active asteroids. *The Astronomical Journal*, 143, 66.

Jewitt, D., Hsieh, H. H., & Agaral, J. (2015) The active asteroids. In P. Michel, F. E. DeMeo, & W. F. Bottke (eds.), *Asteroids IV.* Tucson: University of Arizona Press, pp. 221–242.

Johnson, T. V., & Fanale, F. P. (1973) Optical properties of carbonaceous chondrites and their relationship to asteroids. *Journal of Geophysical Research*, 78, 8507–8518.

Johnson, T. V., Matson, D. L., Veeder, G. J., & Loer, S. J. (1975) Asteroids: Infrared photometry at 1.25, 1.65, and 2.2 microns. *Astrophysical Journal*, 197, 527–531.

Jorda, L., Gaskell, R., Capanna, C., et al. (2016) The global shape, density and rotation of Comet 67P/Churyumov-Gerasimenko from preperihelion Rosetta/OSIRIS observations. *Icarus*, 277, 257–278.

King, T. V. V., Clark, R. N., Calvin, W. M., Sherman, D. M., & Brown, R. H. (1992) Evidence for ammonium-bearing minerals on Ceres. *Science*, 255, 1551–1553.

Konopliv, A. S., Asmar, S. W., Folkner, W. M., et al. (2011) Mars high resolution gravity fields from MRO, Mars seasonal gravity, and other dynamical parameters. *Icarus*, 211, 401–428.

Larson, H. P., & Fink, U. (1975) Infrared spectral observations of asteroid 4 Vesta. *Icarus*, 26, 420–427.

Lazzaro, D., Angeli, C. A., Carvano, J. M., et al. (2004) S^3OS^2: The visible spectroscopic survey of 820 asteroids. *Icarus*, 172, 179–220.

Lazzaro, D., Michtchenko, T., Carvano, J. M., et al. (2000) Discovery of a basaltic asteroid in the outer Main Belt. *Science*, 288, 2033–2035.

Lebofsky, L. A. (1978) Asteroid 1 Ceres: Evidence for water of hydration. *Monthly Notices of the Royal Astronomical Society*, 182, 17–21.

Lebofsky, L. A., Feierberg, M. A., Tokunaga, A. T., et al. (1981) The 1.7 to 4.2 μm spectrum of asteroid 1 Ceres: Evidence for structural water in clay minerals. *Icarus*, 48, 453–459.

Levison, H. F., Bottke, W. F., Gounelle, M., et al. (2009) Contamination of the asteroid belt by primordial trans-Neptunian objects. *Nature*, 460, 364–366.

Li, J.-Y., McFadden, L., Parker, J., et al. (2006) Photometric analysis of 1 Ceres and surface mapping from HST observations. *Icarus*, 182, 143–160.

Licandro, J., Campins, H., Kelley, M., et al. (2011) (65) Cybele: Detection of small silicate grains, water-ice, and organics. *Astronomy and Astrophysics*, 525, id.A34.

Licandro, J., Hargrove, K., Kelley, M., et al. (2012) 5–14 μm Spitzer spectra of Themis family asteroids. *Astronomy and Astrophysics*, 537, A73.

Marchi, S., Ermakov, A. I., Raymond, C. A., et al. (2016) The missing large impact craters on Ceres. *Nature Communications*, 7, 12257.

Marchi, S., McSween, H. Y., O'Brien, D. P., et al. (2012) The violent collisional history of asteroid 4 Vesta. *Science*, 336, 690–694.

Marchis, F., Descamps, P., Hestroffer, D., & Berthier, J. (2005) Discovery of the triple asteroidal system 87 Sylvia. *Nature*, 436, 822–824.

Marchis, F., Enriquez, J. E., Emery, J. P., et al. (2012) Multiple asteroid systems: Dimensions and thermal properties from Spitzer Space Telescope and ground-based observations. *Icarus*, 221, 1130–1161.

Margot, J.-L., Pravec, P., Taylor, P., et al. (2015) Asteroid systems: Binaries, triples, and pairs. In P. Michel, F. E. DeMeo, & W. F. Bottke (eds.), *Asteroids IV*. Tucson: University of Arizona Press, pp. 355–374.

Marsset, M., Broz, M., Vernazza, P., et al. (2020) The violent collisional history of aqueously evolved (2) Pallas. *Nature Astronomy*, 4, 569–576.

Marsset, M., Carry, B., Dumas, C., et al. (2017) 3D shape of asteroid (6) Hebe from VLT/SPHERE imaging: Implications for the origin of ordinary H chondrites. *Astronomy & Astrophysics*, 604, A64.

Marsset, M., Vernazza, P., Birlan, M., et al. (2016) Compositional characterisation of the Themis family. *Astronomy & Astrophysics*, 586, A15.

Masiero, J. R., Grav, T., Mainzer, A. K., et al. (2014) Main-belt asteroids with WISE/NEOWISE: Near-infrared albedos. *The Astrophysical Journal*, 791, 121.

Masiero, J. R., Mainzer, A. K., Grav, T., et al. (2011) Main Belt asteroids with WISE/NEOWISE. I. Preliminary albedos and diameters. *The Astrophysical Journal*, 741, 68.

Matson, D. L. (1971) Infrared Emission from Asteroids at Wavelengths of 8.5, 10.5 and 11.6 Millimicron. PhD thesis, California Institute of Technology.

McCord, T., Adams, J., & Johnson, T. V. (1970) Asteroid Vesta: Spectral reflectivity and compositional implications. *Science*, 168, 1445–1447.

McFadden, L. A., McCord, T. B., & Pieters, C. (1977) Vesta: The first pyroxene band from new spectroscopic measurements. *Icarus*, 31, 439–446.

Merline, W. J., Weidenshilling, S. J., Durda, D. D., et al. (2002) Asteroids do have satellites. In W. F. Bottke Jr., A. Cellino, P. Paolicchi, & R. P. Binzel (eds.), *Asteroids III*, Tucson: University of Arizona Press, pp. 289–312.

Merouane, S., Djouadi, Z., & Le Sergeant d'Hendecourt, L. (2014) Relations between aliphatics and silicate components in 12 stratospheric particles deduced from vibrational spectroscopy. *The Astrophysical Journal*, 780, 174.

Milliken, R. E., & Rivkin, A. S. (2009) Brucite and carbonate assemblages from altered olivine-rich materials on Ceres. *Nature Geoscience*, 2, 258–261.

Monnereau, M., Toplis, M. J., Baratoux, D., & Guignard, J. (2013) Thermal history of the H-chondrite parent body: Implications for metamorphic grade and accretionary time-scale. *Geochimica et Cosmochimica Acta*, 119, 302–321.

Morbidelli, A., Bottke, W. F., Nesvorny, D., & Levison, H. F. (2009) Asteroids were born big. *Icarus*, 204, 558–573.

Morbidelli, A., Levison, H. F., Tsiganis, K., & Gomes, R. (2005) Chaotic capture of Jupiter's Trojan asteroids in the early Solar System. *Nature*, 435, 462–465.

Moskovitz, M., Jedicke, R., Gaidos, E., et al. (2008) The distribution of basaltic asteroids in the Main Belt. *Icarus*, 198, 77–90.

Nesvorny, D. (2015) *Nesvorny HCM Asteroid Families V3.0. EAR-A-VARGBDET-5-NESVORNYFAM-V3.0.* NASA Planetary Data System.

Neveu, M., & Vernazza, P. (2019) IDP-like asteroids formed later than 5 Myr after Ca-Al rich inclusions. *The Astrophysical Journal*, 875, 30.

Nolan, M. C., Magri, C., Howell, E. S., et al. (2013) Shape model and surface properties of the OSIRIS-REx target Asteroid (101955) Bennu from radar and lightcurve observations. *Icarus*, 226, 629–640.

Ockert-Bell, M. E., Clark, B. E., Shepard, M. K., et al. (2010) The composition of M-type asteroids: Synthesis of spectroscopic and radar observations. *Icarus*, 210, 674–692.

Pajuelo, M., Carry, B., Vachier, F., et al. (2018) Physical, spectral, and dynamical properties of asteroid (107) Camilla and its satellites. *Icarus*, 309, 134–161.

Park, R. S., Konopliv, A. S., Bills, B. G., et al. (2016) A partially differentiated interior for (1) Ceres deduced from its gravity field and shape. *Nature*, 537, 515–517.

Park, R. S., Vaughan, A. T., Konopliv, A. S., et al. (2019) High-resolution shape model of Ceres from stereophotoclinometry using Dawn imaging data. *Icarus*, 319, 812–827.

Pinilla-Alonso, N., De Leon, J., Walsh, K. J., et al. (2016) Portrait of the Polana-Eulalia family complex: Surface homogeneity revealed from near-infrared spectroscopy. *Icarus*, 274, 231–248.

Raymond, S. N., & Izidoro, A. (2017) The empty primordial asteroid belt. *Science Advances*, 3, e1701138.

Reddy, V., Dunn, T. L., Thomas, C. A., Moskovitz, N. A., & Burbine, T. H. (2015) Mineralogy and surface composition of asteroids. In P. Michel, F. E. DeMeo, & W. F. Bottke (eds.), *Asteroids IV*. Tucson: University of Arizona Press, pp. 43–64.

Rivkin, A. S. (2012) The fraction of hydrated C-complex asteroids in the asteroid belt from SDSS data. *Icarus*, 221, 744–752.

Rivkin, A. S., Campins, H., Emery, J. P., et al. (2015) Astronomical observations of volatiles on asteroids. In P. Michel, F. E. DeMeo, & W. F. Bottke (eds.), *Asteroids IV*. Tucson: University of Arizona Press, pp. 65–87.

Rivkin, A. S., & Emery, J. P. (2010) Detection of ice and organics on an asteroidal surface. *Nature*, 464, 1322–1323.

Rivkin, A. S., Howell, E. S., & Emery, J. P. (2019) Infrared spectroscopy of large, low-albedo asteroids: Are Ceres and Themis archetypes or outliers? *Journal of Geophysical Research: Planets*, 124, 1393–1409.

Rivkin, A. S., Howell, E. S., Vilas, F., & Lebofsky, L. A. (2002) Hydrated minerals on asteroids: The astronomical record. In W. F. Bottke Jr., A. Cellino, P. Paolicchi, & R. P. Binzel (eds.), *Asteroids III*. Tucson: University of Arizona Press, pp. 235–253.

Rivkin, A. S., Li, J.-Y., Milliken, R. E., et al. (2011) The surface composition of Ceres. *Space Science Reviews*, 163, 95–116.

Rivkin, A. S., McFadden, L., Binzel, R. P., & Sykes, M. (2006a) Rotationally-resolved spectroscopy of Vesta I: 2–4 μm region. *Icarus*, 180, 464–472.

Rivkin, A. S., Volquardsen, E. L., & Clark, B. E. (2006b) The surface composition of Ceres: Discovery of carbonates and iron-rich clays. *Icarus*, 185, 563–567.

Roig, F., & Gil-Hutton, R. (2006) Selecting candidate V-type asteroids from the analysis of the Sloan Digital Sky Survey colors. *Icarus*, 183, 411–419.

Roig, F., Nesvorný, D., Gil-Hutton, R., & Lazzaro, D. (2008) V-type asteroids in the middle Main Belt. *Icarus*, 194, 125–136.

Rubincam, D. P. (2000) Radiative spin-up and spin-down of small asteroids. *Icarus*, 148, 2–11.

Russell, C. T., Raymond, C. A., Ammannito, E., et al. (2016) Dawn arrives at Ceres: Exploration of a small, volatile-rich world. *Science*, 353, 1008–1010.

Russell, C. T., Raymond, C. A., Coradini, A., et al. (2012) Dawn at Vesta: Testing the protoplanetary paradigm. *Science*, 336, 684–686.

Ryan, E. L., & Woodward, C. E. 2010. Rectified asteroid albedos and diameters from IRAS and MSX photometry catalogs. *The Astronomical Journal*, 140, 933–943.

Scheeres, D. J., Britt, D., Carry, B., & Holsapple, K. A. (2015) *Asteroid interiors and morphology*. In P. Michel, F. E. DeMeo, & W. F. Bottke (eds.), *Asteroids IV*. Tucson: University of Arizona Press, pp. 745–766.

Schenk, P., O'Brien, D. P., Marchi, S., et al. (2012) The geologically recent giant impact basins at Vesta's south pole. *Science*, 336, 694.

Schmidt, B. E., Thomas, P. C., Bauer, J. M., et al. (2009) The shape and surface variation of 2 Pallas from the Hubble space telescope. *Science*, 326, 275–278.

Schröder, S. E., Mottola, S., Carsenty, U., et al. (2017) Resolved spectrophotometric properties of the Ceres surface from Dawn Framing Camera images. *Icarus*, 288, 201–225.

Schröder, S. E., Mottola, S., Keller, H. U., et al. (2014) Resolved spectrophotometric properties of the Ceres surface from Dawn Framing Camera images. *Planetary and Space Science*, 103, 66–81.

Shepard, M. K., Taylor, P. A., Nolan, M. C., et al. (2015) A radar survey of M- and X-class asteroids. III. Insights into their composition, hydration state, and structure. *Icarus*, 245, 38–55.

Sunshine, J. M., Bus, S. J., Corrigan, C. M., McCoy, T. J., & Burbine, T. H. (2007) Olivine-dominated asteroids and meteorites: Distinguishing nebular and igneous histories. *Meteoritics & Planetary Science*, 42, 155–170.

Sunshine, J. M., Connolly, H. C., McCoy, T. J., et al. (2008) Ancient asteroids enriched in refractory inclusions. *Science*, 320, 514–516.

Takir, D., & Emery, J. P. (2012) Outer Main Belt asteroids: Identification and distribution of four 3-mum spectral groups. *Icarus*, 219, 641–654.

Takir, D., Emery, J. P., & McSween, H. Y. (2015) Toward an understanding of phyllosilicate mineralogy in the outer main asteroid belt. *Icarus*, 257, 185–193.

Tedesco, E. F., Noah, P. V., Noah, M., & Price, S. D. (2002) The supplemental IRAS minor planet survey. *The Astronomical Journal*, 123, 1056–1085.

Tholen, D. J., & Barucci, M. A. (1989) Asteroid taxonomy. In R. P. Binzel, T. Gehrels, & M. S. Matthews (eds.), *Asteroids II*. Tucson: University of Arizona Press, pp. 298–315.

Thomas C. A., Emery J. P., Trilling D. E., et al. (2014) Physical characterization of warm Spitzer-observed near-Earth objects. *Icarus*, 228, 217–246.

Thomas, P. C., Binzel. R. P., Gaffey, M. J., et al. (1997a) Impact excavation on asteroid 4 Vesta: Hubble space telescope results. *Science*, 277, 1492–1495.

Thomas, P. C., Binzel. R. P., Gaffey, M. J., et al. (1997b) Vesta: Spin pole, size, and shape from HST images. *Icarus*, 128, 88–94.

Thomas, P. C., Parker, J. Wm., McFadden, L., et al. (2005) Differentiation of the asteroid Ceres as revealed by its shape. *Nature*, 437, 224–226.

Tsiganis, K., Gomes, R., Morbidelli, A., & Levison, H. F. (2005) Origin of the orbital architecture of the giant planets of the Solar System. *Nature*, 435, 459–461.

Usui, F., Hasegawa, S., Ootsubo, T., & Onaka, T. (2019) AKARI/IRC near-infrared asteroid spectroscopic survey: AcuA-spec. *Publications of the Astronomical Society of Japan*, 71, 1.

Usui, F., Kuroda, D. Müller, T. G., et al. (2011) Asteroid catalog using AKARI: AKARI/IRC mid-infrared asteroid survey. *Publications of the Astronomical Society of Japan*, 63, 1117–1138.

Vernazza, P., & Beck, P. (2017) Composition of Solar System small bodies. In L. T. Elkins-Tanton, & B. P. Weiss (eds.), *Planetesimals: Early Differentiation and Consequences for Planets*. Cambridge: Cambridge University Press, pp. 269–297.

Vernazza, P., Broz, M., Drouard, D., et al. (2018) The impact crater at the origin of the Julia family detected with VLT/SPHERE? *Astronomy & Astrophysics*, 618, A154.

Vernazza, P., Brunetto, R., Binzel, R. P., et al. (2009) Plausible parent bodies for enstatite chondrites and mesosiderites: Implications for Lutetia's fly-by. *Icarus*, 202, 477–486.

Vernazza, P., Carry, B., Vachier, F., et al. (2019) New satellite around (31) Euphrosyne. *IAU circular 4627*.

Vernazza, P., Castillo-Rogez, J., Beck, P., et al. (2017) Different origins or different evolutions? Decoding the spectral diversity among C-type asteroids. *The Astrophysical Journal*, 153, 72.

Vernazza, P., Ferrais, M., Jorda, L., et al. VLT/SPHERE imaging survey of the largest main-belt asteroids: Final results and synthesis. A&A 654, A56, 2021.

Vernazza, P., Fulvio, D., Brunetto, R., et al. (2013) Paucity of Tagish Lake-like parent bodies in the asteroid belt and among Jupiter trojans. *Icarus*, 225, 517–525.

Vernazza, P., Jorda, L., Sevecek, P., et al. (2020) A basin-free spherical shape as an outcome of a giant impact on asteroid Hygiea. *Nature Astronomy*, 4, 136–141.

Vernazza, P., Lamy, P., Groussin, O., et al. (2011) Asteroid (21) Lutetia as a remnant of Earth's precursor planetesimals. *Icarus*, 216, 650–659.

Vernazza, P., Marsset, B., Beck, P., et al. (2015a) Interplanetary dust particles as samples of icy asteroids. *The Astrophysical Journal*, 806, 204.

Vernazza, P., Marsset, B., Beck, P., et al. (2016) Compositional homogeneity of CM parent bodies. *The Astrophysical Journal*, 152, 54.

Vernazza, P., Mothé-Diniz, T., Barucci, M. A., et al. (2005) Analysis of near-IR spectra of 1 Ceres and 4 Vesta, targets of the Dawn mission. *Astronomy & Astrophysics*, 436, 1113–1121.

Vernazza, P., Zanda, B., Binzel, R. P., et al. (2014) Multiple and fast: The accretion of ordinary chondrite parent bodies. *Astrophysical Journal*, 791, 120.

Vernazza, P., Zanda, B., Nakamura, T., et al. (2015b) The formation and evolution of ordinary chondrite parent bodies. In P. Michel, F. E. DeMeo, & W. F. Bottke (eds.), *Asteroids IV*. Tucson: University of Arizona Press, pp. 617–634.

Veverka, J. F. (1970) Photometric and Polarimetric Studies of Minor Planets and Satellites. PhD thesis, Harvard University.

Viikinkoski, M., Kaasalainen, M., & Durech, J. (2015a) ADAM: A general method for using various data types in asteroid reconstruction. *Astronomy & Astrophysics*, 576, A8.

Viikinkoski, M., Kaasalainen, M., Durech, J., et al. (2015b) VLT/SPHERE- and ALMA-based shape reconstruction of asteroid (3) Juno. *Astronomy & Astrophysics*, 581, L3.

Viikinkoski, M., Vernazza, P., Hanus, J., et al. (2018) (16) Psyche: A mesosiderite-like asteroid? *Astronomy & Astrophysics*, 619, L3.

Vilas, F., & Gaffey, M. J. (1989) Phyllosilicate absorption features in main-belt and outer-belt asteroid reflectance spectra. *Science*, 246, 790–792.

Vokrouhlický, D., Nesvorný, D., & Bottke, W. F. (2003) The vector alignments of asteroid spins by thermal torques. *Nature*, 425, 147–151.

Walsh, K. J., Morbidelli, A., Raymond, S. N., O'Brien, D. P., & Mandell, A. M. (2011) A low mass for Mars from Jupiter's early gas-driven migration. *Nature*, 475, 206–209.

Watanabe, S., HiraBayashi, M., Hirata, N., et al. (2019) Hayabusa2 arrives at the carbonaceous asteroid 162173 Ryugu – A spinning top-shaped rubble pile. *Science*, 364, 268–272.

Williams, J. G. (1979) Proper elements and family memberships of the asteroids. In T. Gehrels (ed.), *Asteroids*. Tucson: University of Arizona Press, pp. 1040–1063.

Yang, B., Hanus, J, Carry, B., et al. (2020) Binary asteroid (31) Euphrosyne: Ice-rich and nearly spherical. *Astronomy & Astrophysics*, 641, A80.

Yang, B., & Jewitt, D. (2010) Identification of magnetite in B-type asteroids. *The Astronomical Journal*, 140, 692–698.

Zappalà, V., Cellino, A., Farinella, P., & Knežević, Z. (1990) Asteroid families. I. Identification by hierarchical clustering and reliability assessment. *The Astronomical Journal*, 100, 2030–2046.

Exploring Vesta and Ceres

CHRISTOPHER T. RUSSELL AND MARC D. RAYMAN

2.1 INTRODUCTION

While Buck Rogers and other science-fiction heroes could explore planet after planet on a single journey into space, our initial exploration of the Solar System consisted of single-targeted orbiters and flybys, with a few multiple-targeted missions. In the 1960s, the United States and the former USSR both sent spacecraft to Venus and Mars to conduct the initial reconnaissance of these distant worlds. NASA's Mariner 2, the first successful interplanetary mission, flew past Venus in 1962, showing the cloud-enshrouded planet. Mariner 4 provided humankind's first close-up views of Mars in 1965, showing a cratered surface quite unlike what many scientists had expected and starkly different from the popular imagination. The first orbiters of those planets were in the 1970s. The Pioneer 10 and 11 spacecraft, launched in 1972 and 1973, traveled to the outer Solar System, and conducted the first measurements of Jupiter in 1973 and 1974, and Saturn in 1974 and 1979. It was obvious that our single-use spacecraft requiring multiple launches to explore our Solar System was a more expensive undertaking than Buck Rogers faced, but the vehicles' propulsion systems in the early space programs had a low specific impulse. Unlike the systems used by Buck Rogers, these spacecraft traveled at a speed much, much slower than the speed of light, and they lacked the capability to orbit multiple destinations. Late in the twentieth century, starting along two different paths that eventually joined, the authors of this article came to appreciate this issue and became involved in an attempt to partially solve this problem, by enabling the exploration by a single spacecraft of the two most massive objects in the main asteroid belt, Vesta and Ceres.

In the 1970s, many planetary scientists were already aware of the promise of solar electric propulsion, and ion propulsion in particular. CTR remembers a 1978 lecture by Janet G. Luhmann of Aerospace Corporation, where she rued the possible pollution of the Earth's magnetosphere with argon, a gas that had been proposed for ion propulsion. Later, CTR himself proposed and completed a NASA-funded study of a four-spacecraft mission that toured the Earth's magnetosphere using solar electric propulsion. This became a proof-of-concept study, since neither argon nor xenon thrusters were space qualified before the last decade of the

twentieth century. These early studies were only paper exercises. We had yet to build hardware.

In 1992, NASA's planetary program was invigorated with the announcement of the Discovery Program, a long-term, PI-led series of missions that allowed innovative ideas to be tested, spawning greater competition. The program began with a workshop in San Juan Capistrano in the fall of that year. NASA's Lewis Research Center (now the Glenn Research Center) participated in the workshop. They had developed and were testing ion engines but had mainly terrestrial missions in mind. The combination of ion engines powered by solar arrays became known as solar electric propulsion, or SEP. Clearly the technology of ion propulsion had even greater potential in the Solar System. CTR approached the Lewis Research Center about planetary applications, and a study group was formed that included Tom McCord, Lucy McFadden, Mark Sykes, and Carle Pieters. This team became the nucleus of a group that, at the first opportunity in 1994, proposed to NASA a lunar orbiter with a cometary encounter.

This proposal would have completed the study of the geology of the moon with multi spectral images at low altitude and hence high resolution, and at the end of the lunar observations, it would fly by a small body that weakly outgassed. It lost to a low-cost chemically-propelled lunar mission, Lunar Prospector, with a similar payload. The study group then submitted two proposals in 1996 that were in competition with each other. Of these two proposals, one would have matched orbits with a small active comet, and the other would have flown to the second most massive body in the asteroid belt, Vesta, and orbited it. Neither were selected.

In 1998, the team decided to pick only one mission to propose, and going to Vesta was clearly the first choice. Vesta was the center of a long-fought controversy regarding the howardite–eucrite–diogenite (HED) meteorites. These meteorites comprise ~5% of the meteorites that have fallen on Earth., and they dominate the class of basaltic achondrite meteorites. Their spectral reflectance closely resembles that of Vesta and a vocal constituency advocated that Vesta was the source of the HED meteorites, while an equally vocal constituency voiced their opposition to the idea of a single HED parent body. If, in fact, we could identify a class of meteorites with a particular asteroid, we would have a powerful tool for understanding the geochemical origin of that body. Geochemistry provides very accurate timing and compositional constraints. The proposal team decided to focus on a single target, but, again, the ion propulsion mission was not selected.

Undaunted, the team decided to propose once again to the Discovery program in 2000. But now, their stars were literally aligned. In the most optimized launch possible for the year 2000

A portion of the research was carried out at the Jet Propulsion Laboratory, California Institute of Technology, under a contract with the National Aeronautics and Space Administration (80NM0018D0004).

proposal, Vesta and Ceres could both be orbited and mapped. This scenario was very much in the manner of a Buck Rogers space voyage. The spacecraft would reach its destination, spiral in and spiral back out, and leave for the next planet. Moreover, Vesta and Ceres were planetary bodies with sizable gravitational fields. They were major players in the origin of the Solar System.

In parallel with this activity in the planetary science program, NASA established the New Millennium Program, dedicated to helping reduce the risk and cost of using new technology, and SEP was determined to be a high priority technology in this program. Marc Rayman, working at JPL, became involved in developing the first mission in this flight technology program, Deep Space 1 (DS1). This mission had the goal of demonstrating flight operations using SEP and learning how to incorporate it into an operational deep-space mission. It would take advantage of SEP's capabilities, accommodate its needs, and quantify its performance. DS1 would assess its interactions with the spacecraft, the instruments, and the environment. While doing this, it would propel the spacecraft along a trajectory that would encounter an asteroid within the 11-month prime mission. Rather than treating it (and the other 11 technologies on DS1) as simple add-ons to a conventional spacecraft, the objective was to use the system as it would be used in a subsequent mission. This motivated a considerable effort in developing new principles in system engineering, spacecraft design, mission planning, and mission operations.

DS1 was launched on 1998 October 24, and spent the next 11 months testing the payload of technologies. The spacecraft encountered (9969) Braille in July 1999. Following the primary mission, DS1 shifted from technology testing to science. Despite what could have been a fatal failure of the spacecraft's star tracker (which was not one of the new technologies), DS1 was able to take advantage of the SEP and some of the other technologies to accomplish a scientifically productive encounter with 19P/Borrelly in September 2001. In addition to testing the hardware and software systems flown on the spacecraft, some of which were subsequently used by the Dawn mission (and many others), DS1 developed the fundamental principles and tools needed to design and operate a mission using SEP, and that were essential to the Dawn mission (Rayman, 2003). Section 2.4 describes some of the significant technical findings that were applicable not only to Dawn but also to other SEP missions.

Dawn and two other missions were chosen as Discovery Program finalists in 2000, and Dawn became one of two chosen for development in 2001. Most appropriately, JPL appointed Marc Rayman to play a key role in its development. Dawn's initial complement of instruments included a gamma ray and neutron spectrometer, a laser altimeter, a visible and IR mapping spectrometer (contributed by Italy), a magnetometer, and a pair of Framing Cameras (contributed by Germany). The radio system was to be used for gravimetric tracking (Russell et al., 2002). Subsequent funding for the laser altimeter was not sufficient, so stereophotogrammetric and stereophotoclinometric analysis of the Framing Camera images were used instead to derive topography (Russell et al., 2004; Raymond et al., 2011). In the end, funds were not provided for the magnetometer. Eventually, Dawn made its way through all the necessary gates and was launched in September 2007.

2.2 THE DAWN MISSION

2.2.1 Objectives

A space mission consists of a coordinated team of scientists, engineers, and administrators, with a set of scientific objectives whose successful outcome depends on the flight of a vehicle to a target, the acquisition of data, and the analysis of the data to obtain scientifically useful information. The Dawn targets were (4) Vesta and (1) Ceres, the two most massive residents of the main asteroid belt, surviving largely intact over the age of the Solar System. Their study was expected to place important constraints on the origin and evolution of the Solar System. As was known prior to Dawn, Vesta is a dry, differentiated body of mean diameter 525 km, with a large impact crater near its south pole (Thomas et al., 1997). The reflectance spectra of the HEDs resemble that of the surface of Vesta, as deduced from telescopic observation (McCord et al., 1970). Confirmation of this identification was expected to go a long way to enable understanding of Vesta's formation and evolution. Ceres, with a mean diameter of 940 km, is quite different from Vesta, even though it is very close in mean distance to the Sun. It is not linked to any meteorite families and appears to have a clay-like surface (Rivkin & Volquardsen, 2010; Rivkin et al., 2011). Telescopic spectra were interpreted as showing hydrated materials covering Ceres' surface (e.g., Milliken & Rivkin, 2009), with OH seen escaping at times above the poles (A'Hearn & Feldman, 1992).

These telescopic and meteoritic observations were used to levy a set of observational requirements for Dawn to measure: density, spin axis orientation, gravity field, topography, abundances of rock forming elements and hydrogen, photometric properties, and visible and infrared spectra of the two bodies' surfaces. These observations were to be performed by the payload and spacecraft systems. A collaboration among various organizations in Germany provided the Framing Cameras, which obtained images for optical navigation as well as the imagery needed for science. Throughout the prime mission, the two cameras were never used simultaneously to minimize the risk of losing both cameras with one anomaly. Neither camera failed during the mission and in the two extended missions this caveat was no longer needed and was suspended, enabling even greater scientific return. A filter wheel was used to provide images in different wavelength ranges that revealed color variations due to varied mineralogy that were used to map the geologic units. The Italian Space Agency, ASI, provided the visible and infrared mapping spectrometer (VIR), which became a veritable workhorse for studying Vesta and Ceres and identifying mineral composition.

The gamma ray and neutron detector (GRaND) identified major rock forming elements and hydrogen (to infer the presence of water) using 21 sensors. The gamma ray spectrometer used a bismuth germanate scintillator and boron-loaded plastic (BLP) scintillators as well as a separate solid-state spectrometer based on cadmium zinc telluride. Neutron spectroscopy was accomplished with the BLP scintillators and with lithium-loaded glass. The instrument was developed by Los Alamos National Laboratory and operated by the Planetary Science Institute.

2.2.2 The Spacecraft

The mechanical design of the Dawn spacecraft was based on Orbital Sciences Corporation's STAR-2 series of spacecraft, and its avionics derived from the LEOStar-2 series. As shown in Figure 2.1, the spacecraft's core was a graphite composite with the hydrazine and xenon tanks mounted inside. The instrument fields of view were all aligned along the +Z axis. The ion propulsion system (IPS) was based on the DS1 system with a 425-kg load of Xe, to be used equally by each of the three thrusters. The IPS accomplished all trajectory control for the entire mission, including travel to Vesta, orbit insertion, maneuvers to each of the 6 science orbits and orbit maintenance, escape from Vesta, travel to Ceres, orbit insertion at Ceres, and maneuvers to each of the 10 science orbits and orbit maintenance. Each thruster had a two-axis gimbal. Only one thruster was used at a time. The attitude control system determined the orientation of the spacecraft and rates of angular motion from a star tracker. A reaction control system used hydrazine to desaturate the reaction wheels and to control the attitude in some safe modes. Hydrazine was never used for trajectory control. The command and data handling system used uplink rates from 7.8 b/s to 2.0 kb/s, and downlink was transmitted at rates from 10 b/s to 124 kb/s.

2.2.3 Dawn Timeline

The spacecraft was launched from Cape Canaveral on a Delta II 7925H-9.5 on 2007 September 27. It arrived at Vesta in July 2011, departed in September 2012, and arrived at Ceres in March 2015, where it operated until October 2018. Figure 2.2 shows Dawn's trajectory and its thrusting history. In this chapter, we will move through this timeline and find that the flight plan was quite different from previous missions. The team built on the lessons learned from DS1 and invented new ones to fly the mission and operate the spacecraft and instruments (Rayman, 2020). It is important to understand how and why the mission was operated as it was. Dawn was a revolutionary mission in many respects. It had much greater capacity for exploration, and at the same time had quite different operational rules. The team learned as the mission progressed, doing new things, and exceeding initial expectations of the amount of data that could be gathered and the new knowledge gained. It was important to learn how to maximize the return from this intricately intercoupled system. In the next section, we discuss the coupling of system resources that on chemically propelled missions are normally independent.

2.2.4 Operating in the Realm of Coupled Resources

Any mission to orbit both Vesta and Ceres that relied on conventional chemical propulsion would not have been possible with a single launch; even a rendezvous with only one of these two bodies would have been unaffordable in the Discovery Program. Dawn's IPS, however, allowed this compelling scientific mission to be undertaken within the available resources.

The IPS enables a coupling of resource margins not possible on purely chemical propulsion (Rayman et al., 2007). The IPS was used for all post-launch trajectory control. The change in velocity of the spacecraft over the course of the mission was 11.5 km/s, comparable to the velocity imparted by the Delta launch vehicle. Both the thrust and the specific impulse of the IPS depend on the power delivered to the thruster. Moreover, the thrust is much lower than for conventional propulsion. As a consequence of these characteristics the IPS couples three key resources: the spacecraft mass,

power, and mission time for thrusting. Because of their coupling, the margins for mass, power, and missed thrust could not be managed independently. In addition, this allowed trades not possible with other missions.

Dawn had to thrust most of its time when not in orbit around its targets. Since such a large amount of thrusting was needed, there had to be a margin to allow for unexpected missed thrusts. This margin varied significantly during the mission. The spacecraft occasionally had to pause thrusting to point its high gain antenna to Earth or to conduct other activities incompatible with optimal thrusting. This required diligent planning for thrusting. Dawn routinely thrust more than 95% of a typical week in which thrusting was required. All of these considerations were needed to calculate the allowable mass of the spacecraft before launch.

2.2.5 Launch and Cruise

Dawn's Delta II liftoff occurred at dawn in picture perfect (and launch perfect) weather conditions. The launch vehicle delivered the flight system to $C_3 = 11.3 \text{ km}^2/\text{s}^2$. Separation from the third stage occurred 61.88 minutes after liftoff. The spacecraft then transitioned to its attitude control system and deployed the solar arrays. The spacecraft found the Sun and oriented itself along the vector to the Sun and spun slowly about it. The spacecraft's X-band traveling wave tube amplifier was turned on 62 minutes after separation and began communications with NASA's Goldstone Deep Space Communications Complex, and mission operations were underway.

The first order of business was to check out the systems needed to operate the spacecraft and sustain long-term ion thrusting. It is misleading to think of thrusting as an IPS function; rather, it is a system function and requires multiple subsystems on the spacecraft to work together. The checkout culminated with a long-duration systems test to verify that the flight and ground systems were ready to execute the thrusting necessary to follow the mission plan. Next, the instruments were checked out. The IPS initially provided an acceleration of 6.5 m/s/day. That gradually increased with decreasing spacecraft mass as xenon was expended, but beyond about 1.9 AU, decreasing solar power due to the increasing heliocentric range required a reduction in the IPS throttle levels. On 2009 February 18, the spacecraft passed 542 km from Mars, gaining a gravitational assist on its way to the asteroid belt. The Mars gravity assist provided a 2.6 km/s velocity change, compared to the 11.5 km/s provided by the IPS over the course of the mission.

After the successful Mars gravity assist, Dawn spent most of the next two and a half years thrusting to rendezvous with Vesta. Cruise operations consisted mainly of thrusting, with weekly communication with the Deep Space Network for downlink, uplink, and radiometric navigation. In June 2010, one of the four reaction wheels failed. Even in a subsequent thorough analysis of the telemetry, no prior indications of problems were found. Only three wheels were needed for normal operations, but since other spacecraft had had failures of wheels of the same design, the Dawn project considered the remaining wheels to be at risk. The operations team immediately embarked on development of new software that would allow it to operate with only two wheels in combination with the reaction control system in case a second wheel failed. In addition, the team quickly worked out methods to complete the cruise to Vesta with all wheels powered off,

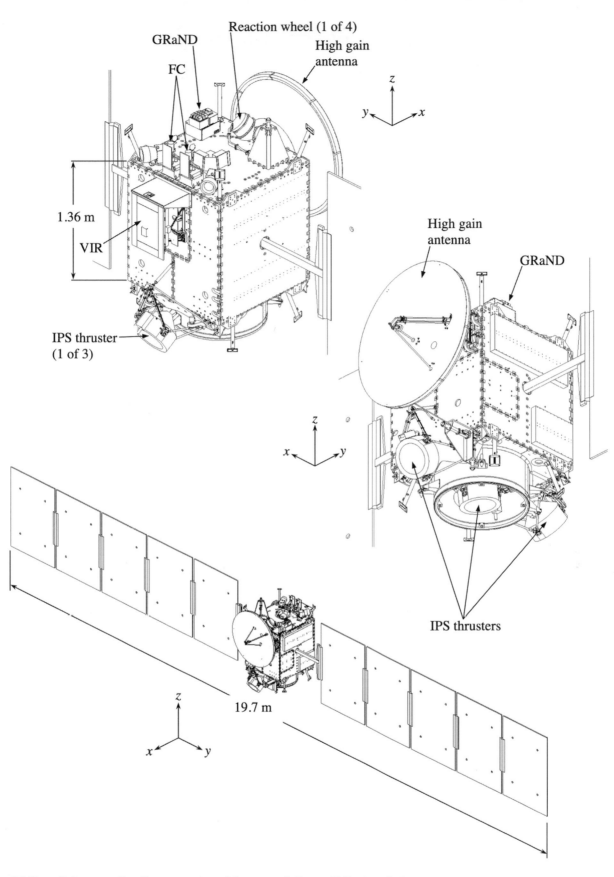

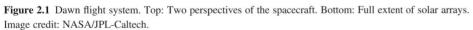

Figure 2.1 Dawn flight system. Top: Two perspectives of the spacecraft. Bottom: Full extent of solar arrays.
Image credit: NASA/JPL-Caltech.

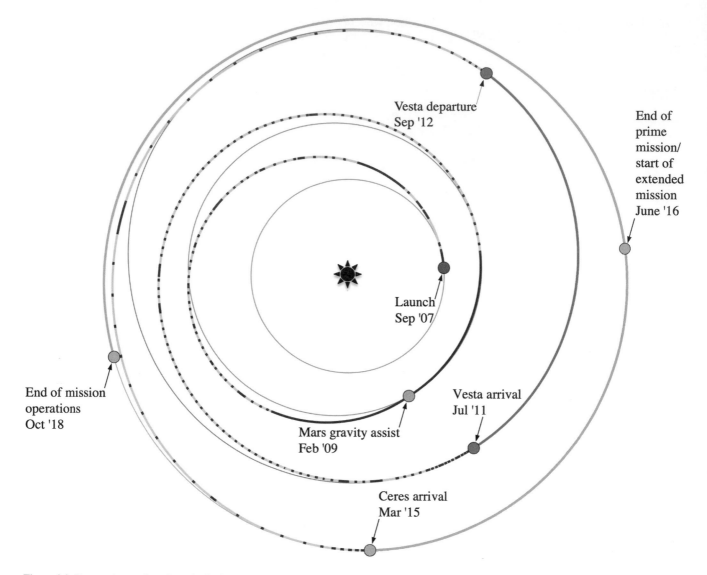

Figure 2.2 Dawn trajectory for prime plus both extended missions. The black portions denote periods of coasting and the gray segments show when the ion propulsion system is thrusting. Thrusting in orbit around Vesta and Ceres is not shown, and coasting periods of less than one day are not shown. Image credit: NASA/JPL-Caltech.

preserving them for operation at Vesta. When the spacecraft was ion thrusting, its attitude control system controlled two of the three spacecraft axes by gimballing the ion engine. Nevertheless, the hydrazine cost of controlling the third axis without wheels and of controlling all three axes when not thrusting was significant, so the project invested considerable effort in reducing hydrazine expenditure. The new software was installed in April 2011, three weeks before the Vesta approach phase began, and the wheels were powered back on at the beginning of that new phase.

2.3 VESTA

Approach operations at Vesta began in May 2011, with the first optical navigation observations of Vesta at a range of 1.2 million km, when Vesta was five pixels across in the Framing Camera (Rayman & Mase, 2014). Optical navigation was essential for

improving the spacecraft–Vesta relative ephemeris. Vesta is the first massive Solar System body a spacecraft was targeted to orbit without first being visited by a flyby spacecraft. Therefore, the approach navigation data were vital to establishing other parameters, such as Vesta's mass, shape, and rotation. In addition to the dedicated optical navigation sessions, there were two periods of observing Vesta with the Framing Camera and VIR throughout its full 5.3-hour rotation. The first of these rotational characterizations (RC1) occurred at 120,000 km, and RC2 was at 37,000 km. The approach images were also used to search for vestan moons.

Unlike missions with conventional propulsion, orbit insertion was not a critical event. As Dawn thrust throughout interplanetary cruise and the approach phase, it was targeting the first science orbit, RC3. On July 16, as thrusting continued, the spacecraft became gravitationally bound to Vesta at an altitude of 16,000 km. Thrusting then was no different from the previous 23,000 hours of thrusting (70% of Dawn's time in space). The project took no special precautions, either on the spacecraft or on the ground.

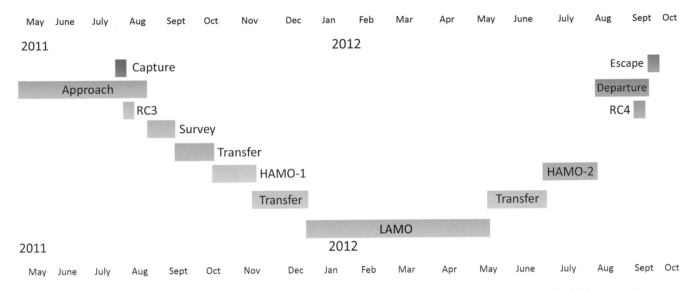

Figure 2.3 Timeline during Vesta operations illustrating the date of each phase. RC3 and RC4 are rotational characterizations 3 and 4 that mapped the entire visible surface at those specific lighting conditions. LAMO is the low altitude mapping orbit, and HAMO is the high altitude mapping orbit. Operations at Ceres initially followed this pattern but became quite different after LAMO was complete.

No extra DSN coverage was scheduled, and orbit capture was not even observed until 25 hours later with the next routine DSN track.

The major science phase began with the RC3 orbit. The sequence of orbits at different altitudes used in the science phase is shown in Figure 2.3. All orbits were polar. The first science observations were obtained in RC3 at an altitude of 5,200 km for about 3 days, roughly half the orbital period. The next orbit was the Survey orbit, at an altitude of 2,735 km, with a period of 69 hours. (Orbit names are based on early plans and no longer reflect the actual objectives or activities conducted.) Seven of these revolutions were completed with the subsolar point at 27°S near solstice. Observations were obtained with the body-mounted instruments pointing to nadir at the illuminated surface. When over the dark side of Vesta, the spacecraft pointed its high gain antenna to Earth to downlink the data. This orbital plan was much more productive than originally planned, with more than 13,000 VIR frames (more than 3.4×10^7 spectra) covering 63% of the surface. The Framing Camera covered 90% of the illuminated surface at 260 m/pixel. Next, the spacecraft spiraled down to 685 km, termed the high altitude mapping orbit (HAMO). Dawn fully mapped Vesta six times in HAMO, each time with the camera pointed at a different angle relative to nadir to provide stereo coverage. In addition, it acquired images in all eight camera filters as well as extensive observations with VIR. More than 7,000 camera images and more than 15,000 spectral frames were obtained by VIR. The observations of the same surface features on multiple photographs were used to solve for a shape model. Figure 2.4 shows five views of the shape model (every 72° of longitude).

The fourth and lowest altitude science orbit was the low altitude mapping orbit, or LAMO. The mean altitude was only 210 km, or 475 km from the center of Vesta, and had a period of 4.3 hours. Since this was the prime orbit for GRaND data acquisition and because its data were independent of surface insolation, GRaND collected data almost continuously. LAMO was also the prime orbit for gravity, so in addition to acquiring radiometric tracking data during communication with Earth, extensive tracking occurred using low gain antennas when the instruments were pointed at nadir.

When Dawn arrived, Vesta was considered to be a protoplanet, a remnant of the very beginning of formation of the Solar System. It was also considered by many, but not all, to be the source of the HED meteorites. When Dawn's spectrometer data, color images, and gravity data arrived at Earth, it was clear that Vesta was planet-like in many ways, and had the same surface composition as the HED meteorites (DeSanctis et al., 2012). It had large mountains and huge craters (Jaumann et al., 2012) and had been heavily battered (Marchi et al., 2012; Schenk et al., 2012). Furthermore, Dawn's gravity data showed an internal structure consistent with an iron core, as predicted by HED geochemical evolution models (Russell et al., 2012). These results are discussed in Section 2.3.1.

Because Dawn had more electrical power than in the original conservative plan, it was able to stay at Vesta longer and still accomplish the ion thrusting required to reach Ceres on schedule. As a result, there was time to extend the duration of LAMO from three months to five months. In addition, the Dawn team added a second HAMO phase during the outbound spiral, providing for much more VIR data as well as camera observations with the Sun significantly closer to the equator than it had been for the first HAMO. In the prelaunch plan, Dawn was to spend seven months at Vesta. With the additional power (which also allowed an earlier arrival), it orbited for 14 months. Images from Dawn's mapping revealed many enigmatic features that proved to be particularly informative, as detailed in Chapter 5. One of the first resolved images of Vesta from RC3 (Figure 2.5(a)) shows the snowman craters and part of the planet-circling fossae. High-resolution LAMO image data shown in Figure 2.5(b) revealed unexpected gullies in the wall of a young crater, Cornelia, and pitted terrain in the crater floor (Figure 2.5(c)).

When Dawn arrived in HAMO-2, the Sun had moved to 7°S latitude and, by the end, to 3°S latitude, so that much of the northern hemisphere had become visible. The final science orbit, RC4, was at an altitude of 6,000 km and collected the final observations of Vesta on August 25–26. With a subspacecraft latitude north of 60°N, and the Sun at 0.75°N, RC4 provided a uniquely good view of the north polar regions. A total of 4,700 images and

Figure 2.4 Vesta shape model with draped image mosaic seen at five different rotational phases. Photojournal Image PIA15678. Credit: NASA/JPL-Caltech/UCLA/MPS/DLR/IDA

nearly nine million VIR spectra were obtained, exceeding all requirements. These results were collected in a series of articles in *Science* magazine (cf. Russell et al., 2012) and in hundreds of subsequent articles in the scientific literature.

2.3.1 Vesta Science

A major scientific controversy leading up to the Dawn mission was whether meteorites falling on Earth in the class of howardite, eucrite, and diogenite were from Vesta. The proponents on both sides of this controversy were quite strongly entrenched. This controversy was quickly resolved when Dawn observed Vesta, as discussed in Chapter 4. The minerals on Vesta's surface closely matched the HED meteorites, providing strong confirmation that Vesta was the parent body.

Vesta is very heavily cratered, and two particularly large impact basins were discovered in the southern hemisphere. The 400-km-diameter Veneneia was formed about two billion years ago, and the overlapping 500-km-diameter Rheasilvia was formed about one billion years ago (Marchi et al., 2012). A large central peak rises to about 25 km above the crater floor. The only known higher

mountain in the Solar System is Olympus Mons on Mars. The impacts that produced the large southern basins did not destroy Vesta, possibly because of its large metallic core of at least 110 km in radius (Russell et al., 2012). The propagation of the energy from the collisions through Vesta's interior also produced more than 90 troughs, the Divalia and Saturnalia Fossae, in two distinct planes orthogonal to the two major impacts (see the basins in Chapter 5). These observations strongly supported the idea that Vesta is a survivor from the earliest days of the Solar System.

2.4 THE LONG VOYAGE RESUMES

As Dawn prepared to depart Vesta for Ceres, a second reaction wheel failed. The team decided to stop using the reaction wheels and return to hydrazine control (and, as before, two-axes of control when ion thrusting). As Dawn spiraled higher, it escaped from Vesta on 2012 September 5. With the second wheel failure, an extremely aggressive hydrazine-conservation campaign was undertaken, significantly altering the plan for the 2.5-year flight to Ceres

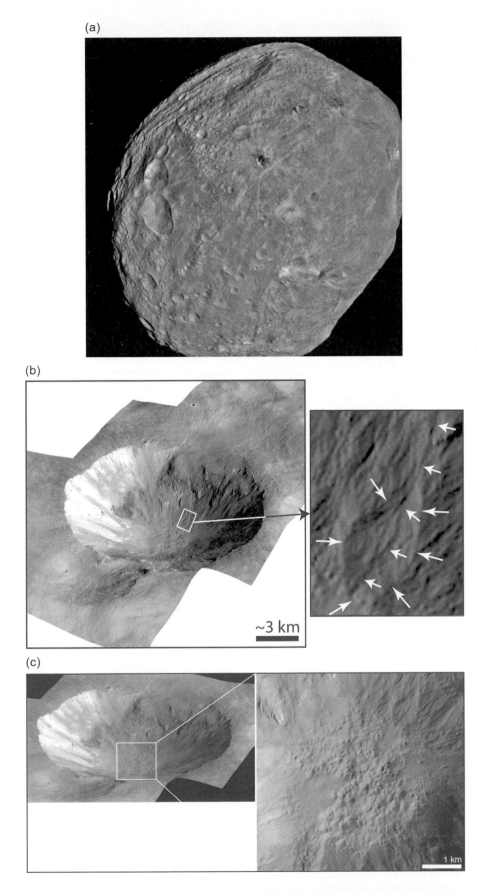

Figure 2.5 (a) Image of Vesta from RC3 orbit. The "snowman" craters, Marcia, Calpurnia, and Minucia, are visible on the left side of the image, and at the top a set of troughs (Saturnalia Fossae), due to ancient massive collisions deep in the southern hemisphere (lower right). The central dark-rayed crater is Cornelia. Photojournal Image PIA14317. Image Credit: NASA/JPL-Caltech/UCLA/MPS/DLR/IDA. (b) Images of gullies (arrows) on the side of Cornelia crater showing evidence for ancient water flows. Photojournal Image PIA19170. Image Credit: NASA/JPL-Caltech/UCLA/MPS/DLR/IDA. (c) Images of pits where volatiles escaped from the floor of Cornelia crater. Photojournal Image PIA16183. Image Credit: NASA/JPL-Caltech/UCLA/MPS/DLR/IDA

and the Ceres operations plan. Without this effort, there would not have been enough hydrazine onboard to accomplish the mission. Among the changes made for the interplanetary phase were to reduce high-gain telecommunication sessions from once a week to once every four weeks. This allowed Dawn to spend more time thrusting and hence controlling attitude with the ion engine, and reduced the significant hydrazine cost of turns to and from the communications attitude. In addition, the turn rate was reduced. The Ceres plan was modified, with a significant reduction in spacecraft turns and other innovations. All these improvements in operation preserved sufficient hydrazine that, by Ceres arrival, a robust plan to acquire all the planned data was in place even if the two remaining wheels failed at Ceres.

2.5 ARRIVAL AT CERES

The operations design used for Vesta was so successful that the Ceres strategy followed that model, albeit with changes to reduce the hydrazine cost (Rayman, 2019). As at Vesta, Figure 2.6(a) and (b) show there were distant, intermediate and low-altitude mapping orbits. Dawn was captured into orbit on 2015 March 6 at an altitude of 60,600 km. The approach phase as well as RC3, Survey, HAMO, and the spiral transfers to each science orbit were conducted with the two operable wheels powered off. The hydrazine cost of attitude control was very sensitive to the orbital altitude, and LAMO would be the most expensive phase. Therefore, the two wheels were saved for the LAMO phase, when they would provide the greatest benefit. Shortly after arrival in the 385-km-altitude LAMO on 2015 December 13, the two wheels were powered on, requiring hydrazine to control only one axis. Apart from some short tests, this was the first use of the capability installed in April 2011, before Vesta operations began.

The start of approach to Ceres was 2014 December 26, and the approach operations concluded on 2015 April 24. This period allowed not only optical navigation but also some science observations by the Framing Camera and VIR. Again, the exploration planned for four orbit phases: RC3, Survey, HAMO, and LAMO. (As there was no intent to depart Ceres, there had been no plan to raise the orbit after LAMO.) This observation program was very similar to that at Vesta, but because Ceres is both larger and more massive than Vesta, observations and orbit transfers required more time. The most fascinating aspect of the approach phase was the

bright spot in Occator crater that was gradually resolved to reveal multiple spots. Figure 2.7 shows these features in a high-resolution image mosaic.

2.5.1 Ceres Science Mapping Orbits

When Dawn arrived at Ceres, the plan was to acquire data in all science orbits through three months of LAMO operations. (The science objectives in LAMO required 1.5 months of data acquisition, and the complexity of operations motivated adding a significant margin.) As it turned out, operations were flawless, and the hydrazine conservation efforts proved so successful that Dawn continued to the end of the originally planned prime mission on 2016 June 30. NASA then extended the mission and LAMO continued for a total of 8.5 months, acquiring not only a wealth of GRaND and gravity data but also fully mapping Ceres at 35 m/pixel and acquiring extensive color and stereo images as well as VIR spectra. Ceres was expected to be a more volatile-rich body than Vesta, and indeed it was. It appears that, at least in one instance during Dawn's stay at Ceres, solar energetic particles released water vapor from the surface that, in turn, interacted with the solar wind (Russell et al., 2016; Villarreal et al., 2017). This mechanism can possibly explain observations of water vapor at Ceres made by the Herschel Space Observatory (Küppers et al., 2014). Dawn's observations showed there was water ice exposed on the surface (Combe et al., 2015), ice in the crust (Prettyman et al., 2017), and even a cryovolcano (Ruesch et al., 2016; Chapter 10). Figure 2.8 shows high-resolution images of this cryovolcano, Ahuna Mons, rendered onto the shape model. Figure 2.9(a) shows a close-up of the bright area in Occator crater and a montage of close-ups of the different

Figure 2.7 Image mosaic of Ceres in orthographic projection at a resolution of ~35 m/pixel, centered on Occator crater. Occator hosts bright deposits called faculae, that occur as a central region known as Cerealia Facula and a series of deposits (Vinalia Faculae) east of the center.
Photojournal Image PIA21906. Image Credit: NASA/JPL-Caltech/UCLA/MPS/DLR/IDA

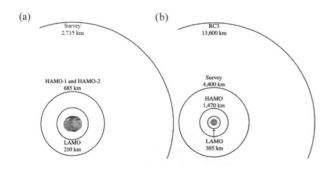

Figure 2.6 Science orbits used in the prime mission. (a) Some of the Vesta orbit altitudes to scale with Vesta (RC3 and RC4, both higher than Survey, are not shown). (b) Four orbital altitudes are shown to scale with Ceres.

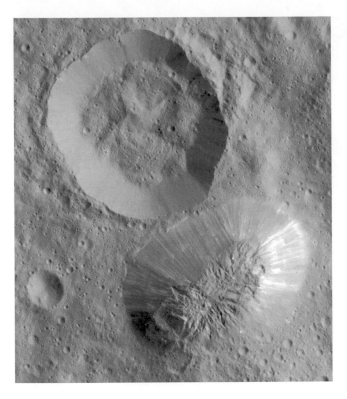

Figure 2.8 Ahuna Mons, a possible cryovolcanic feature on Ceres from a high-resolution mosaic from LAMO images.
Photojournal Image PIA21919. Image Credit: NASA/JPL-Caltech/UCLA/MPS/DLR/IDA

terrain on the crater floor. Figure 2.9(b) shows the remarkable fractured terrain of Nar Sulcus, thought to be evidence of extension (Hughson et al., 2018). Perhaps the biggest surprise of all was the presence of organic molecules (DeSanctis et al., 2017), as discussed in Chapter 8. Following the successful conclusion of the primary mission and noting how limited the remaining hydrazine supply was, a decision was made to return to higher orbits to pursue other objectives at a lower hydrazine cost. This included a high-altitude elliptical orbit, in which GRaND could study the cosmic ray background to reduce the noise and improve the signal-to-noise ratio of the data already gathered.

2.5.2 The Grand Finale

The spacecraft acquired extensive data in a series of extended mission orbits (XMOs) (Rayman, 2019). When another reaction wheel failed in April 2017, in XMO4, just before a unique opposition observation, that observation was conducted successfully with no wheels. (There was no one-wheel mode.) Despite the permanent loss of wheel control, the project devised an ambitious plan to fly to new elliptical orbits, and a second extended mission was approved in October 2017. The spacecraft transferred to XMO6, which was optimized for observing Juling crater, which seemed to have temporal changes in the exposed surface ice. After XMO6, Dawn flew to its final orbit, XMO7 (Figure 2.10), with a periapsis of only 35 km, compared to the previous low altitude of 385 km. This provided a tremendous improvement in spatial resolution for all investigations. The orbit was designed to focus on Occator crater (although other observations were conducted as well), achieving

images of better than 5 m/pixel. The data returned showed a spectacular level of detail, revealing the nature of impact melt, pingo-type mounds, fracture patterns, bright deposits, and mass wasting features in Occator crater that all point to evidence of recent geologic activity (Figure 2.9(a)). Figure 2.11 shows the crater wall and is only about 3 km across. The hydrazine was finally exhausted on 2018 October 31, at which point the spacecraft could no longer control attitude. It thus could not point its solar arrays at the Sun and had insufficient power to operate. It also could not point its antenna at Earth, its instruments at Ceres, or its ion engine in the direction needed to go elsewhere.

2.6 CONCLUSIONS

During the course of its mission at Vesta and Ceres, Dawn's scientific return far exceeded the proposed observation program. The exceptional capability of the spacecraft and the operations team made this mission as productive as it was. The capability to orbit distant massive bodies, without conducting a flyby first, provides a significant leap in knowledge with one mission. Even more, orbiting two contrasting bodies with a single Discovery-class mission provided tremendous scientific return for the investment.

Dawn produced images and reflectance spectra with spatial resolutions unachievable from the vicinity of Earth. Visible-wavelength images at 5 m/pixel on Ceres would require a diffraction-limited aperture well in excess of 20 km. (Even if an optical interferometer could manage such a baseline, the signal would be uselessly weak.) Dawn also acquired images and spectra over ranges of viewing angles and phase angles that are never available to terrestrial observers. What's more, the mission obtained extensive measurements of elemental composition and the mass and higher order gravity fields, data types simply impossible to obtain at astronomical distances. More generally, the opportunity to spend long times in orbit around each target, rather than conducting a rapid flyby, allowed the science and engineering teams not only to conduct comprehensive mapping with all sensors but also to devise new targeted measurements that built on the mission's earlier results.

With the wealth of data acquired at each body, Dawn has shifted Vesta and Ceres from principally residing in the domain of astronomers to that of geologists and geophysicists. Many of the exciting new insights scientists have gained are discussed in the following chapters. But it is immediately clear that, despite both being in the main asteroid belt, they have dramatically different properties. Of course, even telescopic observation showed differences, but Dawn brought those differences into sharper focus. In broad terms, Vesta is now seen in some ways to be more similar to terrestrial planets – strongly differentiated bodies that experienced a hot past – than it is to many of the smaller bodies in the main asteroid belt. Ceres, the largest icy body that has been studied from orbit, is now recognized to have been active in geologically very recent times (and perhaps even in the present).

There is no doubt that with the technical capability Dawn demonstrated, and the fascinating results obtained at its two destinations, the future investigation of the main asteroid belt holds great promise. Eventually, Vesta and Ceres may welcome new missions that will allow still more data types and with greater sensitivity and resolution. These future missions may also include landers, perhaps

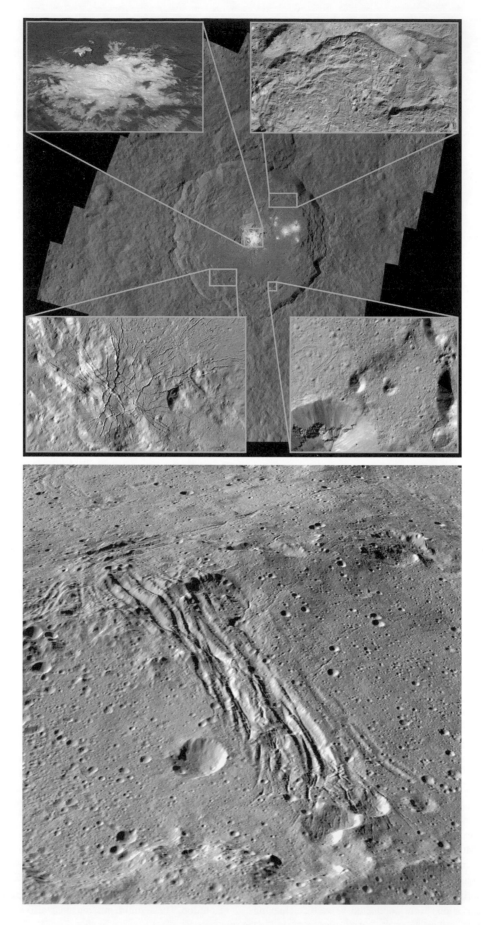

Figure 2.9 (a) Occator crater closeup montage at <10 m/pixel, obtained in XMO7: (Upper left) Perspective view of salt deposits Paola and Ceralia Faculae; (upper right) Northern terminal flow margin of hummocky lobate material, showing conical mounds and varying flow textures; (lower right) Two-hundred

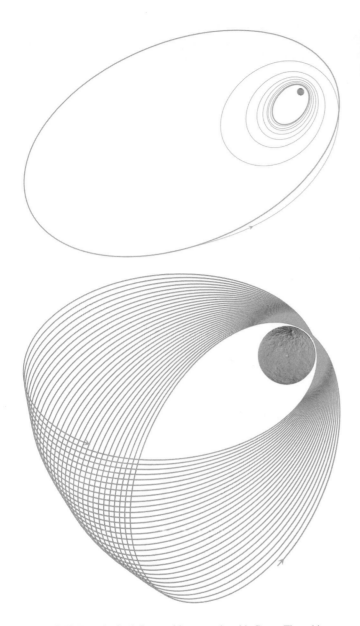

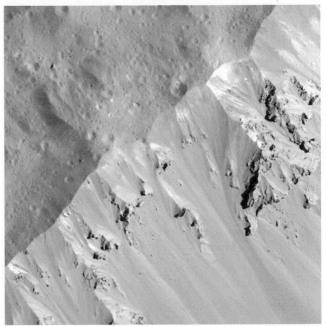

Figure 2.11 A 3-km section of the rim of the north wall of Occator crater, as seen from XMO7. Image acquired from an altitude of 33 km and the image is about 3 km across. Boulders on the slope have slid only partway downslope.
Photojournal Image PIA22550. Image Credit: NASA/JPL-Caltech/UCLA/ MPS/DLR/IDA

with mobility, and perhaps with the capability to bring samples back to Earth.

Although Vesta and Ceres contain a large fraction of the mass in the main asteroid belt, they can reveal only a small part of the story that that part of the Solar System ultimately has to tell. The diversity of bodies in the main asteroid belt is already well known, so there are a great many more types of bodies to visit, and new missions surely will reveal unsuspected differences. Even apart from the differences among the bodies, future discoveries at each body should provide exciting new insights into the physics, chemistry, and geology of the Solar System.

Dawn left a deep legacy of engineering innovation and a vast treasure trove of scientific results. Thanks to its many accomplishments, a new era of Solar System exploration, particularly in the main asteroid belt, has dawned.

REFERENCES

A'Hearn, M. F., & Feldman, P. D. (1992) Water vaporization on Ceres. *Icarus*, 98, 54–60.

Combe, J.-P., McCord, T. B., McFadden, L. A., et al. (2015) Composition of the northern regions of Vesta analyzed by the Dawn mission. *Icarus*, 259, 53–71.

Figure 2.10 Dawn's final Ceres orbits, to scale with Ceres. The orbits are nearly polar, and Dawn revolves counterclockwise in these illustrations. (a) Transfer from the extended mission orbit 5, XMO5, to XMO6. The heavy curves are the initial and final science orbits, and the lighter ones are the transfer. XMO6 was designed to be high enough in the southern hemisphere to provide extensive new coverage with VIR (taking advantage of the improved Sun angle since HAMO) and low enough in the northern hemisphere to allow selected targeted observations with VIR and the Framing Cameras. Image credit: NASA/JPL-Caltech. (b) XMO7. The periapsis moved 1.9° southward with each revolution. Dawn's orbit was in a 1:3 resonance with Ceres' rotation, so the periapsis longitude was nearly fixed. This design provided repeated high-resolution observations with all instruments and gravity measurements of Occator crater, including the spectacular views shown in Figures 2.9(a) and 2.11. When the orbit had shifted much farther south, similar high-resolution observations of the older, larger Urvara crater were conducted. Image credit: NASA/JPL-Caltech

Figure 2.9 (*cont.*) meter scale conical mounds (pingos?) on smooth lobate material; (lower left) Uplifted dome with radial fractures. [K. Hughson, personal communication]. All images credit: NASA/JPL-Caltech/UCLA/MPS/DLR/IDA. (b) Perspective view (with no vertical exaggeration) of the primary set of fractures in Nar Sulcus, interpreted as evidence of significant local extension within the past 100 Myr. The shape implies mechanical properties akin to outer Solar System moons. Image roughly 40 km in width. North is to the left. [K. Hughson, personal communication]. Image credit: NASA/JPL-Caltech/UCLA/ MPS/DLR/IDA.
A black and white version of this figure will appear in some formats. For the color version, refer to the plate section.

DeSanctis, M. C., Ammannito, E., Capria, M. T., et al. (2012) Spectroscopic characterization of mineralogy and its diversity across Vesta. *Science*, 336, 697–700.

DeSanctis, M. C., Ammannito, E., McSween, H. Y., et al. (2017) Localized aliphatic organic material on the surface of Ceres. *Science*, 355, 719–722.

Hughson, K. H. G., Russell, C. T., Schmidt, B. E., et al. (2018) Normal faults on Ceres: Insights into the mechanical properties and thermal evolution of Nar Sulcus. *Geophysical Research Letters*, 46.

Jaumann, R., Williams, D. A., Buczkowski, D. L., et al. (2012) Vesta's shape and morphology. *Science*, 336, 687–690.

Küppers, M., O'Rourke, L., Bockelee-Morvan, D., et al. (2014) Localized sources of water vapour on the dwarf planet (1) Ceres. *Nature*, 505, 525–527.

Marchi, S., McSween, H. Y., O'Brien, D. P., et al. (2012) The violent collisional history of asteroid 4 Vesta. *Science*, 336, 690–694.

McCord, T. B., Adams, J. B., & Johnson, T. V. (1970) Asteroid Vesta: Spectral reflectivity and compositional implications. *Science*, 168, 1445–1447.

Milliken, R. E., & Rivkin, A. S. (2009) Brucite and carbonate assemblages from altered olivine-rich materials on Ceres. *Nature Geoscience*, 2, 258–261.

Prettyman, T. H., Yamashita, N., Toplis, M. J., et al. (2017) Extensive water ice within Ceres' aqueously altered regolith: Evidence from nuclear spectroscopy. *Science*, 355, 55–58.

Rayman, M. D. (2003) The successful conclusion of the Deep Space 1 mission: Important results without a flashy title. *Space Technology*, 23, 185.

Rayman, M. D. (2019) Dawn at Ceres: The first exploration of the first dwarf planet discovered. *Acta Astronautica*, doi:10.1016/j.actaastro.2019.12.017.

Rayman, M. D. (2020) Lessons from the Dawn mission to Ceres and Vesta. *Acta Astronautica*, 176, 233–237.

Rayman, M. D., Fraschetti, T. C., Raymond, C. A., & Russell, C. T. (2007) Coupling of system resource margins through the use of electric propulsion: Implications in preparing for the Dawn mission to Ceres and Vesta. *Acta Astronautica*, 60, 930–938.

Rayman, M. D., & Mase, R. A. (2014) Dawn's exploration of Vesta. *Acta Astronautica*, 94, 159–167.

Raymond, C. A., Jaumann, R., Nathues, A., et al. (2011) The Dawn topography investigation. *Space Science Reviews*, 163, 487–510.

Rivkin, A. S., Li, J.-Y., Milliken, R. E., et al. (2011) The surface composition of Ceres. *Space Science Reviews*, 163, 95–116.

Rivkin, A. S., & Volquardsen, E. L. (2010) Rotationally-resolved spectra of Ceres in the 3 micron region. *Icarus*, 206, 327–333.

Ruesch, O., Platz, T., Schenk, P., et al. (2016) Cryovolcanism on Ceres. *Science*, 353, aaf4286.

Russell, C. T., Coradini, A., Christensen, U., et al. (2004) Dawn: A journey in space and time. *Planetary and Space Science*, 52, 465–489.

Russell, C. T., Coradini, A., Feldman, W. C., et al. (2002) Dawn: A journey to the beginning of the Solar System. *Proceedings of Asteroids, Comets, Meteors*, July 29–August 2, Technical University Berlin, Berlin, Germany, (ESA-SP-500), pp. 63–66.

Russell, C. T., Raymond, C. A., Ammannito, E., et al. (2016) Dawn arrives at Ceres: Exploration of a small, volatile-rich world. *Science*, 353, 1008–1010.

Russell, C. T., Raymond, C. A. Coradini, A., et al. (2012) Dawn at Vesta: Testing the protoplanetary paradigm. *Science*, 336, 684.

Schenk, P., O'Brien, D. P., Marchi, S., et al. (2012) The geologically recent giant impact basins at Vesta's south pole. *Science*, 336, 694–697.

Thomas, P. C., Binzel, R. P., Gaffey, M. J., et al. (1997) Impact excavation on asteroid 4 Vesta: Hubble Space Telescope results. *Science*, 272, 1492–1495.

Villarreal, M. N., Russell, C. T., Luhmann, J. G., et al. (2017) The dependence of the cerean exosphere on solar energetic particle events. *Astrophysical Journal*, 838, L8.

Part II

KEY RESULTS FROM DAWN'S EXPLORATION OF VESTA AND CERES

3

Protoplanet Vesta and HED Meteorites

HARRY Y. MCSWEEN JR. AND RICHARD P. BINZEL

3.1 INTRODUCTION

(4) Vesta, the second most massive asteroid, is one of only a small handful of large, intact protoplanets in the Main Belt. Scientific investigation of Vesta dates back to the time of its discovery, in 1807, when the preeminent astronomer of the day, William Herschel, noted Vesta as having a different color than the three previously discovered asteroids (Cunningham, 2014). While this early distinction may have been subjective, it proved prescient in that twentieth-century astronomers using ground-based and space-based telescopes revealed Vesta to be a differentiated body with a (presumably) ancient volcanic surface. With such an intriguing difference from other large asteroids and strong evidence for a specific meteorite link, Vesta beckoned as a high priority for detailed *in situ* exploration by spacecraft (Binzel, 2012). That exploration was accomplished by the Dawn orbital spacecraft mission (Russell et al., 2012; see Chapter 2). This mission's imagery, and the mineralogical, chemical, and geophysical data it provided, have greatly expanded our understanding of Vesta's origin and geologic evolution.

As noted here, and unlike any other specific asteroid, Vesta is recognized as having been sampled extensively by the howardite–eucrite–diogenite (HED) meteorite suite (see Chapter 1). In fact, the named HED meteorites number more than a thousand and in mass they exceed the sampling of the Moon and Mars (Beck et al., 2015). We will explore why we have so many samples. Having samples, especially in representative numbers, matters. The HEDs were invaluable for calibrating Dawn's instruments and provided a rigorous foundation for interpretation of its data. Laboratory analyses have also provided quantitative information about Vesta that could not otherwise have been obtained by remote sensing.

It is not an overstatement to say that, because of the combination of Dawn's exploration and the extensive sampling afforded by HEDs, Vesta now ranks (along with the Moon and Mars) among the best-understood extra-terrestrial bodies. As a prime example of planetesimals accreted to form the terrestrial planets, Vesta illustrates the complexity of the differentiation and magmatic history experienced by such primordial bodies, and it provides insight into the delivery of volatiles to the inner Solar System.

3.2 HED METEORITES: SAMPLES OF VESTA

The HED meteorites are mafic and ultramafic igneous rocks and impact breccias derived from them. Their petrology and chemistry have been extensively described in numerous reviews (e.g., Mittlefehldt et al., 1998; Keil, 2002; McSween et al., 2011; Mittlefehldt, 2015), so we will only summarize them briefly.

Eucrites are subdivided into basaltic and cumulate varieties (Mayne et al., 2009). Basaltic eucrites occur as fine-grained volcanic rocks and coarser-grained gabbros; cumulate eucrites are plutonic rocks enriched in Mg-rich pyroxene. All eucrites are composed of calcic plagioclase and low-calcium pyroxene (pigeonite or orthopyroxene) with exsolved high-calcium pyroxene (augite) lamellae, and lesser amounts of chromite, ilmenite, troilite, silica, and FeNi metal (kamacite). Most eucrites have been thermally metamorphosed to varying degrees (Takeda & Graham, 1991), and only unmetamorphosed samples (Figure 3.1) retain primary (unrecrystallized) igneous textures and chemically zoned minerals.

Diogenites are mostly orthopyroxenites, although olivine-bearing samples ranging from olivine pyroxenites (10–40% olivine) to harzburgites (>40% olivine) (Beck & McSween, 2010) and rare dunites (>90% olivine) (Beck et al., 2011) have been described. Another less common variety of diogenite, called "Type B," consists of plagioclase pyroxenites. In most diogenites, olivine and chromite occur in only minor proportions, with lesser amounts of troilite, diopside, silica, metal, and phosphates (Mittlefehldt, 1994). A diogenite with a primary igneous texture is shown in Figure 3.1.

Many eucrites and diogenites are monomict breccias. Polymict breccias are mixtures of HED lithologies. Polymict eucrites are composed of basaltic and cumulate eucrite, with <10% diogenite, and howardites contain >10% diogenite mixed with eucrite; however, the distinction can be problematic for coarse-grained rocks studied in thin sections (Mittlefehldt et al., 2013).

Xenocrysts of Mg-rich olivine and orthopyroxene (Lunning et al., 2015) and xenoliths of harzburgite (Hahn et al., 2018) from which the individual crystals were derived have also been found in some howardites. The harzburgite clasts also contain distinctive pyroxene–chromite symplectites. Oxygen isotope analyses demonstrate that they are vestan (Lunning et al., 2015). These uncommon rock and mineral fragments are distinct from the olivine diogenites and are interpreted as samples of Vesta's mantle.

Exogenic carbonaceous chondrite clasts commonly make up a small volume percentage of howardites. Most clasts are similar to CM chondrites (Zolensky et al., 1996) and range in size from identifiable clasts to submicroscopic grains. Clasts of a few other types of foreign material have also been described (Lorenz et al., 2007). The exogenic chondrite component accounts for an observed siderophile element enrichment in many howardites.

Mesosiderites have sometimes been associated with HEDs. These meteorites are composed of eucrite-like clasts and FeNi metal in roughly equal proportions (Benedix et al., 2014). They formed by impact mixing of molten core material and crustal igneous rocks. Although it has been suggested that mesosiderites were formed by a hit-and-run collision of a small differentiating

Figure 3.1 Photomicrographs (crossed polars) of thin sections of a eucrite (EET 90020) composed of pyroxenes and plagioclase, and a diogenite (GRA 98108) composed of orthopyroxenes with minor olivine. These meteorites are unusual because they show primary igneous textures; most eucrites have been metamorphosed and recrystallized, and many HEDs are breccias, having been pulverized and recemented by impact processes on Vesta. The white scale bars are ~2.5 mm.
A black and white version of this figure will appear in some formats. For the color version, refer to the plate section.

body with Vesta (Haba et al., 2019), Dawn has not observed any metal-rich materials on the asteroid itself.

Considerable numbers of bulk-rock analyses of major and trace elements in HEDs are available (e.g., Mittlefehldt & Lindstrom, 2003; Barrat et al., 2007, 2008; Warren et al., 2009). Here we focus on those elements that can be compared with Dawn's remote sensing data or that carry petrogenetic information.

Eucrites have iron-rich compositions similar to those of lunar basalts. Based on chemical compositions (which cluster near a peritectic in the olivine–anorthite–silica system) and crystallization experiments, Stolper (1977) interpreted some eucrites (called the Stannern trend) as primary partial melts, and others (now called Main Group eucrites) as residual melts after fractionation of pyroxene and plagioclase. This interpretation was supported by rare earth element modeling of eucrites (Consolmagno & Drake, 1977). However, models incorporating additional trace element data on new eucrites and other crystallization experiments suggest that the partial-melting model cannot account for eucrite compositions, and instead support their origin as products of fractionation and/or assimilation (e.g., Bartels & Grove, 1991; Mittlefehldt & Lindstrom, 2003; Barrat et al., 2007; McSween et al., 2011). Cumulate eucrites can be modeled as crystals separated from fractionating basaltic eucrite melts, sometimes containing trapped liquid (Treiman, 1997). Eucrites are also depleted in siderophile elements (Righter & Drake, 1997) and show remnant magnetization (Fu et al., 2012), which is taken as evidence for core formation on the HED parent body.

Diogenites have uniform major element compositions but highly variable minor and trace element abundances (e.g., Mittlefehldt, 1994; Shearer et al., 1997; Barrat et al., 2008; Mittlefehldt et al., 2012). These studies concluded that multiple parent magmas were required to produce these cumulates.

The chemical compositions of howardites have been used to calculate the relative proportions of eucrite and diogenite (Warren et al., 2009; Mittlefehldt et al., 2013). The data indicate a eucrite:

diogenite mixing ratio of ~2:1 and suggest only minor amounts of cumulate eucrite; the latter conclusion is at odds with petrologic mapping of howardites (Beck et al., 2012; Lunning et al., 2016). Abundances of siderophile elements and noble gases indicate that only a subset of howardites are actually regolith (surface) breccias (Warren et al., 2009; Cartwright et al., 2014).

The oxygen isotope compositions of HEDs are uniform, except for a small handful of outliers (e.g., Greenwood et al., 2014). This is usually taken as evidence that most, if not all, HEDs are from the same parent body. Nearly uniform Fe:Mn ratios in pyroxenes provide additional evidence for the linkage of eucrites, diogenites, and howardites (Papike et al., 2003). However, outliers in oxygen isotope and FeMn data may suggest that multiple Vesta-like bodies occurred (Scott et al., 2009). Radiogenic isotopes in HEDs have been analyzed extensively for geochronology; these are discussed in Section 3.5.1 and more thoroughly in Chapter 4.

Spectroscopic investigation of Vesta by McCord et al. (1970) showed a strong absorption feature near 0.9 μm that was immediately recognized as being due to a pyroxene-rich surface, thus revealing the basaltic nature of Vesta and a possible relationship to HEDs. Over the visible/near-infrared (VISNIR) spectral range, both Vesta and HEDs show distinctive absorption features centered near 0.9 and 2.0 μm, where the center positions of the bands shift with pyroxene composition (primarily Ca, Mg, and Fe; Adams, 1974). Adding olivine to the pyroxenes serves to broaden the 0.9 μm band and move its center to longer wavelengths (Gaffey, 1976). Because the compositions of pyroxenes in basaltic eucrites and diogenites differ, these lithologies can be distinguished spectrally (Figure 3.2a). Howardites, being mixtures of eucrite and diogenite, sensibly have band positions between the two lithologies (De Sanctis et al., 2012a). Cumulate eucrites have pyroxenes similar in composition to those of howardites, so they are indistinguishable spectrally. Olivine diogenites with 10–30% olivine cannot be spectrally distinguished from the more common orthopyroxenitic diogenites (Beck et al., 2013).

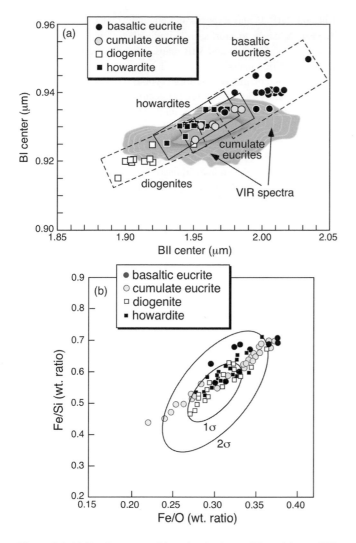

Figure 3.2 (a) Band center positions for the 1 μm (BI) and 2 μm (BII) features distinguish eucrites from diogenites, with howardite and cumulate eucrite bands plotting between them. The cloud represents global VIR pixels for the surface of Vesta (De Sanctis et al., 2013). (b) Global Fe:Si and Fe:O weight ratios (ovals show 1σ and 2σ uncertainties) for Vesta, determined from GRaND measurements (Prettyman et al., 2012), compared to those ratios in HEDs. Adapted from McSween et al. (2013b)
A black and white version of this figure will appear in some formats. For the color version, refer to the plate section.

3.3 VESTA AFTER DAWN

Dawn's instrument payload included two Framing Cameras (FCs), the Visible and Near-Infrared Spectrometer (VIR), and the Gamma Ray and Neutron Detector (GRaND) (Russell et al., 2012) (see Chapter 2). These instruments provided compositional information on Vesta that could be compared with HEDs, as well as images that constrained its geologic history. Precise tracking of the spacecraft's trajectory and velocity provided gravity data for modeling its internal structure.

3.3.1 Geologic Context

The geologic history of Vesta, as inferred from geomorphology, is described in Chapter 5; here we only provide some geologic

context for its compositional and geophysical data. Rotational variability in the spectrum of Vesta was first recognized by Bobrovnikoff (1929), thereby spawning a rich history of observations to interpret its surface in terms of albedo and compositional variations related to its geology. Gaffey (1997) summarized this history and presented a lithologic map derived from ground-based rotationally resolved spectral measurements, identifying regions with howardite, diogenite, and (not subsequently confirmed) olivine mineralogies. Direct low-resolution geologic mapping became possible with Hubble Space Telescope imaging over multiple rotation phases, yielding similar results in identifying a range of surface units present on Vesta, with mineralogic interpretations spanning from eucrites to howardites to diogenites (Binzel et al., 1997).

Vesta's heavily cratered surface is dominated by two huge impact basins near the south pole. The ~500-km-diameter Rheasilvia basin was identified in Hubble Space Telescope images (Thomas et al., 1997; Zellner et al., 1997) and was interpreted as the impact excavation site for the Vesta family of asteroids (see Section 3.4). Rheasilvia is estimated to have formed at ~1 Ga (Marchi et al., 2012) and partly overlaps Veneneia, an older >2 Ga basin. These structures excavated large amounts of material and spread ejecta over at least the southern hemisphere. The geologic map compiled from the Dawn mission and the interpreted timescale for Vesta (Chapter 5) are based on stratigraphy and crater-counting chronology of ejecta from Rheasilvia, Veneneia, and a much younger crater, Marcia (Williams et al., 2014). A prominent system of equatorial troughs is related to the large impacts (Buczkowski et al., 2012). The surface of Vesta is covered by a thick regolith (Jaumann et al., 2012), and no volcanic features have been observed.

3.3.2 Petrology and Chemistry from Spectra of Vesta's Surface

Dawn's VIR measurements of Vesta (De Sanctis et al., 2012a, 2013; Figure 3.2a) confirm its long-established (e.g., Gaffey, 1997) overlap with the spectral signatures of HEDs, centered on howardites. Only a limited number of Dawn's VIR pixels correspond uniquely to eucrite or diogenite, unlike the relative proportions of HEDs in the world's meteorite collections, where eucrite dominates and howardite is subordinate (McSween et al., 2019). This may reflect the differences in spatial scale between meteorite hand samples and VIR footprints (70 m/pixel at the closest approach). The estimated percentages of Vesta's surface mapped as various lithologies (and intermediate lithologies) from VIR spectra (De Sanctis et al., 2013) are: eucrite 22.3%, eucrite-rich howardite 66.3%, howardite 7.1%, diogenitic howardite 4.0%, and diogenite 0.2%.

Beck et al. (2017) used a combination of GRaND data (counts of Fe, fast neutrons, and high-energy gamma rays) to search for *pure* igneous lithologies on Vesta. At the large spatial resolution (200 km FWHM) of this instrument, they found 3% basaltic eucrite but no detectable cumulate eucrite or orthopyroxenitic diogenite. Surprisingly, the most abundant pure igneous rock was Type B diogenite, comprising 11%.

FC color ratios (0.98:0.92 μm), and especially tilt and curvature of the 1 μm band, can also distinguish the HED lithologies (Thangjam et al., 2013). Although less definitive than VIR spectra, the FC compositional map has higher spatial resolution. Color variations on Vesta confirm that the northern hemisphere is dominated by howardite and eucrite, and the southern hemisphere contains more diogenite.

GRaND measured Fe, Si, O, and H abundances on Vesta (Prettyman et al., 2012, 2013), as well as globally averaged values for K and Th (Prettyman et al., 2015). A plot of Fe:O versus Fe:Si for Vesta overlaps those ratios for HEDs (Figure 3.2b), and Vesta's K:Th ratio is similar to that for howardite.

Maps of the surface distributions of HED lithologies on Vesta, based on (a) GRaND neutron absorption data (Prettyman et al., 2013), (b) VIR spectra (Ammannito et al., 2013a), and (c) FC color ratios (Thangjam et al., 2013), are shown in Figure 3.3 (Raymond et al., 2017). GRaND senses up to ~10 cm depth, whereas VIR and FC sample the composition at the surface. The maps also differ in spatial resolution, but all show that the surface is predominantly howardite, with a concentration of eucrite near the equator between

100 and 210° E and deposits of diogenite that extend from the Rheasilvia basin to northern latitudes between 0 and 45° E. However, the GRaND and VIR maps show different distributions of diogenite-rich material.

The unusual plagioclase-bearing Type B diogenite recognized in GRaND data by Beck et al. (2017) occurs near Vesta's north pole region, near the Rheasilvia basin antipode. They suggested that this lithology might have formed through antipodal melting and subsequent magma fractionation (see Chapter 6).

Are the maps in Figure 3.3 really representative of Vesta's crust? The eucrite:diogenite ratios estimated by GRaND and VIR have been compared with those ratios calculated from numbers and masses of eucrite and diogenite in the world's meteorite collections, calculated from chemical-mixing diagrams for howardites, and analyzed in petrologic maps of howardites (McSween et al.,

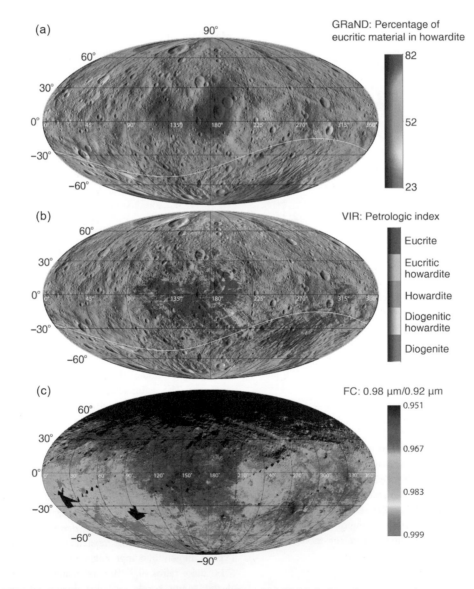

Figure 3.3 Maps of HED lithologies based on data from (a) GRaND, (b) VIR, and (c) FC. Methods used to construct the maps were described by Ammannito et al. (2013a), Prettyman et al. (2013), and Thangjam et al. (2013). The white lines in (a) and (b) indicate the outline of the Rheasilvia basin. From Raymond et al. (2017), with permission

A black and white version of this figure will appear in some formats. For the color version, refer to the plate section.

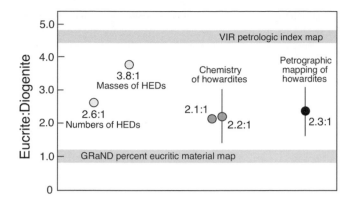

Figure 3.4 Eucrite:diogenite ratios derived from Dawn's GRaND and VIR maps (horizontal bands) compared to ratios calculated from numbers and masses of eucrite and diogenite in the Meteoritical Bulletin Database, from chemical-mixing diagrams for regolithic howardites (two independent estimates), and from petrologic mapping of regolithic howardite thin sections.

Adapted from McSween et al. (2019)

2019). Regolithic howardite is commonly thought to provide the best estimate of the average crustal composition (e.g., Warren et al., 2009; Mittlefehldt et al., 2013). Figure 3.4 shows the Dawn-derived estimates as horizontal bands, which bracket the ratios based on HEDs. The GRaND and VIR-derived mixing-ratio estimates differ significantly.

It is notable that Vesta's compositional maps (Figure 3.3) do not look at all like its geologic map (see Chapter 5). The geologic map is based on geomorphological features and crater ejecta stratigraphy. Some Rheasilvia units tend to correlate with diogenite, but otherwise the surface regolith composition is homogenized relative to the underlying mapped geologic units.

3.3.3 Hydrated Materials on Vesta

One of Dawn's unexpected findings was the discovery of significant hydrogen-bearing materials in a low-albedo region on Vesta. GRaND mapped hydrogen (as water equivalents) by neutron absorption (Prettyman et al., 2019), and VIR provided an analogous map of OH based on the 2.8 μm absorption band (Combe et al., 2015; De Sanctis et al., 2012b). The two maps show similar distributions of hydrogen (see Chapter 6).

The hydrogen-bearing material, occurring near the equatorial eucrite region in Figure 3.3, has been interpreted to be exogenic carbonaceous chondrite delivered by impact. CM-like chondrites contain hydroxyl-bearing phyllosilicates such as serpentine. Such chondrites occur as xenoliths in some howardites, and laboratory spectral analyses of mixtures of CM chondrite and eucrite indicate that admixture of a few percent carbonaceous chondrite can reproduce the spectra of the H-, OH-rich region of Vesta (Reddy et al., 2012). Delivery of this material was tentatively identified with formation of the Veneneia basin, although some dark material might be exogenic dust (De Sanctis et al., 2012b).

3.3.4 Impact Excavation of Vesta's Mantle?

Models of the formation of Vesta's huge, overlapping impact basins indicate excavation to depths of 60–100 km (Jutzi et al.,

2013). This is much deeper than the crustal thickness, and, prior to Dawn's arrival at Vesta, Binzel et al. (1997) predicted that significant ultramafic mantle rock would be exposed in deep basins. Vesta has a chondritic bulk composition (Toplis et al., 2013), and the mantle of any differentiated chondritic body with enough Fe^{2+} to produce eucritic basalts would be dominated by olivine. However, VIR spectra of Rheasilvia curiously revealed little or no olivine (McSween et al., 2013a), although it is challenging to recognize <30% olivine in the presence of pyroxene in diogenite spectra (Beck et al., 2013). Olivine has been identified spectrally elsewhere on Vesta. Ammannito et al. (2013b) reported some olivine occurrences in the northern hemisphere, and Ruesch et al. (2014) mapped a swath of olivine extending from Rheasilvia northward. However, the interpretation of olivine from VIR spectra is ambiguous and still debated, as other interpretations exist (see Chapter 6). There is clearly a dearth of olivine in Vesta's deeply excavated surface and ejecta.

3.3.5 Vesta's Gravity and Internal Structure

Vesta's gravity field is dominated by its shape, complicating its interpretation, although the flattening coefficient J_2 clearly indicates a central core. Core size and density cannot be determined independently, but a mass-balance model with two layers (core, mantle) that matches the J_2 constraint provided an estimate of the core size (Russell et al., 2012). The core radius is 110 ± 3 km for an assumed density of 7,100–7,800 kg/m^3 (as in iron meteorites). A sulfur-rich core would have lower density and a larger size. Such a core composition would approximate a eutectic melt, but is unlikely because of the extensive melting implied by HEDs (Toplis et al., 2013).

The estimated core mass implies an average density for Vesta's silicate fraction (mantle + crust) of 3,100 kg/m^3. This value is low relative to grain densities for mixtures of eucrite, diogenite, and olivine, implying that the mantle + crust layer must have porosity on the order of 10% (Raymond et al., 2017).

The gravity field mapped by Dawn contains anomalies relative to the field calculated, assuming a homogeneous density structure and using Vesta's shape model. The internal structure has been estimated by solving for the densities that minimize residual gravity anomalies for a three-layer (core, mantle, crust) model (Konopliv et al., 2014). That work yields a crust that is 22.4 km thick, with a density of 2,970 kg/m^3 (Ermakov et al., 2014). Such a crust is denser than eucrites and suggests a mixture of eucrite (2,800 kg/m^3) and diogenite (3,050 kg/m^3). The derived mantle density of 3,160 kg/m^3 is low, considering that it should contain appreciable olivine (3,300 kg/m^3).

To study Vesta's gravity anomalies, Park et al. (2014) fixed the mantle density in the same three-layer model to an average value of 3,160 kg/m^3, and they solved for density variations within the crust that minimized the residual gravity. A map of these density anomalies is shown in Chapter 12. Low-density anomalies occur in association with large impact basins and the northern troughs (described in Chapter 5), that is, regions affected by impact excavation and fracturing. High-density anomalies are interpreted to reflect plutons of ultramafic composition emplaced within the crust (Raymond et al., 2017).

3.4 THE VESTA FAMILY AND THE DELIVERY OF HEDS TO EARTH

The discovery of Vesta's spectral match to HED meteorites (McCord et al., 1970) set up a classic debate in planetary science over whether the spectral match was indeed valid and whether there was any pathway for fragments from Vesta to be delivered to Earth. As recounted by Drake (2001), while the spectral match held solid as telescopic measurement precision advanced, the dynamical problem was vexing (Wetherill, 1987). A local source for eucrite delivery was found (Cruikshank et al., 1991) in basaltic asteroids in near-Earth space, but the short lifetime of asteroids in the inner Solar System still demanded a Main Belt source. Consolmagno and Drake (1977) argued that the HED parent body had to be large in order to be sufficiently differentiated and had to be intact, since surface basalts (eucrites) dominated the HED meteorite flux. A spectroscopic survey of the 500 largest asteroids (Zellner et al., 1985) showed that Vesta, and only Vesta, provided a spectral match to HEDs in the Main Belt. Therefore, only Vesta satisfied all the requirements as the HED parent body, even if one could not "get here from there."

Cutting this "Gordian knot" (Gaffey, 1993) required further advancement in detector technology such that smaller Main Belt asteroids (diameters of $\leq$10 km) could be spectroscopically measured. Applying charge-coupled device (CCD) detector technology, Binzel and Xu (1993) found two dozen small asteroids with Vesta-like spectral properties. Most notably, despite surveying hundreds of small asteroids over a wide range of orbits in the inner asteroid belt, the Vesta-like objects (dubbed "vestoids") were found only in a tightly constrained orbital range matching the orbital inclination plane and having the same orbital eccentricity range as Vesta itself. Most importantly, the vestoids spanned the "impossible" orbital range of distances from Vesta to the 3:1 resonance at 2.5 AU, demonstrated by Wisdom (1985) to be a highly efficient delivery route to Earth. The Hubble Space Telescope discovery (Thomas et al., 1997) of a large basin (Rheasilvia) provided the smoking gun as the scar left behind by excavating impact, where the volume of the basin exceeds the sum of all vestoids by about two orders of magnitude. This logic implies that most HEDs were excavated and launched from the Rheasilvia impact basin. However, Type B diogenites located on the opposite side of Vesta (Beck et al., 2017) presumably require an additional ejection event, and Unsulan et al. (2019) opined that a relatively small, young crater was the source of a howardite fall.

Today the inner-belt vestoids are recognized as a "family," a term used to denote asteroids sharing a common parent body. As shown in Figure 3.5, the Vesta family is now an easily recognized major feature in the inner asteroid belt, where the high albedos and unique spectral signature make the family relationship readily apparent in large scale asteroid surveys (Masiero et al., 2013). Relating the Vesta family to the Rheasilvia basin, and its young (~1 Ga) age, explains why Vesta is able to supply a significant percentage of Earth's meteorite flux. Thus, it seems an extraordinary gift of nature that such a unique world as Vesta would be one for which abundant samples "fall" into our laboratories. With the diverse geochemistry of iron meteorites demonstrating many dozens of examples of differentiated asteroids, it is perhaps not surprising that a few other basaltic achondrite examples have been found in different parts of the Main Belt. In the outer belt, (1459) Magnya is an example

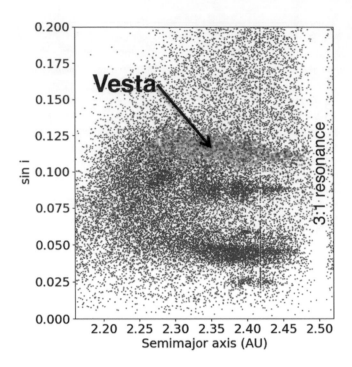

Figure 3.5 Vesta's pathway for delivery of meteorites to Earth was discovered through telescopic spectra initially revealing about two dozen small (D $\leq$10 km) Vesta-like asteroids spreading in the same orbital inclination plane from Vesta to the 3:1 resonance with Jupiter, a dynamical "escape hatch" to the inner Solar System (Binzel & Xu, 1993). Modern space-based infrared surveys easily distinguish the higher albedos of the "Vesta family" asteroids (colored points) showing the vast extent of fragments ejected from Vesta during the formation of the Rheasilvia basin. Adapted from Masiero *et al.* (2013); courtesy J. R. Masiero
A black and white version of this figure will appear in some formats. For the color version, refer to the plate section.

(Lazzaro et al., 2000; see also Chapter 16). As noted in Section 3.2, oxygen isotopes and Fe:Mn analyses also reveal a small number of HEDs that fall outside the main group.

3.5 DISCUSSION: WHAT HAVE WE LEARNED?

The results of the Dawn mission, coupled with decades of research on HED meteorites, have greatly expanded our understanding of the largest surviving differentiated asteroid and, by extension, other protoplanets. This body also clearly reveals how asteroid samples are delivered to Earth and addresses other questions about the kinds of protoplanets that were assembled into the terrestrial planets.

3.5.1 What Have Dawn and Samples Revealed That We Would Not Otherwise Know?

Noting the large number of iron meteorite groups, indicating that extensively melted asteroids were abundant, Wasson (2013) suggested that the HEDs were probably not vestan. McSween et al. (2013b) summarized the evidence strengthening the hypothesis that Vesta is the parent body for HEDs. Vesta's petrologic complexity, detailed spectroscopic characteristics, unusual space weathering, diagnostic geochemical ratios, geochronology, occurrence of exogeneous carbonaceous chondrite, and dimensions of the core, are all consistent with HED observations and constraints. Binzel (2012) similarly concluded the evidence linking these all together comes from the elemental abundance measurements of the GRaND instrument (Prettyman et al., 2012), measurements that can only be obtained by an *in situ* spacecraft. As noted in Section 3.4, the recent excavation of the Rheasilvia basin likely accounts for the excavation and launching of the vestoids, and the nearby orbital escape hatch links Vesta directly to the many HEDs delivered to the Earth.

Cosmic-ray exposure (CRE) ages for HEDs constrain the time spent as small meteoroids in space prior to arrival at Earth. Only two age clusters at ~22 and ~39 Ma are statistically significant (Welten et al., 1997; Cartwright et al., 2013). These CRE spikes likely relate to major impacts that may have destroyed two vestoids and delivered ~80% of HEDs (Herzog, 2007). The larger ~22 Ma cluster delivered all petrologic types of HEDs.

Although the existence of a huge impact basin (Rheasilvia) on Vesta was recognized prior to Dawn (e.g., Thomas et al., 1997), its young (~1 Ga) age was not (Marchi et al., 2012). That age determination required orbital images for the analysis of its crater size-density distribution. The superposition of Rheasilvia on an older basin (>2 Ga Veneneia) was also not known prior to Dawn. The ages of the basins are useful in interpreting argon isotope data from HEDs (Bogard, 2011). Surprisingly, these impacts are not evident in impact-reset ^{40}Ar/^{39}Ar ages, which reflect older events between 3.4 and 4.1 Ga linked to the lunar cataclysm (Marchi et al., 2013).

The spatial scale of Dawn's VIR provides a means of estimating the scale of mixing in the regolith (see Chapter 6). Howardite covers most of the surface, whereas howardites are a relatively minor component of HEDs in meteorite hand samples. Thus, the mixing of eucrite and diogenite in the regolith must occur predominantly as blocks larger than the meteorites. Sampling at both scales, however, indicates considerable variations in the eucrite: diogenite ratio. It is also possible that biased sampling by the Rheasilvia impact event ejected smaller amounts of near-surface howardite, compared to sub-regolith materials. The extent of contamination by foreign chondritic materials was also unexpected. Although it should have been predictable from studies of carbonaceous chondrite xenoliths in howardites, the importance of exogenic contamination of the regolith was not fully appreciated.

Although the occurrence of a large iron core on the HED parent body was predicted from siderophile element depletions in eucrites (Newsom & Drake, 1982; Righter & Drake, 1997) and the former occurrence of a magnetic field as revealed by remnant magnetism (Fu et al., 2012), its size was unknown. The estimated core mass fraction (15–20%) from HEDs compares favorably with the ~18% mass fraction of Vesta's core based on gravitational moment J_2 (Russell et al., 2012) and the combined gravity and shape modeling by Ermakov et al. (2014). The existence of diogenite plutons in the crust (Raymond et al., 2017), as inferred from gravity anomalies (Park et al., 2014), is consistent with trace element studies of diogenites (Mittlefehldt et al., 2012), which require multiple magmas rather than a uniform magmatic source.

Highly siderophile elements should effectively partition into cores, but their abundances in diogenites span five orders of magnitude (Day et al., 2012). The higher abundances are explained by continued minor accretion of chondritic material following metal-silicate equilibration to form Vesta's core.

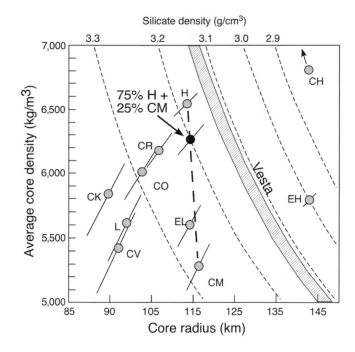

Figure 3.6 Predicted combinations of core radius and density for Vesta, with core estimates based on various kinds of chondritic protoliths. Mixing of 75% H-chondrite and 25% CM-chondrite provides a solution that agrees most closely with Dawn's gravity constraints. Dashed lines are solutions of core size/density that are compatible with values of bulk silicate density in a two-layer model of Vesta.
Modified from Toplis et al. (2013)

A number of estimates for the bulk composition of the HED parent body, based on petrologic and chemical constraints, have been promulgated. Most recently, Toplis et al. (2013) mixed 12 chondritic compositions in various proportions to estimate bulk compositions that satisfied mass-balance and thermodynamic constraints, based on HEDs and Dawn geophysical data. A mixture of 75% H-chondrite and 25% CM-chondrite accounts for the Fe:Mn ratio, oxygen isotope composition, and oxidation state, is most closely consistent with Dawn's core size and density constraints (Figure 3.6), and, when partially melted, can yield eucrite magmas. Without the complete geochemical dataset provided by HEDs, bulk composition models would be highly uncertain.

Although remote sensing of Vesta clearly reveals the presence of basalts and cumulates derived from them, the detailed fractionation patterns of magmas are only revealed by the chemical compositions of HEDs. Major elements are the starting point for experimental petrology (e.g., Stolper, 1977; Bartels & Grove, 1991), and trace elements constrain petrogenetic models (e.g., Mittlefehldt & Lindstrom, 2003; Barrat et al., 2007; Mittlefehldt et al., 2012). These data reveal multiple magmatic pathways and a more complex evolution than normally envisioned for asteroids.

The relative chronology of vestan surfaces has been contested, because different crater production functions were used (O'Brien et al., 2015; Schmedemann et al., 2015). Roig and Nesvorny (2020) argued against rescaling lunar chronology for Vesta, and models using lower impactor flux give the observed number and size distribution for large craters and a ~1 Gy age for Rheasilvia. The crystallization ages of basaltic eucrites, determined by ^{87}Rb–^{87}Sr and ^{147}Sm–^{143}Nd, are as old as ~4.56 Ga (discussed more fully in

Chapter 4). Early melting and crystallization are confirmed by the decay product of short-lived [26]Al (Schiller et al., 2011). Some younger ages of cumulate eucrites and diogenites may indicate a protracted magmatic or cooling history ≥10 Ma. [53]Mn systematics indicate crust–mantle differentiation at ~2 Ma after Solar System formation (Trinquier et al., 2008), and a [182]Hf–[182]W isochron constrains core formation to ~3 Ma (Kleine et al., 2009).

Thermal evolution models for Vesta (Ghosh & McSween, 1998; Gupta & Sahijpal, 2010; Formisano et al., 2013) depend critically on petrologic and chronologic constraints from HEDs. Such models are based on radiogenic heating by short-lived nuclides and indicate accretion within 1–2 Ma of Solar System formation and timescales of >10 Ma for melting, differentiation, and cooling.

Shocked HEDs show evidence of Ar degassing, and their [39]Ar–[40]Ar ages have been reset (Bogard & Garrison, 2010). These ages, ranging between 4.1 and 3.4 Ga, constrain the timing of large impacts. The [39]Ar–[40]Ar ages of impact-melted clasts in howardites fall between 3.8 and 3.3 Ga (Cohen, 2013), essentially the same as the bulk meteorite data. Because these ages overlap the period of the late heavy bombardment, they have been suggested to provide support for such a cataclysm (Marchi et al., 2013). This restricted age range is at odds with crater distributions that suggest Vesta was cratered continuously throughout its history (Marchi et al., 2012).

3.5.2 What Do We Still Not Know?

A major unresolved question is whether Vesta had an early magma ocean. Most models for Vesta's differentiation appeal to a deep magma ocean (Righter & Drake, 1997; Ruzicka et al., 1997; Neumann et al., 2014), despite the fact that all HEDs cannot be explained by magma crystallization on a global scale. Asteroid models that feature very efficient removal of melts from the mantle (Wilson & Keil, 2012) might even preclude a magma ocean. That model would be consistent with the plutons inferred from Vesta's gravity map and would suggest a role for serial magmatism in forming the crust. A hybrid model posits 60–70% equilibrium crystallization of a magma ocean, followed by extraction of residual melt into magma chambers that undergo fractional crystallization and periodic eruption (Mandler & Elkins-Tanton, 2013). This model produces the appropriate eucrite/diogenite ratio, predicts the concentration of olivine in the lower mantle, and is consistent with other Dawn constraints (McSween et al., 2019), although it may not be able to account for diogenite trace element abundances (Barrat & Yamaguchi, 2014). Other models discussed in Chapter 4 also explain the concentration of olivine in the lower mantle, either by crystal accumulation or as residuum from incomplete melting in a shallow magma ocean (Neumann et al., 2014).

Vesta's two huge impact basins, each of which produced a system of troughs following great circles centered on the basins (Buczkowski et al., 2012), demonstrate the profound stresses incurred by this body. It is not at all clear how Vesta remained intact. Perhaps the presence of a massive core acted to strengthen the asteroid against disruption. However, the occurrence of numerous iron meteorites attests to the collisional disruption of other differentiated planetesimals.

Numerical models of the overlapping large impact basins at Vesta's south pole indicate excavation to depths of up to 60–100 km (Jutzi et al., 2013; see also Chapter 16). The thickness of the basaltic crust is estimated at 15 km based on a chondritic model

(Toplis et al., 2013) and 22 km based on a gravity model (Konopliv et al., 2014). The apparent lack of olivine on the floor of Rheasilvia (recalling that olivine diogenite with <30% olivine cannot be detected) and in its ejecta (McSween et al., 2013a) has spawned several explanations for the missing mantle material. Consolmagno et al. (2015) suggested that Vesta has a non-chondritic composition, perhaps produced by collisional processes. Clenet et al. (2014) proposed that Vesta's crust has been significantly thickened by intrusion of diogenite plutons, a suggestion also incompatible with a chondritic bulk composition. A more plausible explanation is sequestration of olivine in the lower mantle (Toplis et al., 2013). Olivine is the earliest liquidus phase for Vesta's magma ocean, crystallizing alone until the fraction of melt remaining was ~60%. Thus, settling and accumulation of olivine in the deep mantle could have occurred during fractional crystallization of a magma ocean (Figure 3.7). Alternatively, olivine could have been concentrated in a bottom residue produced by partial melting to form a shallow magma ocean. The small mantle samples that occur as xenoliths in howardites are more consistent with residues than cumulates (Hahn et al., 2018).

HEDs are usually thought to be anhydrous, and analysis of apatite in eucrites indicates that they are generally fluorine-rich with little OH (Sarafian et al., 2013). However, a few OH-rich

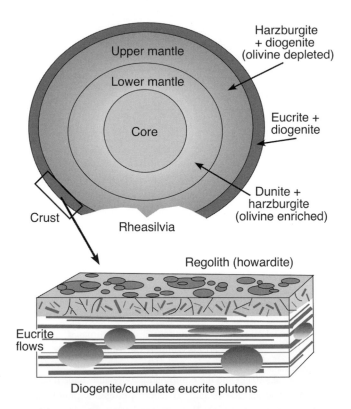

Figure 3.7 Conceptual sketch for the interior structure of Vesta (not to scale), based mostly on the model of Mandler and Elkins-Tanton (2013). The olivine-rich (dunite) lower mantle differs from that model, which posits a lower mantle of harzburgite, but is required to explain the apparent dearth of olivine in deeply excavated craters like Rheasilvia (see Chapter 4). The lower diagram illustrates the structure of the crust, composed of eucrite flows intruded by plutons containing diogenite and cumulate eucrite, and an upper surface regolith of howardite.

A black and white version of this figure will appear in some formats. For the color version, refer to the plate section.

apatites may suggest a role for hydrothermal fluids in the Vestan interior, and Barrat et al. (2011) noted some possible fluid–rock interactions in eucrites. A few craters on Vesta display erosional gullies in their walls, suggested to be debris flows lubricated by fluids released from crater walls (Scully et al., 2015), as well as pits in their floors, interpreted to have formed by escaping subsurface volatiles (Denevi et al., 2012). However, it is unclear whether these geomorphic features are the result of internal fluids or water derived from exogenic carbonaceous chondrite.

3.6 FINAL THOUGHTS: IMPLICATIONS OF VESTA FOR EARLY PLANETESIMALS

Nucleosynthetic isotopic analyses of O, Cr, Ti, Ni, and Mo indicate a compositional dichotomy between carbonaceous and non-carbonaceous meteorites (Scott et al., 2018). HEDs plot clearly within the non-carbonaceous category, demonstrating a link between Vesta and other inner Solar System materials. The possible implications of separated compositional reservoirs for early Solar System history are discussed in Chapter 14.

Vesta, as the sole surviving differentiated rocky protoplanet, can potentially tell us much about the kinds of bodies that accreted to form the terrestrial planets. These bodies had chondritic bulk compositions, although with significant depletions in volatile elements like sodium and potassium. However, the origin of volatile depletions in differentiated bodies is unknown.

The recognition that differentiated meteorites are older than chondrites (Kruijer et al., 2017) has led to the idea that the raw materials for planets were already differentiated. These protoplanets were partially or completely melted by decay of short-lived radionuclides, probably producing magma oceans that allowed efficient core separation. Vesta demonstrates, though, that magmatism was complex, with several stages lasting perhaps for tens of millions of years and producing evolved lithologies. Although models suggest ease of magma ascent and eruption on small bodies, plutons were emplaced into Vesta's crust. Core separation in the terrestrial planets would have been rapid if the accreting planetesimals already had large cores. Given the overlapping timescales for Vesta's magmatic evolution and planetary assembly, it seems likely that some accreting protoplanets might still have been molten and actively differentiating.

The occurrence of exogenic, water-bearing chondrite in Vesta's regolith supports the idea that delivery of volatiles into the inner Solar System by primitive impactors was an important process. However, on Vesta such materials appear to have been incorporated relatively late in its history.

REFERENCES

Adams, J. B. (1974) Visible and near-infrared diffuse reflectance spectra of pyroxenes as applied to remote sensing of solid objects in the solar system. *Journal of Geophysical Research*, 79, 4829–4836.

Ammannito, E., De Sanctis, M. C., Capaccioni, F., et al. (2013a) Vestan lithologies mapped by the visual and infrared spectrometer on Dawn. *Meteoritics & Planetary Science*, 48, 2185–2198.

Ammannito, E., De Sanctis, M. C., Palomba, E., et al. (2013b) Olivine in an unexpected location on Vesta's surface. *Nature*, 504, 122–125.

Barrat, J.-A., & Yamaguchi, A. (2014) Comment on "The origin of eucrites, diogenites, and olivine diogenites: Magma ocean crystallization and shallow magma processes on Vesta" by B. E. Mandler and L. T. Elkins-Tanton. *Meteoritics & Planetary Science*, 49, 468–472.

Barrat, J.-A., Yamaguchi, A., Bunch, T., et al. (2011) Possible fluid–rock interactions of differentiated asteroids recorded in eucritic meteorites. *Geochimica et Cosmochimica Acta*, 75, 3839–3852.

Barrat, J.-A., Yamaguchi, A., Greenwood, R. C., et al. (2007) The Stannern trend eucrites: Contamination of main group eucritic magmas by crustal partial melts. *Geochimica et Cosmochimica Acta*, 71, 4108–4124.

Barrat, J.-A., Yamaguchi, A., Greenwood, R. C., et al. (2008) Geochemistry of diogenites: Still more diversity in their parental melts. *Meteoritics & Planetary Science*, 43, 1759–1775.

Bartels, K. S., & Grove, T. L. (1991) High-pressure experiments on magnesian eucrite compositions: Constraints on magmatic processes in the eucrite parent body. *Proceedings of the Lunar & Planetary Science Conference*, 21, 351–365.

Beck, A. W., Lawrence D. J., Peplowski, P. N., et al. (2015) Using HED meteorites to interpret neutron and gamma-ray data from asteroid 4 Vesta. *Meteoritics & Planetary Science*, 50, 1311–1337.

Beck, A. W., Lawrence, D. J., Peplowski, P. N., et al. (2017) Igneous lithologies on asteroid (4) Vesta mapped using gamma-ray and neutron data. *Icarus*, 286, 35–45.

Beck, A. W., McCoy, T. J., Sunshine, J. M., et al. (2013) Challenges in detecting olivine on the surface of 4 Vesta. *Meteoritics & Planetary Science*, 48, 2155–2165.

Beck, A. W., & McSween, H. Y. (2010) Diogenites as polymict breccias composed of orthopyroxenite and harzburgite. *Meteoritics & Planetary Science*, 45, 850–872.

Beck, A. W., Mittlefehldt, D. W., McSween, H. Y., et al. (2011) MIL 03443, a dunite from asteroid 4 Vesta: Evidence for its classification and cumulate origin. *Meteoritics & Planetary Science*, 46, 1133–1151.

Beck, A. W., Welten, K. C., McSween, H. Y., Viviano, C. E., & Caffee, M. W. (2012) Petrologic and textural diversity among the PCA 02 howardite group, one of the largest pieces of the Vestan surface. *Meteoritics & Planetary Science*, 47, 947–969.

Benedix, G. K., Haack, H., & McCoy, T. J. (2014) Iron and stony-iron meteorites. In A. M. Davis (ed.), *Treatise on Geochemistry*, 2nd ed., Vol. 1. Oxford: Elsevier, pp. 267–285.

Binzel, R. P. (2012) A golden spike for planetary science. *Science*, 338, 203–204.

Binzel, R. P., Gaffey, M. J., Thomas, P., et al. (1997) Geologic mapping of Vesta from 1994 Hubble Space Telescope images. *Icarus*, 128, 95–103.

Binzel, R. P., & Xu, S. (1993) Chips off of asteroid 4 Vesta: Evidence for the parent body of basaltic achondrite meteorites. *Science*, 260, 186–191.

Bobrovnikoff, N. T. (1929) The spectra of minor planets. *Lick Observatirt Bulletin*, 14 (No. 407), 18–27.

Bogard, D. D. (2011) K-Ar ages of meteorites: Clues to parent-body thermal histories. *Chemie der Erde Geochemistry*, 71, 207–226.

Bogard, D. D., & Garrison, D. H. (2010) ^{39}Ar-^{40}Ar ages of eucrites and the thermal history of asteroid 4 Vesta. *Meteoritics & Planetary Science*, 38, 669–710.

Buczkowski, D. L., Wyrick, D. Y., Toplis, M., et al. (2012) Large-scale troughs on Vesta: A signature of planetary tectonics. *Geophysical Research Letters*, 39, L18205.

Cartwright, J. A., Ott, U., & Mittlefehldt, D. W. (2014) The quest for regolithic howardites. Part 2: Surface origins highlighted by noble gases. *Geochimica et Cosmochimica Acta*, 140, 488–508.

Cartwright, J. A., Ott, U., Mittlefehldt, D. W., et al. (2013) The quest for regolithic howardites. Part 1: Two trends uncovered using noble gases. *Geochimica et Cosmochimica Acta*, 105, 395–421.

Clenet, H., Jutzi, M., Barrat, J.-A., et al. (2014) A deep crust mantle boundary in the asteroid 4 Vesta. *Nature*, 511, 303–306.

Cohen, B. (2013) The Vestan cataclysm: Impact-melt clasts in howardites and the bombardment history of 4 Vesta. *Meteoritics & Planetary Science*, 48, 771–785.

Combe, J.-P., McCord, T. B., McFadden, L. A., et al. (2015) Composition of the northern regions of Vesta analyzed by the Dawn mission. *Icarus*, 259, 53–71.

Consolmagno, J. G., & Drake, M. J. (1977) Composition and evolution of eucrite parent body – Evidence from rare earth elements. *Geochimica et Cosmochimica Acta*, 41, 1271–1282.

Consolmagno, J. G., Golabek, G. J., Turrini, D., et al. (2015) Is Vesta an intact and pristine protoplanet? *Icarus*, 254, 190–201.

Cruikshank, D. P., Tholen, D. J., Hartmann, W. K., Bell, J. F., & Brown, R. H. (1991) Three basaltic Earth-approaching asteroids and the source of basaltic meteorites. *Icarus*, 89, 1–13.

Cunningham, C. J. (2014) The First Four Asteroids: A History of Their Impact on English Astronomy in the Early Nineteenth Century. PhD thesis, University of Southern Queensland.

Day, J. M. D., Walker, R. J., Qin, L., & Rumble, D. (2012) Late accretion as a natural consequence of planetary growth. *Nature Geoscience*, 5, 614–617.

De Sanctis, M. C., Ammannito, E., Capria, M. T., et al. (2012a) Spectroscopic characterization of mineralogy and its diversity across Vesta. *Science*, 336, 697–700.

De Sanctis, M. C., Ammannito, E., Capria, M. T., et al. (2013) Vesta's mineralogical composition as revealed by VIR on Dawn. *Meteoritics & Planetary Science*, 48, 2166–2184.

De Sanctis, M. C., Combe, J.-P., Ammannito, E., et al. (2012b) Detection of widespread hydrated materials on Vesta by the VIR imaging spectrometer on board the Dawn mission. *Astrophysical Journal*, 758, L36.

Denevi, B. W., Blewett, D. T., Buczkowski, D. L., et al. (2012) Pitted terrain on Vesta and implications for the presence of volatiles. *Science*, 338, 246–249.

Drake, M. J. (2001) The eucrite/Vesta story. *Meteoritics & Planetary Science*, 36, 501–513.

Ermakov, A. I., Zuber, M. T., Smith, D. E., et al. (2014) Constraints on Vesta's interior structure using gravity and shape models from the Dawn mission. *Icarus*, 240, 146–160.

Formisano, M., Federico, C., Turrini, D., et al. (2013) The heating history of Vesta and the onset of differentiation. *Meteoritics & Planetary Science*, 48, 2316–2332.

Fu, R. R., Weiss, B. P., Shuster, D. L., et al. (2012) An ancient core dynamo in asteroid Vesta. *Science*, 338, 239–241.

Gaffey, M. J. (1976) Spectral reflectance characteristics of the meteorite classes. *Journal of Geophysical Research*, 81, 905–920.

Gaffey, M. J. (1993) Forging an asteroid–meteorite link. *Science*, 260, 167–168.

Gaffey, M. J. (1997) Surface lithologic heterogeneity of asteroid 4 Vesta. *Icarus*, 127, 130–157.

Ghosh, A., & McSween, H. Y. (1998) A thermal model for the differentiation of asteroid 4 Vesta, based on radiogenic heating. *Icarus*, 134, 187–206.

Greenwood, R. C., Barrat, J.-A., Yamaguchi, A., et al. (2014) The oxygen isotope composition of diogenites: Evidence for early global melting on a single, compositionally diverse, HED parent body. *Earth & Planetary Science Letters*, 390, 165–174.

Gupta, G., & Sahijpal, S. (2010) Differentiation of Vesta and the parent bodies of other achondrites. *Journal of Geophysical Research*, 115, E08001.

Haba, M. K., Wotzlaw, J.-W., Lai, Y.-J., et al. (2019) Mesosiderite formation on asteroid 4 Vesta by a hit-and-run collision. *Nature Geoscience*, 12, 510–515.

Hahn, T. M., Lunning, N. G., McSween, H. Y., et al. (2018) Mg-rich harzburgites from Vesta: Mantle residua or cumulates from planetary differentiation? *Meteoritics & Planetary Science*, 53, 514–546.

Herzog, G. F. (2007) Cosmic-ray exposure ages of meteorites. In H. D. Holland, & K. I. Turekian (eds.), *Treatise on Geochemistry*, Vol. 1. Oxford: Pergamon Press, pp. 711–746.

Jaumann, R. J., Williams, D. A., Buczkowski, D. L., et al. (2012) Vesta's shape and morphology. *Science*, 336, 687–690.

Jutzi, M., Asphaug, E., Gillet, P., et al. (2013) The structure of asteroid 4 Vesta as revealed by models of planet-scale collisions. *Nature*, 494, 207–210.

Keil, K. (2002) Geological history of asteroid 4 Vesta: the "smallest terrestrial planet". In W. Bottke, A. Cellino, P. Paolicchi, & R. P. Binzel (eds.), *Asteroids III*. Tucson: University of Arizona Press, pp. 573–584.

Kleine, T., Touboul, M., Bourdon, B., et al. (2009) Hf-W chronology of the accretion and early evolution of asteroids and terrestrial planets. *Geochimica et Cosmochimica Acta*, 73, 5150–5188.

Konopliv, A. S., Asmar, S. W., Park, R. S., et al. (2014) The Vesta gravity field, spin pole and rotation period, landmark positions and ephemeris from the Dawn tracking and optical data. *Icarus*, 240, 103–117.

Kruijer, T. S., Burkhardt, C., Budde, G., et al. (2017) Age of Jupiter inferred from the distinct genetics and formation times of meteorites. *Proceedings of the National Academy of Sciences (USA)*, 114, 6712–6716.

Lazzaro, D., Michtchenco, T., Carvano, J. M., et al. (2000) Discovery of a basaltic asteroid in the outer Main Belt. *Science*, 288, 2033–2035.

Lorenz, K., Nazarov, M., Kurat, G., et al. (2007) Foreign meteoritic material of howardites and polymict eucrites. *Petrology*, 15, 109–125.

Lunning, N. G., McSween, H. Y., Tenner, H. Y., et al. (2015) Olivine and pyroxene from the mantle of asteroid 4 Vesta. *Earth & Planetary Science Letters*, 418, 126–135.

Lunning, N. G., Welten, K. C., McSween, H. Y., et al. (2016) Grosvenor Mountains 95 howardite pairing group: Insights into the surface regolith of asteroid 4 Vesta. *Meteoritics & Planetary Science*, 51, 167–194.

Mandler, B. E., & Elkins-Tanton, L. T. (2013) The origin of eucrites, diogenites, and olivine diogenites: Magma ocean crystallization and shallow magma chamber processes on Vesta. *Meteoritics & Planetary Science*, 48, 2333–2349.

Marchi, S., Bottke, W. F., Cohen, B. A., et al. (2013) High-velocity collisions from the lunar cataclysm recorded in asteroidal meteorites. *Nature Geoscience*, 6, 303–307.

Marchi, S., McSween, H. Y., O'Brien, D. P., et al. (2012) The violent collisional history of asteroid 4 Vesta. *Science*, 336, 690–694.

Masiero, J. R., Mainzer, A. K., Bauer, J. M., et al. (2013) Asteroid family identification using the hierarchical clustering method and WISE/NEOWISE physical properties. *Astrophysical Journal*, 770, 7.

Mayne, R. G., McSween, H. Y., McCoy, T. J., et al. (2009) Petrology of the unbrecciated eucrites. *Geochimica et Cosmochimica Acta*, 73, 794–819.

McCord, T. B., Adams, J. B., & Johnson, T. V. (1970) Asteroid Vesta: Spectral reflectivity and compositional implications. *Science*, 168, 1445–1447.

McSween, H. Y., Ammannito, E., Reddy, V., et al. (2013a) Composition of the Rheasilvia basin, a window into Vesta's interior. *Journal of Geophysical Research*, 118, 335–346.

McSween, H. Y., Binzel, R. P., De Sanctis, M. C., et al. (2013b) Dawn; the Vesta-HED connection; and the geologic context for eucrites, diogenites, and howardites. *Meteoritics & Planetary Science*, 48, 2090–2014.

McSween, H. Y., Mittlefehldt, D. W., Beck, A. W., et al. (2011) HED meteorites and their relationship to the geology of Vesta and the Dawn mission. *Space Science Reviews*, 163, 141–174.

McSween, H. Y., Raymond, C. A., Stolper, E. M., et al. (2019) Differentiation and magmatic history of Vesta: Constraints from

HED meteorites and Dawn spacecraft data. *Geochemistry*, 79, 125526.

Mittlefehldt, D. W. (1994) The genesis of diogenites and HED parent body petrogenesis. *Geochimica et Cosmochimica Acta*, 58, 1537–1552.

Mittlefehldt, D. W. (2015) Asteroid (4) Vesta: I. The howardite-eucrite-diogenite (HED) clan of meteorites. *Chemie der Erde Geochemistry*, 75, 155–183.

Mittlefehldt, D. W., Beck, A. W., Lee, C.-T. A., et al. (2012) Compositional constraints on the genesis of diogenites. *Meteoritics & Planetary Science*, 47, 72–98.

Mittlefehldt, D. W., Herrin, J. S., Quinn, J. E., et al. (2013) Composition and petrology of HED polymict breccias: the regolith of (4) Vesta. *Meteoritics & Planetary Science*, 48, 2105–2134.

Mittlefehldt, D. W., & Lindstrom, M. M. (2003) Geochemistry of eucrites: Genesis of basaltic eucrites, and Hf and Ta as petrogenetic indicators for altered Antarctic eucrites. *Geochimica et Cosmochimica Acta*, 67, 1911–1935.

Mittlefehldt, D. W., McCoy, T. J., Goodrich, C. A., et al. (1998) Non-chondritic meteorites from asteroidal bodies. In J. J. Papike (ed.), *Planetary Materials: Mineralogy & Petrology of Extraterrestrial Materials*. Washington, DC: Mineralogical Society of America, pp. 4-1–4-195.

Neumann, W., Breuer, D., & Spohn, T. (2014) Differentiation of Vesta: Implications for a shallow magma ocean. *Earth & Planetary Science Letters*, 395, 267–280.

Newsom, H. E., & Drake, M. J. (1982) The metal content of the eucrite parent body: Constraints from the partitioning behavior of tungsten. *Geochimica et Cosmochimica Acta*, 46, 2483–2489.

O'Brien, D. P., Marchi, S., Morbidell, A., et al. (2015) Constraining the cratering chronology of Vesta. *Planetary & Space Science*, 103, 131–142.

Papike, J. J., Karner, J. M., & Shearer, C. K. (2003) Determination of planetary basalt parentage: A simple technique using the electron microprobe. *American Mineralogist*, 88, 469–472.

Park, R. S., Konopliv, A. S., Asmar, S. W., et al. (2014) Gravity field expansion in ellipsoidal harmonic and polyhedral internal representations applied to Vesta. *Icarus*, 240, 118–132.

Prettyman, T. H., Mittlefehldt, D. W., Yamashita, N., et al. (2012) Elemental mapping by Dawn reveals exogenic H in Vesta's regolith. *Science*, 338, 242–246.

Prettyman, T. H., Mittlefehldt, D. W., Yamashita, N., et al. (2013) Neutron absorption constraints on the composition of 4 Vesta. *Meteoritics & Planetary Science*, 48, 2211–2236.

Prettyman, T. H., Yamashita, N., Ammannito, E., et al. (2019) Elemental composition and mineralogy of Vesta and Ceres: Distribution and origins of hydrogen-bearing species. *Icarus*, 318, 42–55.

Prettyman, T. H., Yamashita, N., Reedy, R. C., et al. (2015) Concentrations of potassium and thorium within Vesta's regolith. *Icarus*, 259, 39–52.

Raymond, C. A., Russell, C. T., & McSween, H. Y. (2017) Dawn at Vesta: Paradigms and paradoxes. In L. Elkins-Tanton, & B. Weiss (eds.), *Planetesimals*. New York: Cambridge University Press, pp. 321–340.

Reddy, V., Le Corre, L., O'Brien, D. P., et al. (2012) Delivery of dark material to Vesta via carbonaceous chondritic impacts. *Icarus*, 221, 544–559.

Righter, K., & Drake, M. J. (1997) A magma ocean on Vesta: Core formation and petrogenesis of eucrites and diogenites. *Meteoritics & Planetary Science*, 32, 929–944.

Roig, F., & Nesvorny, D. (2020) Modeling the chronologies and size distributions of Ceres and Vesta. *The Astronomical Journal*, 160, 110.

Ruesch, O., Hiesinger, H., De Sanctis, M. C., et al. (2014) Detections and geologic context of local enrichments of olivine on Vesta with VIR/Dawn data. *Journal of Geophysical Research*, 119, 2078–2108.

Russell, C. T., Raymond, C. A., Coradini, A., et al. (2012) Dawn at Vesta: Testing the protoplanetary paradigm. *Science*, 336, 684–686.

Ruzicka, A., Snyder, G. A., & Taylor, L. A. (1997) Vesta as the howardite, eucrite and diogenite parent body: Implications for the size of a core and for large-scale differentiation. *Meteoritics & Planetary Science*, 32, 825–840.

Sarafian, A. R., Roden, M. F., & Patino-Douce, A. E. (2013) The volatile content of Vesta: Clues from apatite in eucrites. *Meteoritics & Planetary Science*, 48, 2135–2154.

Schiller, M., Baker, J., Creech, J., et al. (2011) Rapid timescales for magma ocean crystallization on the howardite-eucrite-diogenite parent body. *Astrophysical Journal Letters*, 740, L22.

Schmedemann, N., Kneissl, T., Ivanov, B. A., et al. (2015) The cratering record, chronology and surface ages of (4) Vesta in comparison to smaller asteroids and ages of HED meteorites. *Planetary & Space Science*, 103, 104–130.

Scott, E. R. D., Greenwood, R. C., Franchi, I. A., & Sanders, I. S. (2009) Oxygen isotopic constraints on the origin and parent bodies of eucrites, diogenites, and howardites. *Geochimica et Cosmochimica Acta*, 73, 5835–5853.

Scott, E. R. D., Krot, A. N., & Sanders, I. S. (2018) Isotopic dichotomy among meteorites and its bearing on the protoplanetary disk. *Astrophysical Journal*, 854, 164.

Scully, J. E. C., Russell, C. T., Yin, A., et al. (2015) Geomorphological evidence for transient water flow on Vesta. *Earth & Planetary Science Letters*, 411, 151–163.

Shearer, C. K., Fowler, G. W., & Papike, J. J. (1997) Petrogenetic models for magmatism on the eucrite parent body: Evidence from orthopyroxene in diogenites. *Meteoritics & Planetary Science*, 32, 877–889.

Stolper, E. M. (1977) Experimental petrology of eucritic meteorites. *Geochimica et Cosmochimica Acta*, 41, 587–681.

Takeda, H., & Graham, A. L. (1991) Degree of equilibration of eucritic pyroxenes and thermal metamorphism of the earliest planetary crust. *Meteoritics*, 26, 129–134.

Thangjam, G., Reddy, V., Le Corre, L., et al. (2013) Lithologic mapping of HED terrains on Vesta using Dawn Framing Camera color data. *Meteoritics & Planetary Science*, 48, 2199–2210.

Thomas, P. C., Binzel, R. P., Gaffey, M. J., et al. (1997) Impact excavation on asteroid 4 Vesta: Hubble Space Telescope results. *Science*, 277, 1492–1495.

Toplis, M. J., Mizzon, H., Monnereau, M., et al. (2013) Chondritic models of 4 Vesta: Implications for geochemical and geophysical properties. *Meteoritics & Planetary Science*, 48, 2300–2315.

Treiman, A. H. (1997) The parent magmas of cumulate eucrites: A mass balance approach. *Meteoritics & Planetary Science*, 32, 138–146.

Trinquier, A., Birck, J. L., Allegre, C. J., et al. (2008) ^{53}Mn–^{53}Cr systematics of the early solar system revisited. *Geochimica et Cosmochimica Acta*, 72, 5146–5163.

Unsulan, O., Jenniskens, P., Yin, Q.-Z., et al. (2019) The Saricicek howardite fall in Turkey: Source crater of HED meteorites on Vesta and impact risk of vestoids. *Meteoritics & Planetary Science*, 54, 953–1008.

Warren, P. H., Kallemeyn, G. W., Huber, H., et al. (2009) Siderophile and other geochemical constraints on mixing relationships among HED-meteoritic breccias. *Geochimica et Cosmochimica Acta*, 73, 5918–5943.

Wasson, J. T. (2013) Vesta and extensively melted asteroids: Why HED meteorites are probably not from Vesta. *Earth & Planetary Science Letters*, 381, 138–146.

Welten, K. C., Lindner, L., Van Der Borg, K., et al. (1997) Cosmic-ray exposure ages of diogenites and the recent collisional history of the howardite, eucrite and diogenite parent body/bodies. *Meteoritics & Planetary Science*, 32, 891–902.

Wetherill, G. W. (1987) Dynamical relations between asteroids, meteorites and Apollo-Amor objects. *Philosophical Transactions of the Royal Society of London, Series A*, 323, 323–336.

Williams, D. A., Jaumann, R., McSween, H. Y., et al. (2014) The chronostratigraphy of protoplanet Vesta. *Icarus*, 244, 158–165.

Wilson, L., & Keil, K. (2012) Volcanic activity on differentiated asteroids: A review and analysis. *Chemie der Erde*, 72, 289–321.

Wisdom, J. (1985) Meteorites may follow a chaotic route to Earth. *Nature*, 315, 731–733.

Zellner, B., Storrs, A. W., Wells, E., et al. (1997). Hubble Space Telescope images of asteroid 4 Vesta. *Icarus*, 128, 83–87.

Zellner, B., Tholen, D. J., & Tedesco, E. F. (1985) The eight-color asteroid survey: Results for 589 minor planets. *Icarus*, 61, 355–416.

Zolensky, M. E., Weisberg, M. K., Buchanan, P. C., et al. (1996) Mineralogy of carbonaceous chondrite clasts in HED meteorites. *Meteoritics & Planetary Science*, 31, 518–537.

4

The Internal Evolution of Vesta

MICHAEL J. TOPLIS AND DORIS BREUER

4.1 INTRODUCTION

The first few million years of the Solar System saw the formation of rocky bodies ranging in size from tens to hundreds of kilometers in diameter (Montmerle et al., 2006). These bodies, made from primitive Solar System materials, were a mixture of high- and low-melting temperature silicates and metal-rich phases including sulfides (Brearley & Jones, 1998). The potential energy associated with collisions between objects of this size is not sufficient to generate large-scale melting, but if accretion occurred early enough in Solar-System history, heat produced by short-lived radioactive elements (e.g. ^{26}Al, ^{60}Fe) had the potential to raise temperatures above the metal–sulfide eutectic and above the solidus of the silicate fraction, on the order of $\sim$1,300 K in both cases. For bodies <1,000 km in diameter, pressure is low enough that, between 1,300 K and the liquidus, partial melting leads to a complex system of a matrix of solid metal and solid silicates containing two immiscible liquids: a molten metal-rich sulfide and a silicate melt. The details will be discussed in this chapter, but, in general terms, if mobility of the liquid phases is possible, then global-scale differentiation may occur, with dense metal-rich liquids having a tendency to sink to form a core, and lighter silicate melts having a tendency to rise to form a differentiated crust.

Despite this apparent simplicity, the physical and chemical evolution of a partially molten body containing both a metal-rich and a silicate fraction is extremely complex. For example, liquid mobility requires that the liquid in question be connected in 3D and that the time scales for liquid movement be shorter than the timescale of cooling of the body. In turn, liquid connectivity is a function of both the wetting behavior of the liquid in question and the fraction of that liquid, while the timescale of extraction is controlled by the viscosities of the solid and liquid phases and grain size. The efficiency of liquid segregation can thus vary from 0 to 100%, and be different for metallic and silicate liquids, depending on the details of the system.

Added to these considerations is the importance of bulk composition, that may contain more or less heat producing elements, more or less semi-volatile elements such as sulfur (that in turn will affect core size and/or density) and more or less volatile elements such as H and N (that may lead to contrasting differentiation histories and very different surface mineralogies). Furthermore, for bodies that experienced magmatic temperatures, the thermodynamics of melting leads to preferential partitioning of heat-producing elements between solid and liquid phases, having the consequence that liquid movement redistributes heat-sources within the body, generating complex retroactions that in turn play a critical role on the physical and chemical evolution of the body (core size and composition, crustal composition and structure). There are thus a huge variety of possibilities for differentiation scenarios in the early Solar System that depend on body size, bulk composition, accretion date, and duration (comparison of Vesta and Ceres being a case in point: e.g., see Chapter 11).

Within this general framework of the question of differentiation in the early Solar System, Vesta is a particularly interesting case study. First, its size is well constrained, simplifying modeling efforts that can concentrate on bodies of relevant size. Second, the rich diversity of HED meteorites that are a convincing match to Vesta (Chapter 3) provide constraints on bulk composition and a unique opportunity to confront predictions of numerical models with petrologic reality. Finally, the Dawn mission (Russell & Raymond, 2012), in addition to confirming the link between Vesta and the HED's, also provides critical constraints on the internal density structure of the asteroid, providing insights into questions of core size and composition (Russell et al., 2012: Chapter 12). In this chapter we will begin by considering petrologic and geochemical constraints on the bulk composition and differentiation timescales of Vesta, before presenting modeling efforts to understand the chemical and physical evolution of this unique preserved rocky protoplanet.

4.2 CONSTRAINTS ON THE BULK COMPOSITION OF VESTA

The meteorite collection in general suggests that the primitive building blocks of the Solar System were "chondritic" in composition. However, diverse classes of chondrite exist, from the volatile-rich Ivuna-class (CI), to metal-poor varieties such as the LL ordinary chondrites (Wasson & Kallemeyn, 1990). Toplis et al. (2013) considered 12 different chondritic groups as potential proxies for bulk Vesta (CI, CV, CO, CM, CK, CR, CH, H, L, LL, EH, EL), noting that none of these groups provide a perfect match to the O-isotope composition of the HED parent body. While this may indicate that bulk Vesta represents an unknown type of chondrite, a useful working hypothesis for constraining bulk composition is that Vesta is a mixture of two or more known groups. Each class of chondrite has a different sulfur and iron content, shows variable extent of volatile depletion, and is characterized by different concentrations of incompatible lithophiles such as the Rare Earth Elements (REE). For fully differentiated bodies, consisting of core–mantle–crust, the relative proportions and densities of each of these three reservoirs will thus be different for different chondritic bulk compositions. In general terms, the core will consist of a mixture of Fe:sulfide and Fe:Ni metal, while the mantle will be dominated by olivine+pyroxene. For Vesta, the "crust" may be assumed to have the composition of the primitive basaltic eucrite, Juvinas (Chapter 3), generated either by partial melting of the bulk silicate fraction (e.g., Jones, 1984; Stolper, 1975), or as a residual liquid of magma-ocean crystallization (e.g. Drake, 2001; Righter & Drake, 1997; Ruzicka et al., 1997; Shearer et al., 1997).

Quantification of the relative proportions and densities of the three principal reservoirs is complicated by the fact that iron may occur in metallic form (in the core) and/or in oxidized form (in the mantle and crust). However, making the assumption that the crust of Vesta has the composition of Juvinas, using the thermodynamics of Fe:Mg partitioning between basaltic liquid (crust) and olivine (mantle), and assuming that the REE are perfectly incompatible (i.e., only in the crust), Toplis et al. (2013) calculated a theoretical internal structure for each of the 12 chondritic compositions tested. Each solution is associated with a different core size, average core density, amount of crust, and bulk composition of the mantle (expressed as a ratio of olivine and pyroxene).

The work of Toplis et al. (2013) shows that solutions corresponding to CI and LL groups predict a negative metal fraction and can thus be excluded. Solutions for enstatite chondrites imply significant oxidation relative to the starting materials and these solutions too were considered unlikely. In the remaining cases, the relative proportion of crust to bulk silicate is typically in the range 15–20%, corresponding to crustal thicknesses of 15–20 km for a porosity-free Vesta-sized body. The mantle is predicted to be largely dominated by olivine (>85%) for carbonaceous chondrites, but to be a roughly equal mixture of olivine and pyroxene for ordinary chondrite precursors. All bulk compositions have a significant core, but the relative proportions of metal and sulfide can be widely different, from the metal dominated case of H-chondrites to the sulfide rich case of CM-chondrites. Using these data, total core size (metal + sulfide) and average core densities can be compared with values derived from the geophysical/gravity data of the Dawn mission, as illustrated in Figure 3.6 of Chapter 3. When making this comparison, it appears that H-chondrites provide the closest match to the geophysical constraints of the Dawn mission, but do not provide a perfect fit.

Although this mass-balance approach provides useful constraints concerning internal structure, it does not address more subtle geochemical issues such as the possibility to generate Juvinas-like liquids at relevant degrees of partial melting/crystallization, nor the capacity to reproduce characteristic elemental and/or isotopic ratios such as Mn:Fe or ^{16}O:^{18}O. Taking all these considerations into account, Toplis et al. (2013) concluded that the best proxy for the bulk composition of Vesta is a mixture of ¾ H chondrite and ¼ CM chondrite, from which Na had been largely lost. One point to note is that for this bulk composition, the eucrite crust has a predicted thickness of ~15 km and the bulk mantle is a ~70:30 mixture of olivine and pyroxene. As such, olivine is expected to be a major mineralogical component of bulk Vesta, despite its paucity at the surface, even in the Rheasilvia basin (McSween et al., 2013).

4.3 CONSTRAINTS ON THE MAGMATIC HISTORY AND DIFFERENTIATION TIME SCALES OF VESTA FROM HED METEORITES

In addition to providing insights into bulk composition, the HED meteorites also provide important constraints on the times and durations of magmatic and metamorphic processes associated with accretion and differentiation. In detail, a number of different radio-chronometers can be used to reconstruct the thermal history of Vesta from eucrites and diogenites, using the fact that different isotopic "clocks" are sensitive to different processes (i.e., core formation, crust–mantle segregation) and that they freeze in information at a variety of "closure temperatures" that correspond to the point at which atomic diffusion of the elements in question ceases in a particular mineral host during cooling (e.g., Dodson, 1973; Ganguly et al., 2007).

Beginning with the question of core formation, Hf:W data can be used to infer the initial ^{182}W:^{184}W ratio, which can in turn be used to date metal silicate separation. Using this method, Kleine et al. (2004) and Touboul et al. (2015) inferred that core formation on Vesta took place within 1 Ma after CAI condensation and that core separation preceded crust formation by ~1 Ma. This sequence of events is supported by the La:W systematics of basaltic eucrites that are consistent with the idea that the eucrite source region was metal-free (Righter & Drake, 1997), although Th:W data indicate that some metal may have been present in the eucrite source (Touboul et al., 2015). In any case, it is of note that large-scale differentiation less than a couple of million years after CAI condensation is remarkably early in solar history, implying that accretion of Vesta had already taken place, even at the time of chondrule-forming events (Villeneuve et al., 2009).

Given such early accretion, there should be evidence of short-lived radioactive elements, and this is effectively the case. Excess ^{26}Mg formed by the decay of ^{26}Al (0.7 Myr half-life) is observed in eucrite plagioclase (e.g., Srinivasan et al., 1999) and there is evidence for the former presence of ^{60}Fe (2.6 Myr half-life) (Shukolyukov & Lugmair, 1993) and ^{53}Mn (3.7 Myr half-life) in basaltic eucrites (Trinquier et al., 2008). The Al:Mg system in particular has proven to be useful for constraining differentiation and crystallization timescales on Vesta (e.g. Schiller et al., 2011; Hublet et al., 2017). These data indicate generation of the HED source region within 1–3 million years of CAI condensation and crystallization of the basaltic eucrites almost immediately, within the time window 2–4 Ma after CAI. On the other hand, high temperature crystallization ages of cumulate eucrites and diogenites are younger by several million years, indicating that the basaltic eucrites are not simply the residues of the observed cumulate lithologies.

A range of other isotopic chronometers have been used to study crystallization ages of the HED lithologies over a wide range of closure temperatures, providing complementary insights into magmatic history. For the case of the basaltic eucrites, Trinquier et al. (2008) used the ^{53}Mn:^{53}Cr system to calculate a global silicate differentiation age of the HED parent body, determining a value of 2.1 ± 1.3 Ma after CAI formation, while Lugmair and Shukolyukov (1998) calculated the age of crystallization of the eucrite Juvinas to be 4 Ma after CAI formation based on Mn:Cr systematics. Whole-rock ^{87}Rb:^{87}Sr isochrons yield an age of 11–23 Ma after CAIs formation for the basaltic eucrites (Smoliar, 1993), these younger ages being consistent with the lower closure temperature (650–800°C) of the Rb:Sr system compared to Mn:Cr or Al:Mg systems. More recently, additional insight into the thermal history of basaltic eucrites is offered by the advent of high precision dating using Pb:Pb and U:Pb systems (Zhou et al., 2013; Iizuka et al., 2015). Crystallization ages of zircons in a series of basaltic eucrites determined in this way (Zhou et al., 2013) point to zircon crystallization ages that cluster around 15 Ma after CAI formation, reaching ages as young as 37 Ma after CAI formation. Large zircon grains in the Agout granulitic eucrite have Pb:Pb ages of 12 Ma after CAI formation and Ti contents that indicate crystallization temperatures of ≥900°C (Iizuka et al., 2015), similar to the equilibration temperatures of pyroxenes from most basaltic eucrites (Yamaguchi et al., 1996). In light of these observations, Iizuka et al. (2015) support the idea that this age reflects the timing of metamorphic zircon growth and pyroxene exsolution during widespread crustal metamorphism on Vesta, explaining the offset between the onset of silicate crystallization and the zircon data. The younger ^{207}Pb:^{206}Pb age of certain zircons (exemplified by Camel Donga: Zhou et al., 2013) are more difficult to explain, but may reflect the age of a subsequent short-term thermal event (i.e., impact) causing local partial melting.

Overall, evidence for global metamorphism is an important feature of certain eucrites that have pyroxene compositions which are highly equilibrated. These samples have been proposed to be produced by a meteoroid impact at or near the floor/wall of a crater (Nyquist et al., 1986) but, in light of more recent data, a more plausible hypothesis for the majority of equilibrated samples is that early formed basaltic eucrites formed at or near the surface were buried under successive lava flows and metamorphosed by heating from the hot interior (Yamaguchi et al., 1996, 2009). This idea finds support from the study of non-brecciated but highly metamorphosed basaltic eucrites using the ^{40}Ar:^{39}Ar system (Iizuka et al., 2019; Jourdan et al., 2020). The Ar:Ar system is of interest as it has a low closure temperature in plagioclase and pyroxene of ~300°C. Thus, combined with data of other chronometers of higher closure temperature, it can be used to constrain not only age, but also cooling rate. Such data imply that equilibrated eucrites took at least 50 million years to cool to 300°C, with low temperature cooling rates on the order of 15–20°C/Ma, consistent with burial beneath lava flows at depths on the order of 15 km (Jourdan et al., 2020).

For crystalline samples, cumulate eucrites are clearly younger than basaltic eucrites, as supported by Hf:W data (Touboul et al., 2015) and self-consistent ^{146}Sm:^{142}Nd isochrons for the cumulate eucrites Binda, Moama, and Moore County, corresponding to ages of 19 ± 15 Ma, 50 ± 18 Ma, and 34 ± 30 Ma after CAI formation, respectively (Boyet et al., 2010). Diogenites are more difficult to date because of their relatively low incompatible element content with respect to eucrites. The ^{87}Rb:^{87}Sr system gives an age of 117 Ma after CAI for the Johnston diogenite (with Rb:Sr closure temperatures in muscovite of 300–500°C, and in biotite of 300°C) (Takahashi & Masudat, 1990). These young ages of cumulate eucrites and diogenites are supported by Al:Mg data (Hublet et al., 2017) and consistent with their formation in deep plutons, where cooling below the closure temperature occurred later than when closer to the surface. It is also of note that the trace element geochemistry of diogenites is extremely variable (Mittlefehldt, 1994; Barrat et al., 2008) and clearly inconsistent with a simple history (i.e., as early formed cumulates of a magma ocean). Indeed, for both the diogenites and the Stannern-trend eucrites (i.e., those for which Mg# and trace element concentration are not correlated), Barrat et al. (2007, 2010) argue that source regions were contaminated by partial melts produced by remobilization of the existing basaltic eucrite crust, consistent with the idea of late emplacement at depth.

Overall, there is thus evidence for very early accretion of Vesta (<1 Ma after CAI formation), leading to core segregation followed by the onset of magmatism within a few Ma. As illustrated in

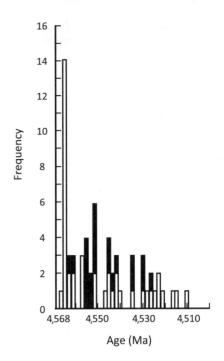

Figure 4.1 Summary of chronological data for basaltic eucrites obtained from isotopic systems of closure temperature in the range 1,200–700°C. Black bars correspond to data obtained from U-Pb dating of zircon, that has a closure temperature of ~900°C, corresponding to the transition from magmatism to metamorphism. White bars correspond to other chronometers. The data that cluster within the first 3–4 Ma of the Solar System are from isotopic systems of high closure temperature that freeze at temperatures >1,000°C (Al:Mg, Hf:W, Mn:Cr, ...), while the data at younger ages are typically ^{207}Pb:^{206}Pb data from pyroxene (closure temperature of 700°C). Figure redrawn from Zhou et al. (2013). The reader is referred to this publication for full details.

Figure 4.1, once the peak of igneous magmatism has passed, Pb:Pb ages of zircon (corresponding to a temperature of 900°C) cluster at 15 ± 7 Ma, representing the start of a long period of metamorphism that took more than 100 Ma to cool to 300°C (as seen by Ar:Ar ages) at the depths of cumulate samples (10–20 km depth). Metamorphism may have been the consequence of burial beneath lava flows but, in any case, the crystallization and cooling ages of cumulate lithologies (including diogenites) are younger than basaltic eucrites. Thus, the latter cannot be residual liquids of the former. Indeed, age constraints and geochemical evidence point to formation of cumulate rocks by intrusion at depth into preexisting basaltic crust. These data provide interesting points of comparison for numerical models of the physical and chemical differentiation of Vesta that will be presented in Section 4.4.

4.4 MODELING THE INTERNAL DIFFERENTIATION OF VESTA

The thermal, physical, and chemical evolution of the HED parent body has been the subject of several studies over the last 20 years. One approach is to focus on the chemical and mineralogical constraints of the HED meteorites to construct models that identify key stages of the physical and chemical evolution of the parent body (e.g., Righter & Drake, 1997; Shearer et al., 1997; Mandler & Elkins-Tanton, 2013). While such models provide useful conceptual frameworks, they do not necessarily capture the full complexity of natural samples (e.g., Barrat & Yamaguchi, 2014) and they lack detailed physical justification. An alternative family of models focuses on the thermal and physical evolution, identifying regions of magma generation and the timing of differentiation (e.g., Ghosh & McSween, 1998; Gupta & Sahijpal, 2010; Šrámek et al., 2012; Formisano et al., 2013). More recent work attempts to combine these approaches, explicitly treating the complex interplay between thermal history, the chemistry of generated liquids, and the physics of melt segregation in a single self-consistent model (Neumann et al., 2014; Mizzon, 2015). In this section, these models will be presented, beginning with an overview of the principal factors affecting planetesimal differentiation and the ranges of relevant parameters.

4.4.1 Accretion, Compaction, and Heat Sources

Planetesimals first form from protoplanetary dust as a porous aggregate that typically consists of a mixture of silicate and metallic phases (e.g., Montmerle et al., 2006). Since most meteorites are well consolidated (Britt & Consolmagno, 2003), this implies that planetesimals underwent some form of sintering from unconsolidated and highly porous to consolidated state. It is generally assumed that there are two successive stages of compaction: cold compaction driven by self-gravitation (e.g., Güttler et al., 2009) and hot pressing (also called sintering), which is associated with the deformation of crystals driven by the effects of temperature and pressure (Rao & Chaklader, 1972; Yomogida & Matsui, 1984). For planetary bodies larger than 10 km, the majority of the interior can be compacted by isostatic cold pressing to a density that roughly corresponds to that of the densest random packing of equally sized spheres, with a corresponding porosity of 0.36–0.4 (Henke et al., 2012).

If the source material heats up, porosity may subsequently be reduced by hot pressing. Such sintering is typically initiated in the deep interior where pressure and temperature are highest. The threshold temperature at which sintering becomes active is about 700–750 K (Yomogida & Matsui, 1984) – a temperature well below the melting temperatures of silicates or sulfides. In this sintered region, porosity rapidly approaches zero. This denser region migrates toward the surface as the interior heats, and the planetesimal shrinks in size as the porosity (and thus the volume) decreases. Assuming an initial porosity in the range 40–60%, calculated evolution of the radial distribution of porosity shows that the radius of a planetesimal shrinks by about 15–25% of its initial value (Figure 4.2; Hevey & Sanders, 2006; Neumann et al., 2012). After compaction and further heating, the planetesimal begins to melt when the silicate solidus and/or the metal–sulfide eutectic is reached.

As discussed in Section 4.1, the most important heat source for melting and hot pressing on protoplanetary bodies of the early Solar System is the radiogenic decay of short-lived isotopes, i.e., ^{26}Al (e.g., Urey, 1955) and ^{60}Fe (e.g., Tachibana & Huss, 2003; Mostefaoui et al., 2005). The presence of ^{26}Al in the early Solar System has been identified in Calcium Aluminum Inclusions (CAI) and chondrules (e.g., Lee et al., 1976; MacPherson et al., 1995; Bizzarro et al., 2004). ^{26}Al has a specific power several orders of magnitude higher than the long-lived radioactive elements uranium, thorium, and potassium, a half-life of ~0.7 Myr, and an initial $^{26}Al:^{27}Al$ ratio of $\geq 5 \times 10^{-5}$ (MacPherson et al., 1995). ^{60}Fe has a comparable decay energy to that of ^{26}Al, but its half-life is four-times longer than that of ^{26}Al (Castillo-Rogez et al., 2009). There is also greater uncertainty concerning the initial $^{60}Fe:^{56}Fe$, with values ranging from 10^{-8} to 10^{-6} (Cameron, 1993; Ogliore et al., 2011; Telus et al., 2012) with the possibility that ^{60}Fe was heterogeneously distributed in the Solar System (Quitté et al., 2010). In

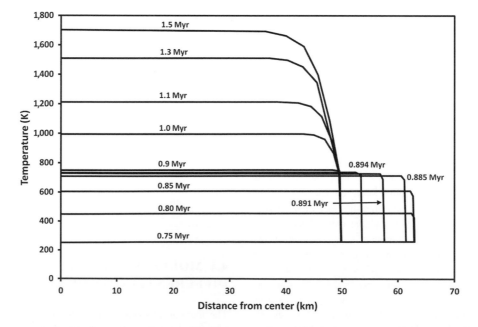

Figure 4.2 Temperature as a function of radius for a planetesimal of initial size of 65 km, formation time is 0.75 Ma after CAI and homogeneous distribution of ^{26}Al. Temperature profiles at different times are shown up to 1.5 Ma after CAI. Between t = 0.88 and t = 0.9 Ma the volume is reduced by compaction to a final radius of 50 km.
Figure from Hevey and Sanders (2006)

any case, the release of energy from these short-lived radionuclides has the potential to heat planetary objects above partial melting temperatures on timescales that are geologically short. In contrast, induction by the magnetically active protosun (Sonnett et al., 1968) is far less effective than ^{26}Al and ^{60}Fe, as shown by induction heating experiments (Marsh et al., 2006). Similarly, impact energy released during the accretion of planetesimals can only cause local heating close to the surface for bodies smaller than $\geq 1,000$ km (Moskovitz, 2009). This is particularly true in the main-belt environment where impacts were typically of low velocity (e.g., Marchi et al., 2013; see also Chapters 15 and 16).

In detail, the maximum temperature reached through radiogenic heating is a strong function of the date of accretion relative to condensation of the CAIs. This is because if accretion takes place earlier, then more radioactive heat-producing nuclei are available. It has been shown that planetesimals with a radius of only ~10 km can partially melt and even differentiate a core if they accrete simultaneously with the CAIs (Hevey & Sanders, 2006; Moskovitz & Gaidos, 2011; Henke et al., 2012; Neumann et al., 2012) (Figure 4.3). For accretion at later times bodies must be larger for melting to occur. However, for chondritic bulk compositions, if the formation time is later than about 2.8–3 Ma after CAI condensation, the energy available from ^{26}Al is no longer sufficient to drive silicate melting of bodies of any size (e.g., Ghosh & McSween, 1998; Sahijpal et al., 2007; Moskovitz & Gaidos, 2011; Neumann et al., 2012; Formisano et al., 2013) constraining the time-window for formation of differentiated bodies. Coming back to the case of accretion that occurs simultaneously with CAI condensation, simple models that only consider heat transport by thermal conduction predict maximum temperatures far above those required for complete melting. However, in reality, such extreme temperatures will not be reached, principally due to the formation and transport of melt. For example, when melting begins, latent heat is consumed, dampening temperature rise. Furthermore, melt migration transports thermal energy toward the surface where it is directly or indirectly lost by interaction with cold surface material (see Section 4.2.2).

4.4.2 Melting and Melt Migration

Given the importance of melt transport, particular attention must be paid to the question of melting and melt migration on differentiating planetesimals. In general terms, partial melting of the metal-rich and silicate-rich subsystems begin at similar temperatures, but at low pressure molten metal and molten silicate do not mix. As such, when sufficient melt is produced, silicate and metallic phases have the potential to separate, driven by density contrasts, denser metals moving downwards to form a core, and lighter silicates moving upwards to form a crust. The interplay between liquids and the solid matrix may lead to variable efficiency of melt segregation. This is particularly true for the metal-rich liquid that is not necessarily connected in 3D (Neri et al., 2019), explaining incomplete metal–silicate separation as observed in primitive achondrites such as acapulcoites and lodranites (McCoy et al., 1997).

In a little more detail, depending on the degree of melting, the differentiation process is either dominated by melt migration through a porous matrix (at lower melt factions) or by metal and silicate separation in a "mushy" or totally molten magma ocean (at high melt fractions). In the latter case, metal may sink as liquid droplets or solid particles in the silicate melt according to Stokes flow. However, this differentiation regime requires a critical and significant melt fraction of silicate such that the overall viscosity of the material is dominated by the liquid phase. This threshold depends on the material and the shapes of crystals, but is typically in the range 30–50% for basaltic systems (e.g., Arzi, 1978; Taylor, 1992; Lejeune & Richet, 1995; Scott & Kohlstedt, 2006).

In the contrasting case of porous flow, an interconnected melt network is nevertheless required because liquid cannot separate from residual crystals when trapped in isolated pockets. Melt connectivity is fundamentally controlled by the relative energies of

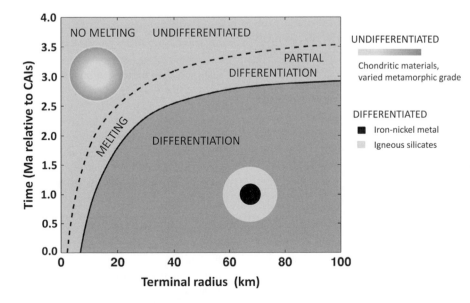

Figure 4.3 Formation time (assuming instantaneous accretion) as a function of the terminal radius of a planetesimal indicating the parameter space for which a planetesimal forms a core (dark grey), melting occurs but no core differentiation (light grey area in between solid and dashed line) and no melting occurs (light grey area above dashed line).
Figure modified from Neumann et al. (2012)

solid–solid and solid–liquid interfaces. In general terms, if solid–liquid interfaces are energetically favorable the melt will have a tendency to wet grain boundaries and be connected while, if energetically unfavorable, melt will remain in isolated pockets (Laporte & Provost, 2000). Macroscopically, the transition from an unconnected to a connected system can be considered in terms of the *dihedral angle*, defined as the angle of the melt wedge measured at the triple junctions of two solid grains and melt. Theoretically, if the dihedral angle is smaller than 60°, a stable interconnected network of melt is formed at all melt fractions (von Bargen & Waff, 1986). This is the case for silicate melts (e.g., Waff & Bulau, 1979) implying that even if the fraction of silicate melt is only a few percent, a connected melt network exists (Taylor et al., 1993). The situation is less clear for molten metal in a silicate(-metal) matrix. Early experimental studies by Takahashi (1983) and Walker and Agee (1988) show unconnected melt pockets of high dihedral angle, but more recent experiments suggest that connected melt networks may be formed under oxidizing conditions and pressures of 2–3 GPa (Terasaki et al., 2008). However, these pressures are much higher than those encountered in planetesimals the size of Vesta. On the other hand, sulfur content of metal-rich melts would also appear to play a role on connectivity, favoring the migration of early formed cotectic Fe,Ni:FeS liquids (Neri et al., 2019). This may explain the loss of FeS in primitive achondrites such as lodranites that initially had ~5% FeS and that experienced ~20% of silicate melting (McCoy et al., 1997).

For porous flow, if there is an interconnected melt fraction as well as a density contrast between the melt and the surrounding solid matrix, melt extraction from the silicate/metal matrix is directly linked to the compaction of the matrix. The ability of the matrix to expel liquid and the rate at which that occurs is determined by the so-called compaction length L_c,

$$L_c = \left(\frac{\eta_m b^2 \phi^n}{\eta_l \tau} \right)^{\frac{1}{2}}$$

where η_m is the matrix viscosity, η_l is the fluid viscosity, b is the grain size, Φ is the degree of partial melting, and n and τ are constants (e.g., Turcotte & Phipps Morgan, 1992). Here, the last three terms can be related to permeability K, which is described by the equation $b^2 \Phi^{(n+1)}/\tau$. For silicate systems of relevance here, $n = 2$ and τ has a value on the order of 270 (e.g., Wark et al., 2003).

If the thickness of the region experiencing melting is much greater than L_c, the percolation velocity can be described by Darcy flow, in which case the relative velocity of melt and matrix is:

$$v_l - v_s = \frac{K \Delta \rho g}{\phi \eta_l} = \frac{b^2 \phi^{n-1} \Delta \rho g}{\tau \eta_l}$$

where $(v_l - v_s)$ is the relative velocity of liquid and solid phases, ϕ is the volume fraction of the melt (= porosity), and $\Delta\rho$ is the density contrast between the solid and the melt. On the contrary, if the thickness of the region over which melting takes place is smaller than or similar to L_c, it is the rate at which compaction can occur that determines the velocity of liquid movement.

Assuming typical values of $\eta_m = 10^{18}$ Pas, $\eta_l = 1$ Pas, $b = 10^{-4}$ m, and $\phi = 0.15$, the compaction length is less than 1 km. For a Vesta-sized body heated by short-lived radioactive isotopes, the melting region will thus have a spatial extent much larger than the compaction length. As such, Darcy flow can be assumed, in which case percolation velocity is independent of the

radius of the body (Neumann et al., 2012; Mizzon, 2015). Predicted characteristic time-scales for liquid movement can vary from 0.1 to 50 Ma, depending strongly on the degree of melting, the melt viscosity, and the permeability (e.g., Mizzon, 2015). Given that permeability is in turn largely controlled by grain-size, the latter parameter may be of particular importance. Indeed, recent work suggests that a threshold grain size on the order of 1 mm is required for significant melt migration to occur on protoplanetary bodies heated by short-lived radionuclides (Lichtenberg et al., 2019).

Overall, given the low dihedral angle of silicate liquids in olivine dominated systems, there is no major obstacle to migration of silicate melt. The case of metal-rich liquids is a little more complex, as a critical liquid fraction may be necessary to ensure connectivity and the role of silicate melt on metal mobility must be assessed (e.g., Bagdassarov et al., 2009). However, the existence of metallic meteorites, experimental and theoretical considerations, and the central mass concentration observed at Vesta (e.g., Russell et al., 2012) all support the possibility for efficient metal segregation in protoplanetary systems.

4.4.3 Redistribution of Heat Sources and Magma Ocean Formation

While the effects of partitioning of heat-producing elements between silicate and metal are important first order factors to take into account (e.g., Formisano et al., 2013), element partitioning between solids and liquids is also of fundamental importance (e.g., Moskovitz & Gaidos, 2011).

In general terms, chemical elements partition between melt and residual minerals during partial melting – incompatible elements are preferentially excluded from the crystals and are concentrated in the melt, while compatible elements remain in the solid residue (Best, 2002). In detail, redistribution between melt and matrix can be quantified by the partition coefficient Di of the element i (Di = concentration in the solid phase / concentration in the liquid phase). This partitioning between liquid and solid silicates was long neglected in modeling efforts of differentiation in the early Solar System but is of utmost importance given that aluminum is highly incompatible during early melting. From a modeling perspective, if relevant values for the distribution coefficient D for ^{26}Al are used (D^{26}Al in the range 0.003–0.02: Kennedy et al., 1993; Pack & Palme, 2003), a huge effect of melt migration on thermal evolution is observed (Figure 4.4, Moskovitz & Gaidos, 2011; Neumann et al., 2014). An alternative approach to using a fixed value of D^{26}Al is to assume an eutectic melting reaction relevant to basalt genesis (Mizzon, 2015). Given that plagioclase is the principal host of Al and that this mineral is the first phase to be lost from the residue during melting, such models imply that all Al is transferred to the liquid at the eutectic temperature, with the same consequence that there is a major redistribution of heat sources driven by melt migration as ^{26}Al-enriched melt rises toward the surface, leaving behind an ^{26}Al-depleted matrix.

Taking into account melt generation and migration, two principal scenarios are possible:
1. If melt migration redistributes heat sources toward the surface on timescales that are short relative to heating, then the residual mantle becomes depleted in those heat sources. The remaining decay energy in the mantle thus becomes insufficient to pursue melting. In this region of the parent body, the local degree of

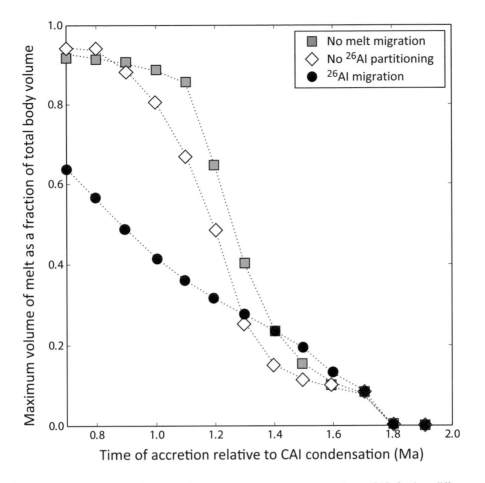

Figure 4.4 Maximum relative volume of the generated melt as a function of the formation time according to CAIs for three different scenarios: without melt transport and homogeneous distribution of the heat sources (grey squares); with melt transport but no preferential partitioning of ^{26}Al between liquid and solid (open diamonds); and with melt transport and preferential enrichment of the melt in ^{26}Al, thus redistribution of the heat source (black circles). Figure adapted from Mizzon (2015)

melting remains low and no global magma ocean is formed. However, the rising melt enriched in ^{26}Al may form a shallow magma ocean near the surface. Melt of this shallow magma ocean can also reach the surface by porous flow or via dykes and form a crust. In addition, the solid surface may lose its cohesion due to thermal expansion, mechanical effects of impacts, or the effect of pressure exerted by the underlying magma ocean, which can lead to surface recycling and efficient cooling (Wilson & Keil, 2012).

2. In the second scenario, slow melt migration leads to a residual mantle that retains sufficient heat sources for melting to continue. In this case, a high degree of local melting throughout the interior of the body can be achieved, resulting in the formation of a deep global magma ocean (Figure 4.5). Although a global magma ocean forms, it should be appreciated that the temperature rise (and thus local degree of melting) is limited by convection driven by low viscosity. This convection leads to an increase of the spatial extent of the magma ocean through partial melting and erosion of the overlying material. This results in effective cooling, such that the melt fraction never rises above a critical value (Figure 4.6), in contrast to early considerations of this problem (e.g., Ghosh & McSween, 1988). When the thermal energy provided by radioactive decay is no longer sufficient to continue melting (i.e., after five or so half-lives) the mantle and the planetesimal begins to cool and crystallize.

In both scenarios, fractions of melt remain distributed throughout the body and can migrate for extended periods of time. The internal thermal structure of the body is complex, with heat sources nevertheless concentrated in a shallow subsurface region, leading to the possibility for protracted periods of magmatism and slow cooling of the lower crust. After cooling below the solidus, no further structural changes due to partial melting and differentiation are expected, except through external influence such as impacts. Indeed, impacts may be of importance throughout the differentiation process (e.g., see Chapter 16) as they have the potential to mix geochemical and petrological reservoirs in ways not accounted for by models of unperturbed evolution.

4.5 MODELS OF VESTA AND COMPARISON WITH PETROLOGICAL AND GEOCHRONOLOGICAL CONSTRAINTS FROM THE HED METEORITES

4.5.1 Principal Features and Results of Recent Numerical Models

The different considerations presented in Section 4.4 have been implemented for Vesta-sized bodies in models of Neumann et al.

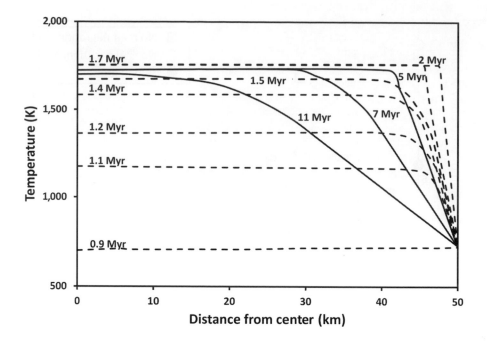

Figure 4.5 Temperature as a function of radius for a planetesimal of initial size of 65 km, formation time is 0.75 Ma after CAI and homogeneous distribution of ^{26}Al. Temperature profiles after compaction from t = 0.9 Ma. Convection starts at a melting degree of 50% at time t = 1.52 Ma. Effective cooling by convection prevents further temperature increase and the partially melted magma ocean expands outwards to about 2 km after the surface. After that the planetesimal slowly cools conductive from the outside.
Figure from Hevey and Sanders (2006)

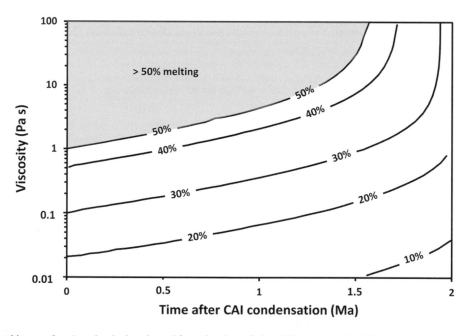

Figure 4.6 Degree of melting as a function of melt viscosity and formation time of a R = 50 km planetesimal. The contours represent 10% intervals of partial melting. Migration of low viscosity depletes ^{26}Al from the deeper mantle and reduces the peak temperature (see also Figure 4.4). For a high melt viscosity and thus slow melt migration the formation of a global magma ocean is possible.
Figure from Moskovitz and Gaidos (2011)

(2014) and Mizzon (2015). These models share the fact that they assume heating primarily by ^{26}Al, that they calculate liquid movement through two-phase flow driven by matrix compaction, and that they explicitly take into account the transport of heat-producing Al with the liquid. Both models assume instantaneous accretion at different times during the first 2 Ma following CAI

condensation and both assume that Vesta is initially an homogeneous mixture of chondritic silicates and metal (of L-chondrite composition for Neumann et al. (2014) and of H-chondrite composition for Mizzon (2015)). In detail, the model of Neumann et al. (2014) takes into account the early stages of the elimination of porosity and it follows the migration of both liquid silicates and liquid metals, providing insights into core formation. On the other hand, the model of Mizzon (2015) concentrates on the evolution of the silicate fraction. It uses the forsterite–enstatite–anorthite phase diagram (Morse, 1980) to model melt production/consumption as a function of temperature and to follow the compositions of melts and residues. The two models thus share a common basis for the thermal and physical evolution, the former providing constraints on core formation and the latter providing insight into the redistribution of different silicate minerals during physical evolution.

Upon heating, phase diagrams indicate that the metal–sulfide subsystem may begin to melt before silicates. However, the model of Neumann et al. (2014) implies that the fraction of metal-rich liquid is insufficient for migration to occur at that time. Migration of FeS melts occurs slightly later, when silicate melting is initiated (see also Neumann et al., 2012). Concerning the silicate fraction, both available models predict that for accretion within the first 1 Ma following CAI condensation, melts rich in ^{26}Al are produced and transported toward the surface on time scales shorter than heat production. As such, heating of the residual mantle is arrested and there is consensus that whole-mantle silicate magma oceans will not form. This important result has the consequence that a fraction of the most refractory components of the deep mantle (i.e., olivine) may never melt, even for the extreme case of accretion that occurs at the same time as CAI condensation.

The early formed silicate liquids are expelled upwards and predicted to accumulate beneath the surface in a melt-rich layer. Note that this layer is not necessarily entirely molten, but probably buffered at a crystal content just below that at which the viscosity of the material is dominated by the liquid and efficient convection is possible. The thickness of this layer varies as a function of model parameters (e.g., melt viscosity) but is predicted to be typically somewhere between 1 and 30 km (resembling a sill in the former case, or a spatially limited magma ocean in the latter). Because this liquid-rich layer concentrates a significant fraction of the available ^{26}Al and it can convect, it rapidly erodes the overlying initially undifferentiated material such that the top of the melt-rich layer is within a few kilometers of the surface and maybe even within a few hundred meters (Figure 4.7). Foundering of the crust and emplacement of basaltic magma at the surface is thus inevitable. At first, pristine undifferentiated material will be digested in the liquid layer but, later, the early-formed basaltic crust may also be recycled. In this near surface region, there is thus competition between internal heating by ^{26}Al in the liquid-rich layer and cooling by convection/recycling. The compositional evolution of the liquid layer is extremely difficult to model during this period, being influenced by thermal erosion of the overlying crust, possible periodic wholesale recycling of that crust (including the effects of impacts of external material), continued input from upwelling melt extracted from the mantle, and precipitation of refractory phases. There is nevertheless a general trend of cooling that results from the decrease in intensity of the ^{26}Al heat-source over time. Models that include mineralogy show that, during this cooling, the most refractory phases (olivine, then pyroxene) precipitate and settle to the base of the liquid layer, while the liquid evolves toward the eutectic composition.

In detail, the onset of differentiation and the extent and duration of magmatic activity is, as expected, primarily a function of accretion date. For example, in the simulations of Neumann et al. (2014), accretion at CAI condensation leads to the formation of a melt-rich

layer in 0.2 Ma that persists for 0.5 Ma while for accretion at CAI+0.5 Ma, the melt-rich layer requires 0.35 Ma to form and it only lasts 0.05 Ma (Figure 4.7). For accretion 1 Ma after CAI condensation, no convecting liquid layer is formed at all. For the simulations of Mizzon (2015), heating times are the same, but cooling by interaction with surface material is less marked and magmatism lasts longer. However, even in this case, the time window is small. For example, for accretion 1 Ma after CAI condensation, final crystallization of the liquid layer is predicted to begin $\sim$3.5 Ma later.

In terms of differentiation into a core–mantle–crust structure, it is of interest to note that the melt-rich layer is predicted to be the first region to be virtually clear of metal. This is followed in time by the underlying residual mantle that will lose its metal to form a core, typically in less than 0.3 Ma (Neumann et al., 2014). In terms of the crust, a near-eutectic liquid is predicted to crystallize from top to bottom, producing an upper crust of roughly equal proportions of pyroxene and plagioclase, similar to normative eucrite compositions (Mizzon, 2015). Below this layer, pyroxene-dominated rocks occur (with or without minor plagioclase and/or olivine; <10%) providing a close match to diogenite-like lithologies. The thickness of the "eucrite" layer is on the order of $\sim$13 km and the pyroxene dominated layer extends down to depths between 40 and 50 km (e.g., Figure 4.8). At greater depths (but above the core), olivine is by far the dominant phase. This olivine can be residual material that never melted and/or early formed cumulates associated with crystallization of the liquid-rich layer. The same final crustal structure is predicted for all dates of accretion less than 1 Ma after CAI condensation. For later accretion, olivine is increasingly present in the near-surface region, reaching the value of the bulk composition when no melting occurs. Timescales of cooling below different isotherms can also be predicted, indicating that up to 100 Ma are required to cool even the upper crust below 300°C (Figure 4.8).

While these calculations provide important first-order insights into differentiation time scales and internal mineralogical structure, it should be appreciated that the models presented here are 1D and cannot capture lateral variations in crustal thickness that may result from the development of compaction waves in 3D (e.g., Wiggins & Spiegelman, 1995). Indeed, in the regions of Vesta unaffected directly by the southern impact basins, Vesta's gravity field implies local variations in crustal thickness and/or density that argue against a simple concentric onion-shell structure (Ermakov et al., 2014; Park et al., 2014; Chapter 12). These geophysical observations thus leave room for lateral variations associated with the formation of diogenite plutons (e.g., Buczkowski et al., 2014; Raymond et al., 2016) and/or local variations in the depths at which lithological boundaries occur.

4.5.2 Comparison with Data from HED Meteorites and Extension to Other Small Bodies

The different modeling results discussed in Section 4.5.1 provide interesting insight into the data from HED meteorites and bring quantitative elements of comparison that help bridge the gap between more conceptual models and petrologic observations (see Chapter 3).

First of all, the modeling clearly requires accretion within the first million years of Solar System history, in agreement with dating of the metal–silicate differentiation of the eucrite source region in

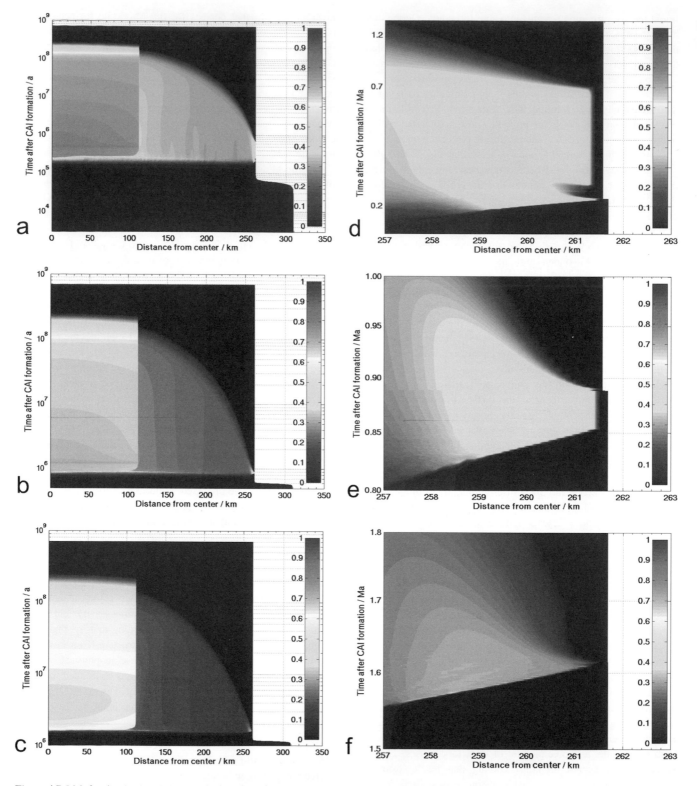

Figure 4.7 Melt fraction (a–c) and close-ups in the sub-surface magma ocean (d–f) of a Vesta-like body which formed at $t_0 = 0$ Ma (a, d), $t_0 = 0.5$ Ma (b, e), and $t_0 = 1$ Ma (c, f) after the CAIs. Figure adapted from Neumann et al. (2014)
A black and white version of this figure will appear in some formats. For the color version, refer to the plate section.

this time window (Kleine et al., 2004) and the possible presence of an early dynamo-driven magnetic field (Fu et al., 2012; Formisano et al., 2016). This is also consistent with geophysical observations at Vesta, that imply efficient core segregation (Russell et al., 2012)

and a viscosity low enough (i.e., a melt fraction high enough) for Vesta to adopt a hydrostatic shape at that time (Fu et al., 2014). Second, models predict that basaltic eucrites should be formed within a couple of million years and be the oldest samples from

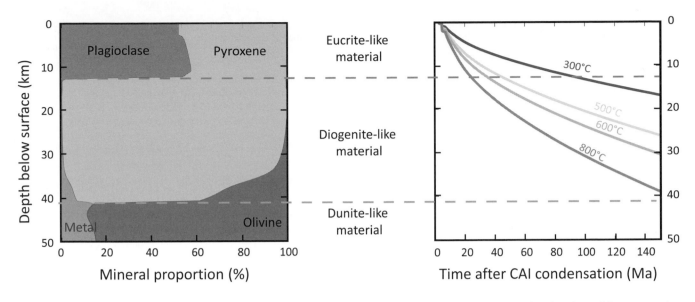

Figure 4.8 Modeled mineralogy and cooling trajectories of the upper portion of Vesta. The left hand panel represents the mineralogy of the upper section of Vesta predicted by the model of Mizzon (2015) for accretion 1 Ma after CAI condensation and full crystallization. Blue = plagioclase; Brown = pyroxene; Green = olivine; Grey = metal. The upper ~13 km has a mineralogy that resembles eucrite, the section between 13 and 40 km has a mineralogy that resembles diogenite, while below that depth olivine is the dominant phase. Note that the model of Mizzon does not include the migration of metal, so its presence in this simulation is an artefact, as indicated by Neumann et al. (2014). The right hand panel represents cooling trajectories as a function of time for the same model. Figure redrawn from Mizzon (2015)
A black and white version of this figure will appear in some formats. For the color version, refer to the plate section.

the vestan crust, in excellent agreement with observations (e.g. Hublet et al., 2017). Third, the numerical models indicate that cumulate rocks from the lower crust (cumulate eucrites and diogenites) should not only be younger, but that they form in a complex environment involving crustal recycling and even late stage magmatic input from fractional melts of the underlying mantle. All these processes may thus act to explain the extreme trace element diversity of these samples that is otherwise difficult to explain (e.g., Barrat et al., 2008). Taken together, the physical models are actually rather consistent with the idea of Mandler and Elkins-Tanton (2013) of an early stage of equilibrium crystallization, followed by a later stage of fractional crystallization (as discussed in Chapter 3). In addition, it is of note that simulations indicate that a vertically stratified mineralogical structure, with olivine sequestered at depth, is the natural consequence of internal heating by ^{26}Al on a Vesta-sized body. This provides a potential explanation for the paucity of olivine measured by Dawn at the surface of Vesta (e.g., Ammannito et al., 2013), even in the Rheasilvia basin (McSween et al., 2013), although hydrodynamical simulations of two successive overlapping head-on impacts at the south-pole of Vesta predict that material as deep as 80 km might be exposed (Jutzi et al., 2013; see also Chapter 16). On the other hand, the possibility of oblique impact excavating to shallower depth, a local transition to an olivine-rich layer that is deeper than the average value of ~50 km, and/or dilution of olivine in the regolith to below the detection limit of ~25 vol.% (Ammannito et al., 2013), may reconcile models and observations, without the need to invoke non-chondritic bulk composition (see Consolmagno et al., 2015; Chapter 3). Finally, the models predict a protracted cooling history at depth, in very good agreement with the extensive metamorphism experienced by eucrites (Yamaguchi et al., 1996) and the timescales of cooling through different closure

temperature and inferred depths of burial inferred for the eucrite crust (e.g., Jourdan et al., 2020). Overall, the numerical models thus appear to provide a self-consistent explanation of the essential features of HED petrology and chronology.

While the models presented here have been tailored for a body the size of Vesta, much of the physical and chemical inferences can be extended to bodies of other size. For example, the fact that melt percolation occurs on time-scales faster than heating (thus limiting the formation of whole-body magma oceans) has been suggested for smaller bodies (Moskovitz & Gaidos, 2011) and will probably be true for bodies of more than 50 km in size. Below this limit, lower gravity and an increase in the relative importance of conductive cooling may affect this conclusion. Finally, another implication of the models explored here is that "eucrite-like" lithologies may be generated on and characteristic of a wide range of large, early-formed asteroidal bodies in the early Solar System. Basaltic mineralogy and composition may thus not be characteristic of a particular parent body, explaining the existence of "anomalous" eucrites such as Pasamonté and the presence of eucrite-like lithologies in the mesosiderites (e.g., McSween et al., 2010) that may well have been formed on parent bodies distinct from Vesta.

REFERENCES

Ammannito, E., De Sanctis, M. C., Palomba, E., et al. (2013) Olivine in an unexpected location on Vesta's surface. *Nature*, 504, 122–125.

Arzi, A. A. (1978) Critical phenomena in the rheology of partially melted rocks. *Tectonophysics*, 44, 173–184.

Bagdassarov, N., Golabek, G. J., Solferino, G., & Schmidt, M. W. (2009) Constraints on the Fe-S melt connectivity in mantle silicates from electrical impedance measurements. *Physics of the Earth and Planetary Interiors*, 177, 139–146.

Barrat, J.-A., & Yamaguchi, A. (2014) Comment on "The origin of eucrites, diogenites, and olivine diogenites: Magma ocean crystallization and shallow magma processes on Vesta" by B. E. Mandler and L. T. Elkins-Tanton. *Meteoritics & Planetary Science*, 49, 468–472.

Barrat, J.-A., Yamaguchi, A., Greenwood, R. C., et al. (2007) The Stannern trend eucrites: Contamination of main group eucritic magmas by crustal partial melts. *Geochimica et Cosmochimica Acta*, 71, 4108–4123.

Barrat, J.-A., Yamaguchi, A., Greenwood, R. C., et al. (2008) Geochemistry of diogenites: Still more diversity in their parental melts. *Meteoritics & Planetary Science*, 43, 1759–1775.

Barrat, J.-A., Yamaguchi, A., Zanda, B., Bollinger, C. & Bohn, M. (2010) Relative chronology of crust formation on asteroid 4-Vesta: Insights from the geochemistry of diogenites. *Geochimica et Cosmochimica Acta*, 74, 6218–6231.

Best. M. G. (2002) *Igneous and Metamorphic Petrology*, 2nd ed. Hoboken, NJ: Wiley-Blackwell.

Bizzarro, M., Baker, J. A., & Haack, H. (2004) Mg isotope evidence for contemporaneous formation of chondrules and refractory inclusions. *Nature*, 431, 275–278.

Boyet, M., Carlson, R. W., & Horan, M. (2010) Old Sm–Nd ages for cumulate eucrites and redetermination of the Solar System initial $^{146}Sm/^{144}Sm$ ratio. *Earth and Planetary Science Letters*, 291, 172–181.

Brearley, A. J., & Jones, R. H. (1998) Chondritic meteorites. In J. J. Papike (ed.), *Planetary Materials. Reviews in Mineralogy*, Vol. 36. Washington, DC: Mineralogical Society of America, pp. 3–39.

Britt, D. T., & Consolmagno, S. J. (2003) Stony meteorite porosities and densities: A review of the data through 2001. *Meteoritics & Planetary Science*, 38, 1161–1180.

Buczkowski, D. L., Wyrick, D. Y., Toplis, M. J., et al. (2014) The unique geomorphology and physical properties of the Vestalia Terra plateau. *Icarus*, 244, 89–103.

Cameron, A. G. W. (1993) Nucleosynthesis and star formation. In E. H. Levy, & J. I. Lunine (eds.), *Protostars and Planets III*. Tucson: University of Arizona Press, p. 47.

Castillo-Rogez, J., Johnson, T. V., Lee, M. H., et al. (2009) ^{26}Al decay: Heat production and a revised age for Iapetus. *Icarus*, 204, 658–662.

Consolmagno, G. J., Golabek, G. J., Turrini, D., et al. (2015). Is Vesta an intact and pristine protoplanet? *Icarus*, 254, 190–201.

Dodson, M. H. (1973) Closure temperature in cooling geochronological and petrological systems. *Contributions to Mineralogy and Petrology*, 40, 259–274.

Drake, M. J. (2001) The eucrite/Vesta story. *Meteoritics & Planetary Science*, 36, 501–513.

Ermakov, A. I., Zuber, M. T., Smith, D. E., et al. (2014) Constraints on Vesta's interior structure using gravity and shape models from the Dawn mission. *Icarus*, 240, 146–160.

Formisano, M., Federico, C., DeAngelis, S., DeSantis, M. C., & Magni, G. (2016) A core dynamo in Vesta? *Monthly Notices of the Royal Astronomical Society*, 458, 695–707.

Formisano, M., Federico, C., Turrini, D., et al. (2013) The heating history of Vesta and the differentiation of Vesta. *Meteoritics & Planetary Science*, 48, 2316–2332.

Fu, R. R., Hager, B. H., Ermakov, A. I., & Zuber, M. T. (2014) Efficient early global relaxation of asteroid Vesta. *Icarus*, 240, 133–145.

Fu, R. R., Weiss, B. P., Schuster, D. L., et al. (2012) An ancient core dynamo in asteroid Vesta. *Science*, 338, 238–241.

Ganguly, J., Ito, M., & Zhang, X. (2007) Cr diffusion in orthopyroxene: Experimental determination, $^{53}Mn–^{53}Cr$ thermochronology, and planetary applications. *Geochimica et Cosmochimica Acta*, 71, 3915–3925.

Ghosh, A., & McSween Jr., H. Y. (1988) A thermal model for the differentiation of asteroid 4 Vesta, based on radiogenic heating. *Icarus*, 134, 187–206.

Gupta, G., & Sahijpal, S. (2010) Differentiation of Vesta and the parent bodies of other achondrites. *Journal of Geophysical Research*, 115, E08001.

Güttler, C., Krause, M., Geretshauser, R., Speith, R., & Blum, J. (2009) The physics of protoplanetesimal dust agglomerates. IV. Towards a dynamical collision model. *The Astrophysical Journal*, 701, 130–141.

Henke, S., Gail, H.-P., Trieloff, M., Schwarz, W. H., & Kleine, T. (2012) Thermal history modelling of the h chondrite parent body. *Astronomy and Astrophysics*, 545, A135.

Hevey, P. J., & Sanders, I. S. (2006) A model for planetesimal meltdown by ^{26}Al and its implications for meteorite parent bodies. *Meteoritics & Planetary Science*, 41, 95–106.

Hublet, G., Debaille, V., Wimpenny, J., & Yin, Q. (2017) Differentiation and magmatic activity in Vesta evidenced by ^{26}Al-^{26}Mg dating in eucrites and diogenites. *Geochimica et Cosmochimica Acta* 218, 73–97.

Iizuka, T., Jourdan, F., Yamaguchi, A., et al. (2019) The geologic history of Vesta inferred from combined $^{207}Pb/^{206}Pb$ and $^{40}Ar/^{39}Ar$ chronology of basaltic eucrites. *Geochimica et Cosmochimica Acta*, 267, 275–299.

Iizuka, T., Yamaguchi, A., Haba, M. K., et al. (2015) Timing of global crustal metamorphism on Vesta as revealed by high-precision U-Pb dating and trace element chemistry. *Earth and Planetary Science Letters*, 409, 182–192.

Jones, J. H. (1984) The composition of the mantle of the eucrite parent body and the origin of eucrites. *Geochimica et Cosmochimica Acta*, 48, 641–648.

Jourdan, F., Kennedy, T., Benedix, G. K., Eroglu, E., & Mayer, C. (2020) Timing of the magmatic activity and upper crustal cooling of differentiated asteroid 4 Vesta. *Geochimica et Cosmochimica Acta*, 273, 205–225.

Jutzi, M., Asphaug, E., Gillet, P., Barrat, J. A., & Benz, W. (2013) The structure of the asteroid 4 Vesta as revealed by models of planet-scale collisions. *Nature*, 494, 207–210.

Kennedy, A. K., Lofgren, G. E., & Wasserburg, G. J. (1993) An experimental study of trace element partitioning between olivine, orthopyroxene, and melt in chondrules: equilibrium values and kinetic effects. *Earth and Planetary Science Letters*, 115, 177–195.

Kleine, T., Mezger, K., Münker, C., Palme, H., & Bischoff, A. (2004) $^{182}Hf–^{182}W$ isotope systematics of chondrites, eucrites, and martian meteorites: Chronology of core formation and early mantle differentiation in Vesta and Mars. *Geochimica et Cosmochimica Acta*, 68, 2935–2946.

Laporte, D., & Provost, A. (2000) The grain-scale distribution of silicate, carbonate and metallosulfide partial melts: A review of theory and experiments. In N. Bagdassarov, D. Laporte, & A. B. Thompson (eds.), *Physics and Chemistry of Partially Molten Rocks*. Dordrecht: Springer, pp. 93–140.

Lee, T., Papanastassiou, D. A., & Wasserburg, G. J. (1976) Demonstration of ^{25}Mg excess in Allende and evidence for ^{26}Al. *Geophysical Research Letters*, 3, 41–44.

Lejeune, A.-M., & Richet, P. (1995) Rheology of crystal bearing silicate melts: An experiment study at high viscosities. *Journal of Geophysical Research*, 100, 4215–4229.

Lichtenberg, T., Keller, T., Katz, R. F., Golabek, G. J., & Gerya, T. V. (2019) Magma ascent in planetesimals: Control by grain size. *Earth and Planetary Science Letters*, 507, 154–165.

Lugmair, G. W., & Shukolyukov, A. (1998) Early Solar System timescales according to ^{53}Mn-^{53}Cr systematics. *Geochimica et Cosmochimica Acta*, 62, 2863–2886.

MacPherson, G. J., Davis, A. M., & Zinner, E. K. (1995) The distribution of aluminium-26 in the early Solar System – A reappraisal. *Meteoritics*, 30, 365–386.

Mandler, B. E., & Elkins-Tanton, L. T. (2013) The origin of eucrites, diogenites, and olivine diogenites: Magma ocean crystallization and shallow magma chamber processes on Vesta. *Meteoritics & Planetary Science*, 48, 2333–2349.

Marchi, S., Bottke, W. F., Cohen, B. A., et al. (2013) High-velocity collisions from the lunar cataclysm recorded in asteroidal meteorites. *Nature Geoscience*, 6, 303–307.

Marsh, C. A., Della-Giustina, D. N., Giacalone, J., & Lauretta, D. S. (2006) Experimental tests of the induction heating hypothesis for planetesimals. *37th Annual Lunar and Planetary Science Conference*, March 13–17, Houston, TX, 2078 (abstract).

McCoy, T. J., Keil, K., Muenow, D. W., & Wilson, L. (1997) Partial melting and melt migration in the acapulcoite-lodranite parent body. *Geochimica and Cosmochimica Acta*, 61, 639–650.

McSween, H. Y., Ammannito, E., Reddy, V., et al. (2013) Composition of the Rheasilvia basin, a window into Vesta's interior. *Journal of Geophysical Research: Planets*, 118, 335–346.

McSween, H. Y., Mittlefehldt, D. W., Beck, A. W., Mayne, R. G., & McCoy, T. J. (2010) HED meteorites and their relationship to the geology of Vesta and the Dawn mission. *Space Science Reviews*, 163, 141–174.

Mittlefehldt, D. W. (1994) The genesis of diogenites and HED parent body petrogenesis. *Geochimica et Cosmochimica Acta*, 58, 1537–1552.

Mizzon, H. (2015) The Magmatic Crust of Vesta. PhD thesis, Universite Toulouse III Paul Sabatier.

Montmerle, T., Augereau, J.-C., Chaussidon, M., et al. (2006) From suns to life: A chronological approach to the history of life on Earth 3. Solar System formation and early evolution: The first 100 million years. *Earth Moon and Planets*, 98, 39–95.

Morse, S. A. (1980) *Basalts and Phase Diagrams: An Introduction to the Quantitative Use of Phase Diagrams in Igneous Petrology*. New York: Springer-Verlag.

Moskovitz, N., & Gaidos, E. (2011) Differentiation of planetesimals and the thermal consequences of melt migration. *Meteoritics & Planetary Science*, 46, 903–918.

Moskovitz, N. A. (2009) Spectroscopic and Theoretical Constraints on the Differentiation of Planetesimals. PhD thesis, University of Hawaii.

Mostefaoui, S., Lugmair, G. W., & Hoppe, P. (2005) ^{60}Fe: A heat source for planetary differentiation from a nearby supernova explosion. *The Astrophysical Journal*, 625, 271–277.

Neri, A., Guignard, J., Monnereau, M., Toplis, M. J., & Quitté, G. (2019) Melt segregation in planetesimals: Constraints from experimentally constrained interfacial energies. *Earth and Planetary Science Letters*, 518, 40–52.

Neumann, W., Breuer, D., & Spohn, T. (2012) Differentiation and core formation in accreting planetesimals. *Astronomy & Astrophysics*, 543, 1–21.

Neumann, W., Breuer, D., & Spohn, T. (2014) Differentiation of Vesta: Implications for a shallow magma ocean. *Earth and Planetary Science Letters*, 395, 267–280.

Nyquist, L. E., Takeda, H., Bansal, B. M., et al. (1986) Rb-Sr and Sm-Nd internal isochron ages of a subophitic basalt clast and matrix sample from the Y75011 eucrite. *Journal of Geophysical Research*, 91, 8137–8150.

Ogliore, R. C., Huss, G. R., & Nagashima, K. (2011) Ratio estimation in SIMS analysis. *Nuclear Instruments and Methods, Physics Research B: Beam Interactions with Materials and Atoms*, 269, 19101918.

Pack, A., & Palme, H. (2003) Partitioning of Ca and Al between forsterite and silicate melt in dynamic systems with implications for the origin of Ca, Al-rich forsterites in primitive meteorites. *Meteoritics & Planetary Science*, 38, 1263–1281.

Park, R. S., Konopliv, A. S., Asmar, S. W., Bills, B. G., & Gaskell, R. W. (2014) Gravity field expansion in ellipsoidal harmonic and polyhedral internal representations applied to Vesta. *Icarus*, 240, 118–132.

Quitté, G., Markowski, A., Latkoczy, C., Gabriel, A., & Pack, A. (2010) Iron-60 heterogeneity and incomplete isotope mixing in the early Solar System. *The Astrophysical Journal*, 720, 1215–1224.

Rao, A. S., & Chaklader, A. C. D. (1972) Plastic flow during hot-pressing. *Journal of the American Ceramic Society*, 55, 596–601.

Raymond, C. A., Russell, C. T., & McSween, H. Y. (2016) Dawn at Vesta: Paradigms and paradoxes. In L. Elkins-Tanton, & B. Weiss (eds.), *Planetesimals*. Cambridge: Cambridge University Press, pp. 321–340.

Righter, K., & Drake, M. J. (1997). A magma ocean on Vesta: Core formation and petrogenesis of eucrites and diogenites. *Meteoritics & Planetary Science*, 32, 929–944.

Russell, C. T., & Raymond, C. A. (2012) The Dawn mission to Vesta and Ceres. *Space Science Reviews*, 163, 3–23.

Russell, C. T., Raymond, C. A., Coradini, A., et al. (2012) Dawn at Vesta: Testing the protoplanetary paradigm. *Science*, 336, 684–686.

Ruzicka, A., Snyder, G. A., & Taylor, L. A. (1997) Vesta as the howardite, eucrite, and diogenite parent body: Implications for the size of a core and for large-scale differentiation. *Meteoritics & Planetary Science*, 32, 825–840.

Sahijpal, S., Soni, P., & Gupta, G. (2007) Numerical simulations of the differentiation of accreting planetesimals with ^{26}Al and ^{60}Fe as the heat sources. *Meteoritics & Planetary Science*, 42, 1529–1548.

Schiller, M., Baker, J., Creech, J., et al. (2011). Rapid timescales for magma ocean crystallization on the Howardites–Eucrite–Diogenite parent body. *The Astrophysical Journal*, 740, L22.

Scott, T., & Kohlstedt, D. L. (2006) The effect of large melt fraction on the deformation behavior of peridotite. *Earth and Planetary Science Letters*, 246, 177–187.

Shearer, C. K., Fowler, G. W., & Papike, J. J. (1997) Petrogenetic models for magmatism on the eucrite parent body: Evidence from orthopyroxene in diogenites. *Meteoritics & Planetary Science*, 32, 877–889.

Shukolyukov, A., & Lugmair, G. W. (1993) Fe-60 in eucrites. *Earth and Planetary Science Letters*, 119, 159–166.

Smoliar, M. I. (1993) A survey of Rb-Sr systematics of eucrites. *Meteoritics*, 28, 105–113.

Sonnett, C. P., Colburn, D. S., & Schwartz, K. (1968) Electrical heating of meteorite parent bodies and planets by dynamo induction from a premain sequence t tauri solar wind. *Nature*, 219, 924–926.

Šrámek, O., Milelli, L., Ricard, Y., & Labrosse, S. (2012) Thermal evolution and differentiation of planetesimals and planetary embryos. *Icarus*, 217, 339 354.

Srinivasan, G., Goswami, J. N., & Bhandari, N. (1999) Al-26 in eucrite Piplia Kalan: Plausible heat source and formation chronology. *Science*, 284, 1348–1350.

Stolper, E. M. (1975) Petrogenesis of eucrite, howardite and diogenite meteorites. *Nature*, 258, 220–222.

Tachibana, S., & Huss, G. R. (2003) The initial abundance of ^{60}Fe in the Solar System. *The Astrophysical Journal Letters*, 588, L41–L44.

Takahashi, E. (1983) Melting of a Yamato L3 chondrite (Y-74191) up to 30 kbar. *National Institute of Polar Research, Memoirs*, Special Issue (ISSN 0386-0744), no. 30.

Takahashi, K., & Masudat, A. (1990) Young ages of two diogenites and their genetic implications. *Nature*, 343, 540–542.

Taylor, G. J. (1992) Core formation in asteroids. *Journal of Geophysical Research*, 97, 14717–14726.

Taylor, G. J., Keil, K., McCoy, T. J., Haack, H., & Scott, E. R. D. (1993) Asteroid differentiation: Pyroclastic volcanism to magma oceans. *Meteoritics*, 28, 34–52.

Telus, M., Huss, G. R., Ogliore, R. C., Nagashima, K., & Tachibana, S. (2012) Recalculation of data for short-lived radionuclide systems using less-biased ratio estimation. *Meteoritics & Planetary Science*, 47, 2013–2030.

Terasaki, H., Frost, D. J., Rubie, D. C., & Langenhorst, F. (2008) Percolative core formation in planetesimals. *Earth and Planetary Science Letters*, 273, 132–137.

Toplis, M. J., Mizzon, H., Monnereau, M., et al. (2013) Chondritic models of 4 Vesta: Implications for geochemical and geophysical properties. *Meteoritics & Planetary Science*, 48, 2300–2315.

Touboul, M., Sprung, P., Aciego, S. M., Bourdon, B., & Kleine, T. (2015) Hf–W chronology of the eucrite parent body. *Geochimica et Cosmochimica Acta*, 156, 106–121.

Trinquier, A., Birck, J. L., Allegre, C. J., Göpel, C., & Ulfbeck, D. (2008) (53)Mn-(53)Cr systematics of the early Solar System revisited. *Geochimica et Cosmochimica Acta*, 72, 5146–5163.

Turcotte, D. L., & Phipps Morgan, J. (1992) The physics of magma migration and mantle flow beneath a mid-ocean ridge. Mantle flow and melt migration beneath oceanic ridges: Models derived from observations in ophiolites. In: J. Phipps Morgan, D. K. Blackman, & J. M. Sinton (eds.), *Mantle Flow and Melt Generation at Mid-Ocean Ridges*, vol. 71, Geophysical Monograph. Washington, DC: American Geophysical Union, pp. 155–182.

Urey, H. C. (1955) The cosmic abundances of potassium, uranium, and thorium and the heat balance of the earth, the moon, and mars. *Proceedings of the National Academy of Sciences (USA)*, 41, 127–144.

Villeneuve, J., Chaussidon, M., & Libourel, G. (2009) Homogeneous distribution of [26]Al in the Solar System from the Mg isotopic composition of chondrules. *Science*, 325, 985–988.

von Bargen, N., & Waff, H. S. (1986) Permeabilities, interfacial areas and curvatures of partially molten systems: results of numerical computations of equilibrium microstructures. *Journal of Geophysical Research*, 91, 9261–9276.

Waff, H. S., & Bulau, J. R. (1979) Equilibrium fluid distribution in an ultramafic partial melt under hydrostatic stress conditions. *Journal of Geophysical Research*, 84, 6109–6114.

Walker, D., & Agee, C. B. (1988) Ureilite compaction. *Meteoritics*, 23, 81–91.

Wark, D. A., Williams, C. A., Watson, E. B., & Price, J. D. (2003). Reassessment of pore shapes in microstructurally equilibrated rocks, with implications for permeability of the upper mantle. *Journal of Geophysical Research*, 108, 2050.

Wasson, J. T., & Kallemeyn, G. W. (1990) Compositions of chondrites. *Philosophical Transactions of the Royal Society of London*, 325, 535–544.

Wiggins, C., & Spiegelman, M. (1995) Magma migration and magmatic solitary waves in 3D. *Geophysical Research Letter*, 22, 1289–1292.

Wilson, L., & Keil, K. (2012) Volcanic activity on differentiated asteroids: A review and analysis. *Chemie der Erde – Geochemistry*, 72, 289–321.

Yamaguchi, A., Barrat, J.-A., Greenwood, R. C., et al. (2009) Crustal partial melting on Vesta: Evidence from highly metamorphosed eucrites. *Geochimica et Cosmochimica Acta*, 73, 7162–7182.

Yamaguchi, A., Taylor, G. J., & Keil, K. (1996) Global crustal metamorphism of the eucrite parent body. *Icarus*, 124, 97–112.

Yomogida, K., & Matsui, T. (1984) Multiple parent bodies of ordinary chondrites. *Earth and Planetary Science Letters*, 68, 34–42.

Zhou, Q., Yin, Q.-Z., Young, E. D., et al. (2013) SIMS Pb–Pb and U-Pb age determination of eucrite zircons at < 5 μm scale and the first 50 Ma of the thermal history of Vesta. *Geochimica et Cosmochimica Acta*, 110, 152–175.

5

Geomorphology of Vesta

DEBRA L. BUCZKOWSKI, RALF JAUMANN, AND SIMONE MARCHI

5.1 INTRODUCTION

5.1.1 History

Vesta has been extensively studied since its discovery in 1807. Prior to 2011, all these studies were based on telescopic observations but even these Earth-based investigations were able to determine several details about Vesta. Multi-spectral data from the Hubble Space Telescope revealed that Vesta is not spherical (Binzel et al., 1997; Zellner et al., 1997). A digital shape model was generated by Thomas et al. (1997) based on Hubble images, and examination of Vesta's irregular shape led to the detection of a large depression, presumably a crater, at the south pole. Several additional depressions detected in the shape model were similarly interpreted to be craters. Surface features were mapped, based on both multispectral (Binzel et al., 1997) and near-infrared (Zellner et al., 2005) Hubble data. These features included low albedo regions that were speculated to perhaps be mare-filled impact craters (Zellner et al., 2005).

NASA's Dawn spacecraft arrived at Vesta on 2011 July 16, and stayed in orbit for 14 months. The three instruments onboard included the Framing Camera (FC) (Sierks et al., 2011), the Visible and Infrared (VIR) Mapping Spectrometer (De Sanctis et al., 2011), and the Gamma Ray and Neutron Detector (GRaND) (Prettyman et al., 2011). These instruments were used to collect imaging, spectroscopic, and elemental abundance data, which were utilized to examine the asteroid's surface. Dawn had three orbital phases at Vesta: a survey orbit at 2,735 km altitude, a high-altitude mapping orbit (HAMO) at 685 km altitude, and a low altitude mapping orbit (LAMO) at 200 km altitude.

The topography of Vesta was first derived from Survey orbit FC data (spatial resolution of ~260 m/pixel) (Jaumann et al., 2012; Preusker et al., 2012). A digital terrain model (Figure 5.1(a)) was created and underwent continuous improvement because the spatial resolution of the FC data steadily increased as the Dawn spacecraft moved closer to Vesta during the three-tier orbital phasing (Preusker et al., 2014). Stereophotogrammetric analysis of Vesta yields an overall relief from –22.3 km to 19.1 km (with respect to a biaxial reference ellipsoid of 285 × 229 km) (Jaumann et al., 2012). Vesta's topography has a much greater range in elevation relative to its radius (15%) than the Moon and Mars (1%) or the Earth (0.3%), but less than highly impacted smaller asteroids like Lutetia (40%).

One of the primary goals of the Dawn mission was to determine if the ~525 km diameter asteroid was in fact a protoplanet, one of the remnants of the material that formed the Solar System (Russell et al., 2012). What Dawn detected on Vesta's surface was used to determine not only that Vesta is indeed a protoplanetary body (Russell et al., 2012), but also an intermediate planetary body, with properties in between those of asteroids and planets (Jaumann et al., 2012; Russell et al., 2012).

5.1.2 The Search for Volcanism on Vesta

Another key motivation for exploring Vesta as a target of the Dawn mission was the tantalizing possibility of volcanism on the asteroid. Several Earth-based spectroscopic studies of Vesta (e.g., McCord et al., 1970; Gaffey, 1997) showed that Vesta has a similar reflectance spectral signature to the howardite–eucrite–diogenite (HED) meteorites (e.g., Drake, 1979, 2001; Consolmagno & Drake, 1977; Takeda, 1997), indicating that the HEDs may be vestan fragments (e.g., Binzel & Xu, 1993).

The HEDs are all igneous in nature. Eucritic meteorites are basaltic rocks (Takeda, 1997; Keil, 2002), diogenites are magnesium-rich orthopyroxenes (Takeda, 1997; Keil, 2002; Beck & McSween, 2010), and howardites are impact breccias comprised of eucrite and diogenite clasts (Takeda, 1997; Keil, 2002). It is now established, utilizing VIR spectroscopic data, that Vesta's crust is dominated by eucritic, howarditic, and diogenetic materials (De Sanctis et al., 2012a), meaning that the HED meteorites can be considered samples of the asteroid, and used to provide a ground-truth for remote sensing observations. Refer to Chapter 3 for more detail on the composition and formation of HED meteorites.

The small crystal size of basaltic eucrites generally indicates that they cooled quickly on the surface of their parent body, likely during some volcanic process. Diogenite crystal size is generally larger than that of basaltic eucrite crystals, suggesting that diogenites experienced slower cooling, consistent with a plutonic origin. This contrast suggests that the crust of the parent body may have a simple layered structure (Takeda, 1979). Surface eucritic basalts overlying deeper diogenites suggest formation during cooling of a global magma ocean. However, Barrat et al. (2010) performed a detailed study of the trace-element chemistry of diogenites, which suggested that their petrogenesis is more complex than that of simple early crystallization products. Instead they proposed that the diogenites possibly represent later-stage plutons injected into an eucrite crust. Utilizing Dawn data, researchers were able to determine that pluton emplacement did indeed play a significant role in the formation of Vesta's crust (Raymond et al., 2017; McSween et al., 2019). Thus, a search for volcanic and plutonic features on Vesta was an important driver for a geomorphological examination of the asteroid.

5.2 LARGE-SCALE GEOMORPHIC FEATURES

5.2.1 Impact Basins

Imaging of Vesta by the Hubble Space Telescope revealed a major depression at the south pole that was suggested to be a giant impact

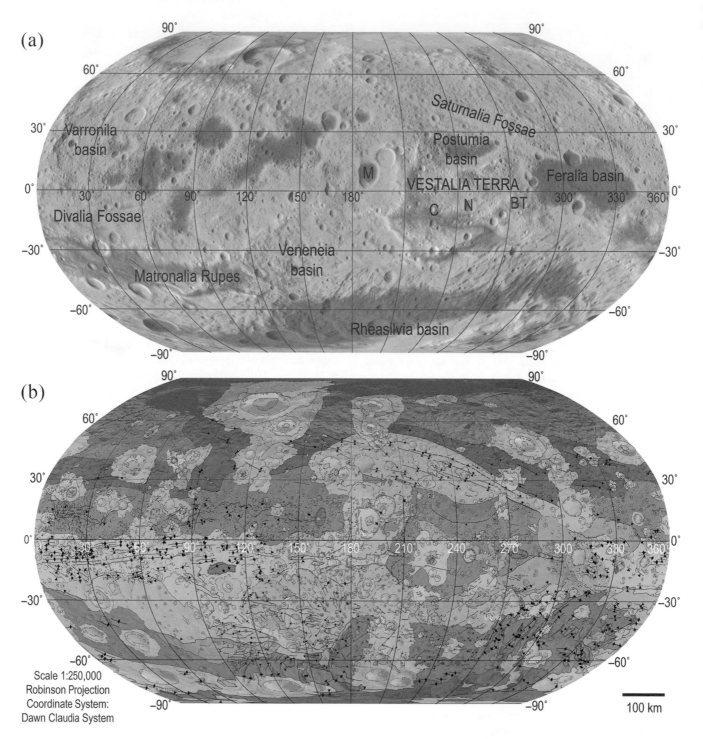

Figure 5.1 (a) Colorized HAMO digital terrain model of Vesta (Preusker et al., 2014), in Robinson projection. GIS processing by David M. Nelson, Regional Planetary Image Facility at Arizona State University. Red shows high regions, while blue represents low. Features of note are labeled. M = Marcia crater; C = Cornelia crater; N = Numisia crater; BT = Brumalia Tholus. (b) The LAMO geologic map of Vesta (Williams et al., 2014a) in Robinson projection. A black and white version of this figure will appear in some formats. For the color version, refer to the plate section.

basin (Thomas et al., 1997). Even at Hubble resolution we could observe that this putative basin was nearly as wide as Vesta in diameter (Thomas et al., 1997). The arrival of Dawn at Vesta discovered that the south polar depression was instead two distinct but overlapping large impact basins, Rheasilvia and Veneneia (Figure 5.2) (Schenk et al., 2012).

The Rheasilvia basin (Figure 5.2), centered at 71.95° S, 86.3° E, is the larger and younger of the two south polar basins. It is ~500 ± 25 km in diameter and 19 ± 6 km deep, making it only slightly smaller than the diameter of Vesta itself. The floor of Rheasilvia is bowl-shaped, characterized by rolling plains (Schenk et al., 2012), and highly deformed by linear and curvilinear ridges and inward

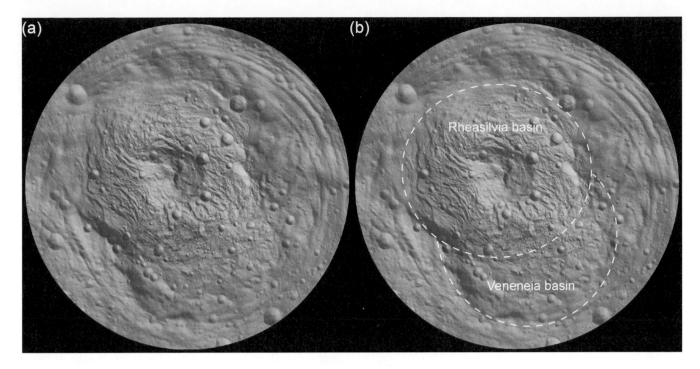

Figure 5.2 (a) Shaded relief HAMO topographic map (resolution 60 m/pixel) of the southern hemisphere of Vesta. The rims of the Rheasilvia and Veneneia basins are visible as steep scarps. The central mound of Rheasilvia is also visible. (b) The margins of the basins are drawn as dashed lines and labeled. The topographic profiles of both basins can be found in Schenk et al. (2012, figure 1).

facing scarps in a pervasive spiral pattern (Otto et al., 2013). Rheasilvia has a wide central massif (180 km at its base and 20–25 km high) from which the scarps and ridges on the basin floor spiral out (Figure 5.2). The basin's outer margin is variable in both shape and elevation (Figure 5.2), being expressed as a broad ridge along the majority of the rim, but as inward-facing scarps in two locations (Schenk et al., 2012; Krohn et al., 2014b). It was predicted that ejecta from Rheasilvia could extend into the northern hemisphere of Vesta (Jutzi & Asphaug, 2011), and there is some mineralogic evidence to support this (Reddy et al., 2012b; Schaefer et al., 2014). However, as of yet, no geomorphic evidence of Rheasilvia ejecta has been identified more than ~100 km from the basin rim (Schenk et al., 2012; Buczkowski et al., 2014; Garry et al., 2014; Schaefer et al., 2014; Williams et al., 2014a).

Vesta has a relatively fast rotation period of 5.3 hours (Russell et al., 2012) and the center of Rheasilvia nearly coincides with its rotational axis (Jaumann et al., 2014). This suggests that Coriolis forces might have deflected radial mass motions of the early crater collapse process (Otto et al., 2013). Slumping toward the center of Rheasilvia exhibits a rotational aspect, with the larger cohesive slump blocks sliding on the concave (spoon-like) shaped surface of the rupture. Parts of the basin are defined by concentric collapse while others are tilted relative to the crater rim of Rheasilvia. The basin also exhibits curved ridges exposed concentric to the crater rim which, in the radial direction, are most likely remnants of the terraced crater rim collapse (Otto et al., 2013).

The Veneneia basin (Figure 5.2) is centered at 47.93° S, 305.68° E, and is partially obscured by the younger Rheasilvia impact (Jaumann et al., 2012). The basin is not quite as large as Rheasilvia, being ~400 ± 25 km in diameter and 12 ± 2 km deep at the point where it is disrupted by the younger basin. Its floor is rolling plains cut by short scarps and ridges; unlike in Rheasilvia,

there is no spiral pattern to the floor structures (Schenk et al., 2012). The Veneneia rim is primarily visible in the topographic data (Figure 5.2), as the disruption caused by the Rheasilvia impact obscured the rim from view in visible data. There is no visible central massif, although it is possible that a pre-existing massif was destroyed by the Rheasilvia impact (Schenk et al., 2012). While no geomorphic evidence of ejecta from Veneneia has been mapped on Vesta (Williams et al., 2014b), the deposition of dark material has been attributed to the Veneneia impact event (McCord et al., 2012; Reddy et al., 2012a; Jaumann et al., 2014) (see Section 5.3.1). Most dark material deposits on Vesta are distributed within 100 km of the northern rim of Veneneia (Jaumann et al., 2014), which is consistent with the observed geomorphic expression of Rheasilvia ejecta (Schenk et al., 2012). Other regions of Veneneia ejecta may be superposed and concealed by overlying Rheasilvia ejecta (Schaefer et al., 2014; Williams et al., 2014a).

These two large basins excavated a large amount of Vesta's surface. The estimated minimum volume excavated by Rheasilvia alone has been estimated to be ~1×10^6 to 3×10^6 km^3, more than half of which was likely lost to space (Schenk et al., 2012). It has been proposed that this excavated rock may be the source of the HED meteorites (Thomas et al., 1997).

If in fact the HEDs were ejected from Vesta during the Rheasilvia impact, as has been suggested, then it might be expected that the ages of the meteorites could serve as a ground truth for the age-dates of the impact event. Ar:Ar dating of 46 eucrites indicate that large impacts on Vesta occurred 3.5–4.1 Ga (Bogard, 2011). This would seem to suggest that the 3.5 Ga age for Rheasilvia derived by Schmedemann et al. (2014) was an accurate prediction. However, similar Ar:Ar dating of howardite impact melt clasts by Cohen (2013) yielded a date of 3.3–3.8 Ga. If the Rheasilvia impact was indeed the source of the HED meteorites, then the basin should have formed after the youngest HED age (Marchi et al., 2013a), which would exclude the Schmedemann et al. (2014) results.

Schenk et al. (2012) pointed out that there is insufficient time for Ar:Ar age resetting to occur during the initial shock of an impact, because an extended time is required to diffuse argon from the rocks. Therefore, most of the thermal alteration recorded in the Ar:Ar ages of the HED meteorites (Bogard, 1995, 2011; Bogard & Garrison, 2003) was more likely to occur in thick slow-cooling units (such as ejecta or basin floor deposits) rather than during the impact event. Since Vesta's surface may be a patchwork of basin-formed units of distinct ages, this means that Rheasilvia could have impacted into pre-existing basins already set to ancient Ar:Ar ages (Marchi et al., 2013a). If Rheasilvia formed into the floor and ejecta deposits of two pre-existing large basins, this could have provided a source of large volumes of thermally altered rocks with ancient resetting ages to become the HED meteorites (Schenk et al., 2012). This would be more consistent with the ~1 Ga age of the Rheasilvia basin predicted by Marchi et al. (2012) and O'Brien et al. (2014). McSween et al. (2013) confirmed this chronological link between the age of the impact event and the HEDs.

Other basins larger than 150 km diameter have also been detected on Vesta (Marchi et al., 2012). Feralia Planitia (Figure 5.1) is a 270 km diameter impact basin centered at 3.03° N, 101.7° E whose western rim marks the eastern boundary of Vestalia Terra. Postumia (Figure 5.1) is a 196 km diameter impact basin centered at 33.84° N, 33.77° E whose southern rim marks the northern boundary of Vestalia Terra (see Section 5.2.3). Both Feralia and Postumia are older than Veneneia (Buczkowski et al., 2014). Varronilla (Figure 5.1) is a heavily degraded 158 km diameter impact basin centered at 29.62° N, 179.58° E; its depth was noted to be shallow for the average vestan crater, possibly due to erosion by the numerous superposed craters (Marchi et al., 2012).

5.2.2 Troughs

Two sets of large-scale linear troughs encircle Vesta (Jaumann et al., 2012). The first set, called the Divalia Fossae (Figure 5.3(a)), are roughly aligned with the equator. Eighty-seven equatorial troughs were mapped, with lengths varying from 19 to 465 km, and widths up to 21.8 km (Buczkowski et al., 2012). The troughs were almost all bounded by steep scarps, with topographic profiles showing that their shape varies from that of classic flat-floored graben (a down-dropped block bounded on both sides by normal faults) to that of half-graben (a structure with a normal fault on only one side). This suggests that there is downward movement along the fault scarps that comprise both the northern and southern walls. Analysis of the largest of the Divalia troughs indicate the graben accommodates at least 5 km of vertical displacement, with the northern fault accommodating the bulk of the fault movement (Buczkowski et al., 2012). The smaller troughs to the south of this largest graben also have greater fault displacement on the northern fault, while the smaller troughs to the north of the largest graben show more displacement on the southern fault. These smaller graben generally only show vertical displacements of up to 2 km (Buczkowski et al., 2012).

Most of the Divalia Fossae structures are troughs, but there are also muted troughs, grooves, and pit crater chains with a similar orientation, suggesting they share a common formation mechanism. Fault plane analysis (Buczkowski et al., 2008) suggests that the formation of the Divalia Fossae structures was triggered by the impact event that formed the Rheasilvia basin, as the poles of the planes described by their fault traces cluster on the Rheasilvia central peak (Jaumann et al., 2012). The Divalia Fossae structures cut the majority (>60%) of the equatorial region of Vesta.

The second set of linear structures, called the Saturnalia Fossae (Figure 5.3(c)), extend to the northwest from the equatorial troughs (Jaumann et al., 2012). There are eight linear structures with this orientation (Buczkowski et al., 2012); the largest is a 39.2 km wide trough that extends north for 366 km from the Lepida crater, while smaller graben and grooves ranging from 31 to 212 km in length are located in the region north of Vestalia Terra. Analysis of topographic profiles of the largest trough indicate that it is a graben that accommodates at least 4 km of vertical displacement, with the southern fault accommodating the bulk of the strain (Buczkowski et al., 2012).

The Saturnalia Fossae are generally shallower than the Divalia Fossae, with rounded edges and infilling on the trough floors (Buczkowski et al., 2012). Both the walls and floor also appear to be more heavily cratered than the equatorial troughs. Fault plane analysis suggests that the formation of the Saturnalia Fossae

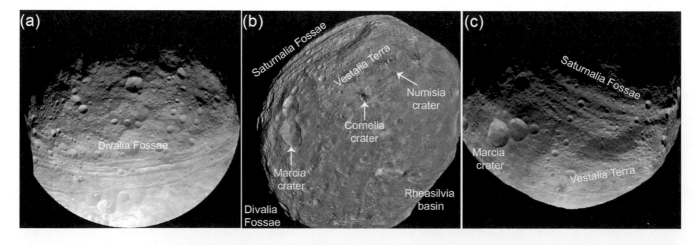

Figure 5.3 Framing Camera clear filter images from RC3 (the late approach phase). Structures of note are labeled. (a) The equatorial region of Vesta, showing the Divalia Fossae. (b) The equatorial region of Vesta, showing Vestalia Terra. While the Divalia Fossae are visible at the margin of the image, they do not cut across Vestalia Terra. Dark material can be observed, even at this scale, in the walls of Cornelia and Numisia craters. (c) The northern hemisphere of Vesta, showing the Saturnalia Fossae.

structures was triggered by the impact event that formed the Veneneia basin (Jaumann et al., 2012). This all suggests that Saturnalia is older than Divalia and the other equatorial structures.

Many researchers have performed numerical models that show that a large impact into a differentiated Vesta could be responsible for the formation of graben roughly circumferential to the impact site (Buczkowski et al., 2012; Ivanov & Melosh, 2013; Bowling et al., 2013a, 2013b; Stickle et al., 2015). This is consistent with the fault plane analysis (Buczkowski et al., 2008) of the Divalia and Saturnalia Fossae (Jaumann et al., 2012), which suggests that the Rheasilvia and Veneneia impacts, respectively, may have been responsible for triggering the formation of the troughs.

Linear structures have previously been identified in a concentric orientation around impact craters on several other asteroids, including Ida (Asphaug et al., 1996), Eros (Prockter et al., 2002; Buczkowski et al., 2008), and Lutetia (Thomas et al., 2012). However, the structures that formed due to impact on these asteroids are all grooves, fractures, and a few shallow troughs, while the geomorphology and displacement profiles of Vesta's equatorial and northern troughs are far more like planetary faults (Buczkowski et al., 2012). The morphological and structural differences in the structural features implies that there must be inherent differences between Vesta and the other asteroids.

The numerical models (Buczkowski et al., 2012; Ivanov & Melosh, 2013; Bowling et al., 2013a, 2013b; Stickle et al., 2015) suggest that the formation of graben instead of grooves on Vesta may be because it is a differentiated body with a mantle and core. The models demonstrate that the density contrast in Vesta's differentiated interior affects the stresses resulting from the Rheasilvia and Veneneia impacts, amplifying and reorienting the stresses resultant from the impact as compared to impact on an undifferentiated body (Buczkowski et al., 2012).

However, the difference could also be due to compositional variations within the asteroid. Vesta's crust is known to be compositionally diverse (e.g., De Sanctis et al., 2012a; Reddy et al., 2012b), while Ida (Sullivan et al., 1996) and Eros (e.g., Trombka et al., 2000) are both compositionally homogeneous. A layered stratigraphy, alternating mechanically strong and weak rock within the crust, would more easily allow for larger scale features, such as graben, to form. Therefore, graben may have formed on Vesta and not the other asteroids simply because Vesta is comprised of multiple crustal layers of differing rock types and strengths (Buczkowski et al., 2012).

5.2.3 Vestalia Terra

Vestalia Terra (Figure 5.1(b)), one of the largest individual features on Vesta, was recognized as a distinct region due to its high topography relative to the rest of the asteroid (Jaumann et al., 2012). An 80,000 km^2 plateau, Vestalia Terra is bounded by steep scarps formed by ancient impact events, including the Rheasilvia, Venenia, Feralia, and Postumia basins (Buczkowski et al., 2014). There are localized topographic variations within Vestalia Terra; while the highest topography on Vesta is in the southwest portion of the plateau, the plateau's interior contains some deep craters and one low intercrater area. However, Vestalia Terra as a whole is topographically high relative to the surrounding terrain, with the bounding scarps being as low as 2.8 km to the west, but 5.4 km to the south, 8.7 km to the east, and as high as 16.5 km to the north.

While the Divalia Fossae structures cut most of the equatorial region of Vesta, they do not cut the Vestalia Terra plateau (Figure 5.3(b)) (Buczkowski et al, 2012). Buczkowski et al. (2014) postulated that Vestalia Terra did not fault during this global-scale stress event because it was comprised of a stronger material than the rest of the equatorial region of Vesta.

5.3 SMALL-SCALE GEOMORPHIC FEATURES

5.3.1 Dark Material

Relatively low-albedo surface deposits appear in the visible wavelength range of Dawn's Framing Camera and VIR spectrometer (Jaumann et al., 2012, 2014; McCord et al., 2012; Reddy et al., 2012a). The spectral characteristics of this dark material resemble that of Vesta's average regolith, indicating that there are no significant mineralogic differences (De Sanctis et al., 2012a; Stephan et al., 2014). There are three types of dark components found in the howardite–eucrite–diogenite meteorites, our putative samples of Vesta's surface (Russell et al., 2012): clasts of carbonaceous chondritic material, impact melt, and impact shock-blackened material (McSween et al., 2011, 2013). Of these, the carbonaceous chondrite component provides the best match with the spectral properties of the dark material (McCord et al., 2012), suggesting that the source of the dark material could be from the impact debris of carbonaceous chondrites that impacted into Vesta, a hypotheses first entertained prior to Dawn's arrival at Vesta (Hasegawa et al., 2003; Prettyman et al., 2011). The highest hydrogen abundances detected by GRaND are associated with broad regions of dark material (Prettyman et al., 2012), while spectral analyses by VIR suggest that debris from carbonaceous chondrites may comprise as much as 60% of the regolith of Vesta (Reddy et al., 2012a). Such a high percentage of carbonaceous materials in Vesta's regolith could indicate that ∼5 wt% water may be present in those areas (Denevi et al., 2012). More information about the composition of the dark material is presented in Chapter 6.

Dark material is distributed unevenly across Vesta's surface, in clusters of different types of exposures (Figure 5.4(a)). Mixed with the regolith and partially excavated by younger impacts, the dark material is exposed as individual layered outcrops in crater walls (e.g., Figure 5.3(b)) or ejecta patches (e.g., Figure 5.4(b)), having been uncovered and broken up by impacts. Dark fans on crater walls and dark deposits on crater floors are the result of gravity-driven mass wasting triggered by steep slopes and impact seismicity. The dark material is mixed with impact ejecta, indicating that it has been processed together with the ejected material (Jaumann et al., 2014). Some small craters display continuous dark ejecta, indicating that the impact excavated the material from beneath the surface. The asymmetric distribution of dark material in impact craters and ejecta suggests non-continuous distribution in the local subsurface (Jaumann et al., 2014). Some dark edifices with positive topography appear to be impact-sculpted hills with dark material distributed over the hill slopes (Jaumann et al., 2012; Williams et al., 2014b). Dark features inside and outside of craters can be arranged as linear outcrops along scarps or as dark streaks perpendicular to the local topography.

On a local scale, some craters expose or are associated with dark material, while others in the immediate vicinity do not show

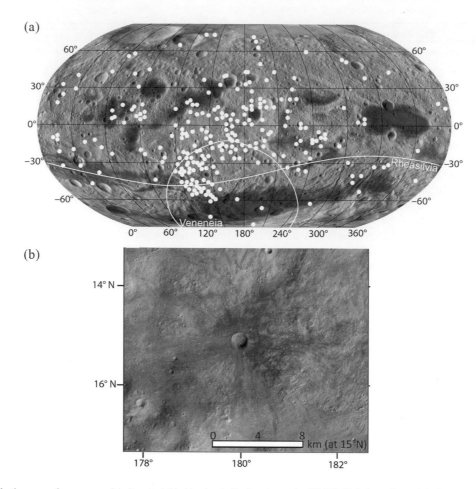

Figure 5.4 (a) Distribution map of exposures of dark material (white dots), displayed over the HAMO digital terrain model of Vesta (Preusker et al., 2014), in Robinson projection. Most dark material is concentrated on the N and NW rim of the Veneneia basin with an extension of about 100 km outside the respective rim (Jaumann et al., 2014). (b) Example dark material exposure. LAMO mosaic (res. 20 m/pixel) of a small crater west of Marcia (179.5°E, 14.5°N). The impact has deposited a dark continuous ejecta blanket and discontinuous radial dark rays, which exhibit clear boundaries with the surrounding regolith.

evidence for dark material. Approximately 500 impact craters with dark material exposed in their walls and rims are found on Vesta (Jaumann et al., 2014). Most of these exposures have the same geological characteristics. Outcrops appear within the first 2,000 m below the rim. In most cases outcrops are correlated with a spur and gully modification of the crater walls. Spurs are composed of bright competent material, whereas the dark material emanates from the less competent gully alcoves. Fans of bright material partly overlay dark ones, and vice versa. Dark outcrops are thin layers of restricted lateral extent, ranging from single spots (a few tens of meters) to short layers of a few kilometers. They are disconnected and separated by areas within the crater wall that are free of dark material. Material accumulating on the crater floor shows a higher albedo than the dark outcrops but lower than the average surrounding regolith, indicating mixing of mass wasting material with dark components. In some cases, the dark material covers the rim above the dark outcrop, darkens the ejecta adjacent to the outcrops, and appears as a dark swirl-like striation emanating from the outcrops outside of the crater. In other instances, the dark ejecta only occurs on one side of the crater. Dark areas that are not directly correlated with larger craters show very small impacts (<1 km), excavating and distributing dark material within limited surface regions (Jaumann et al., 2014).

Multiple factors suggest that dark material is exogenic, from carbon-rich low-velocity impactors, rather than endogenic, from freshly exposed mafic material or melt, exposed or created by impacts (McCord et al., 2012; Reddy et al., 2012a; Jaumann et al., 2014). While the variety of surface exposures of dark material and their different geological correlations with surface features, as well as their uneven distribution (Figure 5.4), indicate a globally inhomogeneous distribution in the subsurface, the dark material seems to be correlated with the rim and ejecta of the older Veneneia south polar basin structure (Jaumann et al., 2014). This regional concentration of dark material associated with a large basin such as Veneneia suggests a direct link between the impact and dark material deposition (Reddy et al., 2012a; Jaumann et al., 2014).

5.3.2 Pitted Terrain

Terrain marked by a distinct pitted morphology was identified on Vesta in the floors, ejecta, and terraces of several craters, including Marcia, Cornelia, Licinia, and Numisia (Denevi et al., 2012). Pits are differentiated from impact craters by their lack of raised rims. These vestan pitted terrains were observed to be morphologically similar to pitted terrains on Mars (Denevi et al., 2012), where pits are thought

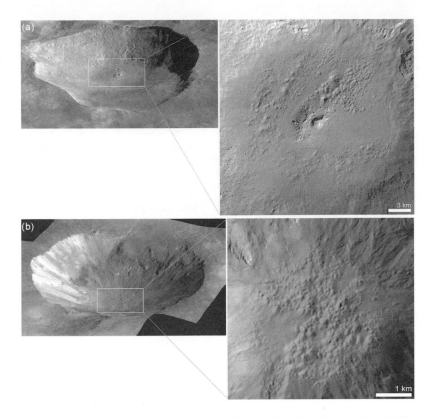

Figure 5.5 (a) Perspective view of a FC mosaic of Marcia crater with a close-up of the pitted terrain on the crater floor. (b) Perspective view of a FC mosaic of Cornelia crater with a close-up of the pitted terrain on the crater floor.
Image credit: NASA/JPL-Caltech/UCLA/MPS/DLR/IDA/JHUAPL

to form through degassing of volatile-bearing material heated by the impact (e.g. Boyce et al., 2012; Tornabene et al., 2012).

More than 200 fresh craters on Mars have occurrences of pitted terrains on their floors, terraces, and ejecta (McEwen et al., 2007; Mouginis-Mark & Garbeil, 2007; Hartmann et al., 2010; Boyce et al., 2012; Tornabene et al., 2012), similar to those observed on Vesta (Denevi et al., 2012). Models for the formation of pitted terrain on Mars generally require the presence of a large amount of volatiles, presumably water ice. These models include (1) the sublimation of ice sometime after the impact event (Hartmann et al., 2010) or (2) rapid degassing of volatiles due to heating by the impact event (Boyce et al., 2012; Tornabene et al., 2012).

The most extensive deposits of pitted terrain on Vesta were observed in Marcia crater (Figure 5.5(a)), where the pits on the crater floor ranged in diameter from 30 m to 1 km; pits on the crater terraces and in the ejecta were smaller, no more than 250 m diameter (Denevi et al., 2012). Pits were observed to be larger and more irregular when formed in regions where the surface materials are thicker. Adjacent pits sometimes overlapped each other and/or coalesced.

Pits identified in Cornelia crater (Figure 5.5(b)) had morphologies similar to those in Marcia, but were only found on the crater floor, not in the ejecta or on the terraces (Denevi et al., 2012). The pits in Licinia crater were degraded, with more superposed impact craters. Identification of pits in Numisia crater was ambiguous; Denevi et al. (2012) determined that a small hummocky area on the crater floor might represent a cluster of small pits.

The morphological similarities between the pitted terrains on Vesta and those on Mars implies that they have a similar formation mechanism. This suggests that there is a substantial amount of

volatiles on Vesta, which is counter to previous evaluations of the asteroid (e.g. Wilson & Keil, 1996; Mittlefehldt et al., 1998). There is a strong association between the vestan pitted terrain and craters that have deposits of low-albedo material in their walls and ejecta (Denevi et al., 2012). This dark material has been shown to be enriched in hydrogen (Prettyman et al., 2012); this hydrogen could be in the form of hydroxyl, which has also been detected on Vesta by VIR, in association with the dark material (De Sanctis et al., 2012b). Impacts into this hydrated dark material would result in devolatilization due to impact heating, as has been modeled for Mars (Denevi et al., 2012).

Denevi et al. (2012) suggested that the occurrence of pitted terrain might have once been more widespread in Vesta's past. This was based on the observation that pits are only currently found in the younger craters of Vesta, leading to the possibility that pits in older craters might have been degraded or buried over time.

5.3.3 Pit Crater Chains

Pit crater chains are a type of geomorphic feature described as lines of circular to elliptical depressions that lack an elevated rim, ejecta deposits, or lava flows (e.g., Ferrill et al., 2004, 2011; Wyrick et al., 2004, 2010; Whitten & Martin, 2019). They have been observed on several planets (Wyrick et al., 2004, 2010), moons (Thomas, 1979; Horstman & Melosh, 1989; Michaud et al., 2008; Wyrick et al., 2010; Martin et al., 2017), and asteroids (Prockter et al., 2002; Veverka et al, 1994; Sullivan et al., 1996; Jaumann et al., 2012; Buczkowski et al., 2016; Scully et al., 2017), including the Earth (e.g., Okubo & Martel, 1998; Ferrill et al., 2011; Whitten & Martin,

2019). Individual pits commonly have a conical shape (Wyrick et al., 2004), but can also be bowl-shaped (Whitten & Martin, 2019) or conical with a flat floor (Wyrick et al., 2004). In many cases, pit crater chains can coalesce into linear troughs, with the individual pits becoming elliptical with the long axis parallel to the chain orientation. Before this coalescence the pits are often bordered by a graben (Wyrick et al., 2004, 2010).

To date, the preferred hypothesis for pit crater formation on small bodies [including Eros (Prockter et al., 2002; Buczkowski et al., 2008), Vesta (Buczkowski et al., 2012), Ceres (Scully et al., 2017), Phobos (Thomas, 1979), and Enceladus (Michaud et al., 2008; Martin et al., 2017)] has been dilational faulting (Ferrill & Morris, 2003) with subsequent regolith drainage. Dilational faulting can occur on normal faults that traverse mechanically strong stratigraphic layers, or where hybrid mode failure (Mode I opening, combined with either Mode II sliding and/or Mode III tearing) occurs under low differential stress (Ferrill & Morris, 2003). Under such conditions, a void is produced beneath fault segments, into which overlying unconsolidated material can drain, producing chains of pit craters (Wyrick et al., 2004). It has also been suggested that pit depth can be used as a proxy for regolith thickness (e.g., Wyrick et al., 2004;

Martin & Kattenhorn, 2013). This method of pit crater chain formation due to dilational faulting has been directly observed on Earth in Iceland (e.g. Ferrill et al., 2004, 2011; Whitten & Martin, 2019).

East–west trending pit crater chains at a variety of scales (Figure 5.6) occur extensively in the northern hemisphere of Vesta (e.g., Denevi et al., 2012; Carsenty et al., 2013). While preliminary mapping of the small-scale (<1 km length) features (Figure 5.6(a)) has been presented (Carsenty et al., 2013), no structural or geomorphic analysis has been performed yet that could indicate how these features formed or explain their orientation. However, three unusually large pit crater chains roughly aligned with the large equatorial troughs have also been detected (Buczkowski et al., 2012). These large pit crater chains – Albalonga Catena (Figure 5.6(b)), Robigalia Catena (Figure 5.6(c)), and an unnamed feature – have merged pits that clearly show signs of collapse, but distinct fault faces are also observed (Buczkowski et al., 2014). It has thus been suggested that these pit crater chains are representative of subsurface faulting (Buczkowski et al., 2012). However, these pit crater chains are significantly different in scale from the other east–west trending features; in terms of both overall length and pit crater diameter and depth, these features are notably larger. If there is a direct correlation

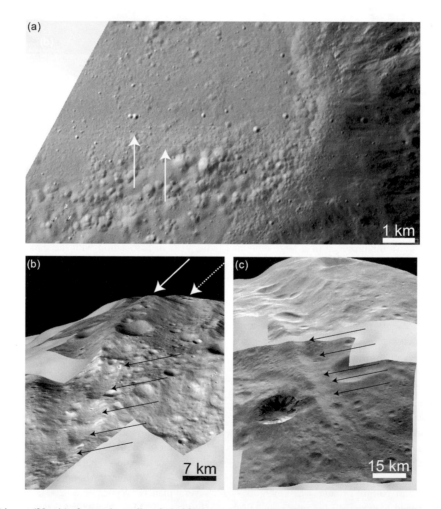

Figure 5.6 (a) FC LAMO image (20 m/p) of example small-scale (<1 km) east–west trending pit crater chains on Vesta. White arrows point to small-scale pit crater chains on the floor of Marcia crater. (b) LAMO mosaic of Albalonga Catena draped over Vesta topography. Black arrows point to the merged pits of the pit crater chain. White arrows point to Brumalia Tholus (left) and Teia crater (right). (c) LAMO mosaic of Robigalia Catena draped over Vesta topography. Black arrows point to the merged pits of the pit crater chain. Cornelia crater is in the lower left.

between pit crater depth and the depth of the regolith that it forms in (e.g., Wyrick et al., 2004; Martin & Kattenhorn, 2013; Whitten & Martin, 2019), this could suggest that the regolith depths in this region are greater than in the rest of the northern hemisphere. However, these differences in scale may also suggest a fundamental difference in the putative faults underlying the pit crater chains.

5.3.4 Mass Wasting

Vesta's surface shows considerable evidence for mass wasting (Jaumann et al., 2012; Otto et al., 2013; Krohn et al., 2014b). Mass wasting events result in subdued impact craters, widen the troughs, and form slumps and landslides, especially within craters. Topography plays a significant role in modification processes on Vesta. Compared to its radius, Vesta has a very intense relief resulting in relatively steep slopes (reaching up to 55°) (Jaumann et al., 2012). Impacts onto such steep surfaces, followed by slope failure, makes resurfacing due to impacts and their associated gravitational and seismic activity an important geologic process on Vesta that significantly alters the morphology of geologic features.

The impact basins of Rheasilvia and Veneneia show significant mass-wasting features that suggest different types of material properties and conditions. Otto et al. (2013) identified six different mass-wasting types in the southern hemisphere, including intra-crater mass-wasting, flow-like movements, creep-like features, rotational slumping, landslides, and curved ridges. The intra-crater mass-wasting features were mainly identified in fresh craters, and included lobate slides, rocky spurs along the rim, boulders inside the crater, dark albedo bands on the crater wall, and talus material. The flow-like and creep-like features are phenomena of shocked and fractured material (Otto et al., 2013).

The youngest slumping area in Rheasilvia occurs along the Matronalia Rupes scarp (Figure 5.7), indicating ongoing erosional degradation (Otto et al., 2013). Landslides are obvious in both the Rheasilvia and Veneneia basins originating from the crater rims, and also from Rheasilvia's central peak. Older landslides are mostly broad features while younger ones, such as those identified on the slope of Matronalia Rupes scarp, exhibit a lobate structure (Figure 5.7) (Otto et al., 2013). The V-shaped pattern at the intersection area of Rheasilvia and Veneneia, and on Rheasilvia's

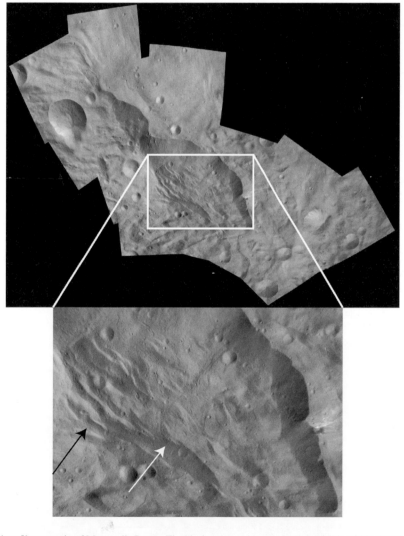

Figure 5.7 Framing Camera clear filter mosaic of Matronalia Rupes. The black arrow points to a large landslide within the Rheasilvia crater, while the white arrow points to a superposed mass movement.
Image credit: NASA / JPL / MPS / DLR / IDA / Daniel Macháček

central peak, can be interpreted as strike–slip faults that are generated during the collapse of Rheasilvia on a bi-layered target with brittle and ductile components (Otto et al., 2013).

5.3.5 Impact Craters

Vesta's surface is dominated by impact craters of all sizes (Jaumann et al., 2012; Marchi et al., 2012, 2013b; Schenk et al., 2012). These craters have characteristics similar to those on smaller asteroids, as well as those on the Moon and Mars, supporting the concept that Vesta is an intermediate body between asteroids and planets (Jaumann et al., 2012; Russell et al., 2012). The degradation history of impact craters on Vesta ranges from fresh to highly degraded, indicating an intensive cratering history over the age of the Solar System (Jaumann et al., 2012; Schenk et al., 2012). A global database of 11,605 craters larger than 0.7 km has been created (Liu et al., 2018). A depth versus diameter analysis of the vestan craters utilizing this database suggests that craters on Vesta are shallower than craters on the Moon or Mars (Marchi et al., 2013b; Liu et al., 2018).

The majority of impact craters on Vesta are roughly circular and bowl-shaped (e.g., Figure 5.5), with crater rims that are approximately the same elevation level all the way around. However, there are a large number of vestan craters that are asymmetrical in both shape and ejecta distribution (e.g., Figure 5.8) (Krohn et al., 2014a). These asymmetric craters formed on Vesta's steep slopes; pronounced collapse and slumping occurred on the upslope sides of the craters, resulting in an extreme asymmetric morphology with a wider upslope crater wall and a narrower downslope wall. In extreme cases, the material from upslope has overrun the downslope rims and flowed out of the craters (Jaumann et al., 2012; Krohn et al., 2014a). Slopes >20° prevent the deposition of ejected material in the uphill direction, and slumping material superimposed the deposit of ejecta on the downhill side. Crater size-frequency measurements of surrounding surfaces, ejecta deposits, and the crater interiors reveal, within measurement and cratering chronology model uncertainties, similar ages, indicating that the deposits were most likely directly formed during the impact and no major post-impact processes modified the crater (Krohn et al., 2014a). Smooth flat regions occur in the interiors of some craters

and in small depressions as pond-like accumulations with well-defined geological contacts that indicate that they are younger than their surroundings (Jaumann et al., 2012). Several scenarios of the origin of these deposits seem plausible, including impact sedimentation, impact melt, dust levitation and transport, seismic shaking, or slumping of fine material.

Gullies have been identified in several vestan craters. Linear gullies, such as those observed in Fabia crater (Figure 5.8(b)) are interpreted to form by the flow of dry granular material (Scully et al., 2015). More unexpectedly, curvilinear gullies, such as those observed in Cornelia and other impact craters that contain pitted terrain, are theorized to possibly form in a transient, debris-flow-like process, where a small amount of fluid was released from volatiles in the pre-existing regolith due to impact heating (Scully et al., 2015).

5.4 THE SEARCH FOR VOLCANIC AND MAGMATIC FEATURES

Because of the igneous nature of the HED meteorites, it has long been suspected that Vesta may have undergone volcanic and/or magmatic activity, certainly in its early history (e.g., Keil, 2002) and possibly even in its more recent past (e.g., Barrat et al., 2010). The eucrites are basalt, suggesting volcanic flows, and the diogenites are plutonic, suggesting magmatic intrusions. Prior to Dawn, several studies assessed the role of igneous intrusion on Vesta, with dikes (Wilson & Keil, 1996), sills (Keil & Wilson, 2012), and plutons (Barrat et al., 2010) all having been proposed. Therefore, a major component of Dawn's mission was a search for volcanic and magmatic features on Vesta.

Numerous lobate flow features were identified around the asteroid, and a study of these lobate flows was performed to assess their formation mechanisms (Williams et al., 2013). These analyses included descriptions of both their morphologies and of their spatial distribution. Williams et al. (2013) determined that there is no unequivocal morphologic evidence of volcanic activity on Vesta. Instead, all the lobate flows were related to impact craters (e.g., impact melt flows) or erosional processes. However, they did not rule out that the evidence of volcanism might have been erased

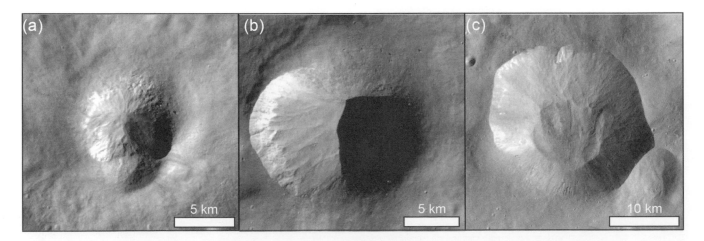

Figure 5.8 Examples of asymmetric craters on Vesta. (a) Teia crater. (b) Fabia crater. (c) Drusilla crater.

by subsequent impact bombardment, since the volcanism likely occurred very early in Vesta's history.

While no evidence of volcanism was found, morphological evidence of magmatic activity on Vesta was identified (Buczkowski et al., 2014). Albalonga Catena is a pit crater chain in eastern Vestalia Terra that phases from being a 76 km long series of merged pits into an elongate hill named Brumalia Tholus that extends 50 km along its ridgeline (Figure 5.5(b)). Westward of the hill, another 70 km linear arrangement of merged pits is visible in the slope data. The orientation of the two sets of merged pits and the axis of Brumalia Tholus suggests that the three features represent a single buried fault (Buczkowski et al., 2014; Wilson et al., 2015). This in turn suggests that the topographic high that emerges along the length of the buried fault most likely formed as some type of magmatic intrusion, in which molten material utilized the subsurface fracture as a conduit to travel toward the surface, intruding into and deforming the surface rock (Figure 5.9).

Laccoliths are a type of magmatic intrusion in which magma pressure is high enough to dome the overlying material without the magma breaking through. If Brumalia Tholus is a laccolith, then its core should be a more plutonic rock like diogenite, rather than basaltic eucrites or brecciated howardites, because when magma cools below the surface it occurs slowly, and larger crystals have time to form.

The ejecta of Teia crater, which impacts the northern face of Brumalia Tholus, have been found to be more diogenitic in composition than the surrounding background terrain (De Sanctis et al., 2014). Since Teia's ejecta is likely sampling Brumalia's core material, its composition supports the assumption that the interior of Brumalia Tholus is diogenitic. This is consistent with the hill being the surface representation of a subsurface magmatic intrusion (Buczkowski et al., 2014).

Albalonga Catena is roughly aligned with the Divalia Fossae (Buczkowski et al., 2012), suggesting that its underlying fault may also have formed during the Rheasilvia impact event (Buczkowski

et al., 2014). However, simple models of the thermal evolution of asteroidal bodies suggest volcanism on Vesta ceased by 10–100 Ma after formation (e.g., Schiller et al., 2010; McSween et al., 2011), long before the time of Rheasilvia impact.

It is possible that the Albalonga faulting occurred due to the Rheasilvia impact event and the putative molten material is diogenitic impact melt associated with the basin's formation. However, previous modeling suggests that melt production on asteroids would be limited due to lower impact velocities (Keil et al., 1997), so impact melt seems an unlikely source of the molten material. Buczkowski et al. (2014) deemed it more likely that the material injected into the fault was sourced by a localized, but still internally driven, melt. For example, the source of the melt could have been the magma plume theorized by Raymond et al. (2013) to have formed the Vestalia Terra plateau, the sill-like intrusions Keil and Wilson (2012) projected to be at the base of the vestan lithosphere, or from one of the diogenitic intrusions into the eucritic crust predicted by Barrat et al. (2010).

Buczkowski et al. (2014) suggested that Brumalia Tholus formed in the following series of steps (Figure 5.9). Sometime in the ancient past the Vestalia Terra sub-surface fractured, forming the Albalonga subsurface fault. Partial melt utilizes the Albalonga fault as a conduit to move toward the surface. Brumalia Tholus forms as this magmatic injection causes surface deformation in the form of laccolith doming. The molten material at the core of Brumalia Tholus cools slowly at depth, forming diogenite. Loose regolith material covering the surface of Vestalia Terra (probably ejecta) collapses into dilational openings along the steep subsurface faults, forming the pit crater chains. Eventually, the Teia impact occurs, exposing the Brumalia diogenitic core material in its ejecta.

5.5 CONCLUSIONS

Vesta's surface is characterized by numerous geomorphic features, including large troughs extending around the equatorial region, enigmatic dark material, considerable mass wasting, and impact craters of all sizes, with a variety of ejecta blankets. Vesta exhibits rugged topography ranging from −22 km to 19 km, relative to a best-fit ellipsoidal shape. Impacts into these steep surfaces are an important geologic process on Vesta, which significantly alter the morphology of geologic features and add to the complexity of interpreting its geologic history.

The importance of impact cratering as a geologic driver on Vesta cannot be understated. A large impact basin (Rheasilvia) was observed at Vesta's south pole by the Dawn spacecraft, as was predicted by images from the Hubble Space Telescope, and an underlying, earlier large basin (Veneneia) was also discovered. The formation of the two distinct sets of large troughs (Divalia Fossae and Saturnalia Fossae), which dominate the global tectonics of Vesta, was likely triggered by the formation of these two south polar basins. Dark material unevenly distributed across Vesta's surface was most likely deposited as ejecta from the Veneneia basin and other carbon-rich low velocity impactors. The ejection of the HED meteorites from Vesta likely occurred during the Rheasilvia impact.

Smaller impact craters on Vesta range from fresh to highly degraded, comparable to the Moon, indicating an extensive

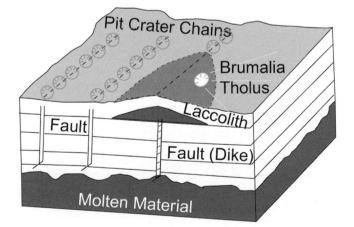

Figure 5.9 Block diagram representing the creation of Brumalia Tholus due to laccolith formation. Although the three faults are oriented in a way that suggests a common formation mechanism, only one extends deeply enough to sample sub-plateau partial melt (dark grey). This fault serves as a conduit for the molten material to rise toward the surface (hatch marks), deforming the surface material above it and creating Brumalia Tholus. Subsequent impact into the hill exposes diogenitic material in the ejecta (white).

cratering history over the age of the Solar System. The pitted terrain and curvilinear gullies found in some of these smaller craters likely formed due to impact cratering processes interacting with pre-existing volatiles on the surface of Vesta.

In contrast to what was expected given the mineralogy of the HEDs, direct surface evidence for volcanic activity on Vesta is lacking. This may be due to volcanism ending early in Vesta's evolution, so that the evidence has been obscured by subsequent resurfacing. However, evidence of subsurface intrusive activity has been detected, and so the presence of magmatic material cannot be excluded.

Although first predicted simply because of its size, analysis of Vesta's geomorphology suggests that it is an intermediate body between asteroids and planets. For example, there is a much greater range in elevation relative to radius on Vesta than there is on planets such as Earth and Mars, but less than there is on smaller asteroids. In addition, impact craters on Vesta have characteristics similar to both those on smaller asteroids as well as those on the Moon and Mars.

REFERENCES

Asphaug, E., Moore, J. M., Morrison, D., et al. (1996) Mechanical and geological effects of impact cratering on Ida. *Icarus*, 120, 158–184.

Barrat, J. A., Yamaguchi, A., Zanda, B., Bollinger, C., & Bohn, M. (2010) Relative chronology of crust formation on asteroid Vesta: Insights form the geochemistry of diogenites. *Geochimica et Cosmochimica Acta*, 74, 6218–6231.

Beck, A. W., & McSween, H. Y. (2010) Diogenites as polymict breccias composed of orthopyroxenite and harzburgite. *Meteoritics & Planetary Science*, 45, 850–872.

Binzel, R. P., Gaffey, M. J., Thomas, B. H., et al. (1997) Geologic mapping of Vesta from 1994 Hubble Space Telescope images, *Icarus*, 128, 95–103.

Binzel, R. P., & Xu, S. (1993) Chips off of asteroid 4 Vesta: Evidence for the parent body of basaltic achondrite meteorites. *Science*, 260, 186–191.

Bogard, D. D. (1995) Impact ages of meteorites: A synthesis. *Meteoritics*, 30, 244.

Bogard, D. D. (2011) K–Ar ages of meteorites: Clues to parent-body thermal histories. *Chemie der Erde – Geochemistry*, 71, 207–226.

Bogard, D. D., & Garrison, D. H. (2003) ^{39}Ar–^{40}Ar ages of eucrites and thermal history of asteroid 4Vesta. *Meteoritics and Planetary Science*, 38, 669–710.

Bowling, T. J., Johnson, B. C., & Melosh, H. J. (2013a) Formation of equatorial graben following the Rheasilvia impact on asteroid 4 Vesta. *44th Lunar & Planetary Science Conference*, March 18–22, Houston, TX, abs. 1673.

Bowling, T. J., Johnson, B. C., Melosh, H. J., et al. (2013b) Antipodal terrains created by the Rheasilvia basin forming impact on asteroid 4 Vesta. *Journal of Geophysical Research*, 118, 1821–1834.

Boyce, J. M., Wilson, L., Mouginis-Mark, P. J., Hamilton, C. W., & Tornabene, L. L. (2012) Origin of small pits in martian impact craters. *Icarus*, 221, 262.

Buczkowski, D. L., Barnouin, O. S., & Prockter, L. M. (2008) 433 Eros lineaments: Global mapping and analysis. *Icarus*, 193, 39–52.

Buczkowski, D. L., Schmidt, B., Williams, D. A., et al. (2016) The geomorphology of Ceres. *Science*, 353, aaf4332.

Buczkowski, D. L., Wyrick, D. Y., Iyer, K. A., et al. (2012) Large-scale troughs on Vesta: A signature of planetary tectonics. *Geophysical Research Letters*, 39, L18205.

Buczkowski, D. L., Wyrick, D. Y., Toplis, M., et al. (2014) The unique geomorphology and physical properties of the Vestalia Terra plateau. *Icarus*, 244, 89–103.

Carsenty, U., Wagner, R. J., Buczkowski, D. L., et al. (2013) The "swarm" – A peculiar crater chain on Vesta. *44th Lunar & Planetary Science Conference*, March 18–22, Houston, TX, abs. 1492.

Cohen, B. A. (2013) The Vestan cataclysm: Impact-melt clasts in howardites and the bombardment history of 4 Vesta. *Meteoritics & Planetary Science*, 48, 771–785.

Consolmagno, G. J., & Drake, M. J. (1977) Composition of the eucrite parent body: Evidence from rare Earth elements. *Geochimica et Cosmochimica Acta*, 41, 1271–1282.

De Sanctis, M. C., Ammannito, E., Buczkowski, D., et al. (2014) Compositional evidence of magmatic activity on Vesta. *Geophysical Research Letters*, 41, 3038–3044.

De Sanctis, M. C., Ammannito, E., Capria, M. T., et al. (2012a) Spectroscopic characterization of mineralogy and its diversity across Vesta. *Science*, 336, 697–700.

De Sanctis, M. C, Combe, J.–Ph., Ammannito, E., et al. (2012b) Detection of widespread hydrated materials on Vesta by the VIR imaging spectrometer on board the Dawn mission. *Astrophysical Journal Letters*, 758, L36.

De Sanctis, M. C., Coradini, A., Ammannito, E., et al. (2011) The VIR spectrometer. *Space Science Reviews*, 163, 329–369.

Denevi, B. W., Blewett, D. T., Buczkowski, D. L., et al. (2012) Pitted terrain on Vesta and implications for the presence of volatiles. *Science*, 338, 246–249.

Drake, M. J. (1979) Geochemical evolution of the eucrite parent body: Possible evolution of Asteroid 4 Vesta? In T. Gehrels, & M. S. Matthews (eds.), *Asteroids*. Tucson: University of Arizona Press, pp. 765–782.

Drake, M. J. (2001) Presidential address: The eucrite/Vesta story. *Meteoritics & Planetary Science*, 36, 501–513.

Ferrill, D. A., & Morris, A. P. (2003) Dilational normal faults. *Journal of Structural Geology*, 25, 183–196.

Ferrill, D. A., Wyrick, D. Y., Morris, A. P., Sims, D. W., & Franklin, N. M. (2004) Dilational fault slip and pit chain formation on Mars, *GSA Today*, 14, 4–12.

Ferrill, D. A., Wyrick, D. Y., & Smart, K. J. (2011) Coseismic, dilational-fault and extension-fracture related pit chain formation in Iceland: Analog for pit chains on Mars. *Lithosphere*, 3, 133–142.

Gaffey, M. J. (1997) Surface lithologic heterogeneity of asteroid 4 Vesta. *Icarus*, 127, 130–157.

Garry, W. B., Williams, D. A., Yingst, R. A., et al. (2014) Geologic mapping of ejecta deposits in Oppia Quadrangle, Asteroid (4) Vesta. *Icarus*, 244, 104–119.

Hartmann, W. K., Quantin, C., Werner, S. C., & Popova, O. (2010) Do young Martian ray craters have ages consistent with the crater count system? *Icarus*, 208, 621.

Hasegawa, S., Murakawa, K., Ishiguro, M., et al. (2003) Evidence of hydrated and/or hydroxylated minerals on the surface of asteroid 4 Vesta. *Geophysical Research Letters*, 30, 2123.

Horstman, K. C., & Melosh, H. J. (1989) Drainage pits in cohesionless materials – Implications for the surface of PHOBOS. *Journal of Geophysical Research*, 94, 12433–12441.

Ivanov, B. A., & Melosh, H. J. (2013) Two-dimensional numerical modeling of the Rheasilvia impact formation. *Journal of Geophysical Research*, 118, 1545–1557. doi:10.1002/jgre.20108

Jaumann, R., Nass, A., Otto, K., et al. (2014) The geological nature of dark material on Vesta and implications for the subsurface structure. *Icarus*, 240, 3–19.

Jaumann, R., Williams, D. A., Buczkowski, D. L., et al. (2012) Vesta's shape and morphology. *Science*, 336, 687–690.

Jutzi, M., & Asphaug, E. (2011) Mega-ejecta on asteroid Vesta. *Geophysical Research Letters*, 38, L01102.

Keil, K. (2002) Geologial history of asteroid 4 Vesta: The "smallest terrestrial planet". In W. F. Bottke, A. Cellino, P. Paolicchi, & R. P. Binzel (eds.), *Asteroids III*. Tucson: University of Arizona Press, pp. 573–584.

Keil, K., Stoffler, D., Love, S. G., & Scott, E. R. D. (1997) Constraints on the role of impact heating and melting in asteroids. *Meteoritics and Planetary Science*, 32, 349–363.

Keil, K., & Wilson, L. (2012) Volcanic eruption and intrusion processes on 4 Vesta: A reappraisal. *43rd Lunar Planetary Science Conference*, Abs. 1127, Houston, TX: Lunar Planetary Institute.

Krohn, K., Jaumann, R., Elbeshausen, D., et al. (2014a) Bimodal craters: Impacts on slopes. *Planetary and Space Science*, 103, 36–56.

Krohn, K., Jaumann, R., Otto, K., et al. (2014b) Mass movement on Vesta at steep scarps and crater rims. *Icarus*, 244, 120–132.

Liu, Z., Yue Z., Michael G., et al. (2018) A global database and statistical analyses of (4) Vesta craters. *Icarus*, 311, 242–257.

Marchi, S., Bottke, W. F., Cohen, B. A., et al. (2013a) High-velocity collisions from the lunar cataclysm recorded in asteroidal meteorites. *Nature Geoscience*, 6, 411.

Marchi, S., Bottke, W. F., O'Brien, D. P., et al. (2013b) Small crater populations on Vesta. *Planetary and Space Science*, 103, 96–103.

Marchi, S., McSween, H. Y., O'Brien, D. P., et al. (2012) The violent collisional history of Asteroid 4 Vesta. *Science*, 336, 690.

Martin, E. S., & Kattenhorn, S. A. (2013) Probing regolith depths on Enceladus by exploring a pit chain proxy. *Lunar and Planetary Science Conference XLIV*, #2047.

Martin, E. S., Kattenhorn, S. A., Collins, G. C., et al. (2017) Pit chains on Enceladus signal the recent tectonic dissection of the ancient cratered terrains. *Icarus*, 294, 209–217.

McCord, T. B., Adams, J. B., & Johnson, T. V. (1970) Asteroid Vesta: Spectral reflectivity and compositional implications. *Science*, 168, 1445–1447.

McCord, T. B., Li, J.-Y., Combe, J.-P., et al. (2012) Dark material on Vesta from the infall of carbonaceous volatile-rich material. *Nature*, 491, 83–86.

McEwen, A. S., Hansen, C. J., Delamere, W. A., et al. (2007) A closer look at water-related geologic activity on Mars. *Science*, 317, 1706.

McSween, H. Y. J., Binzel, R. P., De Sanctis, M. C., et al. (2013) Dawn; the Vesta–HED connection; and the geologic context for eucrites, diogenites, and howardites. *Meteoritics & Planetary Science*, 48, 2090–2104.

McSween, H. Y. J., Mittledfehldt, D. W., Beck, A. W., Mayne, R. G., & McCoy, T. J. (2011) HED meteorites and their relationship to the geology of Vesta and the Dawn mission, *Space Science Review*, 163, 141–174.

McSween, H. Y. J., Raymond, C. A., Stolper, E. M., et al. (2019) Differentiation and magmatic history of Vesta: Constraints from HED meteorites and Dawn spacecraft data. *Geochemistry*, 79, 125526.

Michaud, R. L., Pappalardo, R. T., & Collins, G. C. (2008) Pit chains on Enceladus: A discussion of their origin. *Lunar and Planetary Science*, XXXIX, abs. #1678.

Mittlefehldt, D. W., McCoy, T. J., Goodrich, C. A., & Kracher, A. (1998) Non-chondritic meteorites from asteroidal bodies. In J. J. Papike (ed.), *Planetary Materials*. Washington, DC: Mineralogical Society of America, pp. 4-1–4-195.

Mouginis-Mark, P. J., & Garbeil, H. (2007) Crater geometry and ejecta thickness of the Martian impact crater Tooting. *Meteoritics & Planetary Science*, 42, 1615–1625.

O'Brien, D. P., Marchi, S., Morbidelli, A., et al. (2014) Constraining the cratering chronology of Vesta. *Planetary and Space Science*, 103, 131–142.

Okubo, C. H., & Martel, S. J. (1998) Pit crater formation on Kilauea volcano, Hawaii. *Journal of Volcanology and Geothermal Research*, 86, 1–18.

Otto, K. A., Jaumann, R., Krohn, K., et al. (2013) Mass-wasting features and processes in Vesta's south polar basin Rheasilvia. *Journal of Geophysical Research*, 118, 2279–2294.

Prettyman, T. H., Feldman, W. C., McSween Jr., H. Y., et al. (2011) Dawn's gamma ray and neutron detector. *Space Science Reviews*, 163, 371–459.

Prettyman, T. H., Mittlefehldt, D. W., Yamashita, N., et al. (2012) Elemental mapping by Dawn reveals exogenic H in Vesta's regolith. *Science*, 338, 242–246.

Preusker, F., Scholten, F., Matz, K.-D., et al. (2012) Topography of Vesta from Dawn FC stereo images. *43rd Lunar Planetary Science Conference*, Abs. 2012. Houston, TX: Lunar and Planetary Institute.

Preusker, F., Scholten, F., Matz, K.-D., et al. (2014) Global shape of Vesta from Dawn FC stereo images. *45th Lunar Planetary Science Conference*, abs. 2027. Houston, TX: Lunar and Planetary Institute.

Prockter, L., Thomas, P., Robinson, M., et al. (2002) Surface expressions of structural features on Eros. *Icarus*, 155, 75–93.

Raymond, C. A., Park, R. S., Asmar, S. W., et al. (2013) Vestalia Terra: An ancient mascon in the southern hemisphere of Vesta. *44th Lunar Planetary Science Conference*, Abs. 2882. Houston, TX: Lunar and Planetary Institute.

Raymond, C. A., Russell, C. T., & McSween, H. Y. (2017) Dawn at Vesta: Paradigms and paradoxes. In L. Elkins-Tanton, & B. Weiss (eds.), *Planetesimals: Early Differentiation and Consequences for Planets*. Cambridge: Cambridge University Press, pp. 321–340.

Reddy V., Le Corre, L., O'Brien, D. P., et al. (2012a) Delivery of dark material to Vesta via carbonaceous chondritic impacts. *Icarus*, 221, 544–559.

Reddy, V., Nathues, A., Le Corre, L., et al. (2012b) Color and albedo heterogeneity of Vesta from Dawn. *Science*, 336, 700–704.

Russell, C. T., Raymond, C. A., Coradini, A., et al. (2012) Dawn at Vesta: Testing the protoplanetary paradigm. *Science*, 336, 684–686.

Schaefer, M., Nathues, A., Williams, D. A., et al. (2014) Imprint of the Rheasilvia impact on Vesta – Geologic mapping of quadrangles Gegania and Lucaria. *Icarus*, 244, 60–73.

Schenk, P., O'Brien, D., Marchi, S., et al. (2012) The geologically recent giant impact basins at Vesta's south pole. *Science*, 336, 694.

Schiller, M., Baker, J. A., Bizzaro, M., Creech, J., & Irving, A. J. (2010) Timing and mechanisms of the evolution of the magma ocean on the HED parent body. *73rd Annual Meteorical Society Meeting*, Abst. 5042. New York: Lunar Planetary Institute.

Schmedemann, N., Kneissl, T., Ivanov, B., et al. (2014) The cratering record, chronology and surface ages of (4) Vesta in comparison to smaller asteroids and the ages of the HED meteorites. *Planetary and Space Science*, 103, 104–130.

Scully, J. E. C., Buczkowski, D. L., Schmedemann N., et al. (2017) Evidence for the interior evolution of Ceres from geologic analysis of fractures. *Geophysical Research Letters*, 44, 9564–9572.

Scully, J. E. C., Russell, C. T., Yin, A., et al. (2015) Geomorphical evidence for transient water flow on Vesta. *Earth and Planetary Science Letters*, 411, 151–163.

Sierks, H., Keller, H. U., Jaumann, R., et al. (2011) The Dawn Framing Camera. *Space Science Reviews*, 163, 263–327.

Stephan, K., Jaumann, R., De Sanctis, M. C., et al. (2014) A compositional and geological view of fresh ejecta of small impact craters on Asteroid 4 Vesta. *Journal of Geophysical Research*, 119, 2013JE004388.

Stickle, A. M., Schultz, P. H., & Crawford, D. A. (2015) Subsurface failure in spherical bodies: A formation scenario for linear troughs on Vesta's surface. *Icarus*, 247, 18–34.

Sullivan, R., Greeley, R., Pappalardo, R., et al. (1996) Geology of 243 Ida. *Icarus*, 120, 119–139.

Takeda, H. (1979) A layered-crust model of a howardite parent body. *Icarus*, 40, 455–470.

Takeda, H. (1997) Mineralogical records of early planetary processes on the howardite, eucrite, diogenite parent body with reference to Vesta. *Meteoritic and Planetary Science*, 32, 841–853.

Thomas, N., Barbieri, C., Keller, H. U., et al. (2012) The geomorphology of (21) Lutetia: Results from the OSIRIS imaging system onboard ESA's Rosetta spacecraft. *Planetary and Space Science*, 66, 96–124.

Thomas, P. (1979) Surface features of Phobos and Deimos. *Icarus*, 40, 223–243.

Thomas, P. C., Binzel, R. P., Gaffey, M. J., et al. (1997) Impact excavation on asteroid 4 Vesta: Hubble Space Telescope results. *Science*, 277, 1492–1495.

Tornabene, L. L., Osinski, G. R., McEwen, A. S., et al. (2012) Widespread crater-related pitted materials on Mars: Further evidence for the role of target volatiles during the impact process. *Icarus*, 220, 348.

Trombka, J. I., Squyres, S. W., Bruckner, J., et al. (2000) The elemental composition of asteroid 433 Eros: Results of the NEAR-Shoemaker X-ray spectrometer. *Science*, 289, 2101–2105.

Veverka, J., Thomas, P., Simonelli, D., et al. (1994) Discovery of grooves on Gaspra. *Icarus*, 107, 399–411.

Whitten, J. L., & Martin, E. S. (2019) Icelandic pit chains as planetary analogs: Using morphologic measurements of pit chains to determine regolith thickness. *Journal of Geophysical Research*, 124, 2983–99.

Williams, D. A., Denevi, B. W., Mittlefehldt, D. W., et al. (2014a) The geology of the Marcia quadrangle of asteroid Vesta: Assessing the effects of large, young craters. *Icarus*, 244, 74–88.

Williams, D. A., O'Brien, D. P., Schenk, P. M., et al. (2013) Lobate and flow-like features on asteroid Vesta, *Planetary and Space Science*, 103, 24–35.

Williams, D. A., Yingst, R. A., & Garry, B. (2014b) Introduction: The geologic mapping of Vesta. *Icarus*, 244, 1–12.

Wilson, L., Bland, P., Buczkowski, D., Keil, K., & Krot, S. (2015) Hydrothermal and magmatic fluid flow in asteroids. In P. Michel, F. DeMeo, & W. Bottke (eds.), *Asteroids IV*. Tucson: University of Arizona Press, pp. 553–572.

Wilson, L., & Keil, K. (1996) Volcanic eruptions and intrusions on the asteroid 4 Vesta. *Journal of Geophysical Research*, 101, 18927.

Wyrick, D., Ferrill, D. A., Morris, A. P., Colton, S. L., & Sims, D. W. (2004) Distribution, morphology and origins of Martian pit crater chains. *Journal of Geophysical Research*, 109, E06005.

Wyrick, D. Y., Buczkowski, D. L., Bleamaster, L. F., & Collins, G. C. (2010) Pit crater chains across the Solar System. *41st Lunar Planetary Science Conference*, Abs. 1413. Houston, TX: Lunar and Planetary Institute.

Zellner, B. H., Albrecht, R., Binzel, R. P., et al. (1997) Hubble Space Telescope images of Asteroid 4 Vesta in 1994. *Icarus*, 128, 83–87.

Zellner, N. E. B., Gibbard, S., de Pater, I., Marchis, F., & Gaffey, M. J. (2005) Near-IR imaging of asteroid 4 Vesta. *Icarus*, 177, 190–195.

6

The Surface Composition of Vesta

JEAN-PHILIPPE COMBE AND NAOYUKI YAMASHITA

6.1 INTRODUCTION: DEFINITIONS AND BACKGROUND

The surface of a telluric object such as Vesta is potentially made of components from indigenous or endogenous origins, which subsequently could have undergone alteration processes.

On Vesta, indigenous materials include those from endogenic sources, sometimes from several kilometers in depth, either exposed by impacts or released by tectonic activity. Vesta's surface is mostly composed of mafic silicate mineral pyroxene, which was discovered by near-infrared reflectance spectroscopy from ground telescopic observations (McCord et al., 1970; Gaffey, 1997; De Sanctis et al., 2012). This mineralogy is characteristic of basaltic composition, which implies a fully differentiated body with a crust, a mantle, and a core (Figure 6.1(a)), although questions remain about the structure of the crust and the thickness of the mantle (Figure 6.1(b) and (c); see Chapters 3 and 4). The crust of Vesta has been linked to the howardite, eucrite, and diogenite (HED) meteorites (McCord et al., 1970; McFadden et al., 1977) (Figures 6.2(a–d)). These rock samples are igneous, with characteristics of plutonic and volcanic origin, formed in reducing conditions. Because of this, Vesta was initially expected to have low concentrations of volatile materials to begin with (Hasegawa et al., 2003), and subsequently has undergone further depletion through global melting, differentiation, and regolith gardening by meteoroid impacts. In addition, early spectrophotometry suggested that Vesta has an anhydrous surface (Lebofsky, 1980; Feierberg et al., 1985).

Ground telescopic reflectance spectroscopy of Vesta reported a weak absorption around 3 μm (Hasegawa et al., 2003; Rivkin et al., 2006), suggesting the presence of a hydrous component, either hydroxylated (OH-rich) or hydrated (H_2O-rich) of exogenous origin by carbonaceous chondrite meteorites. This observation is consistent with the link to HED meteorites since hydrated, hydroxylated, or oxyhydroxylated mineral phases are frequently found in carbonaceous chondritic clasts inside of howardites (Zolensky et al., 1996; Gounelle et al., 2003). Exogenous contamination comes from impacts with asteroids or comets. Only the impactors that have a composition different from Vesta may leave foreign materials detectable by remote sensing spectroscopy.

Over time, surface alteration may occur – resulting in changes in the optical characteristics. Minerals exposed to the space environments undergo chemical reactions. Surface processes such as mixing and physical damage due to impacts may also be observed by spectroscopy. Those products and processes are known under the generic term of space weathering. The distinction between exogenous materials and surface alteration can be challenging. For example, one early hypothesis for the hydrous materials on Vesta was formation by solar proton implantation (Hasegawa et al., 2003). However, at first order, the surface of Vesta, characterized by ubiquitous and strong pyroxene absorptions, is considered pristine or not altered (Russell et al., 2012 and references therein).

This chapter compiles the knowledge accumulated from telescopic observations of Vesta, geochemical measurements of the HED meteorites, and the publications from the Dawn mission at Vesta (e.g., Russell et al., 2012; McCord & Scully, 2015; and the whole special issue of *Icarus* 259 on the surface composition of Vesta). In addition, this chapter includes analyses of remote sensing from Dawn that benefits from improved instrument calibrations developed during the Dawn mission at Ceres and from scientific results obtained since Dawn departed from Vesta. In particular, the new analyses are mostly based on data acquired by the Visible and Infrared Mapping Spectrometer (VIR, De Sanctis et al., 2011), in the range 0.26–1.07 μm (visible or VIS) and 1.02–5.10 μm (infrared or IR). All VIR spectra presented in this chapter were processed with the most up-to-date radiometric calibration corrections (Carrozzo et al., 2016; Rousseau et al., 2019). The Dawn mission at Vesta confirmed the early telescopic observations by providing detailed global spatially resolved mapping of the surface composition. This chapter starts with indigenous materials (Section 6.2), with a focus on the HEDs (cf., Chapter 4), followed by components of possible exogenous origin because they differ from the HEDs (Figure 6.2(c)) in Section 6.3 and surficial processes in Section 6.4.

6.2 INDIGENOUS MATERIALS ON VESTA

6.2.1 Distribution of HED Components on Vesta

Vesta is a differentiated protoplanet (Ruzicka et al., 1997). Chapter 4 details the internal structure and evolution from modeling based on the Dawn mission and concludes there is no evidence for a global, deep magma ocean, but supports a shallow magma ocean to produce eucrites and/or delivery of distributed buoyant melts such as dikes or sills, whereas diogenite formed in the lower crust. The high-albedo areas on Vesta are regarded as the most pristine, with compositions ranging from eucritic to diogenitic (Zambon et al., 2014), whereas the lower albedo regions are interpreted as contaminated by exogenous materials (Section 6.3.1) or altered (Section 6.4.1).

We would like to thank the two reviewers, Drs. Hannah Kaplan and Andrea Raponi, as well as the editorial board for their comments, suggestions, and geophysical considerations about Vesta. We are also grateful to Joanna Bastian of the Bear Fight Institute for her contribution in proof-editing.

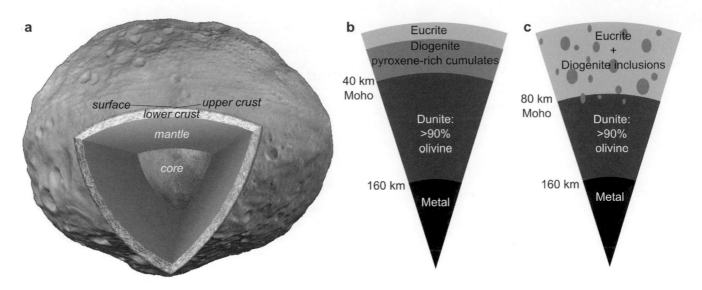

Figure 6.1 Structure and composition of the interior of Vesta derived from Dawn data and informed by the prior geochemical modeling of Righter and Drake (1997) and Ruzicka et al. (1997). The surface is primarily howarditic, with variations between eucrite (common) and diogenite (rare). (a) The image of Vesta is from the Framing Camera (FC) of the Dawn mission; the internal structure is an artist concept from NASA/JPL-Caltech (PIA15510). (b and c) Two models of the internal structure of Vesta by Clenet et al. (2014). (b) A thin differentiated crust with eucrite on top of diogenite and a thick mantle, consistent with internal modeling developed in Chapter 4. (c) A thick, non-differentiated eucritic crust with diogenite inclusions, discussed in Section 6.2.4.

In chronological order of findings, telescopic spectra from ground observations of Vesta exhibit near-infrared absorption features that are similar to those of basaltic achondrites of eucritic composition (McCord et al., 1970; Veeder et al., 1975; Larson, 1977), howarditic composition (Chapman & Salisbury, 1973; Feierberg et al., 1980), and diogenitic composition (Consolmagno, 1979; Takeda, 1979). Therefore, Vesta was suggested to be the source of this class of meteorites (Consolmagno & Drake, 1977; Takeda, 1997; Drake, 2001). This link is supported by nuclear spectroscopy (Prettyman et al., 2012), although it is challenged by some (e.g., Wasson, 2013) based on the abundances of radioelements such as thorium and potassium. However, recent studies show more consistency (Prettyman et al., 2015), and therefore this chapter assumes that Vesta is the parent body of HED meteorites (cf., Chapter 3).

Eucrites are linked to the upper crust of Vesta. They contain low-Ca pyroxene, pigeonite (5–25% Calcium), and plagioclase (anorthite), which is high-Ca. For reference, reflectance spectra of pure minerals are shown in Figure 6.2(e) and (f). Diogenites are igneous, plutonic rocks that originated from the lower crust of Vesta; they are characterized by crystals larger than eucrite and contain magnesium-rich orthopyroxene, with small amounts of plagioclase and olivine. Howardites are regolith breccia mostly constituted of eucrite and diogenite fragments, with occasional carbonaceous chondrules and impact melt. In near infrared spectra, the distinction between eucritic (high-Ca pyroxenes) and diogenitic (low-Ca pyroxenes) compositions appear as a shift of the characteristic Fe^{+2} electronic transition absorption bands located near 1 μm and 2 μm (e.g., Burns, 1970; Adams, 1974, 1975). High-Ca pyroxene eucrites have their absorption bands shifted toward longer wavelengths compared to howardites, whereas low-Ca pyroxene diogenites have their band shifted toward shorter wavelengths (Figure 6.2(a) and (b)).

With visible and near-infrared reflectance spectroscopy (Figure 6.2(a) and (b)), a diogenitic composition is characterized by the center position of the two pyroxene absorption bands at wavelengths shorter than 0.93 and 1.95 μm, respectively (Gaffey, 1976). In VIR spectra (Figure 6.2(c) and (d)), the position of the absorption bands can be measured by a second-order polynomial fit and the calculation of the local minimum (Ammannito et al., 2013). Spectral mixing analysis can also discriminate between the three major surface components of eucrite, diogenite, and carbonaceous-chondritic materials (Zambon et al., 2014; Combe et al., 2015b); Figure 6.3 shows the distribution of the diogenitic component from spectral mixing analysis. In FC spectra, the change of position of the 1 μm absorption band can be sensed by the ratios $R_{0.98 \mu m}/R_{0.92 \mu m}$ and $R_{0.75 \mu m}/R_{0.92 \mu m}$ (Reddy et al., 2012b), where higher ratios indicate a composition richer in diogenite. All these methods provide consistent distributions of diogenite.

6.2.2 Diogenite Excavation from the Rheasilvia Impact

On Vesta, diogenite was detected near the South Pole, especially the central peak of the Rheasilvia impact basin (De Sanctis et al., 2012) and the remnant of the crater wall at Matronalia Rupes (Ammannito et al., 2013, 2015). A broad and elongated diogenitic-rich region that is connected to Matronalia Rupes reaches the northern regions of Vesta and has higher elevations than surrounding terrains (Reddy et al., 2012a); it has been interpreted as an ejecta ray from the Rheasilvia impact (De Sanctis et al., 2012; Reddy et al., 2012a; Ammannito et al., 2013). The elemental observation by Gamma Ray and Neutron Detector (GRaND) on Dawn revealed the low Fe abundance and low neutron absorption in the abovementioned ejecta ray, supporting the hypothesis that the region consists of diogenitic materials to depths of a few decimeters (Prettyman et al., 2013; Yamashita et al., 2013). The strong asymmetry in the distribution of the diogenitic-rich ejecta could likely be explained by a collision that occurred with a high incidence angle (Ammannito et al., 2013). The alternative hypothesis of a buried, large, diogenitic-rich basin has not been supported by any other observations or findings from the Dawn mission

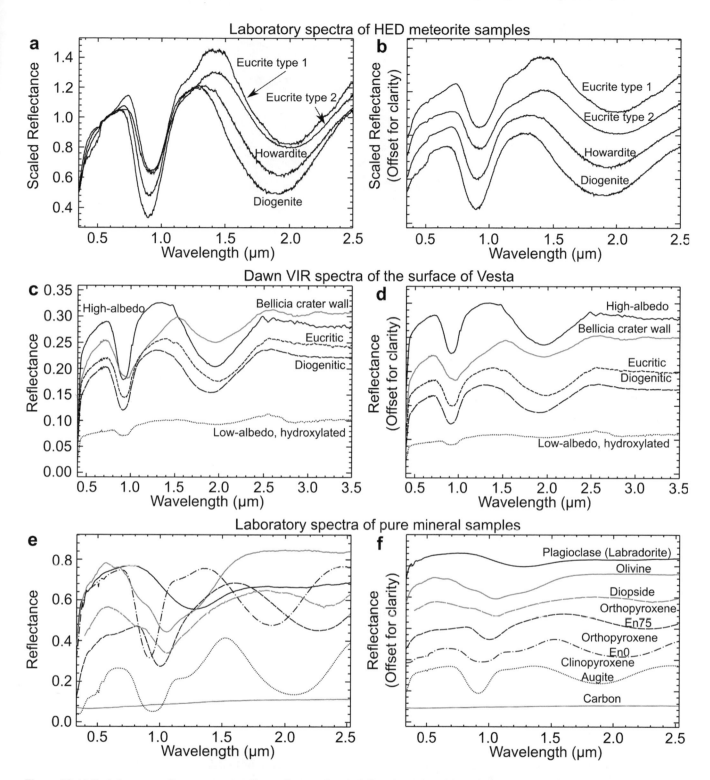

Figure 6.2 (a) Scaled average reflectance spectra of howardites, eucrites, and diogenites (adapted from Gaffey 1976 and following their terminology for eucrites of type 1 and 2). Spectra of eucrite type 1 exhibit an inflexion at 0.65 μm that may be due to absorption, and a more pronounced absorption at 1.2 μm (cf. Sections 6.2.3, 6.2.4, and 6.3.2) than eucrite type 2. (b) Same as (a), offset for clarity. (c) Reflectance spectra of Vesta's surface from the VIR mapping spectrometer of the Dawn mission. (d) Same as (c), offset for clarity. (e) Laboratory reflectance spectra of candidate minerals that may constitute the surface composition of Vesta. Plagioclase labradorite (RELAB PL-ECS-155-Q). Olivine (RELAB AG-TJM-008). Diopside (USGS NMNHR18685). Orthopyroxene En75 (RELAB DL-CMP-022). Orthopyroxene En0 (RELAB DL-CMP-061). Clinopyroxene Augite (RELAB AG-TJM-010). Carbon (RELAB JB-JLB-A05). (f) Same as (e), offset for clarity.

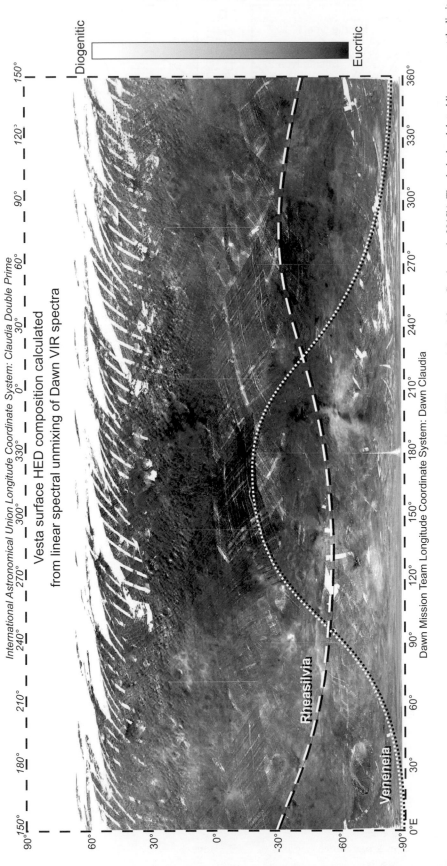

Figure 6.3 Vesta global distribution of eucrite-rich and diogenitic-rich materials by linear spectral unmixing of VIR spectra (adapted from Combe et al., 2015b). The dashed and dotted lines represent the limits of the Rheasilvia and Veneneia impact basins, respectively.

(Ammannito et al., 2013, 2015). Numerical modeling of the distribution of Rheasilvia ejecta by Jutzi et al. (2013) was found to match the location, orientation, and length of the diogenitic-rich ejecta (Combe et al., 2015b).

The detection of a diogenite concentration in the Rheasilvia ejecta and in the remnant of the impact basin, wall, rim, and the central peak supports the global magma ocean model (Ikeda & Takeda, 1985; Righter & Drake, 1997; Ruzicka et al., 1997; Warren, 1997), which implies a thick diogenitic layer below a basaltic eucritic crust (cf., Chapter 4), as illustrated in Figure 6.1(b).

Alternative models of diogenitic-bearing plutons near the crust–mantle boundary or within the lower crust (Figure 6.1(c)) and serial magmatism (Mittlefehldt, 1994; Shearer et al., 1997; Beck & McSween, 2010) derived from measurements of diogenitic trace elements have been suggested based on combined surface composition and density constraints from gravity measurement (cf., Chapters 3 and 12; Raymond et al., 2017). In addition, similar processes may have occurred in the northern regions of Vesta, which is supported by recent studies on type-B diogenites (Beck et al., 2017; and Section 6.2.4), an example of which is the Yamato meteorite that has the composition of a mafic plutonic rock, a plagioclase-bearing pyroxenite.

6.2.3 The Ambiguous Case of Bellicia–Arruntia–Pomponia: Hypothesis of Endogenous Olivine

The Dawn mission revealed that the northern region encompassing craters Bellicia, Arruntia, and Pomponia (40°–90° E, 20°–70° N, see Figure 6.4(a)) is different spectrally from the rest of Vesta because a broad and compound asymmetric absorption band at 1 μm (Figure 6.4(b)) is only detected in that region (Figure 6.4(c)). In addition, a hydroxyl-rich area (Figure 6.4(d)) with a high albedo relative to most of the hydrous surface on Vesta shows a similar distribution (Combe et al., 2015b). Those spectral features led to various interpretations. Endogenous olivine (Ammannito et al., 2013; Ruesch et al., 2014; Palomba et al., 2015; Poulet et al., 2015) was chronologically the first discovered by the Dawn team (this section). The alternative interpretation of pyroxenes (Combe et al., 2015b) in type-B diogenite of endogenous origin (Beck et al., 2015, 2017) is explored in Section 6.2.4. The hypothesis of an exogenous origin (Le Corre et al., 2015; Nathues et al., 2015; Turrini et al., 2016) is described in Section 6.3.1.

The presence of olivine in HED meteorites, 10 ± 3% in eucrite, 8 ± 3% in howardite, 5 ± 1% in diogenite (Batista et al., 2014, and references therein) supports the possible detection of olivine on the surface of Vesta. According to Thangjam et al. (2014), the magma ocean model or the serial magmatism model would be compatible with a hypothetical detection of olivine at the surface of Vesta. However, Clenet et al. (2014) argue that the lack of olivine detection in Rheasilvia indicates a deep crust–mantle boundary that rules out the magma ocean model and favors the serial magmatism model.

According to Ammannito et al. (2013), an exogenous origin of the olivine is unlikely because only low-velocity impacts could have preserved the mineral, which occur mostly with impactors from the Main Belt, and olivine-rich asteroids are rare in the Main Belt. In addition, the study of impact melt breccias in howardites indicates multiple events and that the target body already contained olivine (more detailed are presented in Section 6.2.4). According to

Ruesch et al. (2014), the distribution reported on Vesta and the spatial association with endogenic diogenite rules out an exogeneous origin.

According to Combe et al. (2015b), low-calcium and high-calcium pyroxenes combined such as pigeonite and diopside could explain the VIR spectra of Bellicia, Arruntia, and Pomponia craters, without requiring olivine. Qualitatively, the strong 2 μm absorption band is an argument in favor of pyroxene and not olivine, and claims of olivine detection with FC only (e.g., Thangjam et al., 2016) rely on the spectral range 0.4–1 μm that is not sufficient to clarify the ambiguity.

The timing of those reports is important because those results primarily came from VIR reflectance spectra when the Dawn mission had just left Vesta. Since then, the calibration of VIR has been corrected for Ceres observations in order to be more consistent with telescopic reflectance spectra (Carrozzo et al., 2016), and for variations of the spectral slope in the range of the visible channel as a function of the detector temperature (Rousseau et al., 2019). These corrections affect the shape of spectra between 0.9 and 1.5 μm, which is critical for the identification of olivine, and may explain some of the ambiguities in the mineralogical interpretations of this region. As an attempt to improve the reliability of the results, we have updated the VIR calibration and re-analyzed the spectra of this area.

6.2.4 The Ambiguous Case of Bellicia–Arruntia–Pomponia: Hypothesis of Type-B Diogenite at the Antipode of Rheasilvia

In this section we explore the hypothesis of low-Ca, high-Fe pyroxene in type-B diogenite that is supported by two analyses: (1) an integrated analysis of several types of measurements from the Gamma Ray and Neutron Detector (GRaND) of the Dawn mission by Beck et al. (2015, 2017) and (2) a re-analysis of VIR spectra in the visible and near-infrared (this study) with improved calibration (cf., Section 6.1).

6.2.4.1 Distribution of Type-B Diogenite with GRaND

Beck et al. (2015) demonstrated the possibility of distinguishing igneous lithologies based on a comprehensive HED geochemical dataset, using a combination of neutron absorption (Prettyman et al., 2013), Fe abundance (Yamashita et al., 2013), fast neutron counts (Lawrence et al., 2013), and a compositional parameter (Cp) from High-Energy Gamma Rays (Peplowski et al., 2013). Subsequently, Beck et al. (2017) searched specifically for the following lithologies on Vesta with GRaND data: orthopyroxenitic diogenite, Yamato Type-B diogenite, cumulate eucrite, and basaltic eucrites. Two igneous lithologies were detected and mapped: basaltic eucrite were found on ~3% of the surface at mid-latitudes of the southern hemisphere, and type-B diogenites on ~11% of the surface in a single area centered on crater Pomponia (110° E, 68° N). The two igneous lithologies reported are indicators of endogenous material that have been exposed, or that lay at shallow depth below the surface.

Although basaltic eucritic regions were detectable with GRaND (Beck et al., 2015, 2017), they were not reported in VIR results of eucrite–howardite–diogenite maps (De Sanctis et al., 2012; Ammannito et al., 2013). This is possibly because reflectance spectroscopy is only sensitive to the first micrometers of exposed

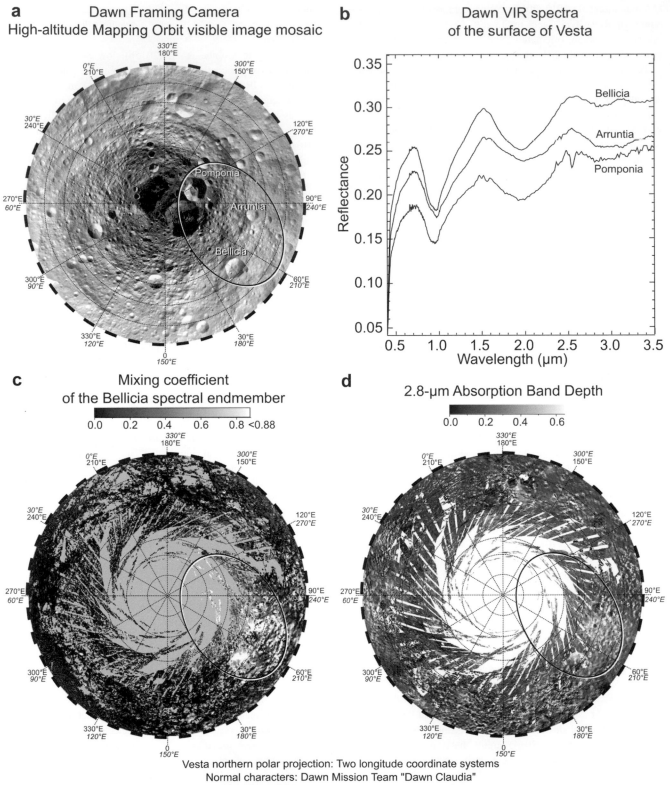

a Dawn Framing Camera
High-altitude Mapping Orbit visible image mosaic

b Dawn VIR spectra
of the surface of Vesta

c Mixing coefficient
of the Bellicia spectral endmember

d 2.8-µm Absorption Band Depth

Vesta northern polar projection: Two longitude coordinate systems
Normal characters: Dawn Mission Team "Dawn Claudia"
Italic: International Astronomical Union "Claudia Double Prime"

Figure 6.4 (a) North Polar map of Vesta from Framing Camera images at 0.75 µm. (b) VIR reflectance spectra showing an inflexion at 1.2 µm in the crater walls of Bellicia, Arruntia, and Pomponia. (c) Distribution of the surface materials similar to Bellicia crater indicated by the ellipse (adapted from Combe et al., 2015b). (d) Distribution of hydrous materials (adapted from Combe et al., 2015b).

materials, and the topmost regolith has undergone heavy impact gardening and mixing (Pieters et al., 2012; Schröder et al., 2013; Zambon et al., 2014; Blewett et al., 2016), whereas GRaND is sensitive to depths up to about a meter (e.g., Prettyman et al., 2019). The type-B diogenite found on crater Pomponia and surrounding terrain is of particular interest because its location may indicate a relationship with the anomalous surface compositions detected in VIR reflectance spectra, despite the multiple interpretations reported to date. The distribution of type-B diogenites is also remarkable because it is correlated with the antipode of Rheasilvia basin and with the impact craters that formed in that region after Rheasilvia, such as Pomponia.

What are the implications for type-B diogenite on Vesta? According to Beck et al. (2015), two formation models can be considered: (1) Takeda and Mori (1985) suggest that type-B diogenite is part of a global single stratigraphic layer (Figure 6.1(b)), deeper than cumulate eucrites and just above orthopyroxenitic diogenites. In this scenario, type-B diogenites in the northern hemisphere of Vesta could only come from Rheasilvia ejecta, because this is the impact that excavated the deepest materials. In contrast, Pomponia was generated by a much less energetic event. The absence of detection of type-B diogenite or pure orthopyroxenitic diogenite in the Rheasilvia basin could mean that the surface of the southern hemisphere was already howarditic when the Veneneia impact was partially overlapped by Rheasilvia. (2) Mittlefehldt and Lindstrom (1993) argue that type-B diogenites formed from a basaltic–eucrite melt in a shallow emplacement in the upper crust, without stratigraphic bound to other lithologies (Figure 6.1(c)), and very localized (i.e., not part of a global layer), and thus have no petrologic relation with orthopyroxenitic diogenite. In this scenario, the finding of type-B diogenite in Pomponia would imply a sampling of a lithology formed in situ. Although the distribution of type-B diogenite on Vesta from Beck et al. (2017) is not sufficient to choose between the two models, the context provided by the absence of olivine in Rheasilvia (McSween et al., 2013; Clenet et al., 2014; see Chapter 3) implies a deep crust–mantle boundary (Clenet et al., 2014) may favor an internal structure of the crust without layers, as in Figure 6.1(c), which is more compatible with a formation of type-B diogenite in the shallow crust without stratigraphic relation to orthopyroxenite diogenite.

6.2.4.2 Low-Ca Pyroxene and Plagioclase Compositions Detected in VIR Spectra

VIR spectra of the northern wall of Bellicia crater (Figure 6.5) have spectral shapes similar to those of Arruntia and Pomponia (Ammannito et al., 2013; Ruesch et al., 2014; Combe et al., 2015b; Palomba et al., 2015; Poulet et al., 2015). They are characterized by a broad asymmetric absorption feature between 0.8 and 1.4 μm, unlike the rest of Vesta, which exhibits a rather symmetric absorption around 1 μm.

Deep absorption bands at 1 and 2 μm are diagnostic of pyroxene compositions (Adams 1974; Hunt 1977; Hazen et al., 1978; Cloutis & Gaffey, 1991). These bands are due to crystal field electronic transitions (e.g., Sunshine et al., 1990; Burns, 1993; Sunshine & Pieters, 1993). In pyroxenes, the tetrahedral structure of SiO_4 forms two octahedral sites M1 and M2 that can host a cation. M1 is a regular octahedron and M2 is a distorted octahedron, larger than M1. Cations in pyroxenes are typically Mg^{2+}, Fe^{2+}, and Ca^{2+}, which form three poles in the classification of pyroxene (Mayne et al., 2009; Klima et al., 2011; Combe et al., 2015b). Mg goes preferentially in the M2 sites, whereas the smaller Fe and Ca go preferentially in the M1 sites.

In low-Ca pyroxenes such as enstatite and ferrosilite, Fe^{2+} in the M1 sites creates a spin-allowed electronic transition absorption at 1.2 μm (Klima et al., 2007, 2008). A similar absorption band exists in olivine (e.g., Sunshine & Pieters, 1998), which may create a confusion with pyroxene, although olivine has an additional absorption due to Fe in M1 sites at wavelengths shorter than the M2 absorption. This 1.2 μm absorption band exists even for low Fe content (enstatite) and intensifies and shifts toward longer wavelengths as Fe abundances increase (Klima et al., 2007, 2008, 2011) toward ferrosilite. The enstatite number (En#) defines the composition with respect to the endmembers enstatite (En100) or ferrosilite (En0). When electronic transition absorption bands from the M1 and M2 sites are broad enough to overlap, they have the visual

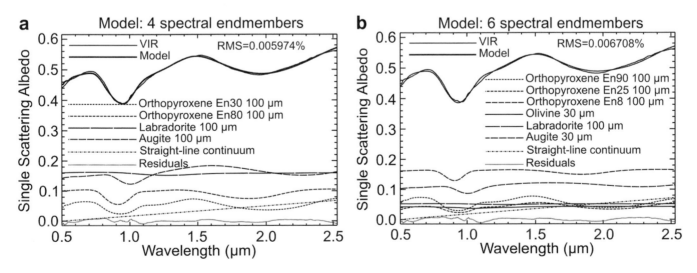

Figure 6.5 Single-scattering spectrum of the wall of crater Bellicia from the Dawn VIR mapping spectrometer (same as in Figure 6.2(c) and (d)) modeled by a linear combination of pure mineral spectra. (a) Model with four different minerals including two orthopyroxenes with different Enstatite number (noted En#), but not olivine. (b) Model with six minerals, including olivine, with lower quality of fit (RMS) than in (a).

appearance of one broad asymmetric absorption skewed toward longer wavelength.

Plagioclase also has a broad and symmetric absorption band centered at 1.25 μm (Conel & Nash, 1970; Bell & Mao, 1973; Adams & Goullaud, 1978), although it is generally shallow. This absorption is due to trace amounts of Fe^{2+} in the plagioclase structure. Increasing FeO content deepens the absorption band at 1.25 μm. In addition, this mineral has high albedo, which means that its spectral signature can be visually masked by materials that have lower albedo, especially the minerals that absorb at the same wavelengths, such as olivine and low-Ca, high-Fe pyroxenes. However, some pyroxene bands can be heavily distorted by plagioclase absorptions (Cheek & Pieters, 2014), and thus can be detected in near-infrared reflectance spectra.

In order to interpret VIR spectra that exhibit this broad and asymmetric absorption band, a spectral mixing analysis based on a radiative transfer model (Hapke, 1981) was performed by Combe (in preparation). The laboratory bidirectional reflectance spectra of pure minerals considered in the analysis included low-Ca pyroxenes in the enstatite–ferrosilite side of the ternary pyroxene classification, olivine, clinopyroxene (augite), plagioclase (labradorite), and carbon. Clinopyroxene, orthopyroxenes, plagioclase, and carbon were found in the RELAB spectral library (Pieters, 1983), and olivine and diopside in the USGS spectral library (Clark et al., 1993). The complex refractive indices of refraction were calculated for each mineral using the method described by Dalton and Pitman (2012) that combines classical dispersion analysis and subtractive Kramers–Krönig analysis. The selection and mixing (Combe et al., 2008) of each mineral spectrum to model the VIR spectra was performed in the space of single scattering albedo. As a result, models that include low-Ca pyroxenes and plagioclase produce the best fit (Figure 6.5(a)), and olivine does not improve the model (Figure 6.5(b)). Another important fact is that it is possible to calculate a good spectral-fitting model without any olivine, whereas models that include olivine must be associated with low-Ca, high-Fe pyroxenes to provide an acceptable fit. These results alone do not rule out the possible presence of olivine, however they are all consistent with low-Ca-poor, high-Fe pyroxenes. Multiple interpretations from spectral mixing analysis is a common occurrence because many free parameters in the radiative transfer model can lead to very different modal composition models and still provide a good quality of fit. As a consequence, a realistic approach would be to provide all the solutions that fit the observations, which can be done with a probabilistic approach (e.g., Lapôtre et al., 2017).

6.2.4.3 Discussion: Possible Relation of Basaltic Eucrite and Type-B Diogenite

The absorption band at 1.2 μm that is observed as an inflexion in VIR spectra of the walls of crater Bellicia, Arruntia, and Pomponia (Figure 6.4(b)), and that is well defined in low-Ca pyroxene with low En# (Figure 6.2(e) and (f)), is also observed in many reflectance spectra of eucrites (Figures 6.2(a), (b) and 6.6). Furthermore, the average telescopic spectrum of Vesta from Burbine et al. (2017) shown in Figure 6.6 exhibits both spectral features. The concave shape at 0.7 μm in the telescopic spectrum is especially similar to the type-1 eucrite in the non-official terminology of Gaffey (1976). Modal mineralogy measurements indicate that these eucrites do contain low-Ca pyroxenes and only small

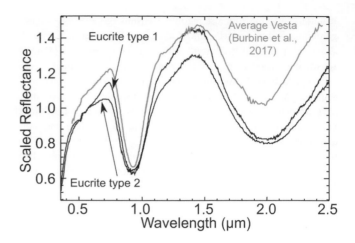

Figure 6.6 Scaled average reflectance spectra of eucrites (same as Figure 6.2(a) and (b) from Gaffey (1976) compared with the telescopic spectrum of Vesta acquired on 2009 November 20, with the Small Main-Belt Asteroid Spectroscopic Survey (SMASS) from Burbine et al. (2017).

abundances of olivine (Mittlefehldt, 2015), which support our hypothesis and the pyroxene-plagioclase compositions without olivine reported in Gaffey (1976).

It is worth noting that the disk-integrated spectrum from Burbine et al. (2017) implies that most of the Vesta's surface is similar to eucrite of type 1. This is a major difference with respect to VIR spectra, which have those spectral characteristics only in the region encompassing the three craters already mentioned. This emphasizes that the radiometric calibration of the instruments must be considered and discussed in the interpretation of those spectra, even after many years of efforts, justifying the revised analysis presented in this chapter.

Telescopic observations of near-Earth asteroid (237442) 1999 TA_{10} with visible near-infrared reflectance spectroscopy established a link with the Vestan family (Reddy et al., 2011). Spectra were acquired on 2010 May 11, using the SpeX instrument (Rayner et al., 2003) on the NASA Infrared Telescope Facility (IRTF) during the close approach. The spectrum exhibits the same absorption band at 1.2 μm that was found in the northern wall of crater Bellicia and in the NWA 5229 meteorite. Its composition was determined to be diogenitic, and thus to be an inner fragment of Vesta. Furthermore, using calibrations developed by Burbine et al. (2009), the calculation of the mean pyroxene chemistry resulted in $Fs_{27-35}Wo_{4-7}$, where Fs (ferrosilite; $FeSiO_3$) measures the iron content and Wo (wollastonite; $CaSiO_3$) measures the calcium content. The low-Ca, and relatively high Fe content is consistent with the detection of enstatite in VIR spectra of crater Bellicia and in the eucritic–basaltic meteorite NWA 5229. The interpretation of the spectrum of (237442) 1999 TA_{10} by Reddy et al. (2011) did not indicate a detection of olivine.

To summarize, on Vesta, a type-B diogenitic composition was detected with GRaND data on a region centered on crater Pomponia, at the antipode of the Rheasilvia basin (Beck et al., 2017). With near-infrared reflectance spectroscopy, low-Ca high-Fe pyroxenes and plagioclase were detected in the same northern region of Vesta (this study), and in basaltic eucrite meteorite NWA 5229 (Combe et al., 2011; Combe, in preparation). These three observations support the formation of type-B diogenite by melting of basaltic eucrites (Mittlefehldt & Lindstrom, 1993) locally in the northern regions of Vesta, possibly triggered by the Rheasilvia impact, and subsequently exposed by post-Rheasilvia impacts such as Pomponia, Bellicia, and Arruntia.

Olivine is not required to explain GRaND, VIR, or FC measurements of Vesta by the Dawn mission. This likely implies that olivine was not excavated in large enough quantities to be detected. The low abundances of olivine in HED meteorites are consistent with a Rheasilvia impact that likely did not reach the mantle of Vesta (see also Chapter 16).

6.3 EXOGENOUS MATERIALS ON VESTA

6.3.1 Low-Albedo, Hydrous Materials from Carbonaceous Chondrite Meteorites

6.3.1.1 Carbonaceous Chondrite in HED Meteorites

The first evidence of exogenous materials on Vesta came from the identification of carbonaceous chondrite fragments found in howardite meteorites (Zolensky & Barrett, 1992; Buchanan et al., 1993; Zolensky et al., 1996; Buchanan & Mittlefehldt, 2003). Another possible indication of a contamination was the detection of a partial absorption band at 2.8 μm, although truncated by absorptions from the Earth's atmosphere, attributed to hydrous materials (i.e., hydroxyl (OH)-bearing materials, Hasegawa et al., 2003, 2004) in both new telescopic spectra of Vesta and the re-analysis of a compilation of older spectra. The delivery of hydrous material by a low-velocity impact was the preferred hypothesis. This was later confirmed by Dawn observations, as detailed in Sections 6.3.1.2–6.3.1.4.

6.3.1.2 Low-Albedo Materials on Vesta Observed by the Dawn Mission

The Dawn mission (Russell & Raymond, 2011) orbited around Vesta from 2011 July 16 to 2012 September 5 (Russell et al., 2012), acquiring more observations supporting exogenous materials on the surface. It started with the discovery of extensive low-albedo regions (Reddy et al., 2012b), as illustrated in Figure 6.7(a). According to Palomba et al. (2014), about 60% of the low-albedo material units are associated with craters or ejecta from impacts and secondary impacts (Figure 6.8(a)). Other low-albedo units form areal and linear features (Figure 6.8(b)) such as flow-like deposits or rays (Figure 6.8(c)) commonly observed on topographic highs (Jaumann et al., 2014). These low-albedo material are characterized by weaker pyroxene absorption bands at 1 and 2 μm than the average surface of Vesta (Palomba et al., 2014). Optical properties of mixed material analyzed in the laboratory indicate that it only requires a few percent of highly absorbing materials mixed with semi-transparent materials to drastically lower the reflectance in a non-proportional way (e.g., Clark, 1983), which is likely the case on Vesta's pyroxene-rich surface (McCord et al., 2012): this makes quantitative interpretations challenging. Geomorphological analyses revealed that, on a global scale, the distribution of low-albedo materials is inhomogeneous on the surface and in the subsurface (Jaumann et al., 2014), and the different depths of the dark material layers (Figure 6.8(d)) may be due to subsequent re-exposure and re-burying by ejecta of multiple impacts. The observations of dark materials on Vesta were conducted to consider several hypotheses: (1) infall of low-albedo objects at low velocity such as carbonaceous chondrite impactors; (2) basalt flows, dikes, or sills on the surface or inside of Vesta that were broken and redistributed by impacts; (3) impact shock or impact melt from major cratering events; (4) pyroclasts from Vesta; and (5) opaque material in eucrites. The geological analysis from Dawn imagery (Reddy et al., 2012b) indicated an association with impact craters such as the ejecta blankets, the crater walls, and rims, which pointed to an exogenic origin from carbon-rich low-velocity impactors rather than endogenic.

6.3.1.3 Correlation between Low-Albedo and Hydroxylated Materials

Near-infrared reflectance spectroscopy revealed an absorption band at 2.8 μm (De Sanctis et al., 2012, 2013; McCord et al., 2012; Combe et al., 2015a) attributed to hydroxyl (OH), anti-correlated with albedo at a global scale (Figure 6.7(b)) and especially concentrated on low-albedo surface features such as craters and spots identified with high-resolutions images (McCord et al., 2012; Reddy et al., 2012b; Turrini et al., 2014), and correlated with the distribution of elemental hydrogen (Prettyman et al., 2012) shown in Figure 6.7(c). In particular, the correlation with OH absorption band is one indicator that rules out impact melt in favor of a carbonaceous chondritic origin (Palomba et al., 2014). The discovery of pitted terrains (Figure 6.8(e)) characteristic of degassing of volatile-bearing material heated by an impact (Denevi et al., 2012) provided independent evidence of hydrated materials, on Vesta, and an indication that some of the impactor materials were preserved locally. On pitted terrains, the abundance of OH or H_2O is a challenge to estimate because pit size is likely controlled by the thickness of the impact-heated deposit and not by the specific volatile content. According to Reddy et al. (2012a), the areas with the lowest albedo on Vesta indicate that carbonaceous material may comprise up to 60% of the regolith, implying that ~5 wt% H_2O may be present in these areas, which may represent an upper limit for Vesta. Overall, these new measurements converged to the theory of the contamination of an achondritic, HED-type of surface by low-albedo, hydroxylated carbonaceous chondritic (Reddy et al., 2012b) from low velocity impacts, and that the composition of those impactors played an important role in the geological evolution of Vesta (Denevi et al., 2012). In particular, the alteration of indigenous surfaces, the retention (Daly & Schultz, 2016), and mixing of the impactor material provides a mechanism by which to transport hydrated and organic-rich materials to Vesta and other bodies. Finally, the interpretation of serpentine from an absorption band centered at 0.72 μm in FC spectra (Nathues et al., 2014) was also reported to be consistent with carbonaceous chondritic materials, because Fe-phyllosilicates are commonly found in carbonaceous chondrite meteorites.

Hydroxyl-free, low-albedo materials exist on Vesta (Figure 6.15(a)). They may be due to carbonaceous chondritic material dehydrated by violent impact, as shock-processed (Palomba et al., 2014), but these must be marginal because the average velocity of impact on Vesta of 4.75 km s^{-1} (O'Brien & Sykes, 2011) is much lower than impacts on objects orbiting closer to the sun, for example, the median impact velocity on the Moon is 20 km s^{-1}.

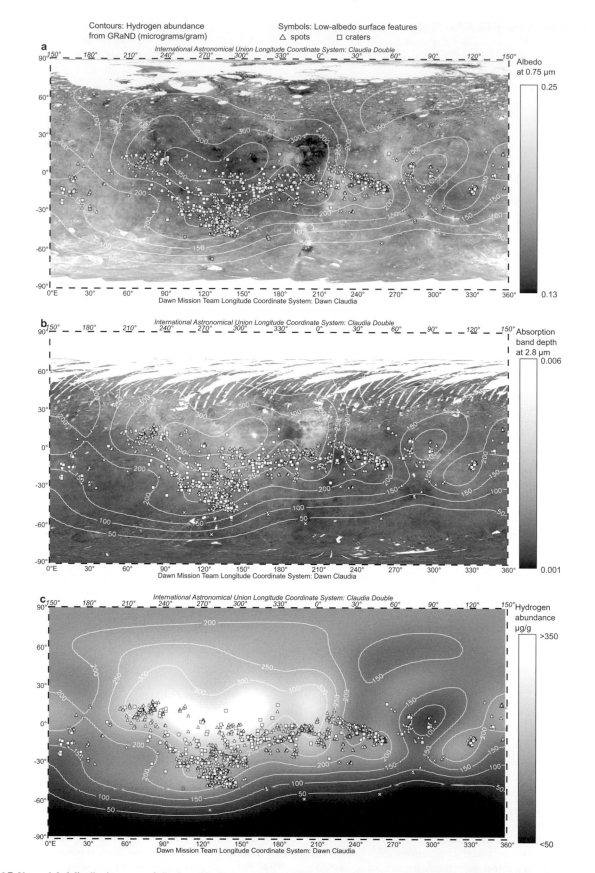

Figure 6.7 Vesta global distribution maps of albedo and hydrous materials. (a) Albedo at 0.75 μm from the Dawn Framing Camera (FC). (b) Absorption band depth at 2.8 μm from the Dawn Visible and InfraRed mapping spectrometer (VIR). (c) Elemental hydrogen abundance from the Gamma Ray and Neutron Detector (GRaND; Prettyman et al., 2012; Prettyman, 2014).

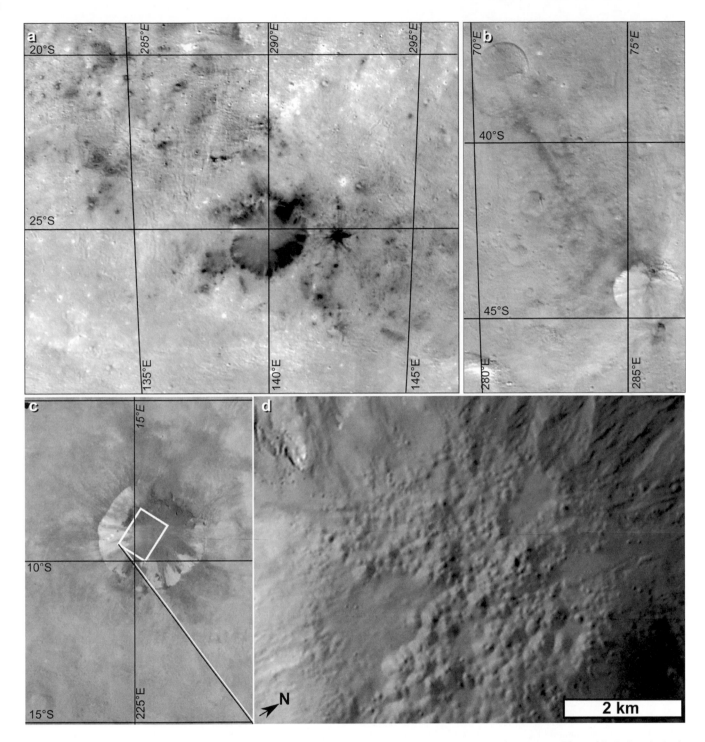

Figure 6.8 Examples of low-albedo surface features on Vesta from a photometrically-corrected global mosaic of FC images (Figure 6.7(a)). Longitudes in italic at the top of each panel are from the International Astronomical Union coordinate system, or Claudia Double Prime. Longitudes in normal font at the bottom of each panel are from the Dawn Mission Team coordinate system, or Dawn Claudia. All panels share the same scale and orientation, except (d). (a) Patches between craters Serena and Fulvia in Mollweide projections centered Fulvia to minimize the distortion. (b) Ejecta ray from crater Sossia (SE) to Urbania (NW) in Mollweide projections centered Sossia to minimize the distortion. (c) Crater Cornelia in equirectangular projection. The rectangle corresponds to the close-up view of pitted terrains in (d). (d) Pitted terrains inside of crater Cornelia from the non-photometrically-corrected FC image FC21B0025747 acquired at low altitude.

Impact melts may explain low-albedo, non-hydrous materials, as observed on HED meteorites (McSween et al., 2013) and on Vesta (Schenk et al., 2012), which is discussed more extensively in Sections 6.4.1 and 6.4.2. Another possible explanation is anhydrous metal rich material, which are found in mesosiderite meteorites (Scott et al., 2014) and were proposed to have originated from Vesta (Greenwood et al., 2006; Scott et al., 2014), although attempts to detect them spectrally with the VIR imaging spectrometer of the Dawn mission did not provide positive results (Palomba et al., 2012).

6.3.1.4 Implications of the Low-Albedo Materials on Vesta on Its Collisional History

The origin of the low-albedo material on Vesta has implications on the collisional history of Vesta and the source of hydrous material in the Solar System. The context of Vesta has to account for the Late Heavy Bombardment, which is a period that supposedly occurred after the main accretion phase, between 4.1 and 3.8 Gyr ago, and that was characterized by an increase in the number of impacts by large objects with the early terrestrial planets in the inner Solar System. This period is often taken as a reference for the evaluation of impact craters fluxes, although the Late Heavy Bombardment theory is still debated (e.g. Mann, 2018). By comparing models of impacts with the crater distribution on Vesta, Turrini et al. (2014) determined that the amounts of dark, hydroxylated materials that were delivered on Vesta before or across the Late Heavy Bombardment are sufficient to match the values estimated by the Dawn mission from surface observations only. In addition, the delivery of the dark material was associated with a continuous flux of impactors over several million years, and the size and impact frequency evolved over time. First, a few (one-to-three) large impact events may have supplied about 30% of the total budget of dark materials (Turrini et al., 2014). Then, once the asteroid belt assumed its present structure, about 300 low-albedo asteroids with diameters between 1 and 10 km could have impacted Vesta during the last 3.5 billion years (McCord et al., 2012). The total estimate of exogenous low-albedo material is about $(3–4) \times 10^{15}$ kg, which may correspond to a 1–2 m thick blanket globally (Turrini et al., 2014) representing $\sim 1.5 \times 10^{-3}$ wt% of the mass of Vesta. For comparison, according to Marchi et al. (2019), the fraction of accreted mass after Vesta's solidification is between 2 and 18 wt%, assuming a mixing depth of 4 km, whereas a modeled contamination of diogenite ranges from 0.4 to 4 wt%, and exogenous contamination measured in diogenites is up to 1–2 wt% (Day et al., 2012). Finally, sub-kilometric meteorites may have delivered the rest of the low-albedo materials, which continued until the present time in the form micrometeorites, with a diminishing mass delivered per impact. However, each impact contributed to the mixing of materials and the formation of the regolith (gardening), generating a several-kilometer thick low albedo layer in the crust (McCord et al., 2012).

6.3.2 The Ambiguous Case of Bellicia–Arruntia–Pomponia: Hypothesis of Exogenous Origin

Any report of exogenous material on Vesta that is not low-albedo and hydrous is an ambiguous case, and thus it is still the topic of debate at the time of this chapter publication. Assuming that certain areas on Vesta have similar composition to olivine-rich diogenites, the detection of olivine by visible and near-infrared spectroscopy in a pyroxene-rich mixture is only possible for olivine abundances >30% (Beck et al., 2013), which is a concentration that may not be represented in the regolith of Vesta due to mixing. Consequently, the regional correlation with hydroxyl-rich materials could be compatible with an exogenous origin.

The uniqueness of the broad and asymmetric 1 μm band implies a mineralogical composition whose origin is not from the pristine surface. Olivine was not expected in this region. Since olivine is supposed to be the main component of the mantle of Vesta

(dunitic), and thus very deep in the interior, it was expected to be found in the cavity, central peak, and ejecta of the two largest impact basins, Veneneia and Rheasilvia (respectively ~400 km and ~500 km diameter). Although these impacts may have excavated all the way through the crust of the South Polar Region, olivine was never detected there (McSween et al., 2013; Ammannito et al., 2013; Clenet et al., 2014). On the other hand, the impacts that formed Bellicia, Arruntia, and Pomponia craters (35 km, 12 km, and 63 km in diameter, respectively) did not excavate as deep inside the crust as Veneneia and Rheasilvia. This unexpected distribution of olivine and the low weight fraction estimated from the near-infrared spectra suggested that the surface typical of Bellicia, Arruntia, and Pomponia could be of exogenous origin (Le Corre et al., 2015; Nathues et al., 2015).

Further argument for exogenous olivine is theorized by Turrini et al. (2016). Potential carriers of olivine on Vesta's surface are A-type (Figure 6.9(a)) and S-type (Figure 6.9(b)) asteroids. The preservation of exogenous olivine on Vesta depends on the time of collision with respect to the Rheasilvia impact and the size of the impactors. A scenario of olivine contamination by only one type of asteroid yields that more than 25% and less than 2% of impacts by A-type and S-type asteroids, respectively, could have created detectable olivine outcrops. More likely, both types of asteroid populations had collisions with Vesta and provided exogenous materials. The young age of the outcrops detected in the Bellicia–Arruntia–Pomponia region indicate that limited removal or dilution by erosion or gardening occurred after their formation or excavation. The lack of outcrops with more than 50% olivine is compatible with no excavation of endogenous olivine and with the contamination by A-type or S-type asteroids.

A counterargument is provided by the comparison between spectra of several olivine-rich asteroids and the VIR spectra of the Bellicia–Arruntia–Pomponia region (Figure 6.9). The spectral data show noticeable differences in the shape of the absorption bands. The best match is obtained with S-type asteroid (15) Eunomia (Figure 6.9(b)), although the center of the 1 μm absorption band in VIR data is deeper, which is more characteristic of pyroxenes than olivine, as discussed in Section 6.2.4.

6.3.3 The Diffuse Surroundings of Craters Oppia and Octavia: Hypothesis of Exogenous Materials

6.3.3.1 Description of the Two Regions

Two diffuse areas near the equator at 290–330° E and 130–160° E, respectively, with their respective longitudes approximately 180° apart, have surfaces characterized by low albedo, a positive spectral slope, or reddening (Figure 6.10 and color Figure 6.15(b)), and slightly weaker pyroxene absorption bands than the background material (Figure 6.11(a)). These two areas have been identified by their distinctive orange appearance in color composites from FC images (Reddy et al., 2012b; Le Corre et al., 2013).

The most distinct of the two, south of crater Oppia, was interpreted by Reddy et al. (2012b) and Le Corre et al. (2013) as diffuse ejecta deposits from the impact that created Oppia on a pre-existing topographic slope, resulting in the asymmetric distribution around the crater.

A spectral comparison with the materials outside of the ejecta (Figure 6.11(a) and (b)) shows that this area has a lower albedo,

Olivine-rich asteroids compared with the Bellicia-Arruntia-Pomponia region of Vesta

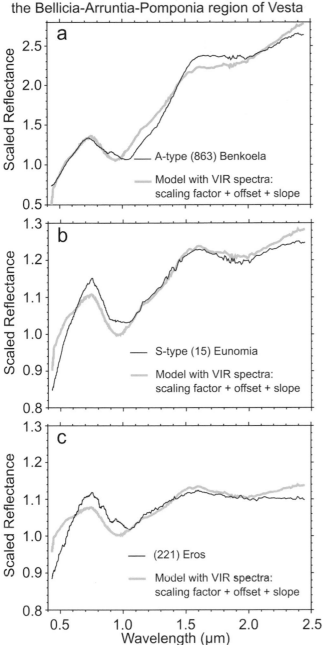

Figure 6.9 Spectra of olivine-rich asteroids (Burbine et al., 2017, in grey) compared with scaled and offset VIR spectra of the Bellicia–Arruntia–Pomponia region (black). (a) A-type asteroid (863) Benkoela has a broader and more complex absorption band at 1 μm, and a more positive spectral slope. (b) S-type asteroid (15) Eunomia has a flatter 1 μm absorption band at the center. (c) (221) Eros has a 1 μm absorption center shifted toward longer wavelengths, a weaker pyroxene absorption band at 2 μm, and a less positive spectral slope.

more positive spectral slopes in the visible spectra (0.4–0.8 μm) and in the near-infrared spectra (0.8–2.5 μm), and weaker pyroxene absorption bands.

The second area is located south of crater Octavia, and it shares a similar spectral character with the first area compared to its surrounding terrain, except with less contrast. The background

material is the main difference between the two craters, as Octavia is on the low-albedo hemisphere of Vesta, whereas Oppia is surrounded with terrains that have averagely high reflectance.

In this chapter, two hypotheses are explored: space weathering (Section 6.4.1) and the contamination by exogenous materials (this section).

6.3.3.2 Comparison with Other Regions of Vesta

The carbonaceous chondritic materials discussed in Section 6.3.1 show a lower albedo, a less positive spectral slope in the visible spectra, a more positive spectral slope in the near-infrared spectra, weaker pyroxene absorption bands, and a stronger hydroxyl absorption band at 2.8 μm than the average of Vesta (De Sanctis et al., 2012; McCord et al., 2012; Combe et al., 2015b). The positive visible spectral slope and the weak absorption band of hydroxyl of the reddish area south of Oppia do not behave like the most common low-albedo materials identified on Vesta as carbonaceous chondritic. However, CM2 type carbonaceous chrontites mixed in the HEDs may contribute to spectral reddening (Cloutis et al., 2013), unlike the other carbonaceous chrontites, and thus could be a plausible cause for the optical properties observed around Oppia and Octavia.

Besides the contamination by infalling meteorites, other hypotheses must be explored, since this area may undergone another process of transformation. Although photometric effects such as the particle phase function may change the albedo, the absorption band strength, and the spectral slope in the visible and near-infrared reflectance, they do not likely explain the relatively red areas considered in this section. Smaller grain sizes increase light scattering, which results in higher albedo and weaker absorption bands, which is opposite of the observations of weaker absorptions and lower albedo.

Besides carbonaceous chondritic materials, grain size variation, and surface irradiation, another hypothesis is impact melt. Although image interpretation revealed that the diffuse orange materials south of Oppia were ejecta and the orange patches west and north of Oppia were impact melt, it was not possible to identify differences from visible spectra alone (Le Corre et al., 2013). Figure 6.10(c) shows the spectral properties of impact melt patches identified in the west and the north of crater Oppia (Le Corre et al., 2013). The spectra of these patches compared to the background low-albedo ejecta from Oppia show a higher albedo, a more positive spectral slope in the visible, a less positive spectral slope in the near-infrared, identically strong pyroxene absorption bands, and a stronger absorption band of hydroxyl. All the spectral properties of the reddish area south of Oppia differ from the impact melt patches west and north of Oppia, except the positive spectral slope in the visible. Because of that, the diffuse ejecta south of Oppia are not likely to contain massive amounts of impact melt materials.

6.3.3.3 C-Type Asteroids as a Possible Source

The low-albedo and relatively strong 2.8 μm absorption (Figure 6.7(a) and (b), respectively) suggests carbonaceous chondrite impactors. In this context, to explain the positive spectral slope in the visible relative to the rest of Vesta, one must demonstrate the spectral diversity of C-type asteroids (Figure 6.12), which may have brought carbonaceous chondritic materials. Among the six

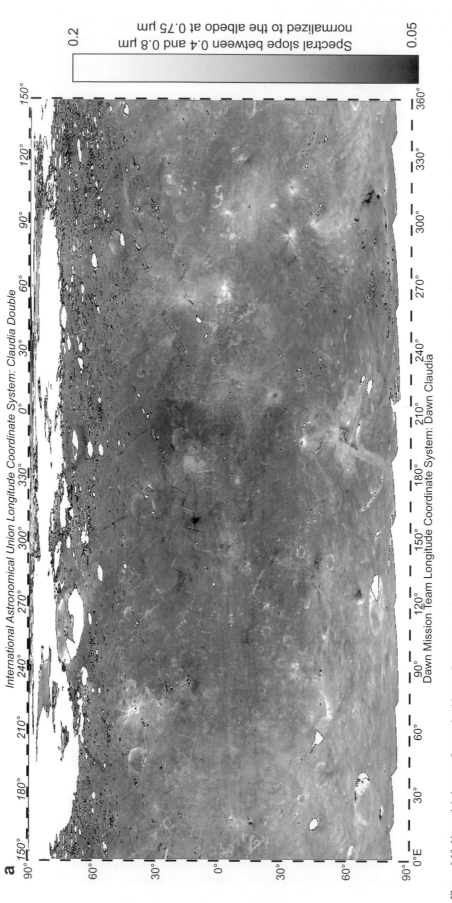

Figure 6.10 Vesta global maps of spectral evidence of space weathering due to solar wind irradiation of surface materials. The dashed ellipses indicate the regions where the visible spectral slope is more positive than average, and the two pyroxene absorption bands are weaker than average, which are two known spectral effects of surface irradiation. On Vesta, no other region exhibits the same characteristics simultaneously (Figure 6.15(b)). (a) Visible spectral slope calculated from FC images. (b) Average of the two absorption band areas characteristic of pyroxene from VIR. Weaker bands appear darker.

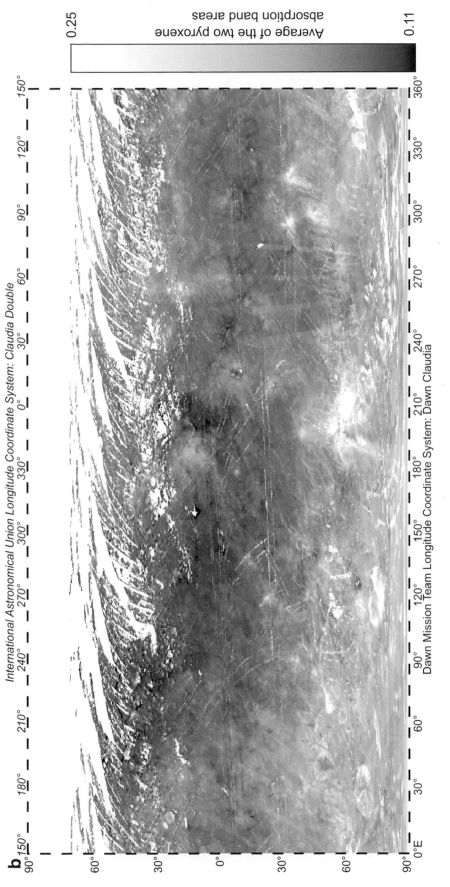

b

International Astronomical Union Longitude Coordinate System: Claudia Double

Dawn Mission Team Longitude Coordinate System: Dawn Claudia

Average of the two pyroxene absorption band areas

0.25

0.11

Figure 6.10 *(cont.)*

95

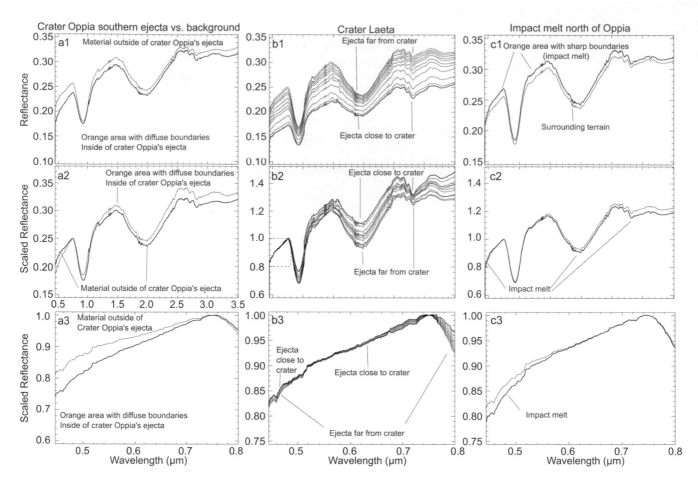

Figure 6.11 Spectral character of potential irradiated areas on Vesta. (a) Vestan orange area with diffuse boundaries south of crater Oppia versus material outside of Oppia ejecta. (b) Vestan carbonaceous chondrite materials from Laeta crater ejecta versus background material. (c) Vestan impact melt north of crater Oppia versus background material constituted of ejecta from crater Oppia.

examples, from Fornasier et al. (2011), the variability in visible spectral slope and spectral curvature in the range 0.4–2.5 μm may be sufficient for objects of this category to have spread materials that remained spectrally distinct after the impact. This was confirmed by more recent of observations in the main asteroid belt (Fornasier et al., 2014; Lantz et al., 2018), as well as in the interplanetary dust particles found in chondrite meteorites (Bradley et al., 1996). In addition, C-type asteroid (10) Hygiea exhibits a similar spectral variability across its surface (Busarev, 2011).

Space weathering is a possible cause for spectral reddening of C-type asteroids (Lantz et al., 2018). Furthermore, laboratory experiments of the Murchison meteorite by Matsuoka et al. (2020) also reveal changes in spectral curvature across the entire range, 0.4–2.5 μm, and weakening of the pyroxene absorption bands, which are observed on Vesta near Octavia and Oppia, and on C-type asteroids.

Alternative arguments exist, however. Although it is located in the low-albedo ejecta from Oppia, the distribution of the positive spectral slope is not clearly correlated with the albedo, as it extends further west and it does not overlap with the rim of Oppia, unlike the low-albedo materials. Consequently, an alternative explanation is needed. If the contamination of Vesta by space weathered C-type asteroids is possible, space weathering of carbonaceous chondritic materials may have occurred directly on Vesta, which is the focus of Section 6.4.1.

6.4 SURFICIAL PROCESSES

6.4.1 The Diffuse Surroundings of Craters Oppia and Octavia: Hypothesis of Space Weathering

These two areas were not studied specifically for the search of space weathering by the Dawn mission (Pieters et al., 2012; Blewett et al., 2016). Indeed, most of Vesta's surface appears pristine. Only a global analysis was able to reveal the two spectral anomalies that resemble the optical effects observed on irradiated samples (Figure 6.13(a)).

6.4.1.1 Spectral Comparison between Vesta and Spatially Weathered Surfaces

Most studies of space weathering are based on lunar samples; they indicate that, on the Moon, the dominant space weathering is caused by melt products such as spherules, agglutinates, and nanophase iron (Noble et al., 2001, 2005; Loeffler et al., 2008), which lowers the reflectance at all the wavelengths of the visible and near-infrared spectra, makes the spectral slope more positive (reddening), and weakens the absorption bands of the pristine mineral components (e.g., Pieters et al., 2000; Taylor et al., 2001). However, as reviewed in Reddy et al. (2013), the surface of asteroids including Vesta exhibit effects of space weathering that differ from the Moon

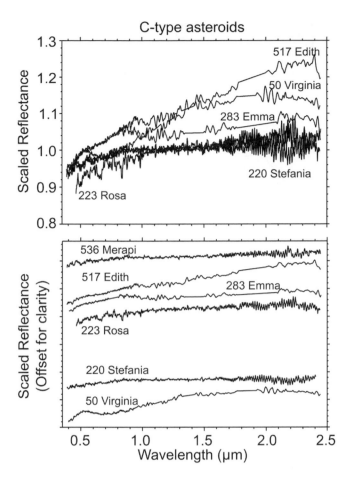

Figure 6.12 Reflectance spectra of C-type asteroids. Adapted from Fornasier et al., 2011).

(Noble et al., 2007; Gaffey, 2010; Pieters et al., 2012; Gaffey et al., 2015), and that arc distinct between one another. For example, space weathering on (243) Ida is associated with unchanged albedo, a decrease of absorption band depth, and an increase of the spectral slope (Veverka et al., 1996), whereas on (433) Eros, it is associated with a lower albedo and no noticeable change of the spectral slope and absorption band depth (Murchie et al., 2002). However, experiments conducted on olivine by Loeffler et al. (2009) concluded that the final effects of the space weathering on reflectance is not necessarily diagnostic of the mechanism, such as vapor deposition or irradiation, although their efficiency over time may be different. This implies that the various types of alterations observed optically may be determined by the surface composition.

The search of space weathering products in HED meteorites conducted by Noble et al. (2011) revealed that howardite Kapoeta contains evidence of melt products, such as spherules, agglutinates, and a possible nanophase bearing rim, suggesting that lunar style space weathering could be active on Vesta. This process normally dominates the spectral properties in a timescale of 10^4–10^6 years (Vernazza et al., 2006). In contrast, Vesta exhibits a ubiquitous high albedo (13–25%) at 0.75 μm and about twice as much (20–40%) at 2.5 μm, a moderately positive spectral slope, and strong pyroxene absorption bands of Fe^{2+} transitions at 1 and 2 μm, which all indicate that the surface is mostly pristine (Vernazza et al., 2006), without obvious effects of irradiation either

qualitatively (Pieters et al., 2012) or quantitatively from radiative transfer modeling (Blewett et al., 2016) assuming nanophase iron metal (npFe0) and optical constants from Cahill et al. (2012).

Irradiation is a likely source of space weathering on Main Belt asteroids (e.g., Brunetto et al., 2015). Irradiation of charged particles from the solar wind of an atmosphereless planetary body alters the surface materials. For example, irradiation of eucrite meteorite (Vernazza et al., 2006) produced obvious darkening, reddening, and reduction of spectral contrast (Figure 6.13(b)). The surface of Vesta is expected to behave similarly if it undergoes irradiation. We therefore investigated this hypothesis by searching FC and VIR reflectance data of Vesta that exhibit the aforementioned spectral characteristics (Figure 6.13(a)). For the same reason, other mafic-rich surfaces of atmosphereless bodies, such as the Moon (Figure 6.13(c)) and asteroid Itokawa (Figure 6.13(d)), are used as a demonstration. They show examples from the literature of natural mafic-rich surfaces where spectral features are similar in character but different in amplitude to the two regions of Vesta that we studied, and that were interpreted to be caused by the same phenomenon: irradiation by charged particles from the solar wind.

6.4.1.2 Implications from Possible Detection of Surface Irradiation on Vesta

The spectral character of the irradiated lunar surfaces has similarities with the spectra collected from the diffuse ejecta south of Oppia, except for the much-subdued contrast with respect to the surrounding terrains. Most of these spectral characteristics could be due to mixtures of low-albedo materials (likely chondritic) and impact melt, both being potentially present on the surface of Vesta at the time and location of the Oppia and Octavia impacts. However, there are differences that would be difficult to explain without irradiation: (1) The low abundance of hydroxyl in the low-albedo, diffuse ejecta of Oppia does not match the strong 2.8 μm absorption band observed in other carbonaceous chondritic materials on Vesta and in the impact melts; (2) The strong positive slope in the visible spectra on the diffuse ejecta of Oppia is the opposite of the less positive spectral slope observed on regions enriched in carbonaceous-chondritic materials (Pieters et al., 2012; Blewett et al., 2016); and (3) A more positive spectral slope at all the wavelengths in the infrared range is unique to irradiated surfaces, both in the laboratory and on the Moon.

The low spectral contrast between the candidate irradiated areas on Vesta and their surrounding terrains was also observed on S-type asteroid (25143) Itokawa, which has the composition of low total iron, low metal chondrite (Nakamura et al., 2011). Visible and near-infrared reflectance spectroscopy of (25143) Itokawa (Hiroi et al., 2006) showed weak, lunar-like space weathering effects (Figure 6.13(d)). Definitive evidence for solar irradiation was later interpreted from regolith samples collected by the Hayabusa mission that landed on Itokawa and returned to Earth. Laboratory analyses of these samples revealed that half the collected grains contained thin rims of nanophase iron (Noguchi et al., 2011), similar to the irradiated lunar samples (Pieters et al., 2012). The weak spectral effects of solar irradiation on Itokawa could be the indication that spectral reddening from nanophase iron may become detectable only if the surface is stable over geological times (Hiroi et al., 2006). Consequently, on Vesta, the nanophase iron-induced reddening could be masked by continuous stirring of the regolith

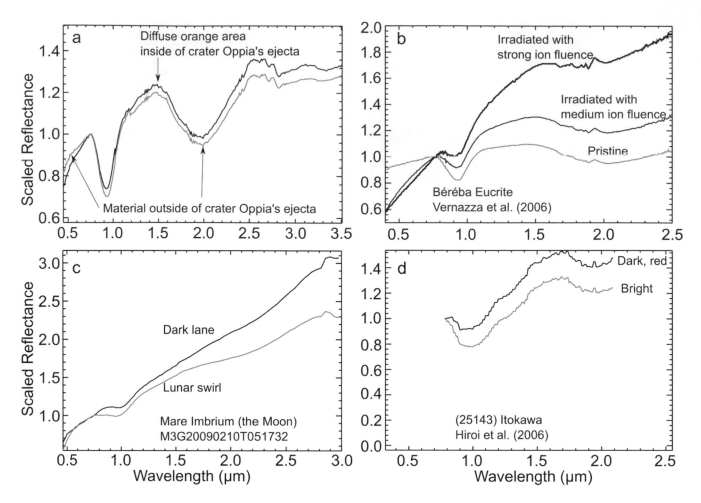

Figure 6.13 Effects of irradiation on reflectance spectra normalized at 0.75 μm. (a) Sample from the Béréba eucrite [12]. Irradiation was performed with ion fluences of 1.6×10^{15} Ar^{++}/cm^2 (medium) and 6.6×10^{15} Ar^{++}/cm^2 (strong). (b) VIR spectra of Vesta near crater Oppia. (c) Lunar mare basalt. (d) Bright and dark red areas of asteroid (25143) Itokawa by the Hayabusa mission (Hiroi et al., 2006).

caused by the high mobility of surface materials that is facilitated by steep slopes and rugged topography (Pieters et al., 2012).

Although our investigation on Vesta is not totally conclusive, it is interesting to note that the reddish area south of crater Octavia is near the antipode of the reddish region south of Oppia. Planetary magnetic fields that exist today are either global dipoles such as the Earth, Mercury, and the giant planets, or have a complex structure recorded as remanent crustal magnetic anomalies such as on the terrestrial oceanic floor, the lunar mare basalts, where the effects can be observed as the lunar swirls and dark lanes, or the crust of Mars (Spohn et al., 1998). All these observations across the Solar System indicate that the surface of a planetary body cannot be entirely shielded by a magnetic field because its polarization is not spatially uniform. For example, the poles create zones of convergence of the incoming charged particles onto the surface, resulting in enhanced weathering. On Vesta, the symmetry of the observed spectral features with respect to the center of the body may be consistent with the hypothesis of a dipolar magnetic field, where Oppia and Octavia could be close to the magnetic poles, and the reddish regions could be the only two areas on Vesta that undergo irradiation by charged particles, while the rest of the surface is shielded and does not experience any irradiation. Given that Vesta does not likely have a dynamo in its metallic core at

the present time, a remanent crustal magnetic field may still exist (Blewett et al., 2016).

The main weakness of the hypothesis of magnetic paleopoles on Vesta is that the global mapping of spectral properties by the Dawn mission is the only indication of possible effects of a remanent crustal magnetic field. No other observation or measurements is available to support this interpretation.

Furthermore, in the low albedo hemisphere of Vesta, the distribution of materials with a positive spectral slope is not limited to the immediate surroundings of Octavia: it forms an ellipse, best illustrated in the color Figure 6.15(b), that includes Lucaria Tholus and the craters Seresa, Octavia, Placia, Eutropia, and Publicia. This geometry might suggest impact-related origin.

6.4.2 Impact Melt

Impact melt was tentatively identified on Vesta by high-resolution color imagery from the Dawn's Framing Camera (Le Corre et al., 2013). Their most prominent optical characteristics are their strongly positive spectral slope in the visible range, their moderately high albedo, and their distribution in multiple patches (Reddy et al., 2012b) that are similar to impact melt identified on the Moon (Le Corre et al., 2013). They mostly occur in a region between

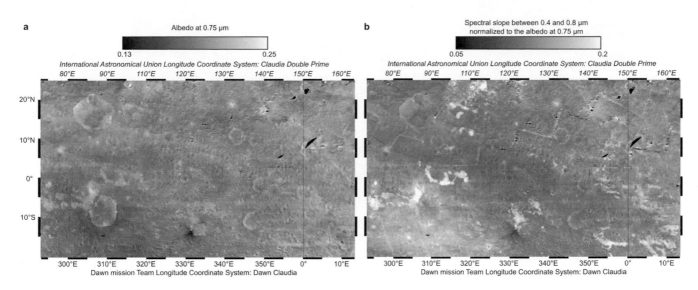

Figure 6.14 Regional maps of Vesta from the Dawn Framing Camera showing impact melt areas. (a) Albedo at 0.75 μm. (b) Spectral slope between 0.438 and 0.75 μm.

70° W and 15° E in the Dawn longitude system, and 20° S and 25° N (Figure 6.14) and are especially concentrated in the west and north of crater Oppia.

Some of these impact melts have a strong absorption band at 2.8 μm, that indicates a higher abundance of hydroxyl, which is unusual on high-albedo or moderately high-albedo areas on Vesta that are typically not rich in carbonaceous chondritic material (De Sanctis et al., 2012; McCord et al., 2012; Combe et al., 2015a; Tosi et al., 2015). Other spectral characteristics of those patches on Vesta include a moderately positive spectral slope in the near-infrared, and pyroxene absorption bands at 1 and 2 μm that have similar depth, position, and width than the surrounding surfaces (Figure 6.11(c)). Similarly, the contours and areas of those patches cannot be distinguished with high-resolution images (Le Corre et al., 2013), since the sharp boundaries appear only in the visible spectral slope and hydroxyl absorption band depth. This apparent lack of distinctive morphological, textural, and optical character of those units may be due to their age, consistent with the formation of Rheasilvia (Le Corre et al., 2013), and which predates the impact of Oppia. Over time, impact gardening from numerous, small meteoritic impacts, and lateral mixing from impact ejecta may have caused the old impact patches to gradually lose their original textural and spectral characteristics, except in the visible and occasionally in the range of the hydroxyl absorption band. Overall, the impact melt is observed on Vesta on a small number of locations, only in the visible spectral slope. Their low frequency of occurrences, limited distribution, and subtle optical characteristics are consistent with the fact that impact melt on Main Belt asteroids is expected to be very minimal due to the low impact speed (Marchi et al., 2013).

6.4.3 Lateral Mixing

Mixing of surface materials is the main contributor of space weathering on Vesta (Pieters et al., 2012). This is consistent with the fact that, in the visible and near-infrared, most of Vesta' surface reflectance spectra can be modeled by liner combinations of diogenite, eucrite, and carbonaceous chondrites (Combe et al., 2015b; Zambon et al., 2016). The type of composition in craters Bellicia

Arruntia and Pomponia is another spectral endmember that cannot be modeled by the other three (Combe et al., 2015b; Zambon et al., 2016) and that was initially reported as olivine (Ammannito et al., 2013; Ruesch et al., 2014; Palomba et al., 2015; Poulet et al., 2015; Zambon et al., 2016, cf. Sections 6.2.3 and 6.3.2) and that could also be interpreted as type-B diogenite (Beck et al., 2015, 2017) and/or basaltic eucrite (Section 6.2.4).

These observations imply that, on Vesta, the surface regolith is - well-mixed and has a composition derived of the host lithologies that include low and high albedo components (Pieters et al., 2012). Three main processes have been identified (Pieters et al., 2012): brecciation or fragmentation, regolith mobility, and fine-scale mixing.

Large and small-scale impacts produce particulate materials and redistribute the surface components. The largest impacts may excavate, expose fragment, and eject minerals that formed in depth, such as the lower crust or the mantle. Rheasilvia ejecta (De Sanctis et al., 2012; Reddy et al., 2012b; Combe et al., 2015b) is the most obvious example (Section 6.2.2), and Pomponia ejecta is also relevant (Section 6.3.2).

On Vesta, which has been enriched by exogenous carbonaceous chondritic material, regolith stirring and gravity-induced mobility cause the dispersion of micrometer-size opaque minerals. Small impacts stir the regolith continuously over extended periods of time. In addition, gravity facilitates the high mobility of the regolith in areas with steep slopes (Pieters et al., 2012). Both processes induce fine-scale mixing. The mixing of high-albedo pristine materials (Li et al., 2012) and low-albedo exogenous materials (De Sanctis et al., 2012; McCord et al., 2012; Palomba et al., 2012; Reddy et al., 2012a) is likely the source of the intermediate albedo materials that covers most of Vesta's surface (Reddy et al., 2013).

6.5 SUMMARY AND CONCLUSIONS

Vesta's surface composition provides insights on its internal structure, geological evolution, and space environment. The bulk igneous composition, the link to the HED meteorites that was first hypothesized in the 1970s, and the differentiation into a crust and

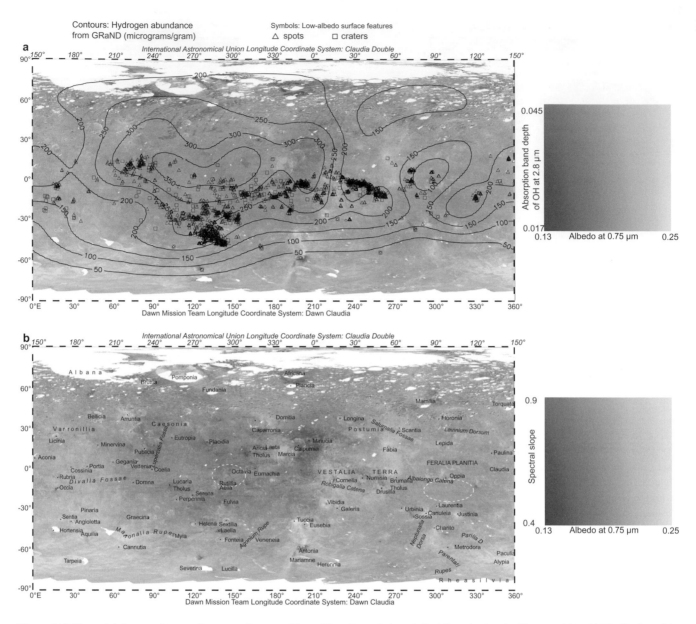

Figure 6.15 Vesta global maps of two surface contaminant candidates. The color scale is optimized for color-impaired human vision. (a) Distribution of the albedo at 0.75 μm and the absorption band depth at 2.8 μm of hydroxyl (OH). The colors from magenta to yellow represent all hydrous components; zones with subdued colors or gray are less or not hydroxylated. Low albedo hydroxylated materials appear in magenta. Low-albedo surfaces, low-albedo craters (Reddy et al., 2012b), and hydrous materials (De Sanctis et al., 2012; Turrini et al., 2014; Combe et al., 2015a) have correlated distributions on Vesta, consistent with contamination with carbonaceous chondrite meteorites. (b) Distribution of the albedo at 0.75 μm and the visible spectral slope between 0.4 and 0.8 μm. The colors from magenta to yellow represent all types of surfaces that exhibit an unusually strong positive visible spectral slope compared to the average of Vesta; areas with subdued colors or gray have a less positive spectral slope. Possible effects of surface irradiation by solar wind particles that combine low albedo and a positive spectral slope (e.g., Vernazza et al., 2006) appear in magenta, such as in the south of crater Oppia (10–20° S, 290–330° E) and the south of crater Octavia (0–10° S, 130–160° E) located with dashed-line ellipses.

A black and white version of this figure will appear in some formats. For the color version, refer to the plate section.

a mantle were confirmed multiple times by telescopic observations and by the Dawn mission. However, several key questions remain the subject of debates.

Understanding the structure and mineralogy of the crust and the thickness of the mantle has become dependent on identifying the nature of the outcrops in a region encompassing the craters Bellicia, Arruntia, and Pomponia. Exogenous olivine is almost ruled out because of the lack of potential olivine-bearing impactors in the present time in the main asteroid belt. Endogenous olivine seems

unlikely because the outcrop distribution is not associated with the largest impact basin Rheasilvia that excavated the deepest in the interior of Vesta. These mutually exclusive hypotheses and observations open the alternative hypothesis of low-Ca, high-Fe pyroxene in type-B diogenite, which could be compatible with both the mineralogy inferred from near-infrared reflectance and elemental abundances measured by gamma-ray spectroscopy. However, this alternative would imply a model of serial magmatism in a thick crust, with diogenite inclusions near the crust–mantle boundary and a relatively thin mantle compared with more standard models. If this is the case, the differentiation of Vesta would be distinct from that of major planets.

Other observations of the surface tend to prove more similarities with larger bodies. Possible optical reddening of the surface near the craters Oppia and Octavia, two equatorial regions near each other's antipode, might be due to space weathering by charged particles from the solar wind. Although the reflectance spectra do not provide a definitive conclusion, if this hypothesis is confirmed it would imply that the charged particles are deviated by a remanent crustal magnetic field, shielding most of Vesta's surface. This would explain why it looks mostly pristine, and the space weathering action concentrated around the magnetic dipole on the surface. This possible scenario would in turn imply that an internal dynamo once existed in the past, due to a convective metallic liquid core, which is more characteristic of larger bodies. Exogenous contamination by red material may not be ruled out, however, and the remarkable location of these two relatively red regions may be only coincidental.

Hydroxylated, carbonaceous–chondritic contamination on low albedo areas is one of the most certain conclusions from the Dawn mission and the analysis of HED meteorites. Hydroxyl infall is of particular interest because hydroxylated minerals, when abundant enough, contribute to a more complex petrological evolution and differentiation by triggering chemical reactions and rock alterations. The various phyllosilicates, carbonates, and organics observed on Ceres are examples of a geological evolution driven by hydroxylated and hydrated minerals in the outer asteroid belt and on a larger-sized body than Vesta (see Chapter 12). As a result, hydrostatic compensation was more efficient on Ceres, which was the reason for classification as a dwarf planet by the International Astronomical Union in 2006. The inner asteroid belt has smaller volatile amounts than the outer asteroid belt, and that may have played a key role in the evolution of Vesta.

Overall, the study of Vesta's surface composition revealed a wealth of information about the formation and evolution of Vesta, including the inner structure and its relationships to the HED meteorites. Among the questions that remain about the history and evolution of Vesta, several them could be overcome at least partially by remote sensing that was not investigated by the Dawn mission. The hypothetical remanent crustal magnetic field could be measured by magnetometry, as it was originally planned for the Dawn mission. In return, those measurements could help verify the possible effects of space weathering by solar irradiation. Other experiments from orbit could complement the observations already performed. For example, thermal infrared could help in discriminating between mineralogic phases of mafic minerals such as the presence of olivine versus low-Ca, high-Fe pyroxenes in craters Belicia, Pomponia, and Arruntia, or the physical characteristics of the regolith.

REFERENCES

Adams, J. B. (1974) Visible and near-infrared diffuse reflectance spectra of pyroxenes as applied to remote sensing of solid objects in the Solar System. *Journal of Geophysical Research* 79, 4829–4836.

Adams, J. B. (1975) Interpretation of visible and near-infrared diffuse reflectance spectra of pyroxenes and other rock-forming minerals. In C. Karr Jr. (ed.), *Infrared and Raman Spectroscopy of Lunar and Terrestrial Minerals*. New York: Academic Press, Inc., pp. 91–116.

Adams, J. B., & Goullaud, L. H. (1978) Plagioclase feldspars: Visible and near infrared diffuse reflectance spectra as applied to remote sensing. *Proceedings of the 9th Lunar and Planetary Science Conference*, March 13–17, Houston, TX, pp. 2901–2909.

Ammannito, E., De Sanctis, M. C., Combe, J.-P, et al. (2015) Compositional variations in the Vestan Rheasilvia basin. *Icarus*, 259, 194–202.

Ammannito, E., De Sanctis, M. C., Palomba, E., et al. (2013) Olivine in an unexpected location on Vesta's surface. *Nature* 504, 122–125.

Batista, S. F. A., Seixas, T. M., Salgueiro da Silva, M. A., & de Albuquerque, R. M. G. (2014) Mineralogy of V-type asteroids as a constraining tool of their past history. *Planetary and Space Science* 104, 295–309.

Beck, A. W., Lawrence, D. J., Peplowski, P. N., et al. (2015) Using HED meteorites to interpret neutron and gamma-ray data from asteroid 4 Vesta. *Meteoritics & Planetary Science* 50, 1311–1337.

Beck, A. W., Lawrence, D. J., Peplowski, P. N., et al. (2017) Igneous lithologies on asteroid (4) Vesta mapped using gamma-ray and neutron data. *Icarus* 286, 35–45.

Beck, A. W., McCoy, T. J., Sunshine, J. M., et al. (2013) Challenges in detecting olivine on the surface of 4 Vesta. *Meteoritics & Planetary Science* 48, 2155–2165.

Beck, A. W., & McSween, H. Y. (2010) Diogenites as polymict breccias composed of or- thopyroxenite and harzburgite. *Meteoritics & Planetary Science* 45, 850–872.

Bell, P. M., & Mao, H. K. (1973) Optical and chemical analysis of iron in Luna 20 plagioclase. *Geochimica et Cosmochimica Acta* 37, 755–758.

Blewett, D. T., Denevi, B. W., Le Corre, L., et al. (2016) Optical space weathering on Vesta: Radiative transfer models and Dawn observations. *Icarus* 265, 161–174.

Bradley, J. P., Keller, L. P., Brownlee, D. E., & Thomas, L. (1996) Reflectance spectroscopy of interplanetary dust particles. *Meteoritics & Planetary Science* 31, 394–402.

Brunetto, R., Loeffler, M. J., Nesvorný, D., et al. (2015) Asteroid surface alteration by space weathering processes. In P. Michel, F. E. DeMeo, & W. F. Bottke (eds.), *Asteroids IV*. Tucson: University of Arizona Press, pp. 597–616.

Buchanan, P. C., & Mittlefehldt, D. W. (2003) Lithic components in the paired howardites EET 87503 and EET 87513: Characterization of the regolith of 4 Vesta. *Antartic Meteorite Research* 16, 128–151.

Buchanan, P. C., Zolensky, M. E., & Reid, A. M. (1993) Carbonaceous chondrite clasts in the howardites Bholghati and EET87513. *Meteoritics* 28, 659–669.

Burbine, T. H., Buchanan, P. C., Dolkar, T., & Binzel, R. P. (2009) Pyroxene mineralogies of near-Earth vestoids. *Meteoritics & Planetary Science* 44, 1331–1341.

Burbine, T. H., DeMeo, F. E., Rivkin, A. S., & Reddy, V. (2017) Evidence for differentiation among asteroid families. In L. T. Elkins-Tanton, & B. P. Weiss (eds.), *Planetesimals: Early Differentiation and Consequences for Planets*. Cambridge: Cambridge University Press, pp. 298–320.

Burns, R. G. (1970) Crystal field spectra and evidence of cation ordering in olivine minerals. *American Mineralogist* 55, 1608–1632.

Burns, R. G. (1993) *Mineralogical Applications of Crystal Field Theory*. Cambridge: Cambridge University Press.

Busarev, V. V. (2011) Asteroids 10 Hygiea, 135 Hertha, and 196 Philomela: Heterogeneity of the material from the reflectance spectra. *Solar System Research* 45, 43–52.

Cahill, J. T. S., Blewett, D. T., Nguyen, N. V., et al. (2012) Determination of iron metal optical constants: Implications for ultraviolet, visible, and near-infrared remote sensing of airless bodies. *Geophysical Research Letters* 39, L10204.

Carrozzo, F. G., Raponi, A., Sanctis, M. C., et al. (2016) Artefacts removal in VIR/DAWN data. *Review of Scientific Instruments* 87, 124501.

Chapman, C. R., & Salisbury, J. W. (1973) Comparisons of meteorite and asteroid spectral reflectivities. *Icarus* 19, 507–552.

Cheek, L. C., & Pieters, C. M. (2014) Reflectance spectroscopy of plagioclase-dominated mineral mixtures: Implications for characterizing lunar anorthosites remotely. *American Mineralogist* 99, 1871–1892.

Clark, R. N. (1983) Spectral properties of mixtures of montmorillonite and dark grains – Implications for remote sensing minerals containing chemically and physically adsorbed water. *Journal of Geophysical Research* 88, 10635–10644.

Clark, R. N., Swayze, G. A., Gallagher, A., King, T. V. V., & Calvin, W. M. (1993) The US Geological Survey, Digital Spectral Library: Version 1: 0.2 to 3.0 μm, US Geological Survey, Open File Report 93-5922.

Clenet, H., Jutzi, M., Barrat, J.-A., et al. (2014) A deep crust-mantle boundary in the asteroid 4 Vesta. *Nature* 511, 303–306.

Cloutis, E. A., & Gaffey, M. J. (1991) Pyroxene spectroscopy revisited – Spectral–compositional correlations and relationship to geothermometry. *Journal of Geophysical Research* 96, 22,809–22,826.

Cloutis, E. A., Izawa, M. R. M., Pompilio, L., et al. (2013) Spectral reflectance properties of HED meteorites + CM2 carbonaceous chondrites: Comparison to HED grain size and compositional variations and implications for the nature of low-albedo features on Asteroid 4 Vesta. *Icarus* 223, 850–877.

Combe J.-Ph. (in preparation) Calcium-poor pyroxenes and plagioclase on the northern regions of Vesta: An alternative interpretation to olivine. *Planetary Science Journal.*

Combe J.-Ph., Ammannito, E., Tosi, F., et al. (2015a) Reflectance properties and hydrated material distribution on Vesta: Global investigation of variations and their relationship using improved calibration of Dawn VIR mapping spectrometer. *Icarus* 259, 21–38.

Combe, J.-Ph., Le Mouélic, S., Launeau, P., Irving, A. J., & McCord, T. B. (2011) Imaging spectrometry of meteorite samples relevant to Vesta and the Moon. *42nd Lunar and Planetary Science Conference*, March 7–11, The Woodlands, Texas. LPI Contribution No. 1608, p. 2449.

Combe, J.-Ph., Le Mouélic, S., Sotin, C., et al. (2008) Analysis of OMEGA/Mars Express data hyperspectral data using a Multiple-Endmember Linear Spectral Unmixing Model (MELSUM): Methodology and first results. *Planetary and Space Science* 56, 951–975.

Combe, J.-Ph., McCord, T. B., McFadden, L. A., et al. (2015b) Composition of the northern regions of Vesta analyzed by the Dawn mission. *Icarus* 259, 53–71.

Conel, J. E., & Nash, D. B. (1970) Spectral reflectance and albedo of Apollo 11 lunar samples: Effects of irradiation and vitrification and comparison with telescopic observations. *Proceedings of the Apollo 11 Lunar Science Conference* 3, 2013–2024.

Consolmagno, G. J. (1979) REE patterns versus the origin of the basaltic achondrites. *Asteroids and Icarus* 40, 522–530.

Consolmagno, G. J., & Drake, M. J. (1977) Composition and evolution of the eucrite parent body: Evidence from rare earth elements. *Geochimica et Cosmochimica Acta* 41, 1271–1282.

Dalton, J. B., & Pitman, K. M. (2012) Low temperature optical constants of some hydrated sulfates relevant to planetary surfaces. *Journal of Geophysical Research* 117.

Daly, R. T., & Schultz, P. H. (2016) Delivering a projectile component to the vestan regolith. *Icarus* 264, 9–19.

Day, J. M. D., Walker, R. J., Qing, L., et al. (2012) Late accretion as a natural consequence of planetary growth. *Nature Geoscience* 9, 614–617.

De Sanctis, M.-C., Ammannito, E., Capria, M.-T., et al. (2013) Vesta's mineralogical composition as revealed by the visible and infrared spectrometer on Dawn. *Meteoritics & Planetary Science* 48, 2166–2184.

De Sanctis, M. C., Combe, J.-P., Ammannito, E., et al. (2012) Detection of widespread hydrated materials on Vesta by the VIR imaging spectrometer on board the Dawn mission. *The Astrophysical Journal Letters* 758, L36.

De Sanctis, M. C., Coradini, A., Ammannito, E., et al. (2011) The VIR spectrometer. *Space Science Reviews* 163, 329–369.

Denevi, B. W., Blewett, D. T., Buczkowski, D. L., et al. (2012) Pitted terrain on Vesta and implications for the presence of volatiles. *Science* 338, 246.

Drake, M. J. (2001) The eucrite/Vesta story. *Meteoritics & Planetary Science* 36, 501–513.

Feierberg, M. H., Larson, H., Fink, U., & Smith, H. (1980) Spectroscopic evidence for at least two achondritic parent bodies. *Geochimica et Cosmochimica Acta* 44, 513–521.

Feierberg, M. H., Lebofsky, L. A., & Tholen, D. J. (1985) The nature of C-class asteroids from 3-μm spectrophotometry. *Icarus* 63, 183–191.

Fornasier, S., Lantz, C., Barucci, M. A., & Lazzarin, M. (2014) Aqueous alteration on Main Belt primitive asteroids: Results from visible spectroscopy. *Icarus* 233, 163–178.

Fornasier, S., Mottola, S., Barucci, M. A., et al. (2011) Photometric observations of asteroid 4 Vesta by the OSIRIS cameras onboard the Rosetta spacecraft. *Astronomy & Astrophysics* 533, 131–146.

Gaffey, M. J. (1976) Spectral reflectance characteristics of the meteorite classes. *Journal of Geophysical Research* 81, 905–920.

Gaffey, M. J. (1997) Surface lithologic heterogeneity of Asteroid 4 Vesta *Icarus* 127, 130–157.

Gaffey, M. J. (2010) Space weathering and the interpretation of asteroid reflectance spectra. *Icarus* 209, 564–574.

Gaffey, M. J., Reddy, V., Fieber-Beyer, S., & Cloutis, E. A. (2015) Asteroid (354) Eleonora: Plucking an odd duck. *Icarus* 250, 623–638.

Gounelle, M. J., Zolensky, M. E., Liou, J.-C., Bland, P. A., & Alard, O. (2003) Mineralogy of carbonaceous chondritic microclasts in howardites: Identification of C2 fossil micrometeorites. *Geochimica et Cosmochimica Acta* 67, 507–527.

Greenwood, R. C., Franchi, I. A., Jambon, A., Barrat, J. A., & Burbine, T. H. (2006) Oxygen isotope variation in stony-iron meteorites. *Science* 313, 1763–1765.

Hapke, B. (1981) Bidirectional reflectance spectroscopy: 1. Theory. *Journal of Geophysical Research* 86, 3039–3054.

Hasegawa, S., Hiroi, T., Ishiguro, M., et al. (2004) Spectroscopic observations of asteroid 4 Vesta from 1.9 to 3.5 microns: Evidence of hydrated and/or hydroxylated minerals. *35th Lunar and Planetary Science Conference*, March 15–19, League City, TX, abstract# 1458.

Hasegawa, S., Murakawa, K., Ishiguro, M., et al. (2003) Evidence of hydrated and/or hydroxylated minerals on the surface of asteroid 4 Vesta. *Geophysical Research Letters* 30, 2123.

Hazen, R. M., Bell, P. M., & Mao, H. K. (1978) Effects of compositional variation on absorption spectra of lunar pyroxenes. *Proceedings of the 9th Lunar and Planetary Science Conference*, March 13–17, Houston, TX, 3. (A79–39253 16-91) New York, Pergamon Press, Inc., pp. 2919–2934.

Hiroi, T., Abe, M., Kitazato, K., et al. (2006) Developing space weathering on the asteroid 25143 Itokawa. *Nature* 443, 56–58.

Hunt, G. R. (1977) Spectral signatures of particulate minerals in the visible and near infrared. *Geophysics* 42, 501–513.

Ikeda, Y., & Takeda, H. (1985) A model for the origin of basaltic achondrites based on the Yamato 7308 howardite. *Journal of Geophysical Research* 90, C649–C663.

Jaumann, R., Nass, A., Otto, K., et al. (2014) The geological nature of dark material on Vesta and implications for the subsurface structure. *Icarus* 240, 3–19.

Jutzi, M., Asphaug, E., Gillet, P., Barrat, J.-A., & Benz, W. (2013) The structure of the asteroid 4 Vesta as revealed by models of planet-scale collisions. *Nature* 494, 207–210.

Klima, R. L., Dyar, M. D., & Pieters, C. M. (2011) Near-infrared spectra of clinopyroxenes: Effects of calcium content and crystal structure. *Meteoritics & Planetary Science* 46, 379–395.

Klima, R. L., Pieters, C. M., & Dyar, M. D. (2007) Spectroscopy of synthetic Mg-Fe pyroxenes I: Spin-allowed and spin-forbidden crystal field bands in the visible and near-infrared. *Meteoritics & Planetary Science* 42, 235–253.

Klima, R. L., Pieters, C. M., & Dyar, M. D. (2008) Characterization of the 1.2 μm M1 pyroxene band: Extracting cooling history from near-IR spectra of pyroxenes and pyroxene-dominated rocks. *Meteoritics & Planetary Science* 43, 1591–1604.

Lantz, C., Binzel, R. P., & DeMeo, F. E. (2018) Space weathering trends on carbonaceous asteroids: A possible explanation for Bennu's blue slope? *Icarus* 302, 10–17.

Lapôtre, M. G. A., Ehlmann, B. L., & Minson, S. E. (2017) A probabilistic approach to remote compositional analysis of planetary surfaces. *Journal of Geophysical Research Planets* 122, 983–1009.

Larson H. P. (1977) Asteroid surface compositions from infrared spectroscopic observations: Results and prospects. In A. H. Delsemme (ed.), *Comet, Asteroids, Meteorites*Toledo, OH: University of Toledo Press, pp. 219–228.

Lawrence, D. J., Peplowski, P. N., Prettyman, T. H., et al. (2013) Constraints on Vesta's elemental composition: Fast neutron measurements by Dawn's gamma ray and neutron detector. *Meteoritics & Planetary Science* 48, 2271–2288.

Le Corre, L., Reddy, V., Sanchez, J. A., et al. (2015) Exploring exogenic sources for the olivine on Asteroid (4) Vesta. *Icarus* 258, 483–499.

Le Corre, L., Reddy, V., Schmedemann, N., et al. (2013) Olivine or impact melt: Nature of the "Orange" material on Vesta from Dawn. *Icarus* 226, 1568–1594.

Lebofsky, L. A. (1980) Infrared reflectance spectra of asteroids: A search for water of hydration. *Astronomical Journal* 85, 573–585.

Li, J.-Y., Mittlefehldt, D. W., Pieters, C. M., et al. (2012) Investigating the origin of bright materials on Vesta: Synthesis, conclusions, and implications. *Lunar and Planetary Science Conference*, 43, Abstract #2381.

Loeffler, M. J., Baragiola, R. A., & Murayama, M. (2008) Laboratory simulations of redeposition of impact ejecta on mineral surfaces. *Icarus* 196, 285–292.

Loeffler, M. J., Dukes, C. A., & Baragiola, R. A. (2009) Irradiation of olivine by 4 keV He$^+$: Simulation of space weathering by the solar wind. *Journal of Geophysics Research* 114, E03003.

Mann, A. (2018) Bashing holes in the tale of Earth's troubled youth. *Nature* 553, 393–395.

Marchi, S., Bottke, W. F., Cohen, B., et al. (2013) High-velocity collisions from the lunar cataclysm recorded in asteroidal meteorites. *Nature Geoscience* 6, 303–307.

Marchi, S., Raponi, A., Prettyman, T. H., et al. (2019) An aqueously altered carbon-rich Ceres. *Nature Astronomy* 3, 140–145.

Matsuoka, M., Nakamura, T., Hiroi, T., et al. (2020) Space weathering simulation with low-energy laser irradiation of Murchison CM chondrite for reproducing micrometeoroid bombardments on C-type asteroids. *The Astrophysical Journal Letters* 890, 1–12.

Mayne, R. G., McSween Jr., H. Y., McCoy, T. J., & Gale, A. (2009) Petrology of the unbrecciated eucrites, *Geochimica et Cosmochimica Acta* 73, 794–819.

McCord, T. B., Adams, J. B., & Johnson, T. V. (1970) Asteroid Vesta: Spectral reflectivity and compositional implications. *Science* 168, 1445–1447.

McCord, T. B., Li, J.-Y., Combe, J.-P., et al. (2012) Dark material on Vesta from the infall of carbonaceous volatile-rich material. *Nature* 491, 83–86.

McCord, T. B., & Scully, J. E. C. (2015) The composition of Vesta from the Dawn mission. *Icarus* 259, 1–9.

McFadden, L. A., McCord, T. B., & Pieters, C. M. (1977) Vesta: The first pyroxene band from new spectroscopic measurements. *Icarus* 31, 439–446.

McSween, H. Y., Ammannito, E., Reddy, V., et al. (2013) Composition of the Rheasilvia basin, a window into Vesta's interior. *Journal of Geophysical Research: Planets* 118, 335–346.

Mittlefehldt, D. W. (1994) The genesis of diogenites and HED parent body petrogenesis. *Geochimica et Cosmochimica Acta* 58, 1537–1552.

Mittlefehldt, D. W. (2015) Asteroid (4) Vesta: I. The howardite–eucrite–diogenite (HED) clan of meteorites. *Chemie der Erde* 75, 155–183.

Mittlefehldt, D. W., & Lindstrom, M. M. (1993) Geochemistry and petrology of a suite of ten Yamato HED meteorites. *Seventeenth Symposium on Antarctic Meteorites. Proceedings of the NIPR Symposium, No. 6*, August 19–21, 1992, National Institute of Polar Research, Tokyo, 268.

Murchie, S., Robinson, M., Clark, B. E., et al. (2002) Color variations on Eros from NEAR multispectral imaging. *Icarus* 155, 145–168.

Nakamura, T., Noguchi, T., Tanaka, M., et al. (2011) Itokawa dust particles: A direct link between S-type asteroids and ordinary chondrites. *Science* 333, 1113.

Nathues, A., Hoffmann, M., Cloutis, E. A., et al. (2014) Detection of serpentine in exogenic carbonaceous chondrite material on Vesta from Dawn FC data. *Icarus* 239, 222–237.

Nathues, A., Hoffmann, M., Schäfer, M., et al. (2015) Exogenic olivine on Vesta from Dawn Framing Camera color data. *Icarus* 258, 467–482.

Noble, S. K., Keller, L. P., & Pieters, C. M. (2011) Evidence of space weathering in regolith breccias II: Asteroidal regolith breccias. *Meteoritics & Planetary Science* 45, 2007–2015.

Noble, S. K., Pieters, C. M., & Keller, L. P. (2005) Evidence of space weathering in regolith breccias I: Lunar regolith breccias. *Meteoritics & Planetary Science* 40, 397–408.

Noble, S. K., Pieters, C. M., & Keller, L. P. (2007) An experimental approach to understanding the optical effects of space weathering. *Icarus* 192, 629–642.

Noble, S. K., Pieters, C. M., Taylor, L. A., et al. (2001) The optical properties of the finest fraction of lunar soil: Implications for space weathering. *Meteoritics & Planetary Science* 36, 31–42.

Noguchi, T., Nakamura, T., Kimura, M., et al. (2011) Incipient space weathering observed on the surface of Itokawa dust particles. *Science* 333, 1121.

O'Brien, D. P., & Sykes, M. V. (2011) The origin and evolution of the asteroid belt – Implications for Vesta and Ceres. *Space Science Reviews* 163, 41–61.

Palomba, E., Combe, J. P., McCord, T. B., et al. (2012) Composition and mineralogy of dark material deposits on Vesta. *43rd Lunar and Planetary Science Conference*, March 19–23, The Woodlands, TX. LPI Contribution No. 1659, id. 1930.

Palomba, E., Longobardo, A., De Sanctis, M., et al. (2014) Composition and mineralogy of dark material units on Vesta. *Icarus*, 240, 58–72.

Palomba, E., Longobardo, A., De Sanctis, M., et al. (2015) Detection of new olivine-rich locations on Vesta. *Icarus* 258, 120–134.

Peplowski, P. N., Lawrence, D. J., Prettyman, T. H., et al. (2013) Compositional variability on the surface of 4 Vesta revealed through GRaND measurements of high-energy gamma rays. *Meteoritics & Planetary Science* 48, 2252–2270.

Pieters, C. M. (1983) Strength of mineral absorption features in the transmitted component of near-infrared reflected light' first results from RELAB. *Journal of Geophysical Research* 88, 9534–9544.

Pieters, C. M., Ammannito, E., Blewett, D. T., et al. (2012) Distinctive space weathering on Vesta from regolith mixing processes. *Nature* 491, 79–82.

Pieters, C. M., Taylor, L. A., & Noble, S. K. (2000) Space weathering on airless bodies: Resolving a mystery with lunar samples. *Meteoritics & Planetary Science* 35, 1101–1107.

Poulet, F., Ruesch, O., Langevin, Y., & Hiesinger, H. (2015) Modal mineralogy of the urface of Vesta: Evidence for ubiquitous olivine and identification of meteorite analogue. *Icarus* 253, 364–377.

Prettyman, T. H. (2014) Dawn GRaND map of hydrogen on Vesta, data set DAWN-A-GRAND-5-VESTA-HYDROGEN-MAP-V1.0. NASA Planetary Data System.

Prettyman, T. H., Mittlefehldt, D. W., Yamashita, N., et al. (2012) Elemental mapping by Dawn reveals exogenic H in Vesta's regolith. *Science* 338, 242.

Prettyman, T. H., Mittlefehldt, D. W., Yamashita, N., et al. (2013) Neutron absorption constraints on the composition of 4 Vesta. *Meteoritics & Planetary Science* 48, 2211–2236.

Prettyman, T. H., Yamahita, Y., Reedy, R. C., et al. (2015) Concentrations of potassium and thorium within Vesta's regolith. *Icarus* 259, 39–52.

Prettyman, T. H., Yamashita, N., Ammannito, E., et al. (2019) Elemental composition and mineralogy of Vesta and Ceres: Distribution and origins of hydrogen-bearing species. *Icarus* 318, 42–55.

Raymond, C. A., Russell, C. T., & McSween, H. Y. (2017) Dawn at Vesta: Paradigms and paradoxes. In L. Elkins-Tanton, & B. Weiss (eds.), *Planetesimals – Differentiation and Consequences for Planets*. Cambridge: Cambridge University Press, pp. 321–340.

Rayner, J. T., Toomey, D. W., Onaka, P. M., et al. (2003) SpeX: A medium-resolution 0.8–5.5 micron spectrograph and imager for the NASA infrared telescope facility. *The Publications of the Astronomical Society of the Pacific* 115, 362–382.

Reddy, V., Le Corre, L., O'Brien, D. P., et al. (2012a) Delivery of dark material to Vesta via carbonaceous chondritic impacts [Erratum: 2013Icar..223.632R]. *Icarus* 221, 544–559.

Reddy, V., Li, J.-Y., Le Corre, L., et al. (2013) Comparing Dawn, Hubble Space Telescope, and ground-based interpretations of (4) Vesta. *Icarus* 226, 1103–1114.

Reddy, V., Nathues, A., & Gaffey, M. J. (2011) First fragment of Asteroid 4 Vesta's mantle detected. *Icarus* 212, 175–179.

Reddy, V., Nathues, A., Le Corre, L., et al. (2012b) Color ad albedo heterogeneity of Vesta from Dawn. *Science* 336, 700–704.

Righter, K., & Drake, M. J. (1997) A magma ocean on Vesta: Core formation and petrogenesis of eucrites and diogenites. *Meteoritics & Planetary Science* 32, 929–944.

Rivkin, A. S., McFadden, L. A., Binzel, R. P., & Sykes, M. (2006) Rotationally-resolved spectroscopy of Vesta I: 2 4 µm region. *Icarus* 180, 464–472.

Rousseau, B., Raponi, A., Ciarniello, M., et al. (2019) Correction of the VIR-visible data set from the Dawn mission. *Review of Scientific Instruments* 90, 123110.

Ruesch, O., Hiesinger, H., De Sanctis, M., et al. (2014) Detections and geologic context of local enrichments in olivine on Vesta with VIR/Dawn data. *Journal of Geophysical Research: Planets* 119, 2078–2108.

Russell, C. T., & Raymond, C. A. (2011) The Dawn mission to Vesta and Ceres. *Space Science Reviews* 163, 3–23.

Russell, C. T., Raymond, C. A., Coradini, A., et al. (2012) Dawn at Vesta: Testing the protoplanetary paradigm. *Science* 336, 684.

Ruzicka, A., Snyder, G. A., & Taylor, L. A. (1997) Vesta as the howardite, eucrite and diogenite parent body: Implications for the size of a core and for large-scale differentiation. *Meteoritics & Planetary Science* 32, 825–840.

Schenk, P., O'Brien, D. P., Marchi, S., et al. (2012) The geologically recent giant impact basins at Vesta's south pole. *Science* 336, 694–697.

Schröder, S. E., Mottola, S., & Keller, H. (2013) Resolved photometry of Vesta reveals physical properties of Crater Regolith. *Planetary and Space Science* 85, 198–213.

Scott, E. R. D., Bottke, W. F., Marchi, S., & Delaney, J. S. (2014) How did mesosiderites form and do they come from Vesta or a Vesta-like body? *45th Lunar and Planetary Science* Conference, March 17–21, The Woodlands, TX, 2260.

Shearer, C. K., Fowler, G. W., & Papike, J. J. (1997) Petrogenetic models for magmatism on the eucrite parent body: Evidence from orthopyroxene in diogenites. *Meteoritics & Planetary Science* 32, 877–889.

Spohn, T., Sohl, F., & Breuer, D. (1998) Mars. *The Astronomy and Astrophysics Review* 8, 181–235.

Sunshine, J. M., & Pieters, C. M. (1993) Estimating modal abundances from the spectra of natural and laboratory pyroxene mixtures using the modified Gaussian model. *Journal of Geophysical Research* 98, 9075–9087.

Sunshine, J. M., & Pieters, C. M. (1998) Determining the composition of olivine from reflectance spectroscopy. *Journal of Geophysical Research* 103, 13675–13688.

Sunshine, J. M., Pieters, C. M. & Pratt, S. F. (1990) Deconvolution of mineral absorption bands: An improved approach. *Journal of Geophysical Research* 95, 6955–6966.

Takeda, H. (1979) A layered-crust model of a Howardite parent body. *Icarus* 40, 455–470.

Takeda, H. (1997) Mineralogical records of early planetary processes on the HED parent body with reference to Vesta. *Meteoritics & Planetary Science* 32, 841–853.

Takeda, H., & Mori, H. (1985) The diogenite–eucrite links and the crystallization history of a crust of their parent body. *Journal of Geophysical Research* 90(Suppl.), C636–C648.

Taylor, L. A., Pieters, C. M., Keller, L. P., Morris, R. V., & McKay, D. S. (2001) Lunar mare soils: Space weathering and the major effects of surface-correlated nanophase Fe. *Journal of Geophysical Research* 106, 27985–28000.

Thangjam, G., Nathues, A., Mengel, K., et al. (2014) Olivine-rich exposures at Bellicia and Arruntia craters on (4) Vesta from Dawn FC. *Meteoritics & Planetary Science* 49, 1831–1850.

Thangjam, G., Nathues, A., Mengel, K., et al. (2016) Three-dimensional spectral analysis of compositional heterogeneity at Arruntia crater on (4) Vesta using Dawn FC. *Icarus* 267, 344–363.

Tosi, F., Frigeri, A., Combe, J.-P., et al. (2015) Mineralogical analysis of the Oppia quadrangle of asteroid (4) Vesta: Evidence for occurrence of moderate-reflectance hydrated minerals. *Icarus* 259, 129–149.

Turrini, D., Combe, J.-P., McCord, T. B., et al. (2014) The contamination of the surface of Vesta by impacts and the delivery of the dark material. *Icarus* 240, 86–102.

Turrini, D., Svetsov, V., Consolmagno, G., Sirono, S., & Pirani, S. (2016) Olivine on Vesta as exogenous contaminants brought by impacts: Constraints from modeling Vesta's collisional history and from impact simulations. *Icarus* 280, 328–339.

Veeder, G. J., Jonson, T. V., & Matson, D L. (1975) Narrowband spectrophotometry of Vesta (abstract). *Bulletin of the American Astronomical Society* 7, 377.

Vernazza, P., Brunetto, R., Strazzulla, G., et al. (2006) Asteroid colors: A novel tool for magnetic field detection? The case of Vesta. *Astronomy & Astrophysics* 451, 43–46.

Veverka, J., Helfenstein, P., Lee, P., et al. (1996) Ida and Dactyl: Spectral and color variations. *Icarus* 120, 66–76.

Warren, P. H. (1997) MgO-FeO mass balance constraints and a more detailed model for the relationship between eucrites and diogenites. *Meteoritics & Planetary Science* 32, 945–963.

Wasson, J. T. (2013) Vesta and extensively melted asteroid: Why HED meteorites are probably not from Vesta. *Earth and Planetary Science Letters* 381, 138–146.

Yamashita, N., Prettyman, T. H., Mittlefehldt, D. W., et al. (2013) Distribution of iron on Vesta. *Meteoritics & Planetary Science* 48, 2237–2251.

Zambon, F., De Sanctis, M., Schröder, S., et al. (2014) Spectral analysis of the bright materials on the asteroid Vesta. *Icarus* 240, 73–85.

Zambon, F., Tosi, F., Carli, C., et al. (2016) Lithologic variation within bright material on Vesta revealed by linear spectral unmixing. *Icarus* 272, 16–31.

Zolensky, M. E., & Barrett, R. (1992) Compositional variations of olivines and pyroxenes in chondritic interplanetary dust particles. *Meteoritics* 27, 312.

Zolensky, M. E., Weisberg, M. K., Buchanan, P. C., & Mittlefehldt, D. W. (1996) Mineralogy of carbonaceous chondrite clasts in HED achondrites and the Moon, *Meteoritics & Planetary Science* 31, 518–537.

Ceres' Surface Composition

MARIA CRISTINA DE SANCTIS AND ANDREA RAPONI

7.1 INTRODUCTION

Ceres is the largest body of the Main Belt and it has been recognized as an unusual "asteroid" due its characteristics. Before Dawn's arrival in 2015, our knowledge of Ceres' properties and surface composition was mainly derived from ground-based and space-based telescopic data (McCord & Castillo-Rogez, 2018, and references therein). The surface composition of Ceres has been a long-standing issue since the first ground based spectrophotometric observations (e.g., Lebofsky et al., 1981; King et al., 1992; Rivkin et al., 2006; Milliken & Rivkin, 2009) and its density, similar to those of the icy satellites, suggested that Ceres was composed of a large amount of water ice.

The first spectral measurements were performed in the early 1970s (Chapman et al., 1973), which led to the recognition that the surface was dark and spectrally neutral over the observed wavelength range, leaving uncertain whether Ceres had a basaltic or carbonaceous surface composition (Chapman & Salisbury, 1973). Measurement in the 0.4–2.5 µm wavelength region (Gaffey, 1976) led to a consensus that Ceres was similar in composition to carbonaceous chondrites. However, the spectrum of Ceres' between 2.9 and 4 µm shows a narrow absorption band centered near 3.06 µm and a set of overlapping bands at 3.3–3.4 µm and 3.8–3.9 µm. These bands have been attributed to different species, like ice frost (Lebofsky et al., 1981), NH_4-bearing minerals (King et al., 1992), irradiated organics and crystalline water ice (Vernazza et al., 2005), carbonates and cronstedtite (Rivkin et al., 2006), and brucite and carbonates (Milliken & Rivkin, 2009). Also, observations in the mid-infrared (IR) spectra have been interpreted in different ways, including the presence of dolomite, magnesite, brucite, and magnetite (Milliken & Rivkin, 2009; Rivkin et al., 2011).

An intriguing aspect is that no meteorite groups have been convincingly linked with Ceres because of spectral mismatches in the 3-µm region, unlike Vesta, which has a well-established connection with HED meteorites. However, Farinella et al. (1993) found that Ceres and Vesta should have similar efficiencies of fragment delivery. Such a discrepancy is confirmed by the discovery of objects in near-Earth space with Vesta-like spectra (Cruikshank et al., 1991; Binzel & Xu, 1993), and the opposite lack of known Ceres-like asteroids (Rivkin et al., 2011). This pointed to a substantial difference between the surface of the two bodies, Ceres being more similar to an icy body: a large impact would create a family with the non-ice fraction too unconsolidated to form coherent objects, and the icy family members approaching the inner Solar System would sublimate away (Rivkin et al., 2011). After the Dawn mission we found a possible confirmation of this picture: Ceres is a body rich in water and water-bearing species (Combe et al., 2016; Prettyman et al., 2017; Carrozzo et al., 2018;

De Sanctis et al., 2020), and there is evidence for aqueous alteration processes being present across its surface and in its interior, as discussed in Section 7.2 and Chapter 11.

The Dawn spacecraft was equipped with a scientific payload capable of distinguishing the various candidate materials and spatially resolve their respective distributions. Dawn's payload is constituted by the Visible and InfraRed mapping spectrometer (VIR, De Sanctis et al., 2011), two redundant Framing Cameras (FCs) with seven wavelength filters (Sierks et al., 2011), and a Gamma Ray and Neutron Detector (GRaND, Prettyman et al., 2011).

The three instruments were sensitive to different radiations, scales, and depths, as they were designed to identify specific mineralogical species and elements, and thus they were complementary in inferring the surface properties of Ceres.

The VIR imaging spectrometer, working in the spectral range 0.4–5 µm, with a spectral sampling of 2–10 nm, represented a breakthrough with respect to the ground-based observations, because of its possibility to obtain spectra of Ceres with unprecedented quality thanks to the huge dataset, without the limitations due to the atmospheric absorptions, and at a resolution from about a hundred meters to a few kilometers. VIR performed reflectance spectroscopy, which is a remote sensing technique to identify selected mineral species, and in some cases the abundance of the species can be quantified via the shape and depth of absorption bands in the spectrum of the outermost layer of the surface (up to several hundred micrometers) (e.g., Hapke, 2012). FC provided multicolor information in the visible range (0.44–0.98 µm) for most of the surface at a spatial resolution of a few tens of meters to a few hundred meters. In addition, FC was a key in the study of the surface photometry, because it provided global shape models (Roatsch et al., 2016) at a spatial resolution that enabled photometric corrections of VIR spectra. GRaND performed nuclear spectroscopy (Prettyman et al., 2011) and was sensitive to Gamma rays and neutrons produced by the interaction of cosmic rays with planetary surfaces, and gamma rays produced by the decay of natural radioelements (K, Th, and U). The intensity and energy distribution of the radiation can be analyzed to infer the concentration of specific elements in the subsurface, up to a few decimeters in depth.

7.2 CERES' AVERAGE SURFACE COMPOSITION

7.2.1 Mineralogy of the Surface

The average reflectance spectrum of Ceres acquired by VIR on Dawn from VIS and IR channels is reported in Figure 7.1, after

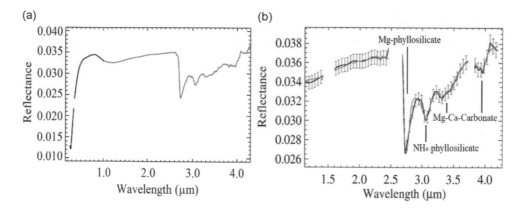

Figure 7.1 (a) Measured average spectrum of Ceres from VIS (black) and IR (gray) channels. (b) Average spectrum of Ceres in the IR channel with error bars of calibration uncertainties (black) and best fit (gray) as derived and modeled in Raponi et al. (2019a), including Mg-phyllosilicates, NH4-phyllosilicates, and Mg-Ca carbonate.

correction for the thermal emission that affects the long wavelengths, as in Raponi et al. (2019a). Both VIS and IR channels are affected by artificial slopes, which has been corrected by Carrozzo et al. (2016) and Rousseau et al. (2020) using a procedure that takes care of the ground-based observations. The ground based corrected VIR data set have been produced using the following steps:

1. Ground observations of Ceres, which are mutually consistent, have been collected (Chapman & Gaffey, 1979; Roettger & Buratti, 1994; Bus & Binzel, 2002a, 2002b; Parker et al., 2002; Lazzaro et al., 2006; Li et al., 2006; Rivkin et al., 2011).
2. Each ground full-disk observation (point 1) was converted in bidirectional reflectance at standard viewing geometry (incidence angle = 30°, emission angle = 0°, phase angle = 30°) by means of Hapke modeling (1993, 2012), according to the photometric parameters derived by Ciarniello et al. (2017)
3. A smooth average spectrum that covers the whole spectral range of the visible and IR channels of the VIR spectrometer has been calculated based on the ground-based spectra (point 2).
4. The average spectrum of VIR-VIS/IR calibrated data has been calculated, after artifact and photometric correction.
5. The ratio between the average spectrum from ground observations (point 3) with the average spectrum obtained from VIR data (point 4) has been calculated. This ratio spectrum is used as a multiplicative correction factor on every single VIR spectrum.

The large dataset produced by the Dawn mission confirmed it as a dark object. The geometric albedo comprised between 0.09 and 0.1 in the VIS (Ciarniello et al., 2017; Schröder et al., 2017; Li et al., 2019), characterized by a steep positive slope between 0.3 and 0.5 μm, and an almost neutral slope at visible wavelengths after 0.5 μm. A broad band centered at about 1.2 μm is visible in the near-infrared spectrum. This broad absorption was also reported in the spectra acquired from telescopic observations and interpreted as magnetite (Larson et al., 1979). This large band also appears in some carbonaceous chondrites (see e.g., Takir et al., 2013). Carbon and magnetite can account for the very low albedo of the cerean surface (De Sanctis et al., 2018; Marchi et al., 2019).

The spectrum beyond 2.6 μm shows several features within a large composite band extending from about 2.6 to 4.2 μm. The distinct absorptions appear at 2.72, 3.05–3.1, 3.3–3.5, and 3.95 μm. The most prominent is a strong and narrow absorption centered at

2.72–2.73 μm, which was not visible from the ground due to the telluric absorption.

The shape of the sharp absorption feature at 2.72–2.73 μm is typical of OH-bearing silicates; H_2O-bearing phases have a much broader absorption band that does not look like the Ceres 2.7 μm band. The position of the 2.7 μm feature suggests the presence of Mg-OH phases (Mg-serpentine or Mg-smectite). In fact, phyllosilicates OH-stretching vibrations, that occur in the 2.7–2.85 μm range, have band centers at different wavelengths for different phyllosilicate species (Farmer, 1974; King & Clark, 1989; Beran, 2002; Bishop et al., 2008).

NH_4-phyllosilicates (King et al., 1992; De Sanctis et al., 2015) were interpreted from an absorption band at about 3.06 μm. Nevertheless, the specific phyllosilicate bearing ammonium was not fully constrained by the observations, partly due to the limited availability of NH_4-phyllosilicate spectra in the literature (Krohn, 1987; Bishop et al., 2002; Berg et al., 2016; Ehlmann et al., 2018). Efforts to identify the ammonium-bearing phyllosilicate have been done recently. Only the expandable phyllosilicates, such as smectite, can incorporate the ammonium ion within their structure by contact with ammonium-rich fluids at relatively low temperature and pressure (Ferrari et al., 2019; De Angelis et al., 2021). In general, data from the smectite samples show that the precise center wavelength of the characteristic ~3.06 μm absorption is variable and is likely related to the degree of hydrogen bonding of NH_4–H_2O complexes. In particular Mg saponites have NH_4-related features that best match the shape and position of the 3.06-μm feature on Ceres (Ehlmann et al., 2018; see Chapter 9).

The band at 3.95 μm and, at least partially, the 3.4–3.5 μm band, have been attributed to carbonates (Rivkin et al., 2006; De Sanctis et al., 2015). Carbonates are characterized by several spectral features with band centers in different positions according to the cations present in the carbonate. By comparing the Ceres average spectrum with the carbonate's absorptions, the predominant carbonates on the surface are most probably dolomite or magnesite.

The dominant carbonate all over the surface is (Ca, Mg)-carbonate, but Na-carbonates (hydrous and anhydrous) have also been detected in various sites (Carrozzo et al., 2018) from spectral absorption at about 4 μm. In these specific areas, the central wavelength of the carbonate absorption is shifted at longer wavelengths (up to 4.1 μm) with respect to the average central wavelength. The most important exposure of Na-carbonate is in

Occator bright facula (De Sanctis et al., 2016) where an amount as large as 80% has been estimated in a few localized areas (De Sanctis et al., 2016; Raponi et al., 2019b). Together with the Na-carbonate in Occator, ammonium salts (NH_4Cl) have been identified thanks to the diagnostic absorption at about 2.2 μm (De Sanctis et al., 2016; Raponi et al., 2019b). Moreover, the high-resolution data over Occator bright material revealed the presence of hydrohalite ($NaCl\ H_2O$) and the hydrous form of halite ($NaCl$) (De Sanctis et al., 2020).

Although the broad signature between 3.3 and 3.6 μm is likely the result of overlapping absorptions due to carbonates and ammoniated phyllosilicates (Beran, 2002; Bishop et al., 2008), it must be noted that organics have a characteristic band within this spectral range (Moroz et al., 1998) and could contribute to the 3.3–3.6 μm cerean absorption (De Sanctis et al., 2017). Indeed, aliphatic organics have been detected in specific locations on the surface and has been estimated that Ceres's surface may contain up to 20 wt% of carbon (De Sanctis et al., 2017, 2019; Marchi et al., 2019).

Finally, apart from mineral phases, a few localized areas clearly show water ice exposures (Combe et al., 2016, 2019; Raponi et al., 2018), normally in shadowed sites or near that.

Thus, the surface mineralogy of Ceres is rather complex, showing different materials in different locations (that will be described in Sections 7.3 and 7.4), but all the identified species indicate the strong unavoidable role of water in the geochemical evolution of the dwarf planet.

7.2.2 Elemental Composition

The GRaND and VIR data jointly constrain characteristics of Ceres' surface composition and geochemical evolution. The data returned by GRaND suggest a high concentration of hydrogen in the Ceres' polar regions (29 wt.% equivalent H_2O) with a decreasing abundance going to lower latitudes (16% at equatorial regions) (Prettyman et al., 2017). The distribution has been interpreted as the presence of subsurface ice as foreseen by ice stability theory, which indicates water ice in the superficial regolith in the polar regions. Also, the concentration of iron follows a water-dilution trend when plotted as a function of regolith hydrogen content, consistent with the presence of subsurface water ice. The equatorial average [Fe] was determined to be 16 ± 1 wt % (Prettyman et al., 2017) and the estimated equatorial concentration for K is 410 ± 40 mg/g, a value that is in between the CI and CM average values (550 and 370 mg/g, respectively) (Prettyman et al., 2017).

The limited amount of Fe determined by GRaND poses an upper limit in the amount of magnetite inferred by VIR, suggesting that carbon is likely also present on the surface (Marchi et al., 2019) in order to explain the low surface albedo. The carbon concentrations are estimated by GRaND measurements to be greater than a few wt.%. A comparison with CC shows that the Ceres superficial carbon content is equal to or more than that found in CI chondrites (Chapter 8). Although the organic matter detected by VIR at Ernutet crater might be widespread, space weathering may have converted the aliphatic organics to graphitized carbon, whose presence was also suggested by UV measurements (Hendrix et al., 2016), that is almost spectrally neutral in the wavelength range explored by VIR.

7.3 REGIONAL DISTRIBUTIONS

The Dawn mission, orbiting Ceres, provided a new, larger, and more varied data set on the mineralogy, molecular and elemental composition, and their distributions in association with surface features and geology (McCord & Castillo-Rogez, 2018).

In particular, VIR spectral imaging capabilities (De Sanctis et al., 2011) determined the associations of mineralogy with surface morphology, linking geo-chemistry with geology. The first mineralogical maps were built on the data acquired during the first orbital phases (Ammannito et al., 2016) and showed the distribution of the phyllosilicates on Ceres (see Figure 7.2). The results indicate a remarkably uniform phyllosilicate composition across the surface. However, macro-regions can be outlined on the base of the variable depth of phyllosilicate absorption (see Figure 7.2). This can be interpreted in different ways, such as changes in abundances in the subsurface composition, local chemical and thermal reactions (e.g., dehydration), and changes with stratigraphy.

The Ceres surface was divided in quadrangles (McCord & Zambon, 2019, and references therein), in analogy with the geological mapping of the dwarf planet (Williams et al., 2018), and the mineralogy was mapped using specific spectral parameters (Frigeri et al., 2019). This methodology has been particularly useful to understand the origin and evolution of geological features present on the surface, associating those features with specific mineralogy, that is, Ahuna Mons (Zambon et al., 2017), Occator bright material (Longobardo et al., 2019a; Raponi et al., 2019b), Ernutet crater (De Sanctis et al., 2017, 2019; Raponi et al., 2019a), exposed water ice, especially in Oxo crater (Combe et al., 2019), and the bluest region in correspondence of Haulani crater (Tosi et al., 2019).

The carbonates distribution was studied by analyzing the 4-μm absorption (Carrozzo et al., 2018), revealing the presence of at least two different carbonates: (Mg,Ca)-carbonate and Na-carbonate (see Figure 7.3). The distribution of the carbonate indicates that the largest part of the Ceres surface is covered by (Mg, Ca)-carbonate, which is also the dominant carbonate species in the average spectrum. Only a few localized locations are rich in Na-carbonates, often associated with very bright material, such as in Occator or Oxo craters (Carrozzo et al., 2018).

7.4 LOCATIONS WITH SPECIFIC MINERALOGIES

7.4.1 Organic-Rich Area Near Crater Ernutet

7.4.1.1 Dawn's Observations

A region of $\sim 1,000$ km^2 close to crater Ernutet (53.4 km diameter, at latitude $\sim 53°$N, longitude 45.5°E), exhibits a prominent 3.4-μm band (Figure 7.4), distinct from the surrounding regions. The pronounced absorption in the 3.3–3.6-μm spectral range does not appreciably affect the remaining parts of the spectrum, with the exception of the spectral slope (Schröder et al., 2017). The main candidates for this band are materials containing C–H bonds, including a variety of organic materials, such as hydrogenated amorphous carbon and complex residues produced by the irradiation of different ices (Hoyle et al., 1982; Greenberg et al., 1995; Pendleton & Allamandola, 2002). The 3.4-μm band is composed of several

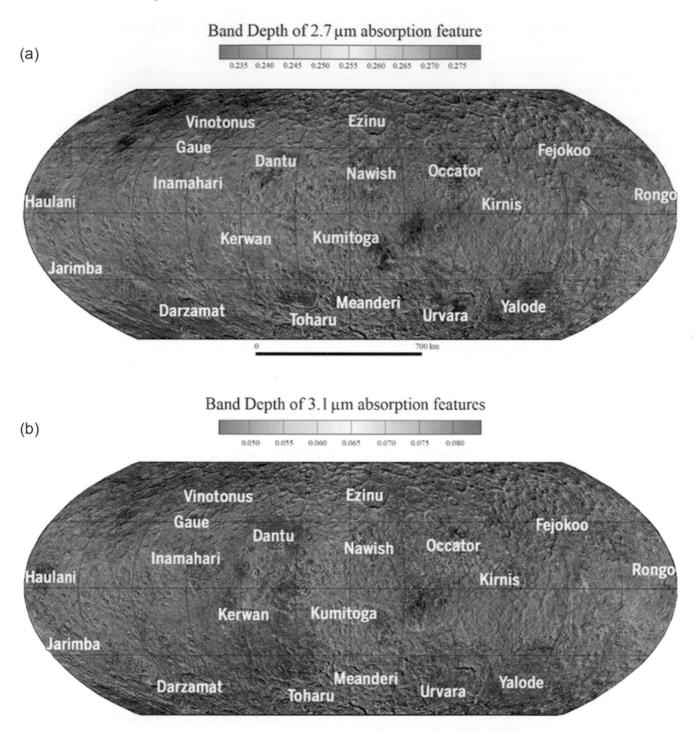

Figure 7.2 Band depth maps. Maps of the peak absorption depths of the (a) 2.7-μm OH-stretch band and (b) 3.1-μm NH4 band. The two maps have a similar global pattern, although they differ in some localized regions such as Urvara. The scale bar is true at the equator (Ammannito et al., 2016). A black and white version of this figure will appear in some formats. For the color version, refer to the plate section.

bands: near 3.38–3.39 μm, 3.40–3.42 μm, and 3.49–3.50 μm. These absorptions are characteristic of the symmetric and antisymmetric stretching frequencies of methyl (CH₃) and methylene (CH₂) functional groups, typical of aliphatic hydrocarbons (Moroz et al., 1998).

The distribution of organic material in the large Ernutet region was mapped using the band depth at 3.4 μm (De Sanctis et al., 2017; Raponi et al., 2019a). In Ernutet, the organic material is found mainly in two broad areas: one is a roughly triangular region in

the northwestern portion of the crater, and one is located on the southwestern part (Figure 7.4). In both cases, organic material is found inside and outside the crater, concentrated in small areas as well as in diffuse regions between the two main concentrations (De Sanctis et al., 2017; Raponi et al., 2019a). Moreover, several other small discrete or diffuse areas are found several tens of kilometers both west and east of Ernutet (Figure 7.4), and in a small area on the rim of Inamahari crater, hundreds of kilometers south of Ernutet.

Band Center of 4.0 µm absorption

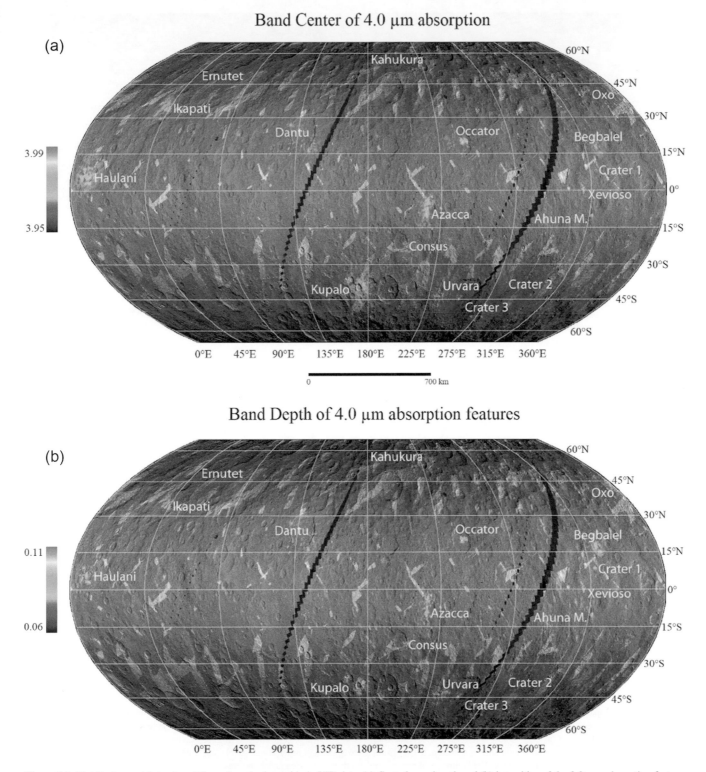

Band Depth of 4.0 µm absorption features

Figure 7.3 Distribution and intensity of the carbonate absorption in VIR data. (a) Central wavelength and (b) intensities of the 3.9-mm absorption feature. The maps are superimposed on the Framing Camera images using a transparency of 25% (Carrozzo et al., 2018).
A black and white version of this figure will appear in some formats. For the color version, refer to the plate section.

7.4.1.2 Analogs in the Solar System

The detection of aliphatic signature has been extensively reported in laboratory samples containing organic material (Orthous-Daunay et al., 2013; Takir et al., 2013). Absorption bands at these wavelengths have been observed remotely for only a few Solar System objects, such as comet 67P/Churyumov-Gerasimenko (Raponi et al., 2020). Other objects such as (24) Themis (Rivkin & Emery, 2010; Campins et al., 2010), (65) Cybele (Licandro et al., 2011), (52) Europa (Takir & Emery, 2012), and Himalia (Brown & Rhoden, 2014) show small absorptions in the same spectral

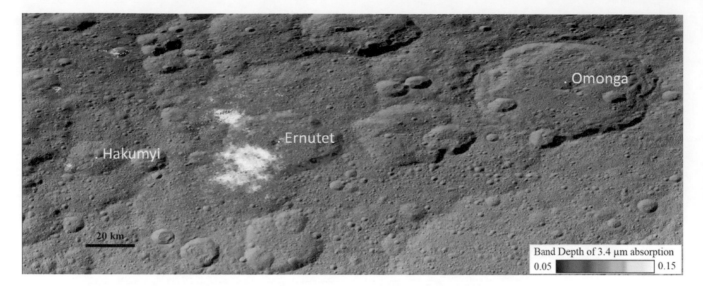

Figure 7.4 Geologic context of the Ernutet region on Ceres that exhibits organic material highlighted by the distribution of the 3.4-μm band depth. Band depths are calculated according to Ammannito et al. (2016) and cylindrically projected on Framing Camera (FC) clear filter mosaic (De Sanctis et al., 2019). A black and white version of this figure will appear in some formats. For the color version, refer to the plate section.

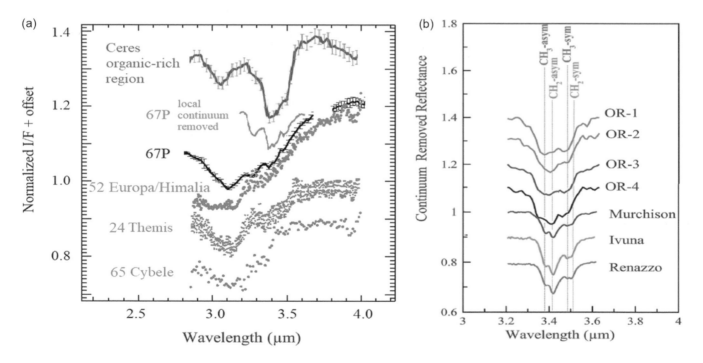

Figure 7.5 (a) Comparison between Ceres organic-rich region spectrum and other bodies of the Solar System. Arbitrary offsets were applied to the spectra. A clear signature of aliphatic organics has been detected on comet 67P, together with an absorption at 3.1 μm. The absorption features centered at 3.1 μm and 3.4 μm are not only observed on Ceres and 67P but also on several asteroids: (52) Europa, (24) Themis, and (65) Cybele. Himalia (not shown here) is an irregular satellite of Jupiter with a strikingly similar spectrum to (52) Europa. (b) Spectra of organic-rich pixels on Ceres surface (OR-1 to OR-4, 450 m/pixel) compared with methyl (CH3) and methylene (CH2) functional groups, and IOM in carbonaceous chondrites. The spectra are offset for clarity.

range, which can be attributed to organic material (Figure 7.5(a)). Notably, all of the abovementioned objects also show an absorption centered at ~3.1 μm, which, similarly to Ceres, in most cases has been attributed to N-H stretching (e.g., Poch et al., 2020, obtained a striking match of the 67P spectrum with laboratory mixtures containing ammonium salts).

A comparison with Insoluble Organic Matter (IOM) extracted from different carbonaceous chondrites shows an overall agreement between spectra of IOM and Ceres organic-rich regions (Figure 7.5(b)) (De Sanctis et al., 2017, 2019).

The non-observation of the CH aromatic stretching band in Ceres at ~3.3 μm spectra do not exclude the presence of aromatic compounds in Ceres OM but instead can give some clues to the cross-linking level of the carbon structure. The CH_2 and CH_3 aliphatic bands of the antisymmetric stretching modes (3.41 and 3.37 μm, respectively) have almost equal intensity, and do not fit with a polymer containing only CH_2 carbons (such as polyisoprene). This may indicate a high abundance of CH_3 groups in the organic matter of Ceres.

The shape of the spectral band between 3.2 and 3.6 μm and the well-defined band shape of the Ceres spectra seem to indicate the occurrence of mainly C–H atoms, and a depletion of oxygen.

Even if the exact composition of the organic matter remains an open question, Ceres' 3.4-μm band shows marked similarities with the organic bands of terrestrial hydrocarbons such as asphaltite and kerite, which are considered to be analogs for asteroidal and cometary organics (Moroz et al., 1998).

7.4.1.3 Quantitative Modeling

Using a nonlinear mixing algorithm (e.g., Raponi et al., 2019a) and considering an average observed spectrum of the organic-rich region, one can find a good match with an intimate mixture of ∼4 to 9% of aliphatic hydrocarbons. The specific value of the abundance of organics retrieved depends on the spectral endmember, that is, kerite or asphaltite (RELAB), used in the fitting procedure. Although the best fit is achieved with 5% of kerite, asphaltite also provides a good result (De Sanctis et al., 2017, 2019) (Figure 7.6). Considering as an end member a shocked asphaltite (AS-LXM-004, RELAB), the estimated abundance is much larger (25%) (De Sanctis et al., 2019). The retrieved abundance can be even larger (45–65%) considering measured samples of Insoluble Organic Matter extracted from Carbonaceous Chondrite (Kaplan et al., 2018): the less pronounced band at 3.4 μm of the laboratory samples can be compensated by a larger abundance retrieved in the spectral fit. However, the aliphatic absorption depth of the analog material used in the best fit cannot be less than the absorption depth measured on Ceres surface. Other types of aliphatic material with less pronounced bands at 3.4 μm, indicating a larger degree of aromaticity (Moroz et al., 1998), produce unsatisfactory fits, indicating that such materials are not suitable to reproduce the observed Ceres spectra (De Sanctis et al., 2019).

7.4.1.4 Other Components Associated with Organics

Using data with a spatial resolution of ∼400 m/pixel, Raponi et al. (2019a) obtained that the region with a higher concentration of organics present a lower amount of Mg-phyllosilicates (5–6%), but a larger amount of NH_4-phyllosilicates (up to 8–9%). A map of carbonates shows a spatial correlation with the larger/southern organic-rich region. From modeling, carbonates are up to 4.5% in the southern part of the floor. The smaller/northern organic-rich region does not present such a good correlation with carbonates. Other than Mg-carbonates, which are ubiquitous on Ceres surface, some spots in the area show the additional presence of Na-carbonates.

A detailed analysis of the spectra at high spatial resolution show some organic-rich spot with a very strong and uncommon 3.06-μm band, attributed to NH_4-phyllosilicates, and another band at 2.99–3.0 μm (De Sanctis et al., 2019). The carrier of this band could be attributed to an additional ammoniated mineral or some amine-bearing compounds.

7.4.1.5 Increased Stability of Organics Mixed with Other Minerals

The observed correlation between carbonate, ammonium-phyllosilicate, and aliphatic compounds might attest to a close relationship between organic and inorganic phases (Ammannito et al., 2016). In altered chondrites, a part of the organic material is observed as finely interspersed within phyllosilicates and carbonates (Le Guillou et al., 2014; Vinogradoff et al., 2017). A mixture with chemical links between organic and inorganic materials could improve the stability of the organic compounds at the surface of Ceres. This is important because irradiation processes (UV, cosmic particles) and space weathering are known to rapidly destroy unprotected organic compounds at the surface of airless bodies (in the span of a few million years) (Clark et al., 2002; Mennella et al., 2003; Pieters & Noble, 2016). In this respect, phyllosilicates can have a noticeable photoprotective effect on the organic compounds because they efficiently absorb organic molecules (Poch et al., 2015; dos Santos et al., 2016).

7.4.1.6 Origins of Organics

The origin of aliphatic organic material in specific locations on Ceres has been debated. While Pieters et al. (2018) suggested a possible exogenous delivery of the organic material on Ceres, Bowling et al. (2020) showed that such scenario is unlikely on the basis of impact modeling.

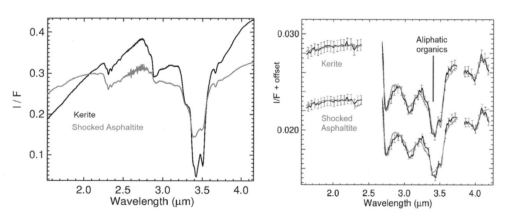

Figure 7.6 Left: Possible analogs of Ceres organic material from RELAB spectral database (De Sanctis et al., 2017). Right: Measured average spectrum of Ernutet organic-rich region with error bars of calibration uncertainties (black), and best fit (gray), as modeled in De Sanctis et al. (2019), including Mg-phyllosilicates, NH4-phyllosilicates, Mg-Ca-carbonate, and the organic materials shown in the left panel.

Analysis and modeling performed with VIR data are consistent with an endogenous scenario (De Sanctis et al., 2019). In fact, the intimate mixing and the correlation between the organic and inorganic phases, the latter being typical of the average Ceres surface, both imply that the two phases have evolved together in the early formation of Ceres. Moreover, the distribution of the organic in the Ernutet area would not be consistent with an exogenous release by meteoritic impact. In particular, this is confirmed by the high abundances of organics retrieved in specific spots, and the observation of some likely extruded materials, not directly linked with the region of Ernutet crater.

Focusing on an endogenous origin of the organic matter in Ceres, De Sanctis et al. (2019) proposed a process for their formation and evolution, based on Ceres' internal evolution (Castillo Rogez et al., 2018; McCord & Castillo-Rogez, 2018): the pervasive hydrothermal alteration of Ceres might have created an environment in the interior conducive to producing organic compounds (Holm et al., 2015). Serpentinization processes are likely to have proceeded inside Ceres (Castillo-Rogez et al., 2018), producing serpentine and other hydrated minerals observed at the surface (De Sanctis et al., 2015; Ammannito et al., 2016) and in combination with another geochemical mechanism, the Fischer-Tropsch-Type process (FTT) (McCollom & Seewald, 2007), hydrocarbons can be formed. At the Ceres scale, FTT reactions may have been very efficient because Ceres is massive enough to have retained H_2 from serpentinization for reactions with CO (Castillo-Rogez et al., 2018). See Chapter 8 for more details.

7.4.2 Carbonate-Rich Area: Cerealia Facula

7.4.2.1 Albedo and Morphology

Several areas with high-albedo material are observed on Ceres surface where significant differences in spectral parameters have been detected, such as slopes, albedo, band depths, and the band center of specific absorption features (Palomba et al., 2019; Stein et al., 2019). The most evident albedo features that stand out from the surrounding terrains are the bright areas, called "Ceralia Facula" and "Vinalia Faculae" (Figure 7.7) in the Occator crater

(15.8–24.9°N and 234.3–244.7°E), a prominent, geologically young crater (diameter 92 km). The Faculae have an albedo 5–10-times higher than the average surface (Li et al., 2016; Ciarniello et al., 2017; Schröder et al., 2017; Longobardo et al., 2019a, 2019b), being brightest (Cerealia Facula) at the center of the crater, including a 10-km-wide central pit (Buczkowski et al., 2016). Dawn's Framing Camera has observed the Cerealia Facula up to a resolution as high as 3.3 m/pixel, revealing a dome (Cerealia Tholus) in the center of the pit that rises 0.4 km above the surrounding terrain (Nathues et al., 2017). Moreover, the faculae appear to be associated with fractures in Occator's floor (Buczkowski et al., 2016). An isolated very bright topographic high (Pasola Facula) stands out on the dark crater floor, close to the tholus. The spectra of the Occator floor, as seen by VIR onboard Dawn, have been photometrically corrected to standard geometry (Ciarniello et al., 2017). The spectral analysis indicates that the reflectance of most of Occator's dark floor (0.03 at 0.55 μm and at 2 μm) outside the faculae is similar to that of regions around the crater. The highest reflectance over Ceres surface is measured in the Cerealia Facula, with a value of 0.26 at 0.55 μm and 0.28 at 2 μm. A transition zone with a decreasing level of reflectance from the bright areas to the surroundings can be recognized, associated with a variation in composition (Figure 7.8).

7.4.2.2 Absorption Bands and Mineral Compositions

The brightest areas show clear spectral differences from the typical crater floor (Figure 7.8). Absorptions near 3.4 μm and ~4 μm increase in the brightest areas and are consistent with enrichments in carbonates. Notably, a shift of the carbonate band center from 3.95 to 4.01 μm has been observed for the brightest regions. De Sanctis et al. (2016) have shown that a changing mineralogy from Mg-carbonate to Na-Carbonate provides the best match with the observed shift.

A complex absorption at 2.20–2.22 μm is observed only in the brightest areas (Figure 7.8). This absorption can be due to a narrow hydroxide (OH) stretching and bending combination vibration in Al-phyllosilicates (Bishop et al., 2008), or ammoniated minerals and salts (De Sanctis et al., 2016), that also have bands at around

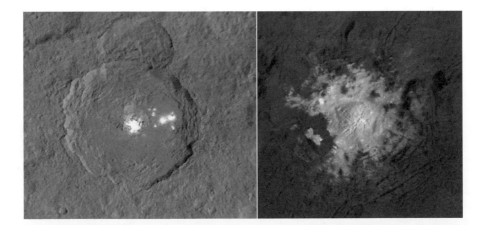

Figure 7.7 Left panel. Framing Camera mosaic (35 m/pixel) obtained with a clear filter during the Low Altitude Mapping Orbit (LAMO) (Roatsch et al., 2016). Right panel: Cerealia Facula obtained by combining Framing Camera images acquired during LAMO phase with three images using spectral filters centered at 438, 550, and 965 nm, during the High-Altitude Mapping Orbit (HAMO) phase.
A black and white version of this figure will appear in some formats. For the color version, refer to the plate section.

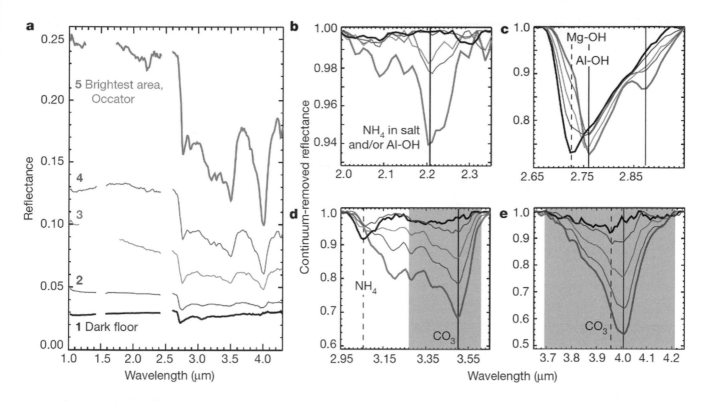

Figure 7.8 Spectra of bright and dark areas in Occator. (a) Reflectance spectra compared from the brightest (top) to darker regions (bottom). Gaps correspond to removed instrumental artefacts or saturated channels. (b–e) Continuum-removed spectra from selected wavelength regions. A 2.2-μm band depth is visible in the spectra of the brightest pixels. The 2.7-μm absorption shifts longward and an absorption at 2.87 μm appears, going from the crater floor to the brightest pixels. The 3.06-μm band depth absorption weakens, while the 3.4-μm and 3.9-μm absorptions strengthen in the brightest areas. The shaded region indicates the band positions and widths for anhydrous carbonates. Dashed lines indicate absorption positions in dark materials and solid lines indicate absorption positions in bright materials (De Sanctis et al., 2016).

2.21 μm (Figure 7.8). Raponi et al. (2019b) have shown that ammonium salts, and more specifically ammonium chloride (NH_4Cl), provide the best match with the observed spectra.

The overall band area between 2.6 and 3.7 μm increases going from the crater floor to the brightest pixels, and several minima at 2.88, 3.2, 3.28, 3.38, and 3.49 μm emerge and become deeper (Figure 7.8).

The 2.7 μm absorption of magnesium (Mg) phyllosilicates typical of Ceres' surface (De Sanctis et al., 2015) shifts from 2.72 to 2.76 μm (Figure 7.8). The 3.06-μm absorption, attributed to ammonia-bearing species (De Sanctis et al., 2015), is clearly present on the crater floor but becomes less evident in brighter terrains and is absent in the brightest areas (Figure 7.8).

Notably, significant heterogeneities of spectral parameters can be observed within the brightest regions (Raponi et al., 2019b; De Sanctis et al., 2020): the regions closer to the dome peak show a stronger 2.2 μm band that does not correspond to a similar increase in the carbonate band depth. Similarly, the band intensities at 4, 3.2, and 3.28 μm do not correlate linearly (Figure 7.9). Oppositely, the bands at 2.2, 3.2, and 3.28 μm show a clear correlation (Figure 7.9), indicating that either the species carrier of the bands is the same or that these bands are due to different species linked to each other. The use of spectral ratios (De Sanctis et al., 2020) highlight the spectral differences associated to the small absorption at 3.2 and 3.28 μm (Figure 7.10): the ratio between areas close to the peak and the flank of Cerealia Facula shows the band at 2.2 μm as well as a large and structured band between 2.76 and 3.7 μm

demonstrating that part of the facula contains a different proportion of species likely associated with the ammonium-salt. Moreover, the overall shape of the spectral ratio indicates the probable hydration. Such a ratio shows a strong similarity with laboratory mixture of sodium bearing species, in particular hydrohalite ($NaCl \cdot 2H_2O$), formed from the freezing of fluids containing sodium, ammonium, carbonate, and chloride ions obtained by Vu et al. (2017).

To obtain information on the abundances of the minerals making up the surface of the Faculae, De Sanctis et al. (2016), Raponi et al. (2019b), and De Sanctis et al. (2020) used a quantitative spectral analysis of the composition using Hapke's radiative transfer model (Hapke, 1993, 2012).

The best fit of the models with the VIR spectra indicate that the sodium carbonate abundance is of the order of 50–80%, making the central bright area in Occator the most concentrated known occurrence of km-scale carbonates beyond the Earth. Other components, which produce clear spectral features, are ammonium chloride, Al phyllosilicates (such as illite), and hydrohalite (De Sanctis et al., 2020).

7.4.2.3 Formation of the Faculae

The distribution of sodium carbonates, hydrohalite, and the other components has been mapped using the abundances retrieved from the spectral modeling of the high-resolution data on Cerealia Facula (De Sanctis et al., 2020). The spectral maps indicate that both

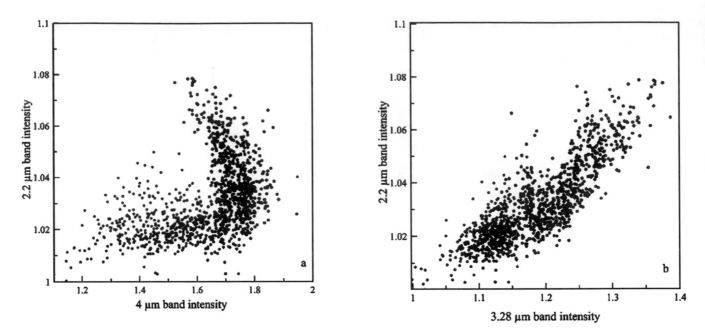

Figure 7.9 Scatter plots of the band intensity. Left panel: plot of 2.2 μm versus 4 μm bands; right panel: plot of 2.2 μm versus 3.28 μm band. The band intensity has been calculated using the band ratios between the band centers and the left shoulders as a proxy of the intensity (De Sanctis et al., 2020).

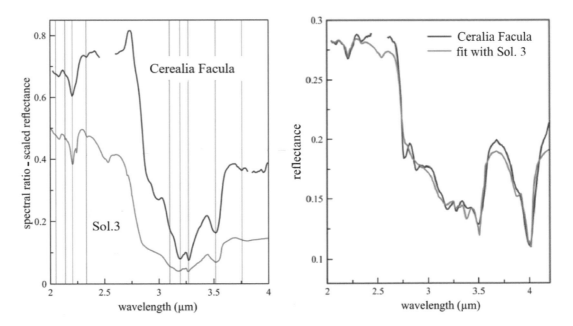

Figure 7.10 Left panel: Cerealia spectral ratio (black) in comparison with laboratory spectra of brines (gray), labeled Sol. 3, with common features highlighted. Right panel: spectral fits of the pixels close to the dome center using Na-carbonate, NH$_4$Cl, illite, and Sol. 3 (De Sanctis et al., 2020).

sodium carbonate and hydrohalite are present on Cerealia Tholus, but the two species are distributed differently. The maximum quantity of hydrohalite is located at the top of the dome, where the carbonate amount is lower, which is the place where the radial fracture patterns are more developed. Thus, the spectral data indicate that the most hydrated and salty material corresponds to the most densely fractured area on the facula located next to the peak of Cerealia Tholus.

The delivery of exogenous materials cannot account for the Occator bright areas, because the spectra are unlike other asteroids and comets. The morphologies of the bright areas argue against a direct impact origin (nearly circular shapes and correlation with fractures), although the central mound of Occator and the presence of bright material along fractures suggests that its emplacement should be related to the impact that formed Occator. Two possible (not necessarily alternative) scenario have been highlighted: (i) the bright material in Occator was formed from local aqueous processing triggered by the impact or (ii) it represents an exposure of deeper material that found its way to the surface via fractures generated by that impact. The interpretation of the most recent results acquired by Dawn flying over Cerealia facula at low altitude indicates that likely both the processes contributed to the emplacement of the bright material (Raymond et al., 2020; Scully et al., 2020) permitting the recent extrusion of these materials from the interior of Ceres.

Carbonates also occur in carbonaceous chondrites but in amounts of only a few volumes per cent. Moreover, no sodium carbonate has been reported in CM and CI chondrites. On Ceres, the occurrence of ammonium salts, as well as the high proportion of carbonates in Occator bright materials, points to a formation mechanism that is distinct from what produces meteoritic carbonates. Ammonium is known to compensate for the acidifying action of dissolved CO_2, keeping the pH high enough for the stability of carbonate or bi-carbonate ions in the fluids. Indeed, a chemical pathway similar to that used to produce industrial Na_2CO_3 on Earth (the Solvay process) may be at work, where dissolved NaCl, CO_2, and ammonia are used to precipitate sodium bicarbonate and ammonium chloride (De Sanctis et al., 2016).

Notably, differences in the mineralogy are observed across Cerealia Facula. The mineral distribution does not seem homogeneous, and it is not perfectly correlated to the brightness of the surface, revealing peculiar regions inside the Facula. From the band area map of the ammonium chloride in Cerealia Facula (e.g., Raponi et al., 2019b), a region with higher concentration of that mineral than the average trend suggests a different emplacement event. Partial superposition between this region and an excess of band areas at 3.20 and 3.28 μm, attributed to hydrohalite (De Sanctis et al., 2020), can be also noted. In particular, the concentration of hydrohalite on peak of the Tholus suggests successive episodes of emplacements of brines, in which the relative abundances of the species were different. Models of cooling of deep reservoirs (Quick et al., 2019) indicate that eutectic temperatures for different solutions would have been reached at diverse times, with the following sequence of crystallization: $NaHCO_3$–ice and Na-carbonate sodium chloride (hydrohalite) and/or ammonium-salts, the former being limited to shallower levels in the subsurface and the latter contributing to late manifestations of cryovolcanism (Quick et al., 2019). Thus, the identification of hydrohalite in Cerelia facula and its concentration on the peak of the dome could infer recent episodes of cryovolcanism with extrusion of hydrated-salt rich brines, implying that aqueous alkaline solutions could even persist in Ceres' subsurface to the present day (see Chapters 10 and 11).

7.4.3 Water Ice, hydrated Carbonates, and Salts

7.4.3.1 Sodium carbonates

Besides Occator (20°N, 240°E), the most representative regions where carbonate exhibits the highest intensity, with absorptions centered between 3.98 and 4.02 mm are: Oxo (42°N, 359°E), Azacca (6°S, 218°E), Kupalo (39°S, 173°E), Ikapati (33°N, 46°E), and Ahuna Mons (10°S, 317°E). Such a band center is consistent with the presence of Na_2CO_3. Only a few regions with strong carbonate absorptions have band centers at 3.95 μm (e.g., western side of Urvara crater and Baltay Catena), consistent with Mg-carbonate (Carrozzo et al., 2018).

Some occurrences of sodium carbonate are related to recently upwelled material from the subsurface, such as Occator's central dome and Ahuna Mons (Zambon et al., 2017), or mounds, such as those close to or inside Haulani, Oxo, Dantu, and Azacca. Other craters such as Haulani and Ikapati also have morphologies, such as floor fractures, that may be indicative of upwelling processes. In other cases, sodium carbonate is found in crater ejecta and floors, such as Kupalo and Haulani (Carrozzo et al., 2018).

7.4.3.2 Hydrated Minerals

Some small areas in specific locations, such as in Oxo (Figure 7.11), Azacca, Kupalo, and Kahukura, are different and cannot be fully fit using Na_2CO_3. The spectra of these small areas show an overall negative spectral slope across the 1.0–4.2 μm range, indicative of changes in the composition, and a larger and broader band in the 2.6–3.8 μm range, suggesting hydration.

Spectral modeling (Carrozzo et al., 2018) indicates that the best fits are obtained using thermonatrite ($Na_2CO_3 \cdot H_2O$), or trona ($NaHCO_3 \cdot Na_2CO_3 \cdot 2H_2O$) (Figure 7.11). Those hydrated carbonates and hydrohalite (detected in the bright material of Occator crater) are not stable on airless surfaces and dehydrate upon exposure to vacuum and irradiation. Dehydration can occur over Myr time scales for hydrated carbonates (Zolotov & Shock, 2001) or even tens of years in the case of hydrohalite (De Sanctis et al., 2020), if no source of hydration is provided. Possible sources of hydration would be the presence of water ice in contact with the carbonates/salts. A second possibility would be a recent/continuous refurbishment of hydrate brines from subsurface reservoirs.

7.4.3.3 Origins of Hydrated Carbonates

The correlation of Na carbonates with some extrusive constructs suggests that at least some Na carbonates are transported to, or near to, the surface by ascending subsurface fluids in several areas of Ceres. On the other hand, Na_2CO_3 and its hydrated forms are found in the same area where H_2O ice has been detected such as in Oxo crater (Combe et al., 2016; Carrozzo et al., 2018; Combe et al., 2019) (Figure 7.11). Hydrated sodium carbonates are also found in Kupalo crater, that, even if devoid of ice, is close to Juling crater, where water ice has been detected, and a water vapor activity has been inferred from a seasonal variation of the ice abundance (Raponi et al., 2018). Water ice has been observed elsewhere on Ceres' surface (Combe et al., 2019), but always in areas where the low temperatures preserve the ice from rapid sublimation (Formisano et al., 2018). The topography and location of Occator bright material does not offer conditions for long-lasting water ice on the surface (Landis et al., 2019; see Chapter 10). Thus, the hydration state of the identified NaCl can be due to very recent or continuous emplacement, implying that brines would still be able to extrude on Ceres' surface. Sodium and ammonium chloride lower the eutectic temperature of the brines. Therefore, it is conceivable that the salt content of Ceres could maintain brine-liquid pockets currently.

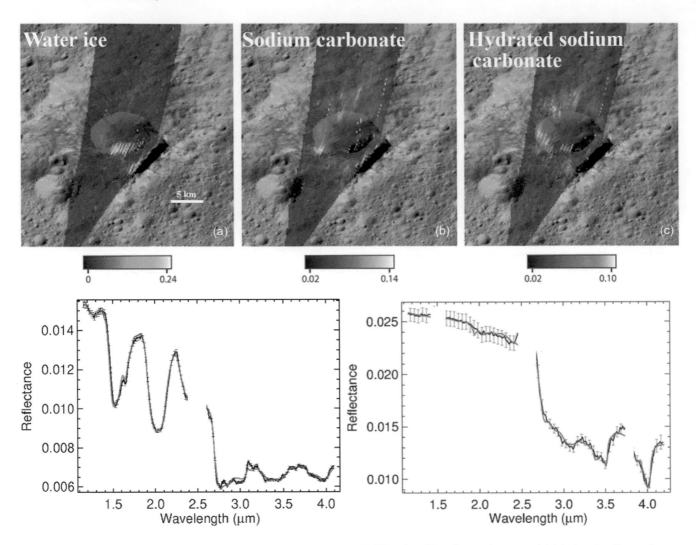

Figure 7.11 Upper panel: distribution of surface compounds in Oxo crater: (a) Water ice, (b) sodium carbonate, and (c) hydrated sodium carbonate. Abundances are derived from the spectral fitting method (Carrozzo et al., 2018). Left lower panel: the observed spectrum and model of ice rich region in Oxo crater (see also Combe et al., 2019). Right lower panel: the spectral fit (red curve) in Oxo using ammonium montmorillonite, dolomite, dark material, sodium carbonate, and hydrated sodium carbonate (Carrozzo et al., 2018).
A black and white version of this figure will appear in some formats. For the color version, refer to the plate section.

7.5 CONCLUSIONS

The surface mineralogy of Ceres reveals three main aspects of the evolution of the dwarf planet:

Organic materials in high abundance in well-defined areas (De Sanctis et al., 2017; Kaplan et al., 2018; Raponi et al., 2019a), the likely presence of widespread carbon on the surface as inferred by GRaND data (Marchi et al., 2019, Chapter 8), and the presence of ammonia into the phyllosilicates (De Sanctis et al., 2015) would indicate that Ceres can be considered as an organic-rich body formed in an environment similar to the one that originated in the Saturn and Jupiter icy satellites.

The spectral features at 3.1 μm, which can be attributed to ammonium ion, and the local feature at 3.4 μm attributed to aliphatic organics, have also been observed on other asteroids, the Jupiter moon Himalia, and comet 67P (see Section 7.4.1.2), revealing a possible common past of the material making up these small bodies.

The global surface composition and the distribution of the observed minerals, especially the extensive presence of phyllosilicates and carbonates, could be explained as a remnant of a global and past pervasive hydrothermal alteration, a consequence of a global ocean in the Ceres past history (De Sanctis et al., 2015, Castillo-Rogez et al., 2020). Unexpectedly, local exposure of hydrated minerals and salts (Carrozzo et al., 2018; De Sanctis et al., 2020), the detection of exposed H_2O ice (Combe et al., 2016, 2019), despite its short lifespan (Landis et al., 2019), changes in water ice content (Raponi et al., 2018), and cryovolcanism evidence (Ruesch et al., 2016) reveal recent or ongoing activity in Ceres' sub-surface, involving liquid water or brines (see also Chapter 11). Sodium salts (NaCl, $NaHCO_3$, and/or Na_2CO_3) identified on Cerealia Facula have also been detected in Enceladus' plume, pointing to Ceres as an object that has experienced aqueous processes in the very recent geological past. These processes are able to form materials similar to those predicted or observed on icy satellites

(Zolotov, 2007; Postberg et al., 2009, 2011; Waite et al., 2009) such as simple organics (Waite et al., 2009) and complex macromolecular organic material (Postberg et al., 2011). Indeed, spectra of Cerealia Facula are unusually positively sloped between 0.4 and 1 µm (Nathues et al., 2019), such as the organic-rich terrains detected at Ernutet (Pieters et al., 2018). At the same time, the mineralogy of Ernutet organic-rich areas is also very complex and rich in Ca-Mg-carbonates and Na-carbonates, which are typical characteristic of the Faculae, pointing to an endogenous origin of the organic matter.

The extent of such organic chemistry remains an open question, together with the difficulty in identify a viable method for transporting the organic material on the surface, where it has been observed (see Chapter 10). Equally difficult is the identification of the source of the modern subsurface fluids. Indeed, Occator is a fresh crater, possibly 100 million years or younger (Buczkowski et al., 2016), so the bright material should be of similar or younger age. However, Cerealia facula was formed at least 18 Ma after the formation of Occator crater (Neesemann et al., 2019), while the modeling of the impact-produced melt chamber suggests complete freezing after ~5 Ma (Bowling et al., 2019) or longer (Raymond et al., 2020). However, an impact-produced melt chamber is not sufficient to explain the most recent activity, which is attributed to a deep brine layer inferred from the thermal modeling and gravity anomalies.

Most of the surface composition of Ceres is made up of a dark material, which has characteristics similar to carbonaceous chondrites. It could originate from the global aqueous alteration of the primordial crust, and/or collisionally accreted material. However, a concentration of carbon more than 5-times higher than in the most C-rich chondrites (Marchi et al., 2019) has been inferred by the GRaND instrument onboard Dawn. Such a high concentration implies that the bulk of Ceres accreted ultra-C-rich materials and/or that C has been concentrated in its crust (see Chapter 8).

Although it is possible that Ceres accreted organic material inherited from the interstellar medium (e.g., Raponi et al., 2020), the pervasive hydrothermal alteration on Ceres might have completely modified this accreted organic matter, and a wide range of organic compounds are expected to form during aqueous alteration (Schulte & Shock, 2004; Vinogradoff et al., 2018). Indeed, organic matter is most abundant in those carbonaceous chondrites that display the greatest amount of inorganic aqueous alteration products as phyllosilicates and carbonates (Pearson et al., 2002), which are known to adsorb organic species and actively participate as catalysts in their syntheses and reactions (Pearson et al., 2002; Williams et al., 2005).

Given this, even if clear infrared signatures of organic matter have only been observed on the local areas of the surface, the mixing between organic matter with the mineral phases could be pervasive and widespread in Ceres, either produced by the past global aqueous alteration or by recent hydrothermal activity, placing Ceres among the most interesting targets from an astrobiological point of view.

REFERENCES

Ammannito, E., De Sanctis, M. C., Ciarniello, M., et al. (2016) Distribution of phyllosilicates on the surface of Ceres. *Science*, 353, aaf4279.

Beran, A. (2002) Infrared spectroscopy of micas. *Reviews in Mineralogy and Geochemistry*, 46, 351–369.

Berg, B. L., Cloutis, E. A., Beck, P., et al. (2016) Reflectance spectroscopy (0.35–8 µm) of ammonium-bearing minerals and qualitative comparison to Ceres-like asteroids. *Icarus*, 265, 218–237.

Binzel, R. P., & Xu, S. (1993) Chips off of Asteroid 4 Vesta: Evidence for the parent body of basaltic achondrite meteorites. *Science*, 260, 186–191.

Bishop, J. L., Banin, A., Mancinelli, R. L., & Klovstad, M. R. (2002) Detection of soluble and fixed NH_4^+ in clay minerals by DTA and IR reflectance spectroscopy: A potential tool for planetary surface exploration. *Planetary and Space Science*, 50, 11.

Bishop, J. L., Lane, M. D., Dyar, M. D., & Brown, A. J. (2008) Reflectance and emission spectroscopy study of four groups of phyllosilicates: Smectites, kaolinite-serpentines, chlorites and micas. *Clay Minerals*, 43, 35–54.

Bowling, T. J., Ciesla, F. J., Davison, T. M., et al. (2019) Post-impact thermal structure and cooling timescales of Occator crater on asteroid 1 Ceres. *Icarus*, 320, 110–118.

Bowling, T. J., Johnson, B. C., Marchi, S., et al. (2020) An endogenic origin of cerean organics. *Earth and Planetary Science Letters*, 534, 116069.

Brown, M. E., & Rhoden, A. R. (2014) The 3 µm spectrum of Jupiter's irregular satellite Himalia. *The Astrophysical Journal*, 793, L44.

Buczkowski, D. L., Schmidt, B. E., Williams, D. A., et al. (2016) The geomorphology of Ceres. *Science*, 353, aaf4332.

Bus, S. J., & Binzel, R. P. (2002a) Phase II of the small main-belt asteroid spectroscopic survey. The observations. *Icarus*, 158, 106.

Bus, S. J., & Binzel, R. P. (2002b) Phase II of the small main-belt asteroid spectroscopic survey. A feature-based taxonomy. *Icarus*, 158, 146.

Campins, H., Hargrove, K., Pinilla-Alonso, N., et al. (2010) Water ice and organics on the surface of the asteroid 24 Themis. *Nature*, 464, 1320–1321.

Carrozzo, F. G., De Sanctis, M. C., Raponi, A., et al. (2018) Nature, formation, and distribution of carbonates on Ceres. *Science Advances*, 4, e1701645.

Carrozzo, F. G., Raponi, A., De Sanctis, M. C., et al. (2016) Artifacts reduction in VIR/Dawn data. *Review of Scientific Instruments*, 87, 124501.

Castillo-Rogez, J. C., Neveu, M., McSween, H. Y., et al. (2018). Insights into Ceres's evolution from surface composition. *Meteoritics & Planetary Science*, 53, 1820.

Castillo-Rogez, J. C., Neveu, M., Scully, J. E. C., et al. (2020) Ceres: Astrobiological target and possible ocean world. *Astrobiology*, 20, 269–291.

Chapman, C. R., & Gaffey, M. J. (1979) Reflectance spectra for 277 asteroids. In T. Gehrels, & M. S. Matthews (eds.), *Asteroids*. Tucson: University of Arizona Press, pp. 655–687.

Chapman, C. R., McCord, T. B., & Johnson, T. V. (1973) Asteroid spectral reflectivities. *The Astronomical Journal*, 78, 126–140.

Chapman, C. R., & Salisbury, J. W. (1973) Comparisons of meteorite and asteroid spectral reflectivities. *Icarus*, 19, 507–522.

Ciarniello, M., De Sanctis, M. C., Ammannito, E., et al. (2017). Spectrophotometric properties of dwarf planet Ceres from the VIR spectrometer on board the Dawn mission. *Astronomy and Astrophysics*, 598, A130.

Clark, B. E., Hapke, B., Pieters, C., & Britt, D. (2002) Asteroid space weathering and regolith evolution. In W. F. Bottke Jr., A. Cellino, P. Paolicchi, & R. P. Binzel (eds.), *Asteroids III*. Tucson: University of Arizona Press, pp. 585–599.

Combe, J.-P., McCord, T. B., Tosi, F., et al. (2016) Detection of local H_2O exposed at the surface of Ceres. *Science*, 353, aaf3010.

Combe, J.-P., Raponi, A., Tosi, F., et al. (2019) Exposed H_2O-rich areas detected on Ceres with the dawn visible and infrared mapping spectrometer. *Icarus*, 318, 22–41.

Cruikshank, D. P., Tholen, D. J., Hartmann, W. K., Bell, J. F., & Brown, R. H. (1991) Three basaltic earth-approaching asteroids and the source of the basaltic meteorites. *Icarus*, 89, 1–13.

De Angelis, S., Ferrari, M., De Sanctis, M. C., et al. (2021) High-temperature VIS-IR spectroscopy of NH4-phyllosilicates. *Journal of Geophysical Research: Planets*, 126, e2020JE006696

De Sanctis, M. C., Ammannito, E., Carrozzo, F. G., et al. (2018) Ceres's global and localized mineralogical composition determined by Dawn's Visible and Infrared Spectrometer (VIR). *Meteoritics & Planetary Science*, 53, 1844.

De Sanctis, M. C., Ammannito, E., McSween, H. Y., et al. (2017) Localized aliphatic organic material on the surface of Ceres. *Science*, 355, 719–722.

De Sanctis, M. C., Ammannito, E., Raponi, A., et al. (2015) Ammoniated phyllosilicates with a likely outer Solar System origin on (1) Ceres. *Nature*, 528, 241–244.

De Sanctis, M. C., Ammannito, E., Raponi, A., et al. (2020) Fresh emplacement of hydrated sodium chloride on Ceres from ascending salty fluids. *Nature Astronomy*, 4, 786–793.

De Sanctis, M. C., Coradini, A., Ammannito, E., et al. (2011) The VIR spectrometer. *Space Science Reviews*, 163, 329–369.

De Sanctis, M. C., Raponi, A., Ammannito, E., et al. (2016) Bright carbonate deposits as evidence of aqueous alteration on (1) Ceres. *Nature*, 536, 54–57.

De Sanctis, M. C., Vinogradoff, V., Raponi, A., et al. (2019) Characteristics of organic matter on Ceres from VIR/Dawn high spatial resolution spectra. *Monthly Notices of the Royal Astronomical Society*, 482, 2407–2421.

dos Santos, R., Patel, M., Cuadros, J., & Martins, Z. (2016) Influence of mineralogy on the preservation of amino acids under simulated Mars conditions. *Icarus*, 277, 342.

Ehlmann, B. L., Hodyss, R., Bristow, T. F., et al. (2018) Ambient and cold-temperature infrared spectra and XRD patterns of ammoniated phyllosilicates and carbonaceous chondrite meteorites relevant to Ceres and other Solar System bodies. *Meteoritics & Planetary Science*, 53, 1884.

Farinella, P., Gonczi, R., Froeschle, C., & Froeschle, C. (1993) The injection of asteroid fragments into resonances. *Icarus*, 101, 174–187.

Farmer, V. C. (1974) The layer silicates. In V. C. Farmer (ed.), *The Infrared Spectra of Minerals, Monograph 4*. London: Mineralogical Society, pp. 331–363.

Ferrari, M., De Angelis, S., De Sanctis, M. C., et al. (2019) Reflectance spectroscopy of ammonium-bearing phyllosilicates. *Presented at EPSC-DPS Joint Meeting 2019*, September 15–20, Geneva, id. EPSC-DPS2019–1864.

Formisano, M., Federico, C., De Sanctis, M. C., et al. (2018) Thermal stability of water ice in Ceres' craters: The case of Juling crater. *Journal of Geophysical Research (Planets)*, 123, 2445–2463.

Frigeri, A., De Sanctis, M. C., Ammannito, E., et al. (2019) The spectral parameter maps of Ceres from NASA/DAWN VIR data. *Icarus*, 318, 14–21.

Gaffey, M. J. (1976) Spectral reflectance characteristics of the meteorite classes. *Journal of Geophysical Research*, 81, 905–920.

Greenberg, J. M., Li, A., Mendoza-Gomez, C. X., et al. (1995) Approaching the interstellar grain organic refractory component. *The Astrophysical Journal*, 455, L177.

Hapke, B. (1993) *Theory of Reflectance and Emittance Spectroscopy*. New York: Cambridge University Press.

Hapke, B. (2012) *Theory of Reflectance and Emittance Spectroscopy*, 2nd ed. New York: Cambridge University Press.

Hendrix, A. R., Vilas, F., & Li, J.-Y. (2016) Ceres: Sulfur deposits and graphitized carbon. *Geophysical Research Letters*, 43, 8920.

Holm, N. G., Oze, C., Mousis, O., Waite, J. H., & Guilbert-Lepoutre, A. (2015) Serpentinization and the formation of H_2 and CH_4 on celestial bodies (planets, moons, comets). *Astrobiology*, 15, 587.

Hoyle, F., Wickramasinghe, N. C., Al-Mufti, S., Olavesen, A. H., & Wickramasinghe, D. T. (1982) Infrared spectroscopy over the 2.9–3.9 μm waveband in biochemistry and astronomy. *Astrophysics and Space Science*, 83, 405–409.

Kaplan, H. H., Milliken, R. E., & Alexander, C. M. O'D. (2018) New constraints on the abundance and composition of organic matter on Ceres. *Geophysical Research Letters*, 45, 5274–5282.

King, T. V. V., & Clark, R. N. (1989) Spectral characteristics of chlorites and Mg-serpentines using high-resolution reflectance spectroscopy. *Journal of Geophysical Research*, 94, 13997–14008.

King, T. V. V., Clark, R. N., Calvin, W. M., Sherman, D. M., & Brown, R. H. (1992) Evidence for ammonium-bearing minerals on Ceres. *Science*, 255, 1551–1553.

Krohn, M. D. (1987) Near-infrared detection of ammonium minerals. *Geophysics*, 52, 924.

Landis, M. E., Byrne, S., Combe, J.-P., et al. (2019) Water vapor contribution to Ceres' exosphere from observed surface ice and postulated ice-exposing impacts. *Journal of Geophysical Research: Planets*, 124, 61–75.

Larson, H. P., Feierberg, M. A., Fink, U., & Smith, H. A. (1979) Remote spectroscopic identification of carbonaceous chondrite mineralogies: Applications to Ceres and Pallas. *Icarus*, 39, 257–271.

Lazzaro, D., Ferraz-Mello, S., & Fernández, J. A. (eds.) (2006) *Asteroids, Comets, Meteors*, IAU Symposium. New York: Cambridge University Press.

Le Guillou, C., Bernard, S., Brearley, A. J., & Remusat, L. (2014) Evolution of organic matter in Orgueil, Murchison and Renazzo during parent body aqueous alteration: In situ investigations. *Geochimica et Cosmochimica Acta*, 131, 368.

Lebofsky, L. A., Feierberg, M. A., Tokunaga, A. T., Larson, H. P., & Johnson, J. R. (1981) The 1.7- to 4.2-μm spectrum of asteroid 1 Ceres: Evidence for structural water in clay minerals. *Icarus*, 48, 453–459.

Li, J.-Y., McFadden, L. A., Parker, J. W., et al. (2006) Photometric analysis of 1 Ceres and surface mapping from HST observations. *Icarus*, 182, 143–160.

Li, J.-Y., Reddy, V., Nathues, A., et al. (2016) Surface albedo and spectral variability of Ceres. *The Astrophysical Journal*, 817, L22.

Li, J.-Y., Schröder, S. E., Mottola, S., et al. (2019) Spectrophotometric modeling and mapping of Ceres. *Icarus*, 322, 144–167.

Licandro, J., Campins, H., Kelley, M., et al. (2011) (65) Cybele: Detection of small silicate grains, water-ice, and organics. *Astronomy and Astrophysics*, 525, A34.

Longobardo, A., Palomba, E., Carrozzo, F. G., et al. (2019a) Mineralogy of the Occator quadrangle. *Icarus*, 318, 205–211.

Longobardo, A., Palomba, E., Galiano, A., et al. (2019b) Photometry of Ceres and Occator faculae as inferred from VIR/Dawn data. *Icarus*, 320, 97–109.

Marchi, S., Raponi, A., Prettyman, T. H., et al. (2019) An aqueously altered carbon-rich Ceres. *Nature Astronomy*, 3, 140–145.

McCollom, T. M., & Seewald, J. S. (2007) Abiotic synthesis of organic compounds in deep-sea hydrothermal environments. *Chemical Reviews*, 107, 382–401.

McCord, T. B., & Castillo-Rogez, J. C. (2018) Ceres's internal evolution: The view after Dawn. *Meteoritics & Planetary Science*, 53, 1778–1792.

McCord, T. B., & Zambon, F. (2019) The surface composition of Ceres from the Dawn mission. *Icarus*, 318, 2.

Mennella, V., Baratta, G. A., Esposito, A., Ferini, G., & Pendleton, Y. J. (2003) The effects of ion irradiation on the evolution of the carrier of the 3.4 micron interstellar absorption band. *The Astrophysical Journal*, 587, 727.

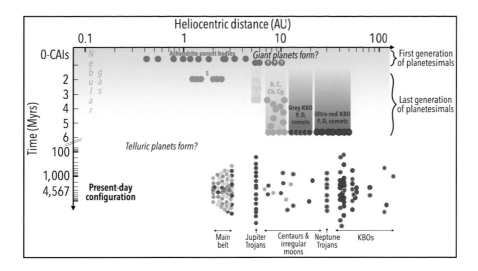

Figure 1.3 Postulated sequence of events tracing the time, place, and duration of formation of small bodies (top) to present-day observed characteristics (bottom; vertical spread reproducing roughly the distribution of orbital inclinations). The accretion duration is shown as gradient boxes ending at the fully formed bodies. Numerical simulations suggest that volatile-rich IDP-like bodies (blue dots; B, C, Cb, Cg, P, D, comets, grey and ultra-red KBOs) accreted their outer layers after 5–6 Myrs (adapted from Neveu & Vernazza, 2019).

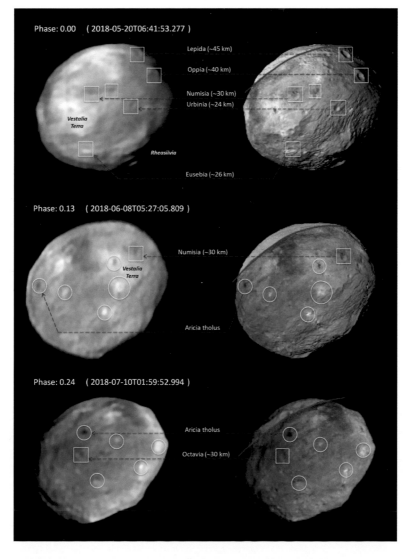

Figure 1.5 Comparison of the VLT/SPHERE deconvolved images of Vesta (left column) with synthetic projections of the Dawn 3D shape model produced with OASIS and with albedo information (right column). No albedo data is available from Dawn for latitudes above 30° N (orange line). The main structures that can be identified in both the VLT/SPHERE images and the synthetic ones are highlighted: craters are embedded in squares and albedo features in circles (from Fetick et al., 2019).

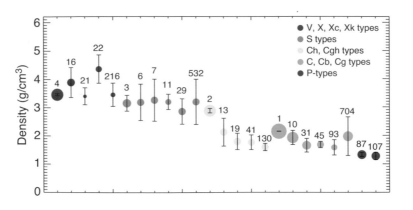

Figure 1.6 Density of some of the largest asteroids. Asteroids are grouped following their spectral classification. The relative sizes of the dots follow the relative diameters of the bodies in logarithmic scale. Error bars are 1-sigma. The science based on these density estimates can be retrieved in Paetzold et al. 2011, Russell et al. 2012, 2016, Viikinkoski et al. 2015b, Marsset et al. 2017, Hanus et al. 2017a, b, Pajuelo et al. 2018, Carry et al. 2019a, b, Ferrais et al. 2020, Hanus et al. 2020, Marsset et al. 2020, Vernazza et al. 2020, Vernazza et al. 2021, Yang et al. 2020 and Dudzinski et al. 2020.

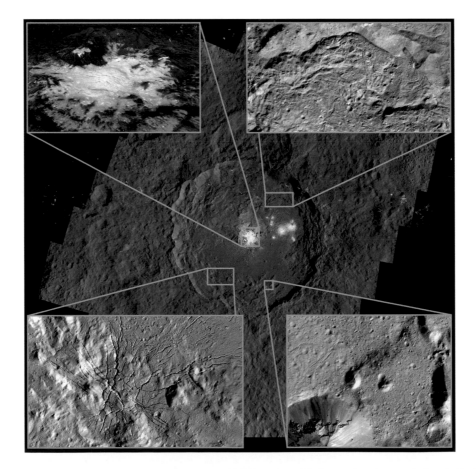

Figure 2.9 (a) Occator crater closeup montage at <10 m/pixel, obtained in XMO7: (Upper left) Perspective view of salt deposits Paola and Ceralia Faculae; (upper right) Northern terminal flow margin of hummocky lobate material, showing conical mounds and varying flow textures; (lower right) Two-hundred meter scale conical mounds (pingos?) on smooth lobate material; (lower left) Uplifted dome with radial fractures. [K. Hughson, personal communication]. All images credit: NASA/JPL-Caltech/UCLA/MPS/DLR/IDA. (b) Perspective view (with no vertical exaggeration) of the primary set of fractures in Nar Sulcus, interpreted as evidence of significant local extension within the past 100 Myr. The shape implies mechanical properties akin to outer Solar System moons. Image roughly 40 km in width. North is to the left. [K. Hughson, personal communication]. Image credit: NASA/JPL-Caltech/UCLA/MPS/DLR/IDA.

Figure 3.1 Photomicrographs (crossed polars) of thin sections of a eucrite (EET 90020) composed of pyroxenes and plagioclase, and a diogenite (GRA 98108) composed of orthopyroxenes with minor olivine. These meteorites are unusual because they show primary igneous textures; most eucrites have been metamorphosed and recrystallized, and many HEDs are breccias, having been pulverized and recemented by impact processes on Vesta. The white scale bars are ~2.5 mm.

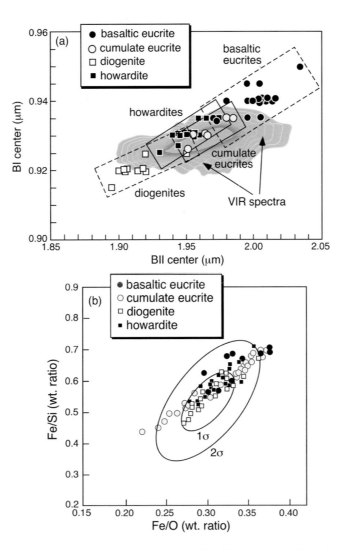

Figure 3.2 (a) Band center positions for the 1 μm (BI) and 2 μm (BII) features distinguish eucrites from diogenites, with howardite and cumulate eucrite bands plotting between them. The cloud represents global VIR pixels for the surface of Vesta (De Sanctis et al., 2013). (b) Global Fe:Si and Fe:O weight ratios (ovals show 1σ and 2σ uncertainties) for Vesta, determined from GRaND measurements (Prettyman et al., 2012), compared to those ratios in HEDs. Adapted from McSween et al. (2013b).

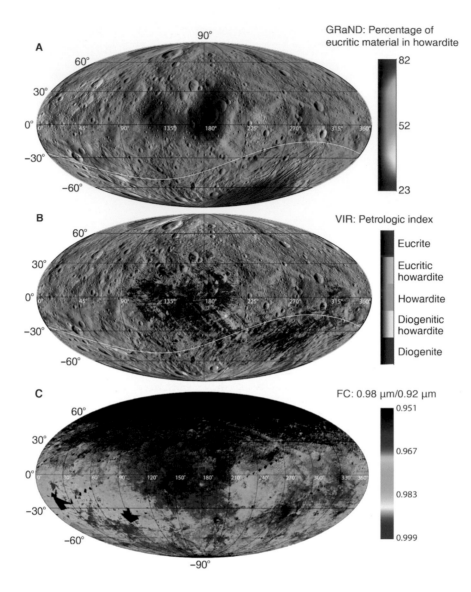

Figure 3.3 Maps of HED lithologies based on data from (a) GRaND, (b) VIR, and (c) FC. Methods used to construct the maps were described by Ammannito et al. (2013a), Prettyman et al. (2013), and Thangjam et al. (2013). The white lines in (a) and (b) indicate the outline of the Rheasilvia basin. From Raymond et al. (2017), with permission.

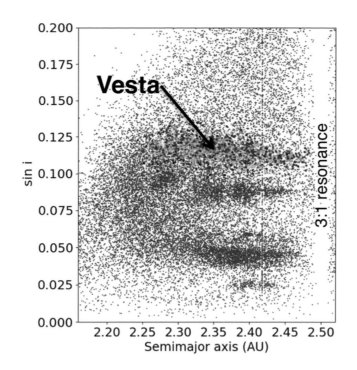

Figure 3.5 Vesta's pathway for delivery of meteorites to Earth was discovered through telescopic spectra initially revealing about two dozen small (D ≤10 km) Vesta-like asteroids spreading in the same orbital inclination plane from Vesta to the 3:1 resonance with Jupiter, a dynamical "escape hatch" to the inner Solar System (Binzel & Xu, 1993). Modern space-based infrared surveys easily distinguish the higher albedos of the "Vesta family" asteroids (colored points) showing the vast extent of fragments ejected from Vesta during the formation of the Rheasilvia basin. Adapted from Masiero *et al.* (2013); courtesy J. R. Masiero.

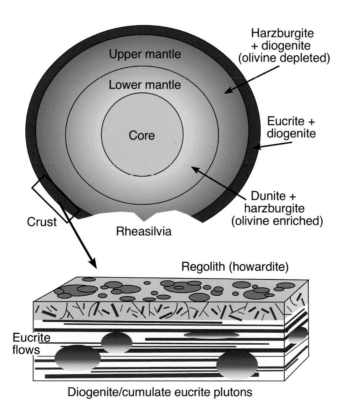

Figure 3.7 Conceptual sketch for the interior structure of Vesta (not to scale), based mostly on the model of Mandler and Elkins-Tanton (2013). The olivine-rich (dunite) lower mantle differs from that model, which posits a lower mantle of harzburgite, but is required to explain the apparent dearth of olivine in deeply excavated craters like Rheasilvia (see Chapter 4). The lower diagram illustrates the structure of the crust, composed of eucrite flows intruded by plutons containing diogenite and cumulate eucrite, and an upper surface regolith of howardite.

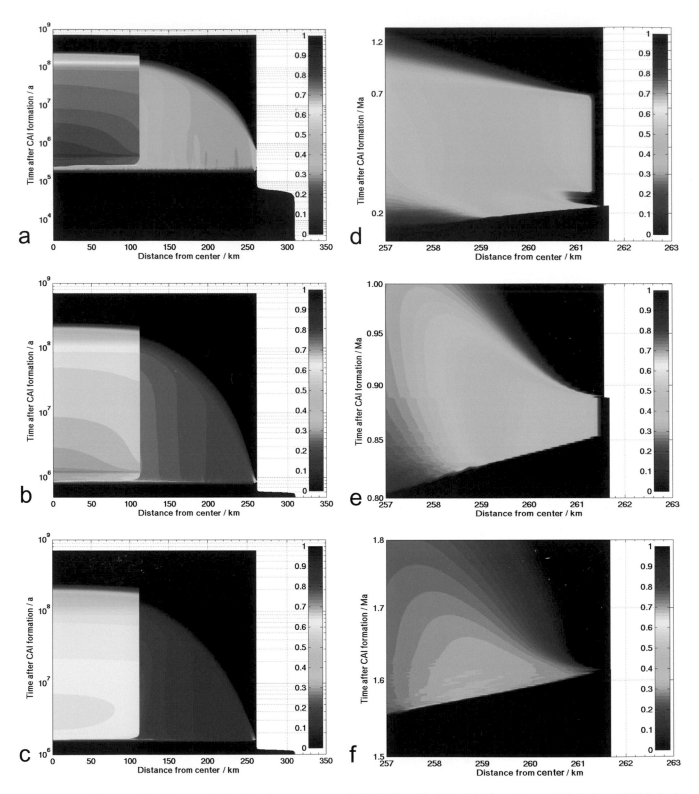

Figure 4.7 Melt fraction (a–c) and close-ups in the sub-surface magma ocean (d–f) of a Vesta-like body which formed at $t_0 = 0$ Ma (a, d), $t_0 = 0.5$ Ma (b, e), and $t_0 = 1$ Ma (c, f) after the CAIs. Figure adapted from Neumann et al. (2014).

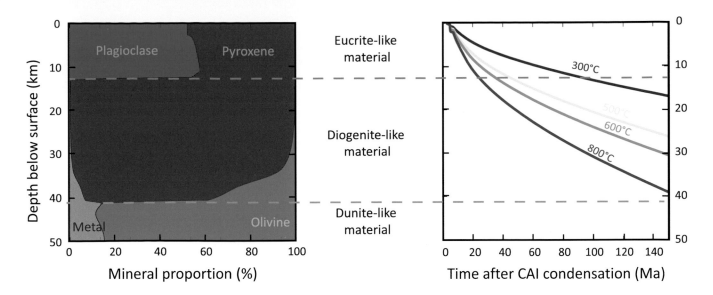

Figure 4.8 Modeled mineralogy and cooling trajectories of the upper portion of Vesta. The left hand panel represents the mineralogy of the upper section of Vesta predicted by the model of Mizzon (2015) for accretion 1 Ma after CAI condensation and full crystallization. Blue = plagioclase; Brown = pyroxene; Green = olivine; Grey = metal. The upper ~13 km has a mineralogy that resembles eucrite, the section between 13 and 40 km has a mineralogy that resembles diogenite, while below that depth olivine is the dominant phase. Note that the model of Mizzon does not include the migration of metal, so its presence in this simulation is an artefact, as indicated by Neumann et al. (2014). The right hand panel represents cooling trajectories as a function of time for the same model. Figure redrawn from Mizzon (2015).

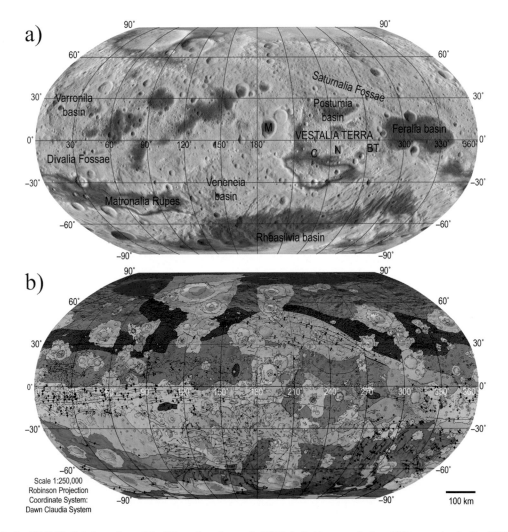

Figure 5.1 (a) Colorized HAMO digital terrain model of Vesta (Preusker et al., 2014), in Robinson projection. GIS processing by David M. Nelson, Regional Planetary Image Facility at Arizona State University. Red shows high regions, while blue represents low. Features of note are labeled. M = Marcia crater; C = Cornelia crater; N = Numisia crater; BT = Brumalia Tholus. (b) The LAMO geologic map of Vesta (Williams et al., 2014a) in Robinson projection.

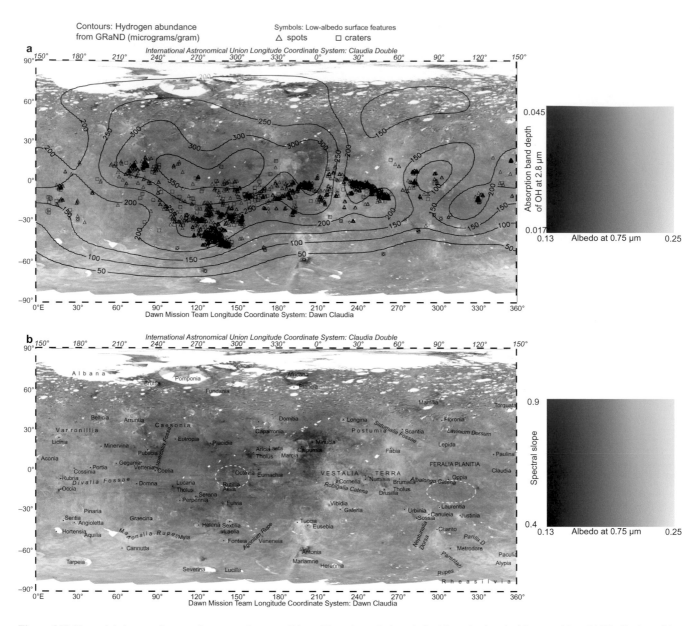

Figure 6.15 Vesta global maps of two surface contaminant candidates. The color scale is optimized for color-impaired human vision. (a) Distribution of the albedo at 0.75 μm and the absorption band depth at 2.8 μm of hydroxyl (OH). The colors from magenta to yellow represent all hydrous components; zones with subdued colors or gray are less or not hydroxylated. Low albedo hydroxylated materials appear in magenta. Low-albedo surfaces, low-albedo craters (Reddy et al., 2012b), and hydrous materials (De Sanctis et al., 2012; Turrini et al., 2014; Combe et al., 2015a) have correlated distributions on Vesta, consistent with contamination with carbonaceous chondrite meteorites. (b) Distribution of the albedo at 0.75 μm and the visible spectral slope between 0.4 and 0.8 μm. The colors from magenta to yellow represent all types of surfaces that exhibit an unusually strong positive visible spectral slope compared to the average of Vesta; areas with subdued colors or gray have a less positive spectral slope. Possible effects of surface irradiation by solar wind particles that combine low albedo and a positive spectral slope (e.g., Vernazza et al., 2006) appear in magenta, such as in the south of crater Oppia (10–20° S, 290–330° E) and the south of crater Octavia (0–10° S, 130–160° E) located with dashed-line ellipses.

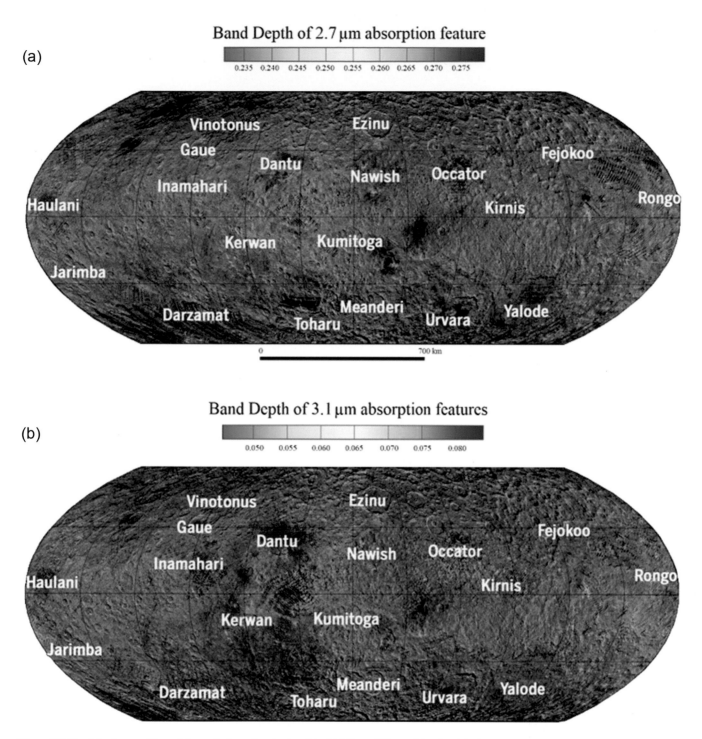

Figure 7.2 Band depth maps. Maps of the peak absorption depths of the (a) 2.7-μm OH-stretch band and (b) 3.1-μm NH4 band. The two maps have a similar global pattern, although they differ in some localized regions such as Urvara. The scale bar is true at the equator (Ammannito et al., 2016).

Band Center of 4.0 μm absorption

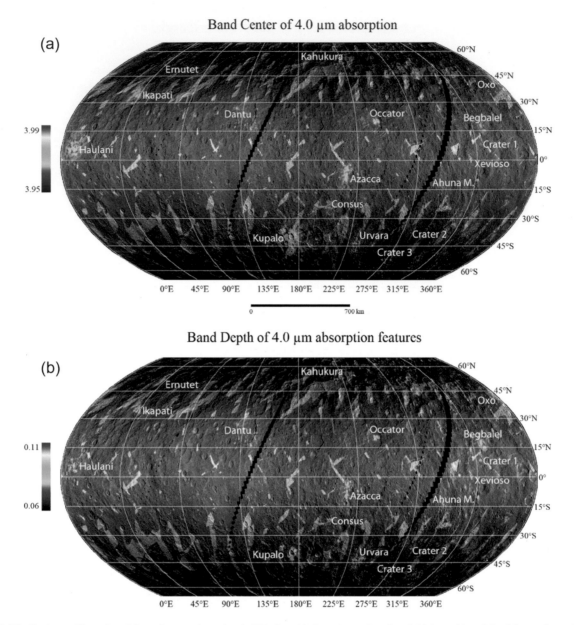

Band Depth of 4.0 μm absorption features

Figure 7.3 Distribution and intensity of the carbonate absorption in VIR data. (a) Central wavelength and (b) intensities of the 3.9-mm absorption feature. The maps are superimposed on the Framing Camera images using a transparency of 25% (Carrozzo et al., 2018).

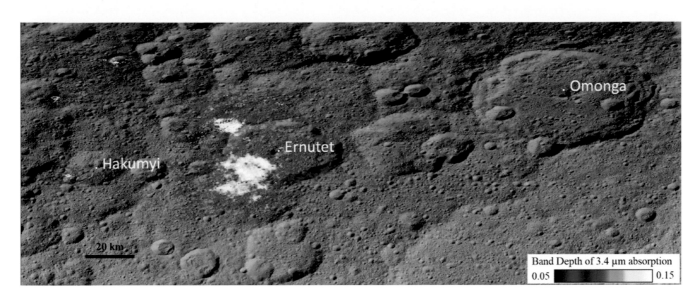

Figure 7.4 Geologic context of the Ernutet region on Ceres that exhibits organic material highlighted by the distribution of the 3.4-μm band depth. Band depths are calculated according to Ammannito et al. (2016) and cylindrically projected on Framing Camera (FC) clear filter mosaic (De Sanctis et al., 2019).

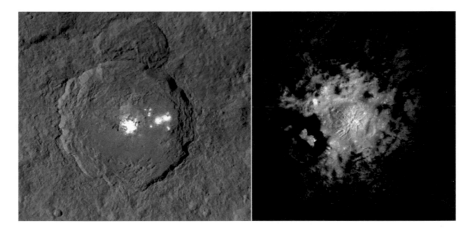

Figure 7.7 Left panel. Framing Camera mosaic (35 m/pixel) obtained with a clear filter during the Low Altitude Mapping Orbit (LAMO) (Roatsch et al., 2016). Right panel: Cerealia Facula obtained by combining Framing Camera images acquired during LAMO phase with three images using spectral filters centered at 438, 550, and 965 nm, during the High-Altitude Mapping Orbit (HAMO) phase.

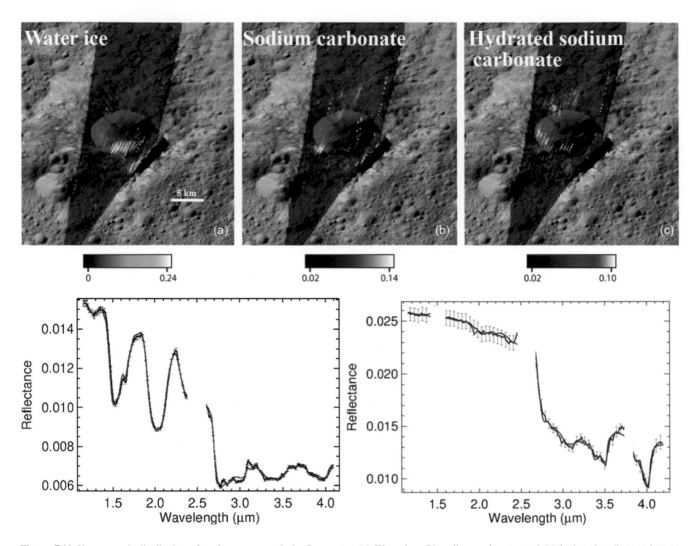

Figure 7.11 Upper panel: distribution of surface compounds in Oxo crater: (a) Water ice, (b) sodium carbonate, and (c) hydrated sodium carbonate. Abundances are derived from the spectral fitting method (Carrozzo et al., 2018). Left lower panel: the observed spectrum and model of ice rich region in Oxo crater (see also Combe et al., 2019). Right lower panel: the spectral fit (red curve) in Oxo using ammonium montmorillonite, dolomite, dark material, sodium carbonate, and hydrated sodium carbonate (Carrozzo et al., 2018).

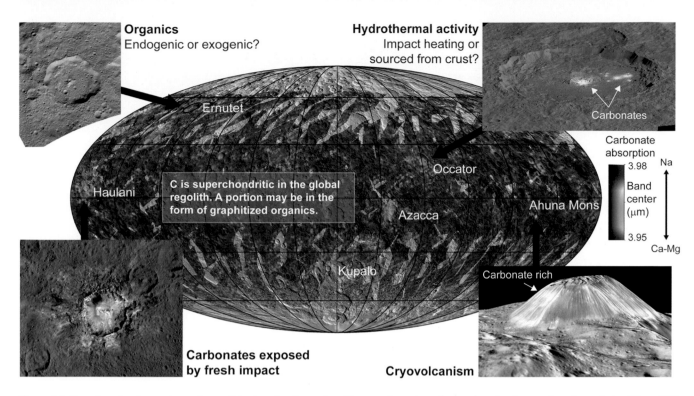

Figure 8.1 Overview of carbon and organics on Ceres based on Dawn data. The map shows the carbonate 4 μm band center from the analysis of Dawn/VIR spectra by Carrozzo et al. (2018). The prime meridian is on the far left of the map. The band center is shifted to longer wavelengths in bright regions, indicating the presence of Na-carbonates, in comparison to Ceres' global regolith, which is rich in Mg-Ca carbonates. Clockwise from the top left, the images show the distribution of organic matter (orange) in and around Ernutet crater (based on Framing Camera, visible spectral slope), the carbonate rich faculae in Occator crater, the largest mountain on Ceres, Ahuna Mons, and carbonates exposed by the impact that formed Haulani crater.
Image credits: NASA/JPL-Caltech/UCLA/MPS/DLR/IDA.

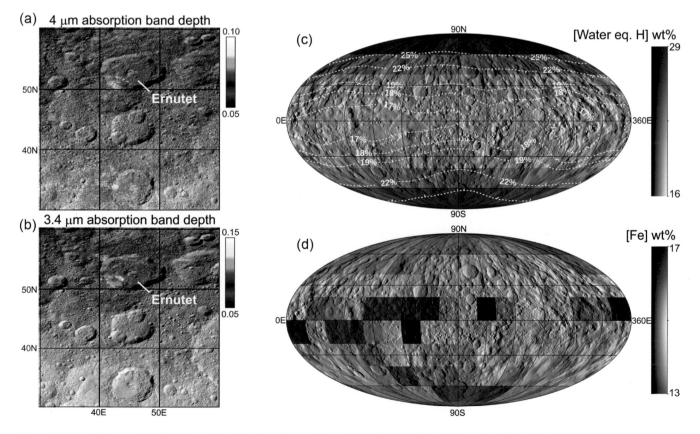

Figure 8.5 Mineral and elemental maps. Maps of the depth of the 3.4- and 4-μm bands, respectively, reveal the distribution of (a) carbonates and (b) aliphatic organic matter in and around Ernutet crater. The maps are from De Sanctis et al. (2017). Maps of water-equivalent H and Fe determined from GRaND data are shown in (c) and (d) (Prettyman et al., 2017, 2019b). The elemental maps are superimposed on shaded relief.

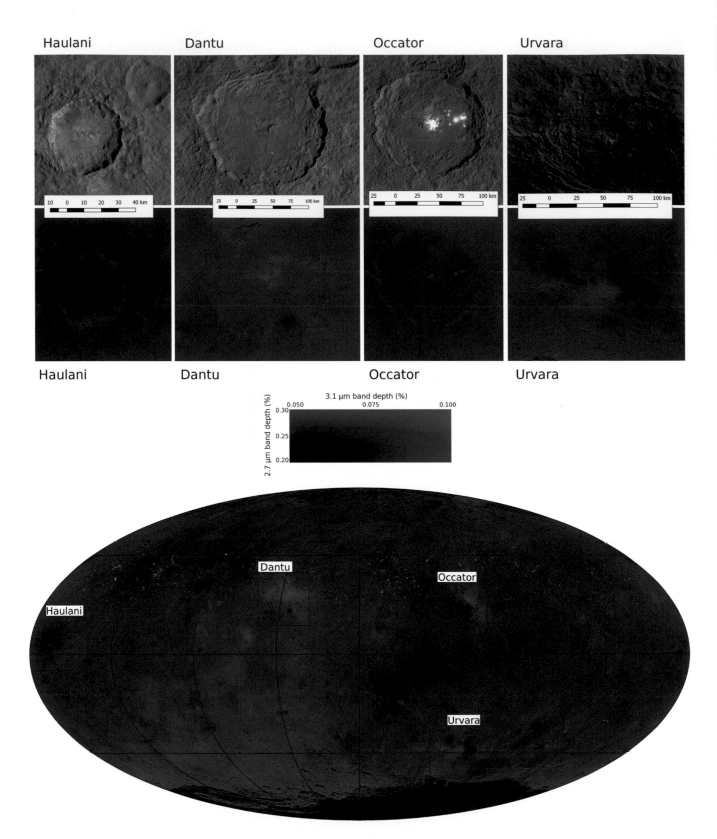

Figure 9.2 Combination of 2.7 μm band depth and 3.1 μm band depth according to the color map in the figure. Values in the map have been obtained merging the acquisition from RC3, survey, and High-Altitude Mapping Orbit Dawn mission phases. On the top there are the close ups of the four craters identified in the map. Data have been projected as described in Ammannito et al. (2016) and corrected for photometric effects as described in Ciarniello et al. (2017).

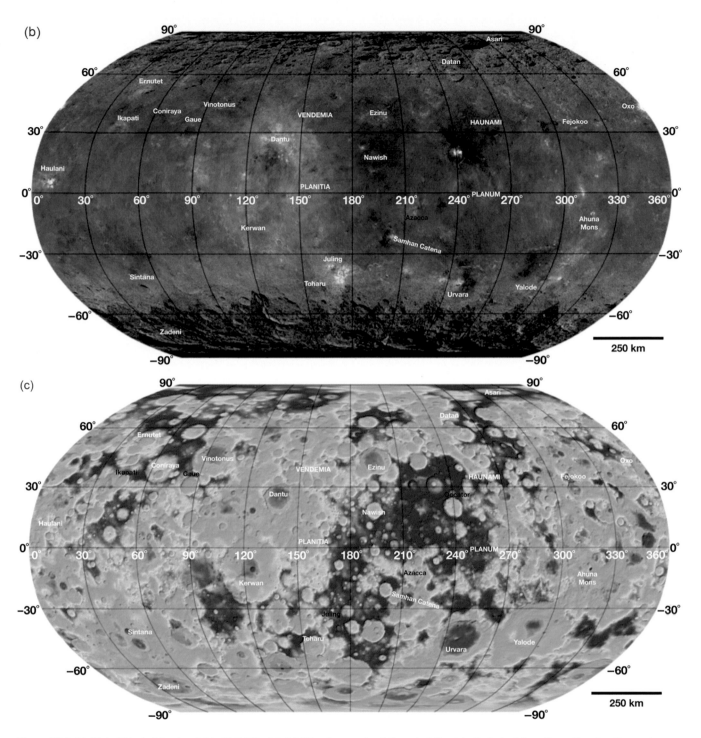

Figure 10.1 (b) High-Altitude Mapping Orbit (HAMO) global RGB color mosaic of Ceres (~140 m/pixel), derived from Dawn Framing Camera images (R = 0.96 μm, G = 0.75 μm, B = 0.44 μm). (c) HAMO SPG global digital terrain model (DTM) of Ceres, color-coded for surface elevation, derived from Dawn Framing Camera images. Mosaics are centered on 0°, 180° in Robinson projection, with a 30° latitude–longitude grid. Major named surface features are indicated.

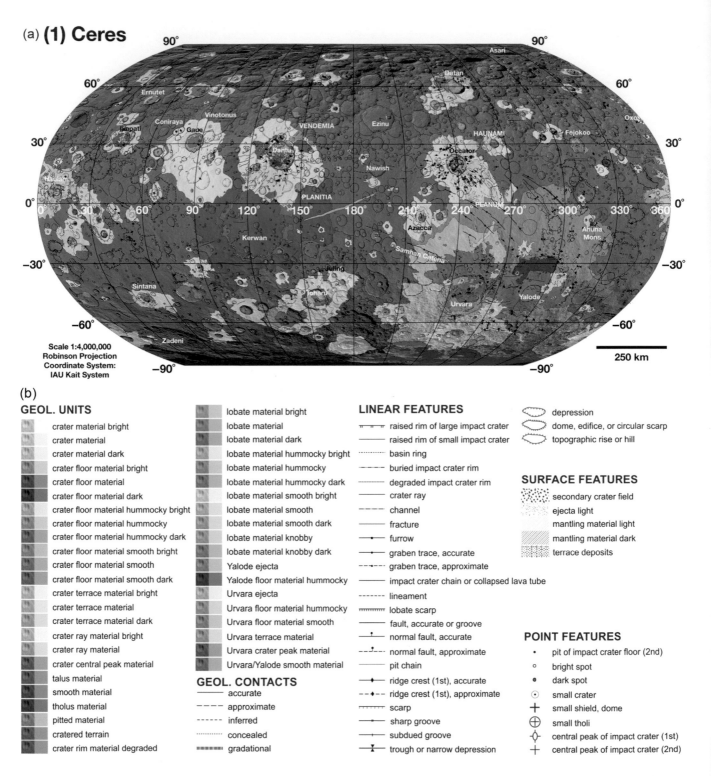

(a) (1) Ceres

Scale 1:4,000,000
Robinson Projection
Coordinate System:
IAU Kait System

250 km

(b)

GEOL. UNITS

- crater material bright
- crater material
- crater material dark
- crater floor material bright
- crater floor material
- crater floor material dark
- crater floor material hummocky bright
- crater floor material hummocky
- crater floor material hummocky dark
- crater floor material smooth bright
- crater floor material smooth
- crater floor material smooth dark
- crater terrace material bright
- crater terrace material
- crater terrace material dark
- crater ray material bright
- crater ray material
- crater central peak material
- talus material
- smooth material
- tholus material
- pitted material
- cratered terrain
- crater rim material degraded

- lobate material bright
- lobate material
- lobate material dark
- lobate material hummocky bright
- lobate material hummocky
- lobate material hummocky dark
- lobate material smooth bright
- lobate material smooth
- lobate material smooth dark
- lobate material knobby
- lobate material knobby dark
- Yalode ejecta
- Yalode floor material hummocky
- Urvara ejecta
- Urvara floor material hummocky
- Urvara floor material smooth
- Urvara terrace material
- Urvara crater peak material
- Urvara/Yalode smooth material

GEOL. CONTACTS
- —— accurate
- – – – approximate
- - - - - inferred
- ········· concealed
- ▪▪▪▪▪▪ gradational

LINEAR FEATURES
- ⊤⊤⊤⊤⊤ raised rim of large impact crater
- —— raised rim of small impact crater
- ········· basin ring
- – – – buried impact crater rim
- ·-·-·-· degraded impact crater rim
- —— crater ray
- – – – channel
- —— fracture
- —•— furrow
- —— graben trace, accurate
- –·–·– graben trace, approximate
- —— impact crater chain or collapsed lava tube
- - - - - lineament
- ⊓⊓⊓⊓⊓ lobate scarp
- —— fault, accurate or groove
- —ᵗ— normal fault, accurate
- ––ᵗ–– normal fault, approximate
- —— pit chain
- —♦— ridge crest (1st), accurate
- –♦– ridge crest (1st), approximate
- ⊤⊤⊤⊤ scarp
- —— sharp groove
- —— subdued groove
- —I— trough or narrow depression

- ⬭ depression
- ⬭ dome, edifice, or circular scarp
- ⬭ topographic rise or hill

SURFACE FEATURES
- secondary crater field
- ejecta light
- mantling material light
- mantling material dark
- terrace deposits

POINT FEATURES
- • pit of impact crater floor (2nd)
- ○ bright spot
- ● dark spot
- ⊙ small crater
- + small shield, dome
- ⊕ small tholi
- ⬦ central peak of impact crater (1st)
- + central peak of impact crater (2nd)

Figure 10.2 (a) Global geologic map of Ceres produced during the Dawn Primary Mission, using Low-Altitude Mapping Orbit (LAMO) images (~35 m/pixel) as the base map. This map is a unification of 15 quadrangle maps, published in a series of papers in *Icarus*, volume 316, December 2018. GIS processing by Andrea Nass, German Aerospace Center (DLR) and David M. Nelson (ASU). (b) Legend for the geological map in (a). GIS processing by Andrea Nass, German Aerospace Center (DLR) and David M. Nelson (ASU).

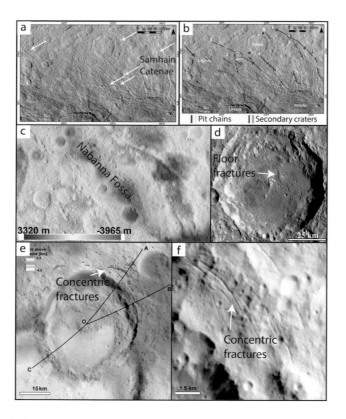

Figure 10.6 Examples of Ceres tectonic features. (a and b) The pit chains of Samhain Catena (red and yellow), with secondary crater chains (pink and magenta), in mapped and unmapped versions. (c) Nabanna Fossa in HAMO topographic data (highest and lowest elevations given in color bar). (d) Floor fractures within Azacca crater. (e and f) Concentric fractures surrounding Ikapati crater in mapped and unmapped versions. Figure adapted from Scully et al. (2017), Buczkowski et al. (2018, 2019), and Otto et al. (2019).

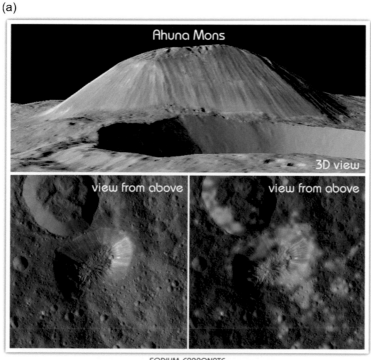

Figure 10.7 (a) Ahuna Mons on Ceres. Top: Perspective view derived from HAMO DTM. Bottom left: LAMO nadir image. Bottom right: LAMO image with overlay of sodium carbonate spectral signature, derived from VIR data. From NASA Photojournal PIA21919. (b) Topographic map of Ceres showing major named domes (tholi) and mountains (mons). Map credit: Preusker et al. (2016).

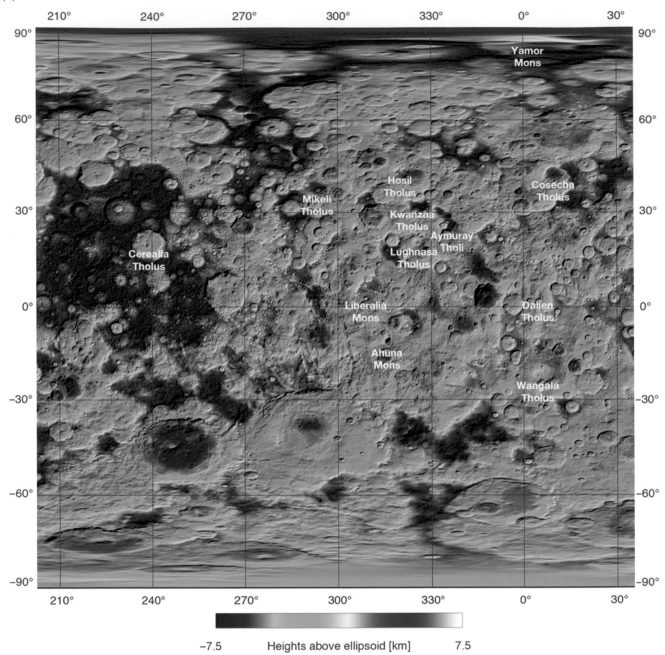

Figure 10.7 (*cont.*)

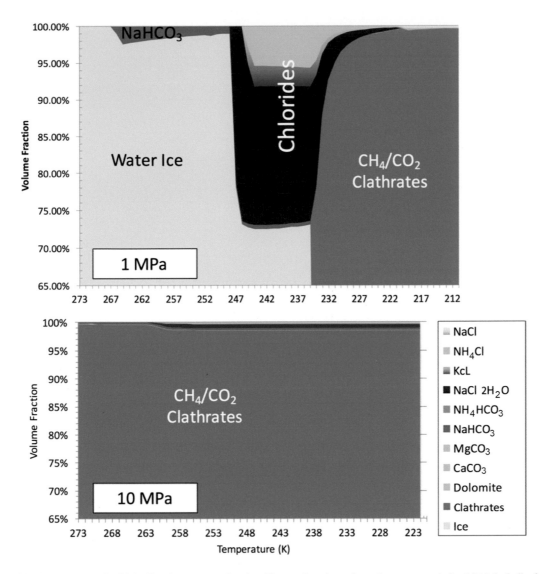

Figure 11.7 Precipitation sequence of solids in Ceres' ocean upon freezing. The top chart shows that at low pressures (a few MPa) the bulk of water freezes in the form of ice with a small fraction (20 vol.%) of salts. The bottom chart shows that under higher pressure, water freezes primarily in the form of clathrate hydrates, and salts are of low abundance (Castillo-Rogez et al., 2018).

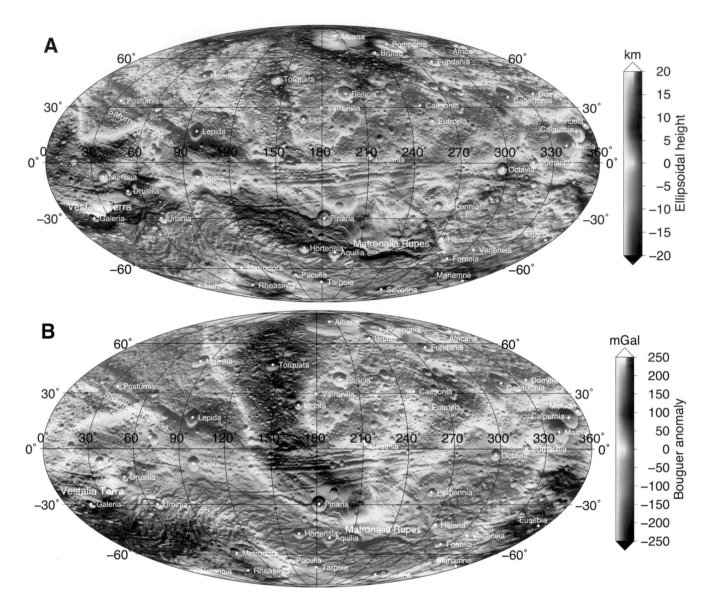

Figure 12.3 (a) Map of Vesta's topography in the Mollweide cartographic projection. (b) Map of Vesta's Bouguer gravity anomaly computed using the VESTA20G gravity model (Konopliv et al., 2014) and SPG shape model (Jaumann et al., 2012). The gravity anomalies are computed on a 293.2 × 266.5 km ellipsoid. The internal structure for computing the Bouguer anomaly was taken from table 3 in Ermakov et al. (2014). The Bouguer gravity anomaly was expanded up to $n = 16$. Here and elsewhere, longitudes increase eastward from 0° to 360°.

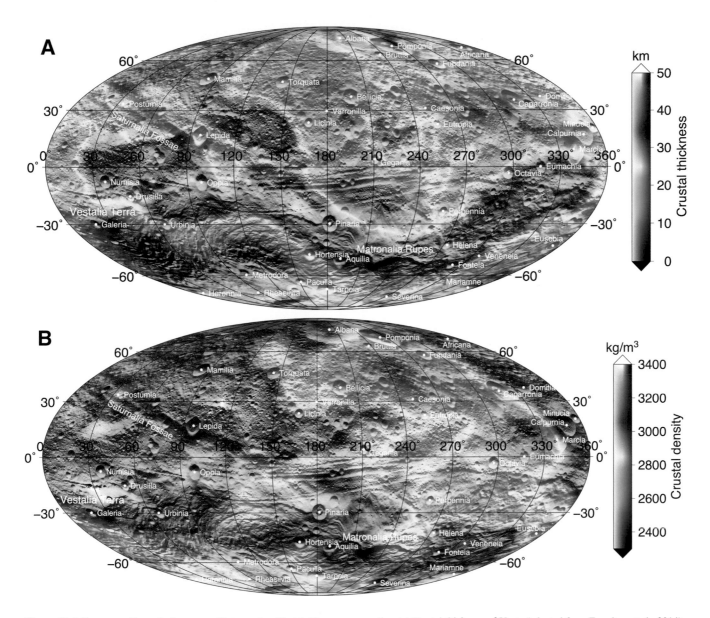

Figure 12.4 Two mutually exclusive ways of interpreting Vesta's Bouguer anomalies. (a) Crustal thickness of Vesta (adapted from Ermakov et al., 2014). Note that this map does not have a degree-1 crustal thickness variation since it was assumed that the degree-1 Bouguer anomaly is due to the core offset. (b) Crustal density variation assuming ellipsoidal core and mantle using the ellipsoidal harmonic model adapted from Park et al. (2014). Note that here the degree-1 term was included in the density inversion and, thus, results in a hemispheric crustal density dichotomy.

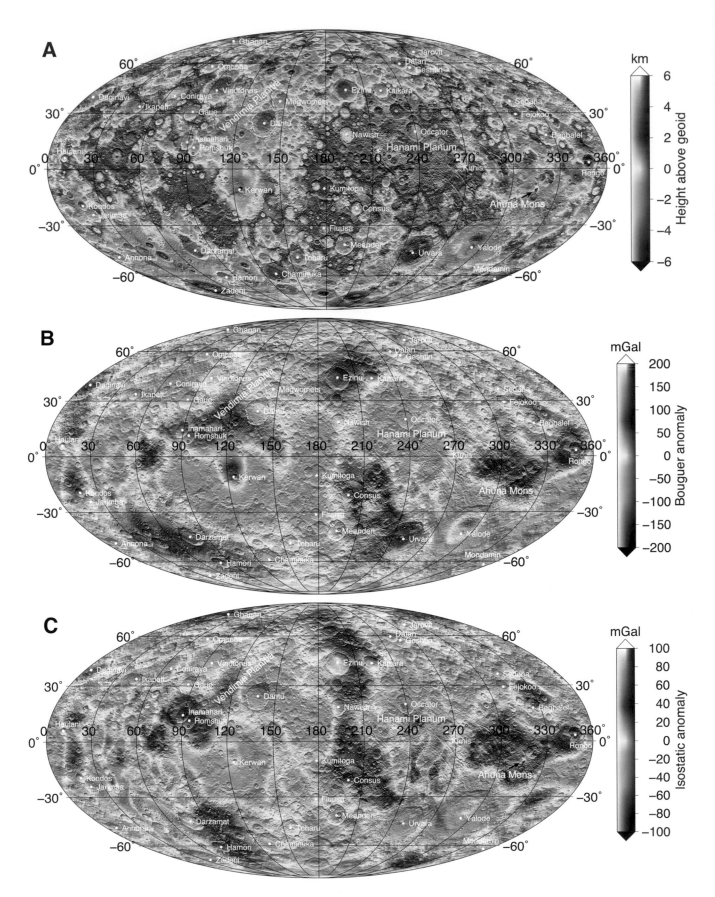

Figure 12.5 Maps of Ceres' (a) topography, (b) Bouguer anomaly, and (c) isostatic anomaly. The CERES70E gravity model (Park et al., 2020) and SPC shape model (Park et al., 2019) were used. The dynamic model of isostasy based on the viscous flow was used for computing the isostatic gravity anomaly for the parameters given in table 4 of Ermakov et al. (2017a). The gravity anomalies were expanded from degree-3 up to the local degree strength (Park et al., 2020) to prevent overwhelming the maps by the sectoral degree-2 anomaly.

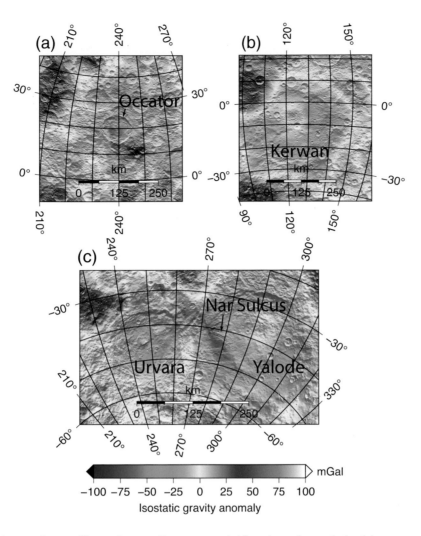

Figure 12.7 Regional isostatic anomaly maps. The gravity anomalies were expanded from degree-3 up to the local degree strength. (a) Hanami Planum and the Occator crater. A negative isostatic anomaly is seen southeast of Occator. (b) Kerwan basin. A bullseye is seen in the basin with a positive anomaly in the center and a negative annulus in the exterior. (c) Urvara and Yalode basins. Similarly to Kerwan, a bullseye pattern is seen in Yalode, whereas smaller Urvara is dominated by a negative isostatic anomaly.

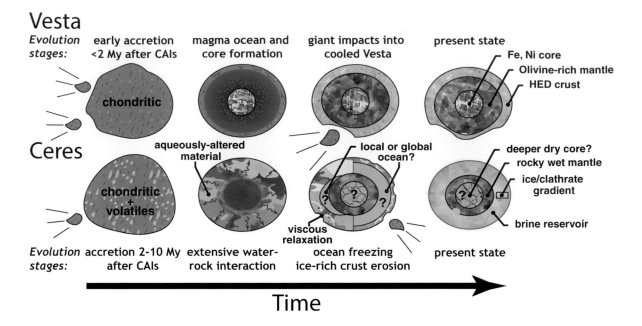

Vesta

Evolution stages:

early accretion <2 My after CAIs | magma ocean and core formation | giant impacts into cooled Vesta | present state

chondritic

- Fe, Ni core
- Olivine-rich mantle
- HED crust

Ceres

aqueously-altered material

chondritic + volatiles

local or global ocean?

viscous relaxation

deeper dry core?
rocky wet mantle
ice/clathrate gradient
brine reservoir

Evolution stages: accretion 2-10 My after CAIs | extensive water-rock interaction | ocean freezing ice-rich crust erosion | present state

Time

Figure 12.8 A comparison of Vesta and Ceres evolutionary scenarios. The drawing is not to scale, and individual phases of the evolution of Vesta are not meant to timewise correspond to the phases of evolution of Ceres.

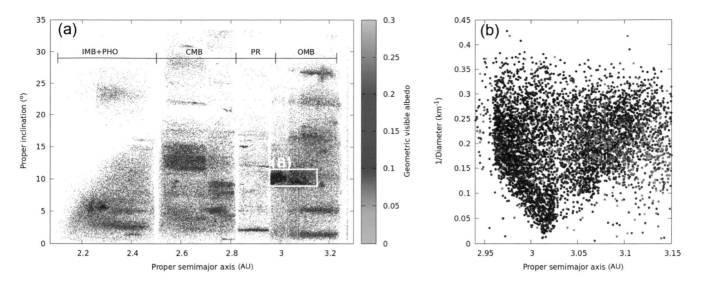

Figure 13.1 The asteroid Main Belt and its collisional families. (a) Proper orbital inclination versus proper orbital semi-major axis of all asteroids with a measured diameter larger or equal to 1 km. Each point represents an asteroid. The different regions of the Main Belt are presented with the following symbols: IMB = Inner Main Belt (below ≈ 18° of proper inclination); PHO = Phocaea group (above ≈ 18° of proper inclination); CMB = Central Main Belt; PR = pristine zone (as named by Brož et al., 2013); OMB = Outer Main Belt. If we extract asteroids with a proper semimajor axis between 2.94 and 3.15 AU, proper inclination between ≈ 9.8° and ≈ 11.2°, proper eccentricity between 0.03 and 0.1 (the white box in (a)), and we plot their inverse diameter as a function of their proper semimajor axis we obtain the plot in (b). This shows a typical V-shape (the Eos family), with its vertex on the x-axis at about 3.014 AU. The inclined inward border of the V-shape is very sharp and is cut at about 1/diameter 0.15 km^{-1} by the 7:3 major mean motion resonance with Jupiter (see text). The outward border is less clear and starts to overlap with background asteroids at 3.07 AU and 0.15 km^{-1}. Source of the data mp3c.oca.eu; download November 2020; For the reference of the proper elements, diameters and albedo see Delbo et al. (2019).

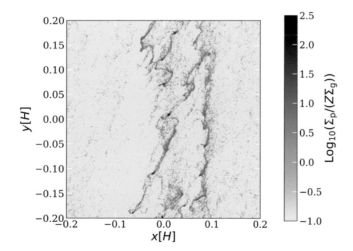

Figure 13.4 3D numerical simulation of planetesimal formation in the presence of streaming instability. Simulations are conducted in the shearing box approximation where x is the radial coordinate and y the azimuthal coordinate. Shown are vertically integrated, normalized particle densities projected onto the disk plane. The realistic conditions in the MB region (see Figure 13.3) are so far not accessible for numerical simulation, thus simulation results have to be extrapolated physical relevant conditions. This snapshot shows that self-gravity can lead to fragmentation of streaming instability filaments and the formation of distinct clumps which are commonly identified as planetesimals. See Gerbig et al. (2020) for a discussion of this particular setup.

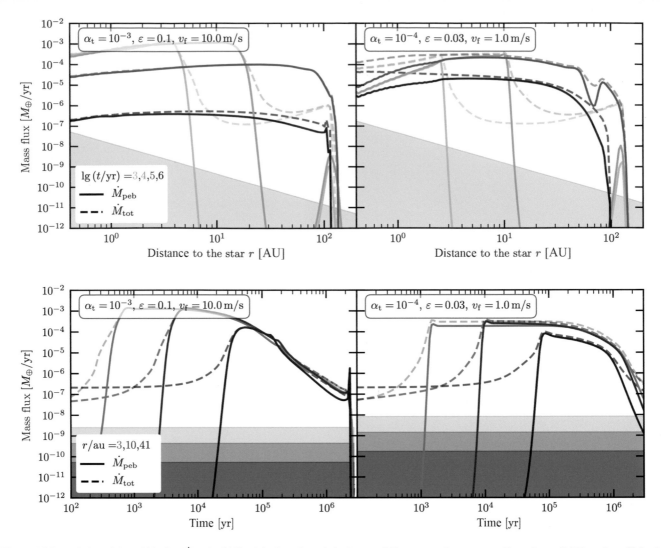

Figure 13.8 Evolution of the pebble flux $\dot{M}_{peb}$ (solid lines) in the solar nebula for two different sets of parameters. Since the availability of a sufficiently massive pebble flux is the key requirement for planetesimal formation (Lenz et al., 2019), the formation of the Asteroid Belt's initial planetesimals (see curves for 3 AU) is expected to last up to ~10^5 to ~10^6 years, which is when the pebble flux starts to decline. The right panels depict the scenario that is preferred by Lenz et al. (2020), with $\epsilon_{d=5H} = 0.03$, $\alpha = 10^{-4}$, and a fragmentation speed of 1 m/s. The plot also shows the total mass flux $\dot{M}_{tot}$, also accounting for smaller dust grains that do not efficiently participate in planetesimal formation. The shaded areas indicate where the mass flux is too low to support the formation of planetesimals. We refer to figure 1 of Lenz et al. (2019) for a detailed discussion of a similar plot.

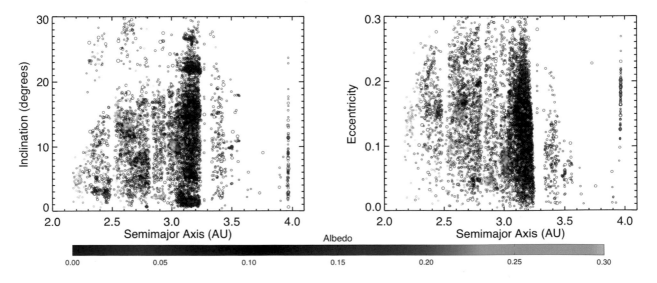

Figure 15.1 Orbital and compositional distribution of 8,237 asteroids larger than 10 km in diameter. The symbol size is proportional to actual size and the colors represent asteroid albedos. Typical albedos of C-types are ~ 5% and of S-types ~20% (Pravec et al., 2012). Asteroids with albedos above 0.3 were given colors corresponding to albedos of 0.3. Asteroid data from mp3c.oca.eu

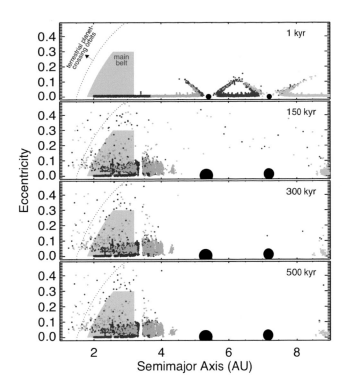

Figure 15.2 Injection of planetesimals into the inner Solar System due to Jupiter and Saturn's rapid gas accretion (adapted from Raymond & Izidoro, 2017a). In this simulation the giant planets' cores (initially $3M_\oplus$) underwent rapid gas accretion from 100–200 kyr (for Jupiter) and 300–400 kyr (for Saturn). The gaseous disk structure responded to the giant planets' growth (from hydrodynamical models of Morbidelli & Crida, 2007). As the giant planets increased in mass, the orbits of nearby planetesimals (whose colors correspond to their starting orbital radii) were destabilized and gravitationally scattered by the giant planets onto eccentric orbits. Under the action of gas drag (assuming $D = 100$ km), some planetesimals were trapped on stable orbits within the Main Belt, originating in a region extending roughly from 4–9 AU. These may be the present-day C-type asteroids. In this simulation giant planet migration was not accounted for, so the 5 AU-wide source region for C-types is the narrowest it could possibly be; when migration is included planetesimals can be implanted into the belt from as far out as 20–30 AU (Raymond & Izidoro, 2017a; Pirani et al., 2019a).

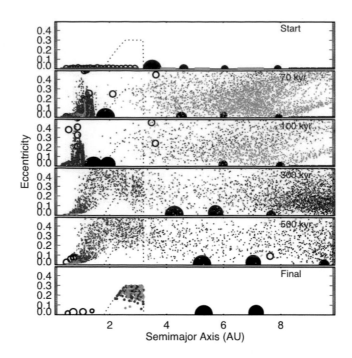

Figure 15.3 Snapshots in the evolution of the Grand Tack model. The simulation starts a few million years into the gaseous disk phase: Jupiter is fully grown, Saturn and the ice giants are large cores, and planetary embryos have formed in the inner Solar System. Jupiter migrates inward through rocky planetesimals and embryos, some of which are shepherded by inner resonances and some of which are scattered outward (see Raymond et al., 2006a; Mandell et al., 2007). Saturn grows and migrates inward, is trapped in 2:3 resonance with Jupiter, and both planets migrate back outward. Despite Jupiter's traversing, the main asteroid belt is populated by S-types (red) originating interior to Jupiter's initial orbit, and C-types from belts originally located between the giant planets' cores (light blue) and from a disk of planetesimals initially exterior to the giant planets (dark blue). The inner belt is dominated by S-types and the outer belt by C-types (Walsh et al., 2011, 2012), consistent with the compositional (Gradie & Tedesco, 1982; DeMeo & Carry, 2014) and orbital (Deienno et al., 2016) structure of the present-day belt.

Milliken, R. E., & Rivkin, A. S. (2009) Brucite and carbonate assemblages from altered olivine-rich materials on Ceres. *Nature Geoscience*, 2, 258–261.

Moroz, L. V., Arnold, G., Korochantsev, A. V., & Wäsch, R. (1998) Natural solid bitumens as possible analogs for cometary and asteroid organics: 1. Reflectance spectroscopy of pure bitumens. *Icarus*, 134, 253–268.

Nathues, A., Platz, T., Thangjam, G., et al. (2017). Evolution of Occator crater on (1) Ceres. *The Astronomical Journal*, 153, 112.

Nathues, A., Platz, T., Thangjam, G., et al. (2019) Occator crater in color at highest spatial resolution. *Icarus*, 320, 24–38.

Neesemann, A., van Gasselt, S., Schmedemann, N., et al. (2019) The various ages of Occator crater, Ceres: Results of a comprehensive synthesis approach. *Icarus*, 320, 60–82.

Orthous-Daunay, F.-R., Quirico, E., Beck, P., et al. (2013) Mid-infrared study of the molecular structure variability of insoluble organic matter from primitive chondrites. *Icarus*, 223, 534–543.

Palomba, E., Longobardo, A., De Sanctis, M. C., et al. (2019) Compositional differences among bright spots on the Ceres surface. *Icarus*, 320, 202–212.

Parker, J. W., Stern, S. A., Thomas, P. C., et al. (2002) Analysis of the first disk-resolved images of Ceres from ultraviolet observations with the Hubble Space Telescope. *The Astronomical Journal*, 123, 549.

Pearson, V. K., Sephton, M. A., Kearsley, A. T., et al. (2002) Clay mineral-organic matter relationships in the early Solar System. *Meteoritics & Planetary Science*, 37, 1829–1833.

Pendleton, Y. J., & Allamandola, L. J. (2002) The organic refractory material in the diffuse interstellar medium: Mid-infrared spectroscopic constraints. *The Astrophysical Journal Supplement Series*, 138, 75–98.

Pieters, C. M., Nathues, A., Thangjam, G., et al. (2018) Geologic constraints on the origin of red organic-rich material on Ceres. *Meteoritics & Planetary Science*, 53, 1983–1998.

Pieters, C. M., & Noble, S. K. (2016) Space weathering on airless bodies. *Journal of Geophysical Research (Planets)*, 121, 1865.

Poch, O., Istiqomah, I., Quirico, E., et al. (2020) Ammonium salts are a reservoir of nitrogen on a cometary nucleus and possibly on some asteroids. *Science*, 367, aaw7462.

Poch, O., Jaber, M., Stalport, F., et al. (2015) Effect of nontronite smectite clay on the chemical evolution of several organic molecules under simulated martian surface ultraviolet radiation conditions. *Astrobiology*, 15, 221.

Postberg, F., Kempf, S., Schmidt, J., et al. (2009) Sodium salts in E-ring ice grains from an ocean below the surface of Enceladus. *Nature*, 459, 1098–1101.

Postberg, F., Schmidt, J., Hillier, J., Kempf, S., & Srama, R. (2011) A salt-water reservoir as the source of a compositionally stratified plume on Enceladus. *Nature*, 474, 620–622.

Prettyman, T. H., Feldman, W. C., McSween, H. Y., et al. (2011) Dawn's gamma ray and neutron detector. *Space Science Reviews*, 163, 371–459.

Prettyman, T. H., Yamashita, N., Toplis, M. J., et al. (2017) Extensive water ice within Ceres' aqueously altered regolith: Evidence from nuclear spectroscopy. *Science*, 355, 55–59.

Quick, L. C., Buczkowski, D. L., Ruesch, O., et al. (2019) A possible brine reservoir beneath Occator crater: Thermal and compositional evolution and formation of the Cerealia dome and Vinalia Faculae. *Icarus*, 320, 119–135.

Raponi, A., Carrozzo, F. G., Zambon, F., et al. (2019a) Mineralogical mapping of Coniraya quadrangle of the dwarf planet Ceres. *Icarus*, 318, 99–110.

Raponi, A., Ciarniello, M., Capaccioni, F., et al. (2020) Infrared detection of aliphatic organics on a cometary nucleus. *Nature Astronomy*, 4, 500.

Raponi, A., De Sanctis, M. C., Carrozzo, F. G., et al. (2019b) Mineralogy of Occator crater on Ceres and insight into its evolution from the properties of carbonates, phyllosilicates, and chlorides. *Icarus*, 320, 83–96.

Raponi, A., De Sanctis, M. C., Frigeri, A., et al. (2018) Variations in the amount of water ice on Ceres' surface suggest a seasonal water cycle. *Science Advances*, 4, eaao3757.

Raymond, C. A., Ermakov, A. I., Castillo-Rogez, J. C., et al. (2020). Impact-driven mobilization of deep crustal brines on dwarf planet Ceres. *Nature Astronomy*, 4, 741–747.

Rivkin, A. S., & Emery, J. P. (2010) Detection of ice and organics on an asteroidal surface. *Nature*, 464, 1322.

Rivkin, A. S., Li, J.-Y., Milliken, R. E., et al. (2011) The surface composition of Ceres. *Space Science Reviews*, 163, 95–116.

Rivkin, A. S., Volquardsen, E. L., & Clark, B. E. (2006) The surface composition of Ceres: Discovery of carbonates and iron-rich clays. *Icarus*, 185, 563–567.

Roatsch, T., Kersten, E., Matz, K.-D., et al. (2016) Ceres survey atlas derived from Dawn Framing Camera images. *Planetary and Space Science*, 121, 115–120.

Roettger, E. E., & Buratti, B. J. (1994) Ultraviolet spectra and geometric albedos of 45 asteroids. *Icarus*, 112, 496.

Rousseau, B., De Sanctis, M. C., Raponi, A., et al. (2020) Correction of the VIR-visible dataset from the Dawn mission at Vesta. *Review of Scientific Instruments*, 91, 123102.

Ruesch, O., Platz, T., Schenk, P., et al. (2016) Cryovolcanism on Ceres. *Science*, 353, aaf4286.

Schröder, S. E., Mottola, S., Carsenty, U., et al. (2017) Resolved spectro-photometric properties of the Ceres surface from Dawn Framing Camera images. *Icarus*, 288, 201–225.

Schulte, M., & Shock, E. (2004) Coupled organic synthesis and mineral alteration on meteorite parent bodies. *Meteoritics & Planetary Science*, 39, 1577–1590.

Scully, J. E. C., Schenk, P. M., Castillo-Rogez, J. C., et al. (2020) The varied sources of faculae-forming brines in Ceres' Occator crater emplaced via hydrothermal brine effusion. *Nature Communications*, 11, 3680.

Sierks, H., Keller, H. U., Jaumann, R., et al. (2011) The Dawn Framing Camera. *Space Science Reviews*, 163, 263–327.

Stein, N. T., Ehlmann, B. L., Palomba, E., et al. (2019) The formation and evolution of bright spots on Ceres. *Icarus*, 320, 188–201.

Takir, D., & Emery, J. P. (2012) Outer Main Belt asteroids: Identification and distribution of four 3-µm spectral groups. *Icarus*, 219, 641–654.

Takir, D., Emery, J. P., McSween, H. Y., et al. (2013) Nature and degree of aqueous alteration in CM and CI carbonaceous chondrites. *Meteoritics & Planetary Science*, 48, 1618–1637.

Tosi, F., Carrozzo, F. G., Zambon, F., et al. (2019) Mineralogical analysis of the Ac-H-6 Haulani quadrangle of the dwarf planet Ceres. *Icarus*, 318, 170–187.

Vernazza, P., Mothé-Diniz, T., Barucci, M. A., et al. (2005) Analysis of near-IR spectra of 1 Ceres and 4 Vesta, targets of the Dawn mission. *Astronomy and Astrophysics*, 436, 1113–1121.

Vinogradoff, V., Bernard, S., Le Guillou, C., & Remusat, L. (2018) Evolution of interstellar organic compounds under asteroidal hydro-thermal conditions. *Icarus*, 305, 358–370.

Vinogradoff, V., Le Guillou, C., Bernard, S., et al. (2017) Paris vs. Murchison: Impact of hydrothermal alteration on organic matter in CM chondrites. *Geochimica et Cosmochimica Acta*, 212, 234.

Vu, T. H., Hodyss, R., Johnson, P. V., & Choukroun, M. (2017) Preferential formation of sodium salts from frozen sodium-ammonium-chloride-carbonate brines – Implications for Ceres' bright spots. *Planetary and Space Science*, 141, 73–77.

Waite, Jr., J. H., Lewis, W. S., Magee, B. A., et al. (2009) Liquid water on Enceladus from observations of ammonia and ^{40}Ar in the plume. *Nature*, 460, 487–490.

Williams, D. A., Buczkowski, D. L., Mest, S. C., et al. (2018) Introduction: The geologic mapping of Ceres. *Icarus*, 316, 1–13.

Williams, L. B., Canfield, B., Voglesonger, K. M., & Holloway, J. R. (2005) Organic molecules formed in a "primordial womb". *Geology*, 33, 913–916.

Zambon, F., Raponi, A., Tosi, F., et al. (2017) Spectral analysis of Ahuna Mons from Dawn mission's visible-infrared spectrometer. *Geophysical Research Letters*, 44, 97–104.

Zolotov, M. Y. (2007) An oceanic composition on early and today's Enceladus. *Geophysical Research Letters*, 34, L23203.

Zolotov, M. Y., & Shock, E. L. (2001) Composition and stability of salts on the surface of Europa and their oceanic origin. *Journal of Geophysical Research*, 106, 32815–32827.

Carbon and Organic Matter on Ceres

THOMAS PRETTYMAN, MARIA CRISTINA DE SANCTIS, AND SIMONE MARCHI

8.1 INTRODUCTION

Carbon is central to astrobiology and influenced the development of water-rich planetesimals in the early Solar System. As a candidate ocean world with outer Solar System origins, exploration of Ceres has the potential to provide new insights into prebiotic chemistry (Castillo-Rogez et al., 2020). In this chapter, we summarize observations of carbon and organics on dwarf planet Ceres, including the results of the NASA Dawn mission and recent telescopic observations and their interpretation. We place these results in context with astrophysical processes that produced carbon and organics in nebular materials that condensed to form Ceres. Moreover, we consider the possibility that accreted organic matter could have been altered and/or synthesized within Ceres' interior. The discussion is framed by results of studies of meteorites and other Solar System materials, for which the aqueously altered CI carbonaceous chondrites are Ceres' closest analog.

Early orbital observations by Dawn show that Ceres' dark, global surface contains aqueous alteration assemblages, including Mg-rich phyllosilicates and carbonates as well as unidentified opaque phases (e.g., De Sanctis et al., 2015). That Mg-phyllosilicates are pervasive suggests surface materials have experienced a high degree of alteration (McSween et al., 2017). Dotting the surface are bright salt-rich deposits, some of which are associated with fresh impacts, such as Occator and Haulani craters (e.g., Carrozzo et al., 2018) (Figure 8.1). The bright spots may have formed via impact-triggered hydrothermal processes and/or the release of subsurface carbonate-rich brines sourced from the base of the crust (Bowling et al., 2019; Quick et al., 2019). Studies of Occator crater suggest that both mechanisms contributed to the formation of the faculae (Raymond et al., 2020; Scully et al., 2020). The surface mineralogy shows that materials that accreted to form Ceres were profoundly altered by water–rock interactions at low temperature within the interior (De Sanctis et al., 2016; Prettyman et al., 2017; Castillo-Rogez et al., 2018). Ceres shows endogenic activity: The largest mountain on Ceres, Ahuna Mons, is a candidate cryovolcano (Ruesch et al., 2016). The extruded

cryomagma was likely a slurry of brine and solid particles sourced from a mantle plume (Ruesch et al., 2019).

The geophysical evidence for Ceres' water-rich origins and implications for interior structure are described in detail elsewhere (Chapters 11 and 12). The geochemical inferences for the structure and chemical composition of Ceres interior support a complex evolutionary history in which carbon has played an important role. Dawn gravity data and geomorphology are consistent with a low density ($\sim$1.3 g/cm^3), approximately 40-km thick crust, likely consisting of water ice, phyllosilicates, salts, and hydrate clathrates overlying a rocky mantle (Bland et al., 2016; Ermakov et al., 2017; Fu et al., 2017). The crust is supported by a weak layer that contains residual brines, perhaps the remnants of an ancient subsurface ocean.

The inferred presence of ammonium in clay minerals indicates Ceres' crust differs compositionally from CI chondrites and implies an outer Solar System origin (De Sanctis et al., 2015). Ceres probably formed in the outer Solar System and was implanted into the Main Belt along with other water-rich bodies in response to the growth of the giant planets (e.g., Vokrouhlický et al., 2016; Raymond & Izidoro, 2017; refer to Chapters 15 and 17). As such, Ceres likely grew from nebular grains containing silicates, organics, and ices, including water, carbon dioxide, methane, and ammonia. The timing of formation and/or water-to-silicate ratio of the accreted materials resulted in melting of ice within the interior, perhaps from heat generated by the decay of the short-lived radionuclide ^{26}Al, as proposed for the parent bodies of the carbonaceous chondrites (Grimm & McSween, 1993; Castillo-Rogez et al., 2018). The melting of ice initiated serpentinization, which led to the physically differentiated, aqueously altered body that we see today. Conditions within the interior would have modified the accreted carbon, producing carbonates and possibly altering organic compounds to synthesize more complex species (e.g., Holm et al., 2015). As discussed in Section 8.2.3, the latter requires high hydrogen partial pressure that could have accompanied serpentinization deep within Ceres interior.

Key observations of carbon and organics include the aforementioned geophysical constraints from Dawn, surface studies with Dawn's instrument suite, which includes a Framing Camera (FC) (Sierks et al., 2011), Visible to Infrared Mapping Spectrometer (VIR) (De Sanctis et al., 2011), and the Gamma Ray and Neutron Detector (GRaND) (Prettyman et al., 2011), and telescopic observations of Ceres at ultraviolet (UV) wavelengths (Hendrix et al., 2016a). The geophysical data along with GRaND measurements provide constraints on the total amount of carbon in Ceres' regolith (Marchi et al., 2019; Prettyman et al., 2019b). The UV data hint at the widespread presence of graphitized carbon species on Ceres'

This work was supported by the NASA SSERVI Toolbox for Research and Exploration (TREX) project and by the Discovery Data Analysis Program (Grant Number 80NSSC20K1153). We are grateful to Amanda Hendrix for providing the UV data and model shown in Figure 8.3 and to Eleonora Ammannito for providing the maps shown in Figure 8.4. The GRaND data, including the elemental maps presented here, are available from the NASA Planetary Data System (https://sbn.psi.edu/pds/resource/dawn/dawngrandPDS4.html). Critical reviews by H. Y. McSween and an anonymous reviewer greatly improved the manuscript.

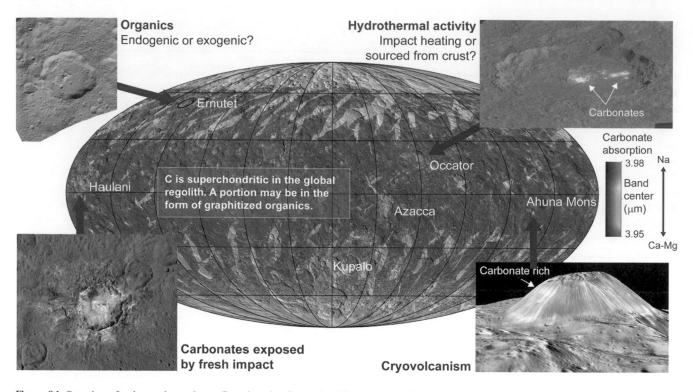

Figure 8.1 Overview of carbon and organics on Ceres based on Dawn data. The map shows the carbonate 4 μm band center from the analysis of Dawn/VIR spectra by Carrozzo et al. (2018). The prime meridian is on the far left of the map. The band center is shifted to longer wavelengths in bright regions, indicating the presence of Na-carbonates, in comparison to Ceres' global regolith, which is rich in Mg-Ca carbonates. Clockwise from the top left, the images show the distribution of organic matter (orange) in and around Ernutet crater (based on Framing Camera, visible spectral slope), the carbonate rich faculae in Occator crater, the largest mountain on Ceres, Ahuna Mons, and carbonates exposed by the impact that formed Haulani crater. Image credits: NASA/JPL-Caltech/UCLA/MPS/DLR/IDA

A black and white version of this figure will appear in some formats. For the color version, refer to the plate section.

surface, possibly the result of space weathering of organic matter (Hendrix et al., 2016a). The VIR data enabled detailed characterization of carbonates (e.g., Carrozzo et al., 2018) and the detection of aliphatic organics within and around Ernutet crater (De Sanctis et al., 2017). Multifilter imagery (FC) of the characteristic red slope of the organic materials detected by VIR provides high spatial resolution context for the organics, suggesting possible mechanisms for emplacement (Pieters et al., 2018; Bowling et al., 2020).

Ceres carbon-rich surface results from a combination of impacts and complex processes that occurred within Ceres' interior, including low-temperature aqueous alteration, physical differentiation via ice–rock fractionation, and modification of the accreted carbon species during serpentinization. In this chapter, we provide a review of current evidence, including sources of carbon, parent body processes, remote sensing observations, and their interpretation.

8.2 CARBON AND ORGANICS: SOURCES AND PROCESSES

8.2.1 Carbon Chemistry in the Interstellar Medium and Solar Nebula

Our Solar System condensed from a giant molecular cloud, likely in the vicinity of massive stars (Armitage, 2020). In this scenario, shock compression by ultraviolet radiation from a nearby supernova triggered gravitational collapse of the cloud to form the early Sun and protoplanetary disk from which the planets accreted (e.g., Bally et al., 2000). Nearby supernovae produced short-lived radionuclides such as ^{26}Al that were injected into the solar nebula (Sahijpal et al., 1998), providing a heat source that influenced the evolution of planetary embryos.

Organic matter has its origins in the interstellar medium and protoplanetary disks. Within the dense, cold cloud of gas surrounding a protostar (<30 K), gas phase chemistry can produce simple molecules and carbon chains. Carbon-bearing species made by this process include CO, C_2H_2, C_2H_4, and HCN (Ehrenfreund & Charnley, 2000). In addition, atoms and molecules collide with and adhere to fine silicate particles, resulting in the formation of icy mantles (Charnley, 1994; Ehrenfreund & Charnley, 2000). Gas-grain chemistry at cryotemperatures produces CO_2 and CH_3OH as well as more complex organic species (Hasegawa et al., 1992). Ethanol, acetaldehyde, ketene, propanal, isocyanic acid, and formamide are made on the surface of grains (Ehrenfreund & Charnley, 2000). As the cloud is heated by the young star, the dust grains warm (>100 K), releasing ice and volatiles. Gas phase chemistry within hot cores may produce yet more complex organic molecules, with additional speciation resulting from exposure to UV radiation and cosmic rays (e.g., production of hydrogenated, amorphous carbon). At high temperatures, Fischer-Tropsch type reactions can form carbonaceous coatings on grains, resulting in the production of macromolecular carbon (Nuth et al., 2008). Large molecules, including fullerenes (C_{60} and C_{70}), have been observed in young planetary nebulae (Cami et al., 2010). As a result of these processes, a rich carbon chemistry was available for assimilation into planetary embryos and later delivery to the inner Solar System.

8.2.2 Carbon in Analog Materials: Comets, IDPs, and Meteorites

Comets are agglomerations of ice and dust that condensed at low temperature from the solar nebula, most likely in the outer Solar System, beyond the orbits of Neptune and Uranus (e.g., Brandt, 2014). Particles collected from comet 81P/Wild 2 by the Stardust mission include chondritic refractory minerals that must have formed close to the Sun, indicating large-scale mixing of nonvolatile components within the solar nebula (Brownlee et al., 2006). Thermal processing and chemical alteration within the icy planetesimals from which comets are derived is thought to be negligible. Consequently, comets are among the most primitive objects in the Solar System.

Telescopic observations of comets and in situ characterization of dust by spacecraft show that comets are carbon-rich, containing organic compounds, including high molecular weight organic molecules like the insoluble organic matter found in meteorites (e.g., Jessberger et al., 1988; Fray et al., 2016). Interplanetary dust particles (IDPs) have been collected in Earth's stratosphere and from Antarctic snow. Some anhydrous, pyroxene-rich chondritic IDPs have significantly higher abundances of carbon than the carbonaceous chondrite meteorites (Thomas et al., 1993). These are thought to be good candidates for cometary dust (e.g., Fomenkova et al., 1994; Duprat et al., 2010); however, thermal processing of dust particles during their passage through the atmosphere can alter their carbon chemistry (Riebe et al., 2020). Organic matter is found in similar abundances and types in anhydrous chondritic IDPs and those containing hydrated silicates formed by aqueous processing on water-rich asteroids (Flynn et al., 2003). This implies that much of the prebiotic matter in the Solar System was formed non-aqueously within the solar nebula and/or the interstellar medium. Investigations of comets and IDPs connect carbon and organics in the interstellar medium and solar nebula to meteorites. The latter constitute large samples that can be subjected to detailed laboratory analyses with reduced susceptibility to the effects of atmospheric processing.

Meteorites can be subdivided into three broad categories: stones, stony-irons, and irons (e.g., Weisberg et al., 2006). Stony meteorites include chondrites and achondrites. Achondrites have experienced igneous processing (melting and recrystallization) within their parent bodies, whereas chondrites have not been modified by melting. A portion of chondrites contain spherical silicate chondrules with calcium and aluminum rich inclusions (CAIs), the earliest Solar System condensates. The most primitive meteorites are the carbonaceous chondrites. These are further subdivided into groups based on their chemistry and petrology. The Renazzo (CR), Mighei (CM), and Ivuna (CI) groups have experienced varying degrees of aqueous alteration on their parent bodies. Stable nucleosynthetic isotope anomalies indicate that the parent bodies of carbonaceous chondrites formed from a separate reservoir, distinct from most other meteorites (Warren, 2011; Scott et al., 2018). This may reflect partitioning of the inner and outer Solar Systems by a long-lived pressure maximum that triggered the formation of Jupiter (Brasser & Mojzsis, 2020). Rapid in situ growth of Jupiter subsequently scattered carbonaceous asteroids into the Main Belt (Vokrouhlický et al., 2016; Kretke et al., 2017; Raymond & Izidoro, 2017). For more details, see Chapter 14.

The CI chondrites are the most altered meteorites, with high concentrations of H and C, an abundance of hydrated minerals, carbonates, and organic matter (Figure 8.2(a)) (McSween et al., 2017). They lack chondrules, which may have been destroyed by exposure to liquid water at low temperatures within the interior of their parent body. With few exceptions, the abundances of rock-forming elements in CI chondrites are nearly identical to that of the solar photosphere, thought to be representative of the composition of the solar nebula (Anders & Grevesse, 1989). This implies that alteration of the accreted materials was isochemical – with constant elemental composition – and occurred within a closed system on the CI parent body.

Petrologic studies of CI chondrites indicate metasomatism (compositional changes due to the loss or gain of elements) of soluble elements usually occurred on scales less than a few 100 micrometers, possibly due to low permeability resulting from the fine grain size of the accreted primordial dust (Bland et al., 2009). In the absence of liquid transport on macroscopic scales, small parent bodies are required to explain the narrow temperature range for which alteration occurred within CI and CM chondrites.

Bland and Travis (2017) propose an alternative model in which "mud," consisting of accreted chondrules, fine grained silicates, and water, was convected over large spatial scales without chemical fractionation, resulting in a chemically homogeneous body. Mud convection processes could allow a primitive composition to be realized for bodies as large as Ceres, provided a portion of the interior was not lithified by other geologic processes. This condition was likely met for Ceres. The Dawn data and geochemical modeling show that Ceres' interior is chemically and physically layered, with a mantle consisting of aqueously altered rock (Ermakov et al., 2017; Fu et al., 2017; Prettyman et al., 2017; Castillo-Rogez et al., 2018; Travis et al., 2018). Although both Ceres and the meteorites experienced similar evolutionary processes, the aqueously altered carbonaceous chondrites are less altered and thus are an imperfect analog for Ceres.

Of relevance to carbon and organics on Ceres are studies of the meteorites Zag and Monahans, H ordinary chondrites (OC) that underwent thermal metamorphism within their parent bodies (Rubin et al., 2002). Zag contains xenolithic clasts with carbonaceous chondrite-like materials and halite (NaCl) crystals with inclusions containing brines (Chan et al., 2018; Kebukawa et al., 2019). The halite crystals also contain soluble organic matter, including polyaromatic hydrocarbons (PAH) and heterocycles. The clasts are thought to result from the accretion of a primitive asteroid with the OC parent body. The halite crystals, which are rare in carbonaceous chondrites, were likely formed in a hydrothermal system on their parent body, under conditions similar to that expected within Ceres (Chan et al., 2018). Organic molecules, including amino acids, are associated with the included brines, providing further evidence that asteroid interiors can host environments suitable for prebiotic chemistry.

8.2.3 Production/Modification of Carbonaceous Materials on Meteorite Parent Bodies

In carbonaceous chondrites, carbon is found primarily in organic matter, to a lesser degree in carbonates (Figure 8.2(a)), and in trace quantities as exotic forms, such as silicon carbide, diamond, and graphite (e.g., Sephton, 2002). Much of the organic matter is thought to have originated from the accreted interstellar and nebular

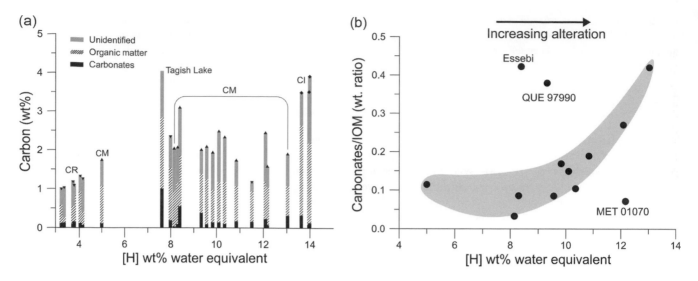

Figure 8.2 (a) The concentration of carbon in CR, CM, and CI carbonaceous chondrites is shown as a function of hydrogen concentration, both determined using mass spectrometry by Alexander et al. (2014, 2015). The amount of carbon in organics and carbonates is shown. The components do not sum to the bulk concentration, indicating a portion of carbon-bearing species were not identified. Hydrogen in carbonaceous chondrites is incorporated into the mineral lattice. As such, hydrogen concentration is a proxy for the degree of aqueous alteration. (b) For the CM chondrites, the proportion of carbonates to insoluble organic matter increases with hydrogen content. This alteration trend implies that organic matter was oxidized to produce carbonates within the parent body of the CM chondrites (after McSween et al., 2017).

materials; although modification and/or synthesis of organics on the parent body is expected (e.g., Pizzarello et al., 2013). The organic matter can be subdivided into organic compounds that can be dissolved by organic solvents ($\sim$30%) and insoluble organic matter, containing a kerogen-like, macromolecular material ($\sim$70%). In the well-studied Murchison meteorite, a CM carbonaceous chondrite, an insoluble component is accompanied by aromatic and aliphatic hydrocarbons, amino acids and precursors, carboxylic acids, alcohols, aldehydes, ketones, sugars, amines and amides, and nitrogen and sulfur heterocycles (Cronin & Chang, 1993; Sephton, 2002).

For the Murchison meteorite, Schulte and Shock (2004) argue that at least a portion – if not all – of the soluble organic matter could result from coupled aqueous alteration and abiotic synthesis of aliphatic organics. The accreted silicate, metals, and amorphous phases react with liquid water to produce aqueously altered minerals, progressing from Fe-serpentine to Mg-serpentine to clays, accompanied by the production of magnetite, tochilinite, sulfides, and sulfates (McSween et al., 2017). Serpentinization is exothermic and once initiated results in melting of additional ice, providing a positive feedback mechanism that continues the process. Many of the reactions produce gaseous hydrogen (e.g., Holm et al., 2015). For example,

$$3Fe_2SiO_4 + 2H_2O \rightarrow 2Fe_3O_4 + 3SiO_2 + 2H_2$$
$$\text{(fayalite)} \qquad\qquad \text{(magnetite)}$$

The thermal energy and potential for high hydrogen fugacity resulting from serpentinization could provide the conditions necessary for synthesis of complex organics. Provided simple organic compounds are present in the accreted material, more complex aliphatic molecules can be synthesized (Schulte & Shock, 2004). For example,

$$H_2O + H_2 + CH_2O + \; HCN \rightarrow CH_3COOH + NH_3$$
$$\text{(formaldehyde)} \qquad\qquad \text{(acetic acid)}$$

Studies of progressive alteration in CR-, CM-, and CI-chondrites by Le Guillou et al. (2014) suggest clay-mediated reactions could

modify the originally accreted organics, supporting the potential role of clay minerals in prebiotic evolution (Garvie & Buseck, 2007). Pizzarello et al. (2013) suggest that hydrothermal alteration of insoluble organic matter under oxidizing conditions within parent bodies could account for the diverse inventory of organic compounds found in carbonaceous chondrites, potentially benefiting molecular evolution early in Earth's history.

Carbonates are secondary mineral precipitates formed by the oxidation of organic matter and/or other volatile carbon species, such as CH_4 and CO_2 in an aqueous environment. Oxidative decarboxylation of organic matter leads to the production of CO_2, which combines with water and minerals to form carbonates. Evidence for oxidation of organic carbon can be found in CM chondrites (Figure 8.2(b)) for which the ratio of carbonates-to-organics increases as more hydrogen is incorporated into phyllosilicates (McSween et al., 2017). In addition, a study of organocarbonate associations in carbonaceous chondrites shows evidence for oxidizing fluids barren of organic matter along with more evolved fluids rich in organics (Chan et al., 2017), indicating temporal changes to the alteration chemistry. Carbonate mineralogy provides an indication of the degree of aqueous alteration, with calcium carbonates replaced by more complex carbonates (e.g., dolomite) as alteration progresses (Zolensky et al., 1997). CM carbonaceous chondrites primarily contain calcite, with minor amounts of dolomite; whereas the more altered CI carbonaceous chondrites contain abundant dolomite as well as breunnerite, siderite, and calcite (Endreß & Bischoff, 1996; De Leuw et al., 2010).

8.3 CERES OBSERVATIONS

8.3.1 Telescopic UV

Diagnostic spectral features for carbonaceous species in the UV are produced by solid-state electronic transitions associated with strong

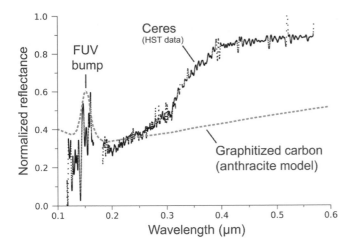

Figure 8.3 Ceres' reflectance spectrum acquired by the Hubble Space Telescope (HST) is compared to a model for anthracite, which contains graphitized carbon (Hendrix et al., 2016b). The presence of the "FUV bump" may result from processing of organic matter on the surface of Ceres by ultraviolet light and ionizing radiation (see text).

covalent bonds between carbon atoms in organic molecules (Hendrix et al., 2016b). These result in absorption features at about 0.088 μm (σ bond) and 0.22 μm (π bond), which influence the shape of the reflectance spectrum in the UV. Organic molecules exposed to ionizing radiation, including solar energetic charged particles and UV light, undergo graphitization in which bonds are broken and hydrogen is removed, resulting in the production of anthracite-like species, amorphous carbon, and ultimately graphite. As graphitization progresses, the 0.22 μm absorption feature increases in depth and shifts to longer wavelengths and an upturn appears in the far UV wavelengths (0.1–0.15 μm).

These features are present in spectra of Ceres acquired by the International Ultraviolet Explorer and Hubble Space Telescope (HST) (Hendrix et al., 2016a, 2016b). The observed bump in the far UV matches that of anthracite, which suggests that organic matter that has been processed by exposure to ultraviolet light and ionizing radiation is pervasive on Ceres' surface (Figure 8.3). Ceres' globally low visible albedo and flat spectral shape is consistent with the presence of weathered hydrocarbons (Hendrix et al., 2016b). Long-term exposure of anhydrous and hydrous Na-carbonates to vacuum could result in a 0.28 μm absorption feature observed in spectra of Ceres acquired by HST (Bu et al., 2019).

8.3.2 Carbonates on Ceres

Anhydrous carbonates produce distinct absorption features in the near IR (3.3–3.5 μm and 3.8–4.0 μm). These were detected in reflectance spectra acquired by the NASA Infrared Telescope Facility on Mauna Kea, Hawaii, spanning a complete rotation of Ceres, indicating carbonates are prevalent on Ceres' surface (Rivkin et al., 2006). Globally-averaged reflectance spectra acquired by VIR (0.4- to 5-μm) at Ceres revealed a strong absorption centered on 3.95 μm with (Mg,Ca)-carbonates providing a good fit to the data (De Sanctis et al., 2015). If the spectral mixing fractions determined by De Sanctis et al. (2015) are interpreted as mineral volume fractions, then the amount of C in carbonates is about 1 wt.% for Ceres global average. Except for Tagish Lake, the

VIR-inferred concentration of C in carbonates exceeds that found in carbonaceous chondrite meteorites (Figure 8.2(a)).

During Dawn's primary mission to Ceres, near global coverage was achieved by VIR with about 1 km/pixel during the Survey mapping orbit (Survey), about 365 m/pixel in a high-altitude mapping orbit (HAMO), and about 100 m/pixel in a low-altitude mapping orbit (LAMO) (De Sanctis et al., 2018). These orbits enabled detailed characterization of bright and dark materials on Ceres. Carrozzo et al. (2018) globally mapped the distribution of carbonates primarily with data acquired at HAMO altitudes (Figure 8.1). In many of the high albedo regions, including Occator's faculae and Ahuna Mons, the 3.960 μm band center is shifted to longer wavelengths consistent with the presence of natrite (Na_2CO_3) (e.g., Figure 8.4(a)). In addition, spectral fits to the 1.0- to 4.2-μm region are consistent with the presence of partially hydrated Na-carbonate species, such as thermonatrite ($Na_2CO_3 \cdot H_2O$) and trona ($NaHCO_3 \cdot Na_2CO_3 \cdot 2H_2O$). These highly localized hydrated species are associated with natrite and in some cases near surface exposures of water ice (e.g., Oxo crater). Considering the short lifetime (weeks to years) of hydrated Na-carbonates in vacuum at temperatures found on Ceres' surface (e.g., Bu et al., 2019), the detections are consistent with ongoing emplacement and exposure of hydrated salts (De Sanctis et al., 2016; Carrozzo et al., 2018). With high resolution (LAMO) data, Raponi et al. (2019) found that the faculae in Occator crater primarily consist of Al-phyllosilicates, natrite, and ammonium chloride (see Chapter 7).

The VIR-derived mineralogy provides constraints on the water–salt system from which the bright deposits originated, inferred by Zolotov (2017) to be a cold solution with a pH of ~10 and moderate salinity that primarily contained Na^+, CO_3^{-2}, HCO_3^-, NH_4^+, NH_3, and Cl^-. This alkaline solution precipitates NaCl and its hydrates (cf. discussion of halite crystals in the Zag and Monahans meteorites, Section 8.2.2). De Sanctis et al. (2016) note that the eutectic temperature of such fluids is higher than that expected for Ceres' crust (Castillo-Rogez & McCord, 2010). As such, impact heating would be required to mobilize them.

De Sanctis et al. (2020) reported the detection of hydrated sodium chloride (hydrohalite) within Cerealia Facula in Occator crater based on an analysis of VIR spectra acquired with high spatial resolution in Dawn's final mission phase. Given estimates of the rate of dehydration of hydrohalite, they conclude that chloride salts were emplaced recently, long after fluids produced by impact heating would have frozen. The heterogeneous composition and geologic context of Cerealia Facula indicate episodic emplacement of brines. Recent emplacement of hydrohalite suggest brines rich in sodium chloride may still be circulating deep within Ceres' crust and may contribute to ongoing cryovolcanism.

8.3.3 Detection of Organics near Ernutet Crater

The 3.3–3.6 μm spectral region includes absorptions characteristic of stretching frequencies for methyl and methylene functional groups associated with aliphatic hydrocarbons (Kaplan & Milliken, 2016; De Sanctis et al., 2017, 2018) (Figure 8.4(a)). This signature was detected within a ~1,000 km^2 region around Ernutet crater and tentatively in a small area of a few km^2 on the rim of Inamhari crater. The strongest absorption is in the immediate vicinity of the crater in a region containing high concentrations of Na-carbonates (Figures 8.5(a) and (b)), although neither the organic

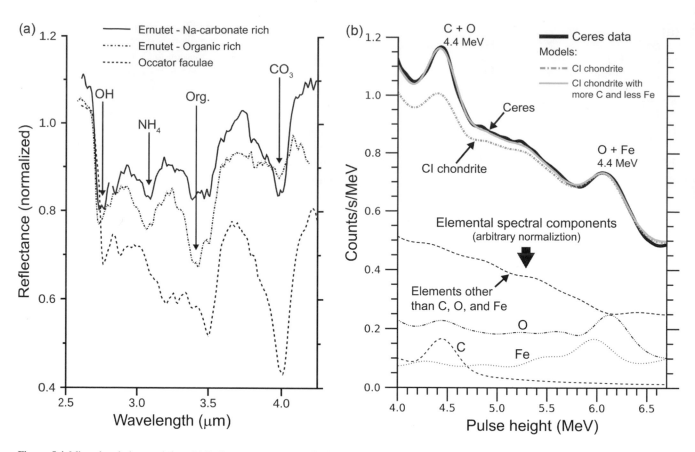

Figure 8.4 Mineral and elemental data. (a) Reflectance spectra acquired by VIR in the near IR reveal absorption features associated with OH- and NH_4-bearing phyllosilicates, organics (org.) and carbonates (De Sanctis et al., 2017). (b) LAMO-averaged gamma ray difference spectrum for measurements taken within 20 degrees of the equator is compared to models (Prettyman et al., 2018). The spectral region contains contributions from O, C, and Fe. These include continua and gamma ray peaks produced by neutron capture with ^{56}Fe (5.9- and 6.0-MeV) and by fast neutron reactions with ^{16}O (6.1 MeV and 4.4 MeV) and ^{12}C (4.4 MeV).

nor carbonate signature correspond to a specific crater. The signatures are patchy and span the southwest region of Ernutet, spilling into an adjacent, older crater. The organic signature is also associated with elevated concentrations of ammoniated phyllosilicates.

De Sanctis et al. (2017) argue that the correspondence between the organics and the NH_4- and CO_3-bearing minerals, which are formed elsewhere on Ceres via extrusion of cryovolcanic fluids and/or hydrothermal processes, strongly support an endogenic origin for the organics. Regions containing organics have a red slope at visible wavelengths that contrasts with the surroundings. This enables high-resolution studies of their geologic context using Framing Camera color filter imagery (e.g., Figure 8.1, image top left). Pieters et al. (2018) suggest that the observed pattern of organics could result from secondary impacts, although a candidate for the primary impact has not been identified. As an alternative, they propose that the observed pattern could result from recent exposures of an organic-rich subsurface layer by impacts; however, this hypothesis is not fully supported given the geologic context of the exposed organics.

Additional support for an endogenic origin comes from numerical impact simulations by Bowling et al. (2020), who show that delivery of aliphatic organics by impacts is very inefficient. They suggest that the impact that formed Ernutet excavated organic-rich crustal material, bringing it closer to the surface where it could be exposed by mass wasting and subsequent small

impacts. An endogenic origin is consistent with the synthesis of aliphatic organics within hydrothermal systems on carbonaceous protoplanets and organo-carbonate relationships identified in meteorites (Section 8.2.3).

8.3.4 Constraints on Regolith Carbon Content from GRaND

Dawn's Gamma-Ray and Neutron Detector (GRaND) globally mapped Ceres' regolith composition on spatial scales of ~400 km (Prettyman et al., 2017, 2019b). The instrument is sensitive to bulk regolith elemental composition to depths of a few decimeters. GRaND can detect carbon when it is present at wt.% levels; however, quantification of carbon is model dependent.

Gamma rays are produced by the steady interaction of galactic cosmic ray secondaries, primarily neutrons, with nuclei that make up the regolith (e.g., Prettyman et al., 2019a). Neutron inelastic scattering produces the first excited state of ^{12}C, which de-excites by the emission of a 4.4 MeV gamma ray (Figure 8.4(b)) (Prettyman et al., 2019a); however, the same gamma ray can be made by nonelastic scattering, primarily with oxygen, for example, $^{16}O + n \rightarrow {}^{12}C + {}^{4}He + n + \gamma$, or $^{16}O(n,n\alpha\gamma)$ in shorthand notation. This reaction occurs for neutrons with kinetic energies greater than 7.2 MeV. The prominent 6.1 MeV gamma ray observed in planetary spectra originates primarily from inelastic neutron scattering with oxygen: $^{16}O(n,n\gamma)$. If the energy distribution of fast neutrons

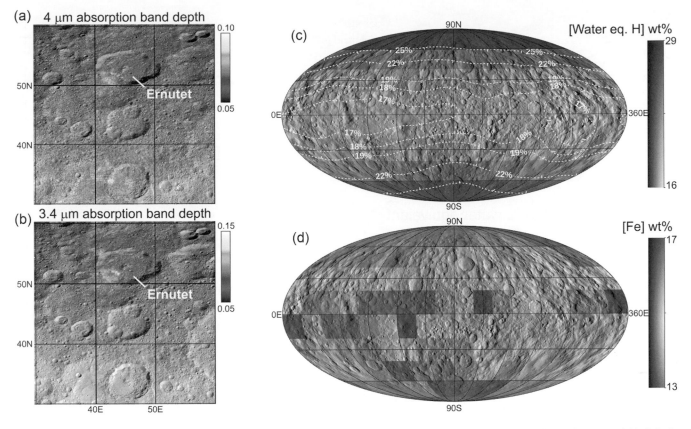

Figure 8.5 Mineral and elemental maps. Maps of the depth of the 3.4- and 4-μm bands, respectively, reveal the distribution of (a) carbonates and (b) aliphatic organic matter in and around Ernutet crater. The maps are from De Sanctis et al. (2017). Maps of water-equivalent H and Fe determined from GRaND data are shown in (c) and (d) (Prettyman et al., 2017, 2019b). The elemental maps are superimposed on shaded relief.
A black and white version of this figure will appear in some formats. For the color version, refer to the plate section.

is known, then contributions from ^{16}O to the 4.4 MeV can be determined and removed, leaving only contributions from ^{12}C. For GRaND, a complicating matter is that the aforementioned reactions occur both in the planetary surface and in the instrument housing and spacecraft structure, which also contains oxygen and carbon. Fast neutrons leaking away from Ceres interact with the instrument housing, producing 4.4- and 6.1-MeV gamma-rays.

An assessment of Ceres' carbon content was made by comparing gamma-ray spectra of Ceres with Vesta acquired with nearly identical geometry and galactic cosmic ray energy distributions (Prettyman et al., 2017). For Vesta, the concentration of C is small, ~0.1 wt.% assuming the 250 μg/g average hydrogen concentration measured by GRaND was delivered by CM-like impactors with negligible loss in hydrogen (Prettyman et al., 2012). As such, the 4.4 MeV peak was almost entirely due to fast neutron interactions with oxygen in Vesta's regolith and oxygen and carbon in the instrument and spacecraft. The fast neutron leakage flux was observed to be nearly three times higher at Vesta than at Ceres. This accounts for a decrease in the 6.1 MeV O peak observed at Ceres. If Ceres regolith were free of carbon, then the intensity of the 4.4-MeV peak should also have decreased; however, the intensity of this peak increased relative to Vesta. The data show that carbon is pervasive in Ceres' regolith, consistent with UV and VIR observations.

Quantitative analyses of carbon were made by forward modeling of gamma-ray spectra and neutron counting data. The gamma-ray background from galactic cosmic ray interactions with the instrument and spacecraft was measured at high altitude and subtracted from the LAMO data to form a difference spectrum (Figure 8.4(b)). The difference spectrum is sensitive only to gamma-rays and neutrons originating from Ceres. The data are compared to a model of the gamma ray spectrum assuming a CI carbonaceous chondrite composition along with elemental spectral components for Fe, C, and O. The model includes gamma-rays induced by neutron interactions with the instrument. The data can be fitted by adjusting the spectral components to decrease the contribution from Fe and increase the contribution of C relative to CI chondrites (~3.5 wt.% C). The analysis indicates Ceres' bulk regolith contains between 7- and 20-wt.% C (Prettyman et al., 2018), consistent with analyses of neutron counting data (Prettyman et al., 2017) that suggest C concentrations are bracketed by that of the CI chondrites and cometary dust.

Prettyman et al. (2019b) estimated the amount of carbon on Ceres using a mixing model to fit the concentration of hydrogen and iron at equatorial latitudes (±20° latitude) (Figures 8.5(c) and (d)). In this region, ice stability modeling predicts water ice to be absent or in low concentrations at depths sensed by GRaND (Prettyman et al., 2017). If so, then hydrogen should primarily be in the form of hydrated minerals and – if present – hydrocarbons. The equatorial region is directly comparable to the aqueously altered meteorites, for which hydrogen concentration is a proxy for the degree of aqueous alteration (Figure 8.2(b) and McSween et al., 2017).

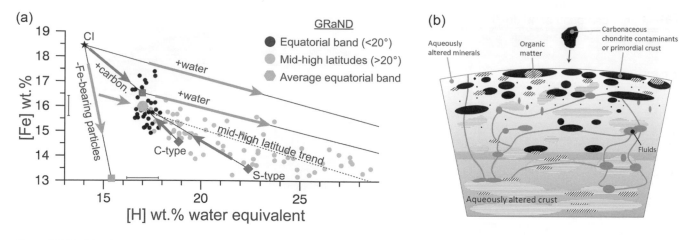

Figure 8.6 Models of crustal composition. (a) A scatter plot of elemental concentrations from Figure 8.5 shows that Ceres' ice-free equatorial regolith contains more H and less Fe than average CI chondrites. Water dilution trends are indicated by arrows (+water). Removal of heavy, anhydrous Fe-bearing particles (magnetite & troilite) via sedimentation (-Fe-bearing minerals) cannot explain the difference between the CI composition and Ceres regolith, unless a hydrous species is also enriched in the regolith. The addition of carbonaceous material (+carbon), consisting of carbonates and organics, can explain the difference. Estimates of Ceres' pristine crustal composition (diamonds) are provided for an impactor flux of C-type and S-type asteroids, with high retention efficiency (50%), from table S4 of Prettyman et al. (2017). (b) A schematic diagram by Marchi et al. (2019) depicts contamination of the outermost ∼4 km of the aqueously altered crust by carbonaceous chondrite impactors.

Ceres' average equatorial hydrogen content is higher than that of the CI chondrites, whereas the concentration of Fe is lower (Figure 8.6(a)). The observed concentration of Fe placed limits on the amount of opaque, Fe-rich minerals, such as magnetite and troilite, in the regolith (about 6 vol.%). De Sanctis et al. (2015) included magnetite as a dark component in spectral fits of global near-IR reflectance data along with amorphous carbon. The estimated volume fraction of magnetite based on spectral fitting was 60% to 90%, which is incompatible with the GRaND observations. While magnetite was useful for modeling VIR spectral data, the darkening agent could not possibly be anhydrous, Fe-bearing minerals. This leaves graphitized organic matter as a viable alternative (McSween et al., 2017; Prettyman et al., 2017), consistent with inferences from UV observations (Hendrix et al., 2016a, 2016b). Prettyman et al. (2019b) found that a mixture of carbonates and organic matter could fit the equatorial elemental data, resulting in values for the carbon content of Ceres regolith ranging from 8- to 14-wt%, in agreement with forward modeling studies of gamma-ray and neutron data.

8.4 SYNTHESIS

The telescopic UV and Dawn observations, together with laboratory studies, can be synthesized to form a coherent picture of the chemical makeup of Ceres' surface and regolith. The UV observations suggest graphitized organic species, partially depleted in hydrogen, are pervasive on Ceres surface. The Dawn/VIR investigation found carbonates everywhere on Ceres' surface. Based on spectral mixing fractions and mineral compositional models, carbonates contribute about 1 wt.% C to the global average (Prettyman et al., 2019b). Aliphatic organics were clearly detected in one location, in the vicinity of Ernutet crater; however, the dark phase that dominates Ceres spectrum in the VIR spectral range is consistent with amorphous carbon as well as magnetite or other opaque

phases (De Sanctis et al., 2015, 2018). GRaND elemental analyses indicate carbon is pervasive, with concentrations more than the CI chondrites. As such, there is an apparent disconnect between the low carbon concentrations inferred from near-IR mineralogy and much higher concentrations determined by GRaND.

A possible explanation for the apparent discrepancy is that the dark opaque phase is composed of amorphous carbon. Laboratory studies show that aliphatic absorptions are not detectable via measurements of the 3.4 mm band depth when the H/C atom ratio is less than about 0.4 (Kaplan et al., 2018, 2019). Above this detection limit, sensitivity increases with the H/C ratio. The sensitivity of near-IR measurements to graphitized organics places broad bounds on the concentration of organics in locations where organics were detected by VIR (spectral abundance of 45%–65%) (Kaplan et al., 2018). Furthermore, exposure of surficial organics to ionizing radiation results in the removal of hydrogen, forming kerogen-like species (Hendrix et al., 2016b). As such, organics may be pervasive on Ceres, but not easily detectable in the near-IR.

Marchi et al. (2019) set out to reconcile near IR measurements by VIR and elemental measurements by GRaND for the global, ice-free regolith. This was accomplished via an iterative approach, using a spectral mixing model to fit VIR data with relevant mineral assemblages, including mixtures of carbonaceous chondrite compositions and amorphous carbon with Ceres-specific minerals, including Mg-phyllosilicates, NH$_4$-clays, (Mg,Ca)-carbonates, and magnetite. The H/C atom ratio of amorphous carbon was selected to be ∼0.6 H/C to simulate kerogen-like species near the detection limit for VIR. Elemental weight fractions were estimated given the density and composition of the minerals and the fitted spectral mixing ratios, which were interpreted as volume fractions. The VIR-estimated elemental ratios were compared with measurements by GRaND of ice-free concentrations of H, Fe, K, and C.

An important result of this study was the discovery of feasible mineral models compatible with both VIR and GRaND data; although the solutions are not unique (see Table 8.1 for an example composition). The analysis indicates that a mixture of >30 vol.%

Table 8.1 *Possible composition of Ceres' global regolith (from Marchi et al., 2019).*

	Source/formula	Vol. (%)	H (wt.%)	C (wt.%)	K (μg/g)	Fe (wt.%)	Density (g/cm^3)
GRaND	Prettyman et al. (2017)		1.9 ± 0.2	> CI	414 ± 40	16 ± 1	–
Ceres model	Marchi et al. (2019)		1.97	20.2	334.2	16.9	2.34
CI chondrite[a]	Lodders and Fegley (1998)	60	1.57	3.61	576.0	18.4	2.26
Magnetite	Fe_3O_4	2.5	–	–	–	72.4	5.17
Am. carbon	H/C = 0.6	22	4.79	95.21	–	–	2
Antigorite	$Mg_{2.25}Fe_{0.75}(Si_2O_5)(OH)_4$	2	1.34	–	–	13.9	2.61
NH$_4$-annite[b]	$NH_4Fe_3AlSi_3O_{10}(OH)_2$	4	1.23	–	0.8	34.1	3.30
NH$_4$-mont.[c]	$(N_{0.3}H_{1.2}Al_2Si_4O_{12}H_2)_{0.37}$ $(Na_{0.2}Ca_{0.1}Al_2Si_4O_{12}H_2)_{0.63}$	8.5	0.67	–	–	–	2.35
Dolomite	$CaMg(CO_3)_2$	1	0.00	13.03	–	–	2.85

[a] Hydrogen concentration adjusted to be consistent with that found by mass spectrometry (Alexander et al., 2012).

[b] Ammoniated annite was blended with a small amount of K-annite (10 ppm). Most of the K in the resulting mixture is from the CI chondrite composition.

[c] For ammoniated montmorillonite, a cation exchange capacity of 100 millimoles cation per 100 g dry clay was assumed.

CI- or CM-like material and minerals specific to Ceres is needed to match the data. Similar concentrations of amorphous carbon are also required, primarily to match the ice-free regolith hydrogen concentration measured by GRaND. The best fit model resulted in ~20 wt.% C, which if valid would suggest extreme C fractionation occurred within Ceres' interior. Of significance, both VIR and GRaND analyses (see Section 8.3.4) – independently and together – allow superchondritic amounts of C to be present in Ceres' global regolith. For models that fitted the VIR data, consistent with elemental constraints from GRaND, the bulk of carbon is in the form of organics and graphitized species, with relatively small contributions from carbonates (~1 wt.%).

Kurokawa et al. (2020) carried out an independent analysis of VIR infrared spectra using a probabilistic approach that included constraints on elemental composition from GRaND (H, C, K, and Fe). Carbon was constrained between 8 and 14 wt.% per Prettyman et al. (2019b). Kurokawa et al. found that the data could be fitted by 40–70 wt.% carbonaceous chondrite materials mixed with amorphous carbon and aqueous alteration products. Similar results were obtained for the global regolith for an analysis of VIR visible and infrared spectra by Raponi et al. (2021). Moreover, Raponi et al. (2021) found evidence for reddening at visible wavelengths of bright materials within Occator crater, suggesting that organic matter could be present in the brines that erupted onto the surface. If so, conditions may have been right for the survival or synthesis of organic matter within Ceres (see Sections 8.2.2 and 8.2.3).

Based on a laboratory study of near-infrared spectra for carbonaceous chondrite meteorites, Beck et al. (2021) developed a spectral metric using a combination of the 2.75 μm and 2.80 μm band depths to determine equivalent water content. They applied their "hygrometer" to VIR spectra acquired at Ceres and estimated that 1.22 wt.% of hydrogen on Ceres' surface is in the form of organic matter. This resulted in a range of possible C concentrations (~9- to 20-wt.%), depending on the H/C ratio, which varies between soluble and insoluble components.

8.5 INTERPRETATION

8.5.1 Exogenic Contamination of Ceres' Crust by Carbonaceous Impactors

The high concentration of CI-chondrite material inferred by combining VIR and GRaND data by Marchi et al. (2019) supports pervasive exogenic contamination by carbonaceous impactors following Ceres' injection into the main asteroid belt. Collisional models indicate that the outermost few kilometers of the crust could be contaminated by impact debris (Figure 8.6(b)). Consistent with this hypothesis, vertical rail gun experiments suggest that Ceres' volatile crust could retain much of the infalling material (Daly & Schultz, 2015). Significant contamination by anhydrous silicates present in the Main Belt can be ruled out based on Ceres' surface mineralogy and inferred crustal composition.

If the parent bodies of the carbonaceous chondrites formed in the same region of the outer Solar System, then these small bodies and their fragments would have been the building blocks for Ceres. Fragments of these planetesimals could also have contaminated Ceres' crust following differentiation. In the supplementary materials for Prettyman et al. (2017), the hydrogen and iron content of Ceres' pristine (uncontaminated), ice-free regolith was estimated given measured concentrations and simulations of exogenic pollution (Figure 8.6(a)). Ceres' pristine regolith would have had higher hydrogen concentrations prior to the accretion of impactors with compositions like the carbonaceous chondrite meteorites. Given the mineralogy of Ceres, this is only possible if the aqueously altered crust was much richer in carbon and hydrogen than the carbonaceous chondrites (Figure 8.6(a)). If so, then organic matter is likely a significant crustal component. As discussed in Section 8.2, organo-carbonate relationships provide constraints on hydrothermal processes that produced the crust. Consequently, determining how endogenic C is partitioned between carbonates and organic matter is the subject of ongoing research.

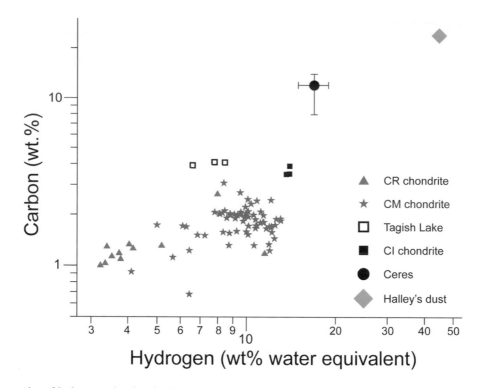

Figure 8.7 The concentration of hydrogen and carbon in Ceres' ice-free regolith is compared with analog materials, including carbonaceous chondrites (Alexander et al., 2012) and dust from comet Halley (Jessberger et al., 1988). The concentration of hydrogen in Ceres' regolith was determined by neutron spectroscopy (Prettyman et al., 2017). The concentration of carbon was estimated by Prettyman et al. (2019b), wherein organic matter was mixed with a representative CI chondrite composition to fit the ice-free iron and hydrogen concentrations measured by GRaND. The error bar gives the range of carbon estimated from the data.

8.5.2 Implications for Ceres' Origin and Hydrothermal Evolution

The concentration of hydrogen and carbon determined by GRaND is compared to that of the aqueously altered meteorites and cometary dust in Figure 8.7. While Ceres surface mineralogy is similar to that of the carbonaceous chondrites, there are key differences, including the presence of ammoniated clay minerals (De Sanctis et al., 2015). The latter supports Ceres outer Solar System origins, which could explain its comet-like surface composition (Beck et al., 2021). In addition, ice-free concentrations of H, C, and Fe differ significantly from the CI carbonaceous chondrites. Superchondritic concentrations of H and C with subchondritic concentrations of Fe are also consistent with Ceres as an icy differentiated body. Concentrations of carbon in excess of the CI chondrites imply ice-rock fractionation – and fractionation of C-bearing species – occurred within Ceres' interior. Prettyman et al. (2017) suggest that density driven separation of Fe-rich phases from accreted primitive materials along with convective upwelling of brines and ice could lead to a surface depleted in Fe and enriched in H and C. This is consistent with geophysical observations of a bulk crustal composition that is enriched in volatiles, including clathrate hydrates (Castillo-Rogez et al., 2020). As such, Ceres' crustal composition results from a combination of factors, including its outer Solar System origins, its collisional history, and its interior evolution.

8.6 SUMMARY AND FUTURE WORK

Telescopic observations and data acquired by Dawn demonstrate that Ceres' surface is carbon rich, with concentrations in excess of the CI chondrite meteorites. Organic matter, detected in localized regions, is inferred to be pervasive in high concentrations within Ceres' global regolith. Ceres' surface composition is the result of its outer Solar System origins, collisional history, and interior evolution, which may have included multiple phases of hydrothermal alteration. The presence of carbon and organics in the aqueous environment likely present throughout Ceres' history may have resulted in the synthesis of complex organic molecules. As such, Ceres is an attractive target for future missions focused on understanding ocean worlds and astrobiology.

Continued progress toward understanding the nature and origin of carbon on Ceres can be made using available remote sensing data and laboratory studies. Open questions include whether the provenance of carbon-rich crustal materials is primarily endogenic or exogenic, how endogenic C is partitioned between carbonates and organic matter, and how the composition of the organic matter differs from that found in meteorites. Studies of organo-carbonate relationships will provide insights into the hydrothermal conditions within Ceres' interior and whether synthesis of organic matter could have occurred.

Considerable work remains to test models of Ceres' surface composition, representing various hypotheses for the formation of Cere's crust, against the remote sensing data. Accurate determination of the abundance and distribution of C on the surface is needed to answer many of the outstanding questions. To this end, we are analyzing C using high sensitivity and high-spatial-resolution data acquired by GRaND in eccentric orbits during Dawn's final mission phase. With this data set, which includes coverage of Occator crater, more accurate analyses can be carried out on spatial scales comparable to surface geologic units. This will enable focused regional studies that combine VIR and GRaND data, providing further insights into the role of C in Ceres' geochemical evolution.

REFERENCES

Alexander, C. M. O'd., Bowden, R., Fogel, M. L., et al. (2012) The provenances of asteroids, and their contributions to the volatile inventory of the terrestrial planets. *Science*, 337, 721–723.

Alexander, C. M. O'd., Bowden, R., Fogel, M. L., & Howard, K. T. (2015) Carbonate abundances and isotopic compositions in chondrites. *Meteoritics & Planetary Science*, 50, 810–833.

Alexander, C. M. O'd., Cody, G. D., Kebukawa, Y., et al. (2014) Elemental, isotopic, and structural changes in Tagish Lake insoluble organic matter produced by parent body processes. *Meteoritics & Planetary Science*, 49, 503–525.

Anders, E., & Grevesse, N. (1989) Abundances of the elements: Meteoritic and solar. *Geochimica et Cosmochimica Acta*, 53, 197–214.

Armitage, P. J. (2020) *Astrophysics of Planet Formation*. Cambridge: Cambridge University Press

Bally, J., O'Dell, C. R., & McCaughrean, M. J. (2000) Disks, microjets, windblown bubbles, and outflows in the Orion nebula. *The Astronomical Journal*, 119, 2919–2959.

Beck, P., Eschrig, J., Potin, S., et al. (2021) "Water" abundance at the surface of C-complex main-belt asteroids. *Icarus*, 357, 114125.

Bland, M. T., Raymond, C. A., Schenk, P. M., et al. (2016) Composition and structure of the shallow subsurface of Ceres revealed by crater morphology. *Nature Geoscience*, 9, 538.

Bland, P. A., Jackson, M. D., Coker, R. F., et al. (2009) Why aqueous alteration in asteroids was isochemical: High porosity≠high permeability. *Earth and Planetary Science Letters*, 287, 559–568.

Bland, P. A., & Travis, B. J. (2017) Giant convecting mud balls of the early Solar System. *Science Advances*, 3, e1602514.

Bowling, T. J., Ciesla, F. J., Davison, T. M., et al. (2019) Post-impact thermal structure and cooling timescales of Occator crater on asteroid 1 Ceres. *Icarus*, 320, 110–118.

Bowling, T. J., Johnson, B. C., Marchi, S., et al. (2020) An endogenic origin of cerean organics. *Earth and Planetary Science Letters*, 534, 116069.

Brandt, J. C. (2014) Physics and chemistry of comets. In T. Spohn, D. Breuer, & T. Johnson (eds.), *Encyclopedia of the Solar System*, 3rd ed. Amsterdam: Elsevier, pp. 683–703.

Brasser, R., & Mojszis, S. J. (2020) The partitioning of the inner and outer Solar System by a structured protoplanetary disk. *Nature Astronomy*, 4, 492–499.

Brownlee, D., Tsou, P., Aléon, J., et al. (2006) Comet 81P/Wild 2 under a microscope. *Science*, 314, 1711.

Bu, C., Rodriguez Lopez, G., Dukes, C. A., et al. (2019) Stability of hydrated carbonates on Ceres. *Icarus*, 320, 136–149.

Cami, J., Bernard-Salas, J., Peeters, E., & Malek, S. E. (2010) Detection of C60 and C70 in a young planetary nebula. *Science*, 329, 1180.

Carrozzo, F. G., De Sanctis, M. C., Raponi, A., et al. (2018) Nature, formation, and distribution of carbonates on Ceres. *Science Advances*, 4, e1701645.

Castillo-Rogez, J. C., & McCord, T. B. (2010) Ceres' evolution and present state constrained by shape data. *Icarus*, 205, 443–459.

Castillo-Rogez, J. C., Neveu, M., McSween, H. Y., et al. (2018) Insights into Ceres's evolution from surface composition. *Meteoritics & Planetary Science*, 53, 1820–1843.

Castillo-Rogez, J. C., Neveu, M., Scully, J. E. C., et al. (2020) Ceres: Astrobiological target and possible ocean world. *Astrobiology*, 20, 269–291.

Chan, Q. H. S., Zolensky, M. E., Bodnar, R. J., Farley, C., & Cheung, J. C. H. (2017) Investigation of organo-carbonate associations in carbonaceous chondrites by Raman spectroscopy. *Geochimica et Cosmochimica Acta*, 201, 392–409.

Chan, Q. H. S., Zolensky, M. E., Kebukawa, Y., et al. (2018) Organic matter in extraterrestrial water-bearing salt crystals. *Science Advances*, 4, eaao3521.

Charnley, S. B. (1994) Chemistry of star-forming cores. *AIP Conference Proceedings*, 312, 155–159.

Cronin, J. R., & Chang, S. (1993) Organic matter in meteorites: Molecular and isotopic analyses of the Murchison meteorite. In J. M. Greenberg, C. X. Mendoza-Gómez, & V. Pirronello (eds.), *The Chemistry of Life's Origins* Dordrecht: Springer Netherlands, pp. 209–258.

Daly, R. T., & Schultz, P. H. (2015) Predictions for impactor contamination on Ceres based on hypervelocity impact experiments. *Geophysical Research Letters*, 42, 7890–7898.

De Leuw, S., Rubin, A. E., & Wasson, J. T. (2010) Carbonates in CM chondrites: Complex formational histories and comparison to carbonates in CI chondrites. *Meteoritics & Planetary Science*, 45, 513–530.

De Sanctis, M. C., Ammannito, E., McSween, H. Y., et al. (2017) Localized aliphatic organic material on the surface of Ceres. *Science*, 355, 719–722.

De Sanctis, M. C., Ammannito, E., Raponi, A., et al. (2015) Ammoniated phyllosilicates with a likely outer Solar System origin on (1) Ceres. *Nature*, 528, 241–244.

De Sanctis, M. C., Ammannito, E., Raponi, A., et al. (2020) Fresh emplacement of hydrated sodium chloride on Ceres from ascending salty fluids. *Nature Astronomy*, 4, 786–793.

De Sanctis, M. C., Coradini, A., Ammannito, E., et al. (2011) The VIR spectrometer. *Space Science Reviews*, 163, 329–369.

De Sanctis, M. C., Raponi, A., Ammannito, E., et al. (2016) Bright carbonate deposits as evidence of aqueous alteration on (1) Ceres. *Nature*, 536, 54–57.

De Sanctis, M. C., Vinogradoff, V., Raponi, A., et al. (2018) Characteristics of organic matter on Ceres from VIR/Dawn high spatial resolution spectra. *Monthly Notices of the Royal Astronomical Society*, 482, 2407–2421.

Duprat, J., Dobrică, E., Engrand, C., et al. (2010) Extreme deuterium excesses in ultracarbonaceous micrometeorites from central Antarctic snow. *Science*, 328, 742.

Ehrenfreund, P., & Charnley, S. B. (2000) Organic molecules in the interstellar medium, comets, and meteorites: A voyage from dark clouds to the early Earth. *Annual Review of Astronomy and Astrophysics*, 38, 427–483.

Endreß, M., & Bischoff, A. (1996) Carbonates in CI chondrites: Clues to parent body evolution. *Geochimica et Cosmochimica Acta*, 60, 489–507.

Ermakov, A. I., Fu, R. R., Castillo-Rogez, J. C., et al. (2017) Constraints on Ceres' internal structure and evolution from its shape and gravity measured by the Dawn spacecraft. *Journal of Geophysical Research: Planets*, 122, 2267–2293.

Flynn, G. J., Keller, L. P., Feser, M., Wirick, S., & Jacobsen, C. (2003) The origin of organic matter in the Solar System: evidence from the interplanetary dust particles. *Geochimica et Cosmochimica Acta*, 67, 4791–4806.

Fomenkova, M. N., Chang, S., & Mukhin, L. M. (1994) Carbonaceous components in the comet Halley dust. *Geochimica et Cosmochimica Acta*, 58, 4503–4512.

Fray, N., Bardyn, A., Cottin, H., et al. (2016) High-molecular-weight organic matter in the particles of comet 67P/Churyumov–Gerasimenko. *Nature*, 538, 72–74.

Fu, R. R., Ermakov, A. I., Marchi, S., et al. (2017) The interior structure of Ceres as revealed by surface topography. *Earth and Planetary Science Letters*, 476, 153–164.

Garvie, L. A. J., & Buseck, P. R. (2007) Prebiotic carbon in clays from Orgueil and Ivuna (CI), and Tagish Lake (C2 ungrouped) meteorites. *Meteoritics & Planetary Science*, 42, 2111–2117.

Grimm, R. E., & McSween, H. Y. (1993) Heliocentric zoning of the asteroid belt by aluminum-26 heating. *Science*, 259, 653.

Hasegawa, T. I., Herbst, E., & Leung, C. M. (1992) Models of gas-grain chemistry in dense interstellar clouds with complex organic molecules. *The Astrophysical Journal Supplement Series*, 82, 167–195.

Hendrix, A. R., Vilas, F., & Li, J.-Y. (2016a) Ceres: Sulfur deposits and graphitized carbon. *Geophysical Research Letters*, 43, 8920–8927.

Hendrix, A. R., Vilas, F., & Li, J.-Y. (2016b) The UV signature of carbon in the Solar System. *Meteoritics & Planetary Science*, 51, 105–115.

Holm, N. G., Oze, C., Mousis, O., Waite, J. H., & Guilbert-Lepoutre, A. (2015) Serpentinization and the formation of H2 and CH4 on celestial bodies (planets, moons, comets). *Astrobiology*, 15, 587–600.

Jessberger, E. K., Christoforidis, A., & Kissel, J. (1988) Aspects of the major element composition of Halley's dust. *Nature*, 332, 691–695.

Kaplan, H. H., & Milliken, R. E. (2016) Reflectance spectroscopy for organic detection and quantification in clay-bearing samples: Effects of albedo, clay type, and water content. *Clays and Clay Minerals*, 64, 167–184.

Kaplan, H. H., Milliken, R. E., & Alexander, C. M. O. D. (2018) New constraints on the abundance and composition of organic matter on Ceres. *Geophysical Research Letters*, 45, 5274–5282.

Kaplan, H. H., Milliken, R. E., Alexander, C. M. O. D., & Herd, C. D. K. (2019) Reflectance spectroscopy of insoluble organic matter (IOM) and carbonaceous meteorites. *Meteoritics & Planetary Science*, 54, 1051–1068.

Kebukawa, Y., Ito, M., Zolensky, M. E., et al. (2019) A novel organic-rich meteoritic clast from the outer Solar System. *Scientific Reports*, 9, 3169.

Kretke, K. A., Bottke, W. F., Levinson, H. F., & Kring, D. A. (2017) Mixing of the asteroid belt due to the formation of the giant planets. *Accretion: Building New Worlds, LPI Topical Conference*, August 15–18, Lunar and Planetary Institute, Houston, TX, #2027.

Kurokawa, H., Ehlmann, B. L., De Sanctis, M. C., et al. (2020) A probabilistic approach to determination of Ceres' average surface composition from Dawn VIR and GRaND data. *Journal of Geophysical Research: Planets*, n/a, e2020JE006606.

Le Guillou, C., Bernard, S., Brearley, A. J., & Remusat, L. (2014) Evolution of organic matter in Orgueil, Murchison and Renazzo during parent body aqueous alteration: In situ investigations. *Geochimica et Cosmochimica Acta*, 131, 368–392.

Lodders, K., & Fegley, B., Jr. (1998) *The Planetary Scientist's Companion*. Oxford: Oxford University Press on Demand.

Marchi, S., Raponi, A., Prettyman, T. H., et al. (2019) An aqueously altered carbon-rich Ceres. *Nature Astronomy*, 3, 140–145.

McSween, H. Y., Emery, J. P., Rivkin, A. S., et al. (2017) Carbonaceous chondrites as analogs for the composition and alteration of Ceres. *Meteoritics & Planetary Science*, 53, 1793–1804.

Nuth, J. A., Johnson, N. M., & Manning, S. (2008) A self-perpetuating catalyst for the production of complex organic molecules in protostellar nebulae. *Proceedings of the International Astronomical Union*, 4, 403–408.

Pieters, C. M., Nathues, A., Thangjam, G., et al. (2018) Geologic constraints on the origin of red organic-rich material on Ceres. *Meteoritics & Planetary Science*, 53, 1983–1998.

Pizzarello, S., Davidowski, S. K., Holland, G. P., & Williams, L. B. (2013) Processing of meteoritic organic materials as a possible analog of early molecular evolution in planetary environments. *Proceedings of the National Academy of Sciences (USA)*, 110, 15614.

Prettyman, T. H., Englert, P. A. J., & Yamashita, N. (2019a) Neutron, gamma-ray, and X-ray spectroscopy: Theory and applications. In J. L. Bishop, J. F. Bell, & J.E. Moersch (eds.), *Remote Compositional Analysis*. Cambridge: Cambridge University Press, pp. 191–238.

Prettyman, T. H., Feldman, W. C., McSween, H. Y., Jr., et al. (2011) Dawn's gamma ray and neutron detector. *Space Science Reviews*, 163, 371–459.

Prettyman, T. H., Mittlefehldt, D. W., Yamashita, N., et al. (2012) Elemental mapping by Dawn reveals exogenic H in Vesta's regolith. *Science*, 338, 242–246.

Prettyman, T. H., Yamashita, N., Ammannito, E., et al. (2019b) Elemental composition and mineralogy of Vesta and Ceres: Distribution and origins of hydrogen-bearing species. *Icarus*, 318, 42–55.

Prettyman, T. H., Yamashita, N., & McSween, H. Y. (2018) Carbon on Ceres: Implications for origins and interior evolution. *49th Lunar and Planetary Science Conference*, March 19–23, The Woodlands, TX, #1151.

Prettyman, T. H., Yamashita, N., Toplis, M. J., et al. (2017) Extensive water ice within Ceres' aqueously altered regolith: Evidence from nuclear spectroscopy. *Science*, 355, 55–59.

Quick, L. C., Buczkowski, D. L., Ruesch, O., et al. (2019) A possible brine reservoir beneath Occator crater: Thermal and compositional evolution and formation of the Cerealia dome and Vinalia Faculae. *Icarus*, 320, 119–135.

Raponi, A., De Sanctis, M. C., Carrozzo, F. G., et al. (2019) Mineralogy of Occator crater on Ceres and insight into its evolution from the properties of carbonates, phyllosilicates, and chlorides. *Icarus*, 320, 83–96.

Raponi, A., De Sanctis, M. C., Carrozzo, F. G., et al. (2021) Organic material on Ceres: Insights from visible and infrared space observations. *Life*, 11, 9.

Raymond, C. A., Ermakov, A. I., Castillo-Rogez, J. C., et al. (2020) Impact-driven mobilization of deep crustal brines on dwarf planet Ceres. *Nature Astronomy*, 4, 741–747.

Raymond, S. N., & Izidoro, A. (2017) Origin of water in the inner solar system: Planetesimals scattered inward during Jupiter and Saturn's rapid gas accretion. *Icarus*, 297, 134–148.

Riebe, M. E. I., Foustoukos, D. I., Alexander, C. M. O. D., et al. (2020) The effects of atmospheric entry heating on organic matter in interplanetary dust particles and micrometeorites. *Earth and Planetary Science Letters*, 540, 116266.

Rivkin, A. S., Volquardsen, E. L., & Clark, B. E. (2006) The surface composition of Ceres: Discovery of carbonates and iron-rich clays. *Icarus*, 185, 563–567.

Rubin, A. E., Zolensky, M. E., & Bodnar, R. J. (2002) The halite-bearing Zag and Monahans (1998) meteorite breccias: Shock metamorphism, thermal metamorphism and aqueous alteration on the H-chondrite parent body. *Meteoritics & Planetary Science*, 37, 125–141.

Ruesch, O., Genova, A., Neumann, W., et al. (2019) Slurry extrusion on Ceres from a convective mud-bearing mantle. *Nature Geoscience*, 12, 505–509.

Ruesch, O., Platz, T., Schenk, P., et al. (2016) Cryovolcanism on Ceres. *Science*, 353, aaf4286.

Sahijpal, S., Goswami, J. N., Davis, A. M., Grossman, L., & Lewis, R. S. (1998) A stellar origin for the short-lived nuclides in the early solar system. *Nature*, 391, 559–561.

Schulte, M., & Shock, E. (2004) Coupled organic synthesis and mineral alteration on meteorite parent bodies. *Meteoritics & Planetary Science*, 39, 1577–1590.

Scott, E. R. D., Krot, A. N., & Sanders, I. S. (2018) Isotopic dichotomy among meteorites and its bearing on the protoplanetary disk. *The Astrophysical Journal*, 854, 164.

Scully, J. E. C., Schenk, P. M., Castillo-Rogez, J. C., et al. (2020) The varied sources of faculae-forming brines in Ceres' Occator crater emplaced via hydrothermal brine effusion. *Nature Communications*, 11, 3680.

Sephton, M. A. (2002) Organic compounds in carbonaceous meteorites. *Natural Product Reports*, 19, 292–311.

Sierks, H., Keller, H. U., Jaumann, R., et al. (2011) The Dawn Framing Camera. *Space Science Reviews*, 163, 263–327.

Thomas, K. L., Blanford, G. E., Keller, L. P., Klöck, W., & McKay, D. S. (1993) Carbon abundance and silicate mineralogy of anhydrous interplanetary dust particles. *Geochimica et Cosmochimica Acta*, 57, 1551–1566.

Travis, B. J., Bland, P. A., Feldman, W. C., & Sykes, M. V. (2018) Hydrothermal dynamics in a CM-based model of Ceres. *Meteoritics & Planetary Science*, 53, 2008–2032.

Vokrouhlický, D., Bottke, W. F., & Nesvorný, D. (2016) Capture of trans-Neptunian planetesimals in the main asteroid belt. *The Astronomical Journal*, 152, 39.

Warren, P. H. (2011) Stable-isotopic anomalies and the accretionary assemblage of the Earth and Mars: A subordinate role for carbonaceous chondrites. *Earth and Planetary Science Letters*, 311, 93–100.

Weisberg, M. K., McCoy, T. J., & Krot, A. N. (2006) Systematics and evaluation of meteorite classification. In D. S. Lauretta, &

H. Y. McSween, Jr. (eds.), *Meteorites and the Early Solar System II*. Tucson: University of Arizona Press, pp. 19–52.

Zolensky, M. E., Mittlefehldt, D. W., Lipschutz, M. E., et al. (1997) CM chondrites exhibit the complete petrologic range from type 2 to 1. *Geochimica et Cosmochimica Acta*, 61, 5099–5115.

Zolotov, M. Y. (2017) Aqueous origins of bright salt deposits on Ceres. *Icarus*, 296, 289–304.

9

Ammonia on Ceres

ELEONORA AMMANNITO AND BETHANY EHLMANN

9.1 INTRODUCTION

Even before the arrival of the Dawn spacecraft, Ceres attracted attention due to its infrared spectral properties, which were distinct from most other asteroids (see Chapter 1). First speculated based on telescopic data (King et al., 1992), subsequent data from the Dawn spacecraft confirmed the presence of ammonium on Ceres. Ammonium has been identified within near-ubiquitous dark materials, and in salts in a few localized bright faculae in the interiors of craters, as we describe further in this chapter (De Sanctis et al., 2015, 2016; Ammannito et al., 2016; Raponi et al., 2019; see also Chapter 7).

The presence of ammonium on Ceres is significant because it implies the availability of ammonia during its evolution. Ammonia is most likely derived from ices stable today only in the outer Solar System. Ammonia is important as a freezing point depressant in fluids and is potentially important in supplying nitrogen for prebiotic organic chemical reactions that occurred on primitive asteroids. The processes driving the availability of ammonia at the asteroid belt and inward are only partially understood. Today, the "snow line" of NH_3 ice is at >10 AU, at or just beyond the distance of Saturn. Inward of Saturn, ammonia ice species are modeled to have been lost to sublimation over relatively short timescales (Brown et al., 2011). If this is the case, it could render ammonia largely unavailable during accretion and early processes on asteroid parent bodies and planets. On the other hand, strong binding of NH_3 with H_2O may have caused NH_3 availability to closely track the H_2O snowline in the main asteroid belt, rendering it more available than predicted by considering NH_3 only in isolation (Dodson-Robinson et al., 2009). Thus, the discovery of ammoniated species on Ceres and deciphering their origins is key to understanding Ceres' history including accretion, differentiation, and its dynamical history. More broadly, understanding the processes that led to the presence of ammonium on Ceres provides important information on the aqueous environments in the early Solar System and the origins and dynamical histories of the large outer Main Belt asteroids.

In this chapter, we briefly review the significance of ammonia and then describe what was known or speculated about ammoniated species on Ceres prior to Dawn's arrival. We then review the findings of the Dawn mission, in particular the detection and mapping of ammoniated phases by the Visible and Infrared spectrometer (VIR): which species host ammonia/ammonium, their abundance and spatial distribution. We then discuss the potential origins and implications of ammonia, drawing on laboratory studies and modeling efforts. Finally, we summarize the key findings and the outstanding questions that remain for future investigation.

9.2 AMMONIA AND AMMONIUM SPECIES

In the Solar System today, ammonia, NH_3, is most common in ices. NH_3 ices are present in the outer Solar System, from the Saturnian system outward. Ammonia ices are found on the surfaces of the Uranian moon Miranda (Bauer et al., 2002), Pluto (Dalle Ore et al., 2019), Pluto's Moon Charon (Brown & Calvin, 2000), and possibly as minor phases on Kuiper Belt Objects (e.g., Jewitt & Luu, 2004). Ammonia is also observed in carbonaceous chondrite meteorites at up to parts per thousand concentrations within select phases (e.g., Pizzarello & Williams, 2012).

Ammonium – a cation with the chemical formula NH_4^+ – can substitute for potassium and other alkali elements in crystal lattices to form ammonium-bearing minerals (Krohn & Altaner, 1987). Although it is uncommon on Earth, ammonium can be found in mineral groups including silicates, halides, and carbonates. This ammonium is often sourced from the decomposition of nitrogen-bearing compounds in organic matter, which are dissolved in hydrothermal or metamorphic fluids and later incorporated into minerals. Volcanically- or hydrothermally-sourced NH_3 gases, NH_3 ice, or NH_4-bearing fluids also supply ammonium in minerals. Ammoniated alkali feldspars and ammoniated illites and micas are the most common mineral products on Earth where the NH_4^+ is substituted for alkali cations in the lattice at the time of formation. NH_4^+ can also be hosted in the interlayer cations in phyllosilicates, such as smectites and illites, and, unlike lattice NH_4^+, interlayer NH_4^+ is exchangeable and can be gained or lost during interaction with fluids.

In Solar System and laboratory studies, infrared spectroscopy provides a means of detecting and assessing the nature of both ammonia and ammonium. The free NH_4^+ ion has IR-active internal modes $v3$ at 3,145 cm^{-1} ($\lambda = 3.18$ μm) and $v4$ at 1,400 cm^{-1} ($\lambda = 7.14$ μm) which are, respectively, an anti-symmetric N–H stretch and an anti-symmetric H–N–H deformation (Mookherjee et al., 2002). IR reflectance studies of ammonium-bearing minerals and rocks (e.g., Farmer & Russell, 1967; Krohn & Altaner, 1987; Bishop et al., 2002; Berg et al., 2016; Ehlmann et al., 2018, Ferrari et al., 2019) show spectral features related to the presence of nitrogen complexes at 1.56, 2.05, 2.12, 3.06, 3.25, 3.55, 4.2, 5.7, and 7.0 μm (Robinson, 1974; Bruno & Svoronos, 1989; Figure 9.1). The nature of the bonding environment, including unit cell size and the configuration of hydrogen bonding, will cause the various absorption bands to shift in position and strength (Silverstein et al., 1991).

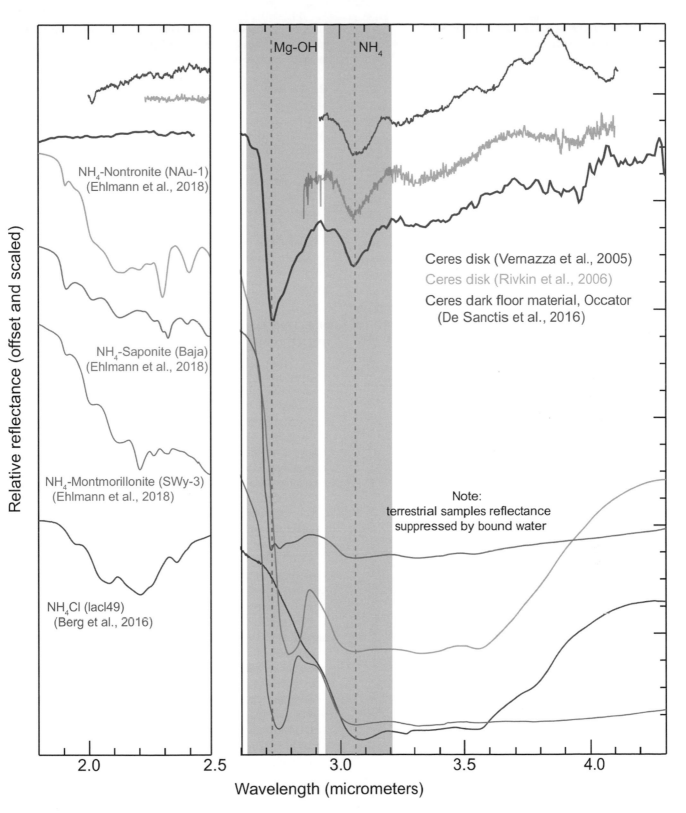

Figure 9.1 Ceres spectra show a prominent feature at 2.73 μm (Mg-OH) and at 3.06 μm (NH₄). Terrestrial data measured in vacuum show similar features in NH₄-exchanged phyllosilicates at −100°C and NH₄-bearing salts. The x-axis is split because the scaling of spectra is different in the two panels in order to make key spectral features discernible. The "Ceres dark floor material" spectrum is from Dawn's VIR, while the "Ceres" disk spectra are from Earth-based telescopes. Gaps in telescopic data are due to atmospheric absorptions in Earth's atmosphere. Note that, relative to airless bodies, even terrestrial samples in vacuum have more bound H₂O, leading to broad suppression of reflectance near 3.0-4.0 μm relative to Ceres' reflectance, though specific absorptions are still clearly discernible.

9.3 AMMONIA AT CERES? THE STATE OF KNOWLEDGE

Ceres attracted attention as soon as shortwave infrared surveys of the Main Belt could be conducted beyond 2.5 μm. Ceres and at least three other asteroids had unusual absorptions centered near 3.05 μm that were not characteristic of hydrous silicates nor ice (e.g., Takir & Emery, 2012; Rivkin et al., 2019), observations since further borne out by observations outside of Earth's atmosphere (Usui et al., 2019; Chapter 1; Figure 9.1). Originally interpreted as a thin film of water ice frost (Lebofsky, 1978; Lebofsky et al., 1981), King et al. (1992) combined telescopic observations, laboratory measurements, and theoretical calculations to propose that the peculiar ~3.1 μm absorption feature in the spectrum of Ceres is generated by ammoniated saponite, an Mg-smectite, on its surface. The interpretation was, however, debated and alternatives proposed, including a mixture of irradiated organics and crystalline water ice (Vernazza et al., 2005), iron rich clays (Rivkin et al., 2006), or brucite (Milliken & Rivkin, 2009). The possibility that ammonium-bearing mineral species are present on the surface of Ceres posed attention on the fact that ammonia was available during its evolution. Such possibility was recognized by several authors at the time due to its role in freezing point depression of brines and consequent implications for geophysical evolution as well as its instability in the modern Main Belt (e.g., McCord & Sotin, 2005; McKinnon, 2008; Castillo-Rogez & McCord, 2010).

The unprecedented quality of the spectra acquired by the Visible and Infrared spectrometer (VIR, De Sanctis et al., 2011) on board the Dawn spacecraft (Russell & Raymond, 2011), in combination with the accessibility of the 2.6–2.8 μm spectral range – which is hidden in ground based telescopic observation by the water vapor in earth's atmosphere – allowed confirmation of the King et al. (1992) "ammoniated phyllosilicate" interpretation for the 3.1-μm absorption feature, while all the other proposed interpretations have been found incompatible with the VIR spectra (De Sanctis et al., 2015). Indeed, the presence of the principal hydration feature at 2.73 μm confirmed specifically the association with a Mg phyllosilicate (see Chapter 7).

Considering a wider VIR spectral range from 1.0 to 4.2 μm, the mineral assemblage most consistent with VIR average observations is made of a darkening component(s), Mg-phyllosilicates, ammoniated clays, and Mg-Ca carbonates (De Sanctis et al., 2015, 2018). The darkening phase responsible for Ceres' exceptionally low albedo (ranging from 0.085 to 0.094; Reddy et al., 2015; Li et al., 2016; Ciarniello et al., 2017) is spectrally neutral and likely a combination of organic carbon, magnetite, and other iron sulfides or oxides.

Spatially resolved, 100-meter scale VIR data were acquired to map the composition of the entire Ceres surface. With relatively constant surface physical properties based on relatively homogenous thermal inertia (Rognini et al., 2020), after applying a photometric correction to the VIR dataset (Ciarniello et al., 2017), the main factor that contributes to the strength of an absorption feature is the relative abundance of the components in the assemblage. The spectral center of the metal–OH phyllosilicate absorption at 2.7 μm is sensitive to the specific chemical formulation of the mineral (Bishop et al., 2008), and the spectral position of the 3.1-μm feature is a function of the species that is hosting the ammonium (Berg et al., 2016). These data show that the mineral assemblage of dark material, phyllosilicates, ammoniated clays, and carbonates is widespread across the surface of Ceres and remarkably chemically homogeneous across the Ceres surface (Ammannito et al., 2016; Carrozzo et al., 2018; De Sanctis et al., 2018; Raponi et al., 2019), though there are minor variations in the strengths and relative strength of the 2.7 μm and 3.1 μm features (Figures 9.2 and 9.3).

The constancy of the 2.73 μm absorption position indicates Mg^{2+} is the dominant cation for phyllosilicates across the whole surface, with the exception of one location, in the brighter material comprising Cerealia facula within Occator crater (Raponi et al., 2019) and possibly Kupalo (De Sanctis et al., 2019). There, a 2.76 μm absorption may indicate Al^{3+} phyllosilicates. Similarly, because the position of the ammonium related feature at 3.1 μm is related to the hosting species, its constancy at 3.05–3.07 μm is an indication that most of the ammonium has been embedded in the identical species or a combination of species (Figure 9.3).

9.4 THE NATURE OF THE AMMONIATED PHYLLOSILICATE ASSEMBLAGE

Readily available infrared reflectance spectral libraries have ammoniated feldspars, ammoniated salts, and ammoniated Al-phyllosilicates, such as NH_4-annite and NH_4-montmorillonite (e.g., Srasra et al., 1994; Bishop et al., 2002; Berg et al., 2016; NASA Keck RELAB database (www.planetary.brown.edu/relab/)[1]). Among these species, using radiative transfer unmixing models (e.g., De Sanctis et al., 2015), NH_4-annite and/or NH_4 montmorillonite have been identified as the most consistent with the VIR spectrum (De Sanctis et al., 2015). However, spectra of ammoniated phyllosilicates more consistent with the composition of meteoritic phyllosilicates are not present in existing, publicly available spectral libraries. In addition, the potential for ammoniation of carbonaceous meteorite materials, a likely source material for the regolith of Ceres (e.g., Prettyman et al., 2017), has not been systematically assessed. To overcome these issues, several laboratory experiments have been done in the last few years to check the capabilities of relevant clays to incorporate NH_4^+ and eventually to study the spectral characteristic of the ammoniated assemblage.

Ehlmann et al. (2018) selected some Ceres-relevant analog materials, including both phyllosilicates and carbonaceous chondrite meteorites, to evaluate which phases ammoniate; which ammoniated phases exhibit absorptions with similar positions, widths, and shapes to those observed for Ceres; and if/how the NH_4-related absorptions change at different temperatures under vacuum. They found that the Al-, Fe-, and Mg-smectite clay mineral samples readily accommodated ammonium in the clay interlayer at room temperature, while phyllosilicates lacking an exchangeable interlayer like serpentines and chlorite did not accept the ammonium. With the same procedure, carbonaceous chondrite material, ranging from CM chondrites rich in serpentine to CI and C2 chondrites with appreciable saponite, did not readily incorporate the ammonium, although a few exhibited ambiguous spectral changes. The different behavior of different classes of phyllosilicates to the ammoniation process has been confirmed by Ferrari

[1] Website last accessed 2021 September 13.

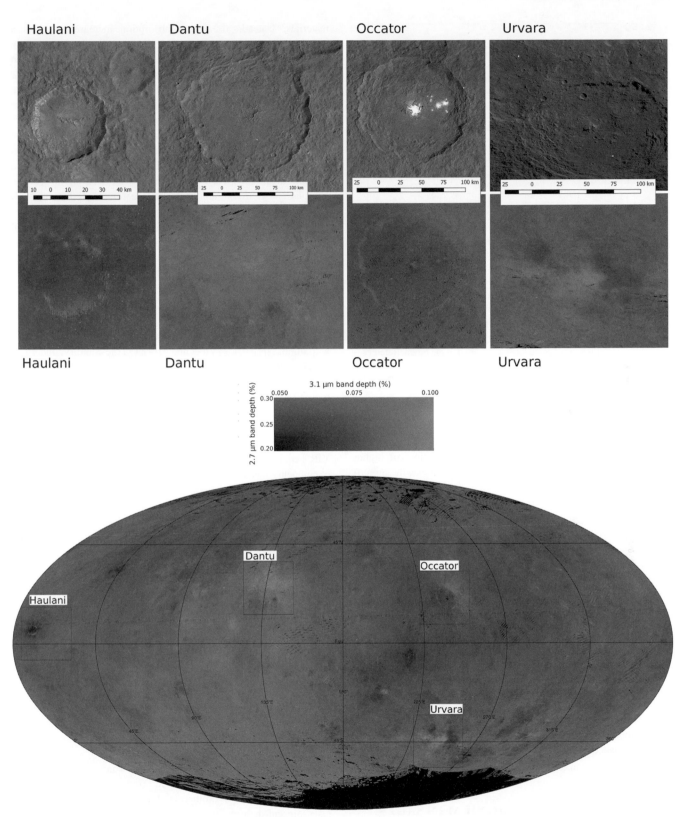

Figure 9.2 Combination of 2.7 μm band depth and 3.1 μm band depth according to the color map in the figure. Values in the map have been obtained merging the acquisition from RC3, survey, and High-Altitude Mapping Orbit Dawn mission phases. On the top there are the close ups of the four craters identified in the map. Data have been projected as described in Ammannito et al. (2016) and corrected for photometric effects as described in Ciarniello et al. (2017)

A black and white version of this figure will appear in some formats. For the color version, refer to the plate section.

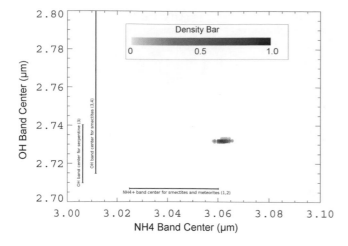

Figure 9.3 Plot of the band centers of 2.7 versus 3.1 μm absorption feature. Color has been assigned according to the density bar where values have been normalized to the max value for the same dataset visualized in Figure 9.2. Occator faculae values do not show in the plot because there is no 3.06 μm absorption band detected. The ranges of the axis have been chosen to account for the possible variability of the band centers in different hosting species of the represented ions, as reported in the figure. References for the ranges: (1) Ehlmann et al. (2018), (2) Ferrari et al (2019), (3) Bishop et al. (2008), (4) Fox et al. (2021).

et al. (2019), who found that only the expandable phyllosilicates can incorporate the ammonium ion within their structure by contact with ammonium-rich fluids at low temperature and pressure, although it cannot be excluded that, under conditions of higher temperature or pressure, all phyllosilicates can allocate the ammonium (Vedder, 1965; Itihara & Honma, 1979; Honma, 1996; Busigny et al., 2003; Papineau et al., 2005). Their clays with the NH_4^+ related absorption, however, show the OH stretch absorption with values slightly higher than the 2.73 μm of Ceres' average spectrum (i.e., nontronites at 2.78 μm, illite-smectite, montmorillonite, and hectorite at 2.75 μm). Thus, presently available data permit two explanations for associated ammonium and phyllosilicates on Ceres: (1) the presence of NH_4-smectites like nontronite, montmorillonite, or hectorite could be accompanied by another phase with OH stretching at 2.73 μm, for example, an Mg-phyllosilicate like serpentine used in radiative transfer models and found in carbonaceous chondrites (De Sanctis et al., 2015), or (2) the ammoniated phyllosilicate could be ammoniated Mg-saponite (King et al., 1992; Ehlmann et al., 2018) without requiring another phyllosilicate.

Regardless of the ammoniated phyllosilicate used, radiative transfer modeling shows abundances <10 wt.% of the ammoniated phase (De Sanctis et al., 2015; Kurokawa et al., 2020).

9.5 EVIDENCE FOR AMMONIUM IN SALTS ON CERES

The only location where VIR was able to undoubtedly detect a different, nonphyllosilicate bonding environment for the ammonium is the bright material of the Cerealia Facula within Occator crater. In this location, in fact, the 3.05 μm feature diminishes in strength as the albedo increases and is absent in the brightest

materials. Instead, a 2.76-μm absorption increases as does a broad and narrow absorption near 2.2 μm. While the 2.76-μm absorption and the narrow 2.2-μm absorption could be explained by Al phyllosilicates, the superimposed broad absorption near 2.2 μm requires an ammonium chloride or carbonate salt (De Sanctis et al., 2016). High resolution data show minor 2.08 μm and 2.35 μm absorptions adjacent to the 2.2 μm one that identify the NH_4 salt as NH_4Cl with a concentration of a few volume weight percent (Raponi et al., 2019). In this instance, ammonium is in a distinctly different mineral assemblage. The bright material in Cerealia facula have Na_2CO_3 (natrite), NH_4Cl, Al-phyllosilicates, and $NaCl_2H_2O$ (hydrohalite) responsible for absorptions at 3.11, 3.20, and 3.28 μm (De Sanctis et al., 2020). Bright faculae all over Ceres may have similar salt assemblages, though the coarser spatial scale of data elsewhere relative to the size of the faculae precludes definitive identification of minor NH_4Cl and $NaCl_2H_2O$ within the Na_2CO_3 enriched deposits (Figure 9.4).

Laboratory studies of the freezing of brines containing sodium, ammonium, chloride, and carbonate ions show that, in slow freezing conditions, Na_2CO_3 and $NaCl·2H_2O$ (hydrohalite) form preferentially, even in ammonium-dominated solutions, along with minor NH_4HCO_3 (Vu et al., 2017). On the contrary, under fast freezing conditions, the dominant specie is NH_4Cl rather than $NaCl·2H_2O$ which, however, is still present if the initial concentration of sodium and chloride ions in the brine is higher (Thomas et al., 2019). De Sanctis et al. (2020) noted that the spectra obtained during the laboratory experiments from solutions with an excess of sodium and chloride ions resembled very nicely the spectra that VIR acquired over Cerealia Facula. The specific assemblage of components measured by VIR, including NH_4Cl, along with the distribution of those components in association with the geologic features of the region suggest that cryovolcanic activity with extrusion of hydrated salt rich brines is the most likely formation scenario for the Cerealia Facula.

9.6 GEOLOGIC SETTING OF AMMONIATED MATERIALS

The average absorption band depth of the 3.1 μm absorption of ammoniated phyllosilicates is ~0.08. A special case are the faculae within Occator crater where the overall shape of the spectrum in the 3 μm region does not allow clear identification of the 3.06 μm band, which is consistent with the absence or low concentrations of ammoniated phyllosilicates. Ammoniated salts have instead been detected in these spatially restricted bright materials in the Cerealia Facula. This suggests a clear association with brines. The materials post-date the formation of Occator crater, which is between 1 and 60 Ma (Neesemann et al., 2019, and references therein), and were either formed from hydrothermal fluids ascending from a deep reservoir and then mainly transported in solid state to the surface or formed from upwelling brines after impact (Raymond et al., 2020; Schenk et al., 2020; Scully et al., 2020).

The 3.1-μm absorption of ammoniated phyllosilicates is 14% higher than average in Ceres' eastern hemisphere and correlates with the 800-km depression called Vendimia Planitia interpreted as an ancient basin (Marchi et al., 2016). Either these heterogeneities occur laterally across Ceres' surface or indicate a higher abundance

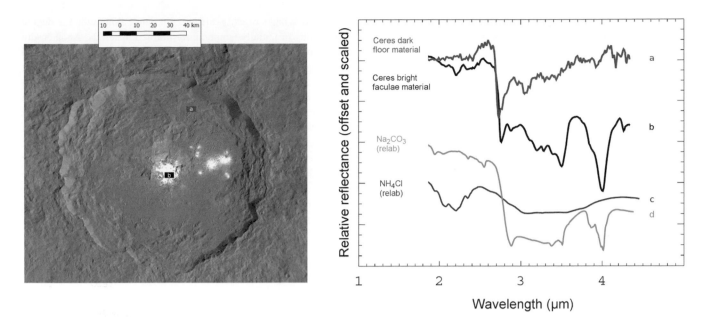

Figure 9.4 Left: Framing Camera image of Occator crater with locations of Dawn VIR spectra shown (black box a, grey box b). Right: NH₄Cl rich spectrum from Cerealia Facula (b) and average spectrum from Occator floor (a), respectively, from the black and grey boxes on the left. Reference spectra of NH₄Cl (c) and Na₂CO₃ (d) from the Brown University RELAB database.

of ammonium-bearing phyllosilicates in deeper regions of Ceres' crust (Stephan et al., 2019). The ammoniated phyllosilicate absorption is most prominent near the central peak of Urvara crater, with an average increase of about 25% in the depth of the absorption at 3.06 μm with respect to the average and the ejecta of the young Dantu crater (20% increase), while Haulani crater is the most depleted region with a decrease of the strength of the feature of about 30%.

Overall, in spite of these minor variations, the ammoniated clays on Ceres are widespread and associated with diverse geological features and spanning ancient cratered terrain to fresh ejecta of young craters. Furthermore, the spectral properties of the ammoniated materials do not change appreciably with age, from recent crater ejecta (<1 Ma) to ancient basin floors (>3 Ga). Thus, it is clear that the mineral assemblage of dark materials with the ammoniated clays is ubiquitous both at the surface and at down to at least a few kilometers depth, comprising the bulk composition of the uppermost Ceres crust.

9.7 AMMONIA AND ITS CONSEQUENCES FOR CERES

Ammonia is significant for astrobiology, geophysical evolution, and understanding the origin of Ceres. Carbon in Ceres surface materials is significantly enriched relative to the carbonaceous chondrites (Chapter 8). Additionally, ammonia adds to the richness of organic geochemical reactions and is important in driving the complex set of prebiotic chemical reactions necessary for developing precursors for life. For example, nucleobases are synthesized in carbonaceous chondrites via ammonium cyanide chemistry (Callahan et al., 2011). Insoluble organic matter in meteorites has been observed to release up to 1 wt.% ammonia, and certain CR-2

meteorites have more ammonia and amino acids than hydrocarbons and carboxylic acids (e.g., Pizzarello & Williams, 2012). Ceres does not appear to be represented in the meteorite collection, therefore, although some aliphatic organics have been detected on its surface (De Sanctis et al., 2018), the nature of its organic chemistry will await data from a future landed mission. From a historical geological perspective, the presence of ammoniated phyllosilicates on Ceres places an upper bound on temperatures experienced as <~400 K, else ammonium smectites observed become unstable (Chourabi & Fripiat, 1981).

Whether or not Ceres accreted with ammonia has long-term consequences for its geophysical and geological evolution because the accretion of ammonia would decrease the eutectic temperature of brincs to as low as 176 K and promote formation of a liquid brines, cryovolcanism, and a subsurface ocean (Kargel, 1992; Castillo-Rogez & McCord, 2010). The widespread occurrence of ammoniated materials in surface terrains of all depths and ages, including brine products, suggests ammonia is one of the bulk materials of Ceres rather than a later-delivered product. Indeed, the presence of a "weak layer" in the upper mantle/lower crust inferred by Fu et al. (2017) may suggest existence of subsurface liquids on Ceres to the present, facilitated by ammonia's freezing point depression. Additionally, as summarized by Castillo-Rogez et al. (2020), bright spots in select faculae may have formed by low temperature brine extrusion and/or diapirism. Modeling of the freezing of Ceres' ocean (Zolotov, 2017; Castillo-Rogez et al., 2018) reproduces the sodium bicarbonate and ammonium salts and carbonates found in these surface bright materials, some of which are excavated from depth (Stein et al., 2019). Thus, it seems clear that ammonia is central to the ongoing geologic activity involving salts and/or brines continuing at Ceres today to form such striking landforms as Cerealia facula in Occator crater and Ahuna Mons. Such a scenario is supported by measurements over the central region of Occator crater during the second extension of the Dawn spacecraft operations. The identification of $NaCl_2H_2O$ (hydrohalite) in combination with NH_4Cl (ammonium chloride) and Na_2CO_3 (natrite) in the Cerealia Facula and the uneven distribution of such components is strongly suggestive of recent and possibly continuous episodes of cryovolcanism as formation mechanism for the Facula itself (De Sanctis et al., 2020; Chapter 7).

In addition, as ammonia is typically found in outer Solar System objects, its presence on Ceres has implications as a key tracer of the accretional environment in which Ceres formed. Recent modeling by Kurokawa et al. (2020) shows that the synthesis of ammonium phyllosilicates is only explained if Ceres had ammonia at concentrations intermediate between pure water ice and cometary materials at water:rock ratios >1 and $T < 70°C$. Scenarios include the possibility that Ceres was implanted in the Main Belt from beyond the NH_3 snow line, perhaps driven by dynamical instabilities from giant planet migration, or that large deliveries of outer Solar System ices were brought into the Main Belt (Part III). Indeed, the Mg phyllosilicates, sodium carbonates, and ammoniated salts point to a Ceres water chemistry dominated by sodium, ammonium, chloride, and carbonate, and similar to that found on Enceladus (Matson et al., 2007; Waite et al., 2009). The larger Ceres and smaller Enceladus may be close cousins in a family of ocean worlds: Enceladus is the warmer version of Ceres, tidally heated to sustain its global subsurface ocean to the present, whereas Ceres' "ocean" is likely restricted to the pore volumes briny waters deep in the silicate crust/mantle (Castillo-Rogez et al., 2020; De Sanctis et al., 2020).

9.8 CONCLUSIONS AND OUTSTANDING QUESTIONS

Ceres has ammoniated phyllosilicates in the widespread dark materials that are ubiquitous both over the Ceres surface and down to depths of at least a few kilometers, excavated by impact craters. The 3.05-μm absorption is associated specifically with ammoniated smectites and co-occurs with a 2.73 μm absorption characteristic of Mg-OH. Thus, either the phase is ammoniated Mg-saponite or a mixture in roughly constant proportions of another ammonium smectite with Mg serpentine. The ammoniated phyllosilicates are present in at least a few wt.% and do not change in spectral properties with either geologic unit or surface exposure age within the dark materials. Bright materials on Ceres host ammonium salts, including specifically NH_4Cl, in at least one location at Cerealia Facula, in association with Na carbonate-enriched deposits and brine or diapir processes. Ammonia thus is a key phase for Ceres' evolution, playing an important role in its organic chemistry (Chapter 8), dictating the stability of water and the evolution of subsurface brines, and serving as a pointer to a dynamical evolution of the outer Main Belt that includes the substantial delivery of outer Solar System volatiles like ammonia to the inner Solar System.

There are, however, several aspects linked to the presence of ammonia on Ceres that remain to be fully resolved. For instance, what is the nature of a phyllosilicate carrier of ammonium in the ubiquitous dark materials: a single ammoniated Mg-saponite or a mixture of ammoniated smectite with Mg-serpentine? The answer to this question is important for establishing the temperatures and

water:rock ratios of early aqueous alteration along with the degree of mixing of initially temperature-stratified materials. Another key point is whether ammonia supports brines on Ceres that persist to the present-day in the subsurface. The geological and geophysical data suggest low temperature brines, but these remain to be absolutely confirmed by further geophysical data, such as nutation or magnetometer measurements. In addition, did Ceres arrive from the Main Belt as an ammonia-rich body or were ammonia-rich materials delivered to the Main Belt during early accretion and differentiation of the large asteroids? The ever growing number of bodies in the outer Main Belt with ammoniated materials like Ceres (Usui et al., 2019; Chapter 1) points to ammonia as a fundamental tracer of the dynamical evolution of our early Solar System. A combination of modeling and observation is required to establish implications for Solar System evolution and the delivery of volatiles inward.

REFERENCES

Ammannito, E., DeSanctis, M. C., Ciarniello, M., et al. (2016) Distribution of phyllosilicates on the surface of Ceres. *Science*, 353, aaf4279.

Bauer, J. M., Roush, R. L., Geballe, T. R., et al. (2002) The near-infrared spectrum of Miranda. *Icarus*, 158, 178–190.

Berg, B. L., Cloutis, E. A., Beck, P., et al. (2016) Reflectance spectroscopy (0.35–8 μm) of ammonium-bearing minerals and qualitative comparison to Ceres-like asteroid. *Icarus*, 265, 218–237.

Bishop, J. L., Banina, A., Mancinelli, R. L., & Klovstad, M. R. (2002) Detection of soluble and fixed NH4+ in clay minerals by DTA and IR reflectance spectroscopy: A potential tool for planetary surface exploration. *Planetary and Space Science*, 50, 11–19.

Bishop, J. L., Lane, M. D., Dyar, M. D., & Brown, A. J. (2008) Reflectance and emission spectroscopy study of four groups of phyllosilicates: Smectites, kaolinite-serpentines, chlorites and micas. *Clay Minerals*, 43, 35–54.

Brown, M. E., & Calvin, W. M. (2000) Evidence for crystalline water and ammonia ices on Pluto's satellite Charon. *Science*, 287, 107–109.

Brown, M. E., Schaller, E. L., & Fraser, W. C. (2011) A hypothesis for the color diversity of the Kuiper Belt. *Astrophysics Journal Letters*, 739, L60–L64.

Bruno, T. J., & Svoronos, P. D. N. (1989) *CRC Handbook of Basic Tables for Chemical Analysis*. Boca Raton, FL: CRC Press.

Busigny, V., Cartigny, P., Philippot, P., & Javoy, M. (2003) Ammonium quantification in muscovite by infrared spectroscopy. *Chemical Geology*, 198, 21–31.

Callahan, M. P., Smith, K., Cleaves, H., et al. (2011) Carbonaceous meteorites contain a wide range of extraterrestrial nucleobases. *Proceedings of the National Acadamy of Sciences (USA)*, 108, 13995–13998.

Carrozzo, F. G., De Sanctis, M. C., Raponi, A., et al. (2018) Nature, formation, and distribution of carbonates on Ceres. *Science Advances*, 4, e1701645.

Castillo-Rogez, J. C., & McCord, T. B. (2010) Ceres' evolution and present state constrained by shape data. *Icarus*, 205, 443–459.

Castillo-Rogez, J., Neveu, M., McSween, H. Y., et al. (2018) Insights into Ceres's evolution from surface composition. *Meteoritics & Planetary Science*, 53, 1820–1843.

Castillo-Rogez, J. C., Neveu, M., Scully, J. E. C., et al. (2020) Ceres: Astrobiological target and possible ocean world. *Astrobiology*, 20, 269–291.

Ciarniello, M., De Sanctis, M. C., Ammannito, E., et al. (2017) Spectrophotometric properties of dwarf planet Ceres from the VIR spectrometer on board the Dawn mission. *Astronomy & Astrophysics*, 598, A130.

Chourabi, B., & Fripiat, J. J. (1981). Determination of tetrahedral substitutions and interlayer surface heterogeneity from vibrational spectra of ammonium in smectites. *Clays and Clay Minerals*, 29, 260–268.

Dalle Ore, C. M., Cruikshank, D. P., Protopapa, S. et al. (2019) Detection of ammonia on Pluto's surface in a region of geologically recent tectonism. *Science Advances*, 5, eaav5731.

De Sanctis, M. C., Ammannito, E., Carrozzo, F. G., et al. (2018) Ceres's global and localized mineralogical composition determined by Dawn's Visible and Infrared Spectrometer (VIR). *Meteoritics & Planetary Science*, 53, 1844–1865.

De Sanctis, M. C., Ammannito, E., Raponi, A., et al. (2015) Ammoniated phyllosilicates with a likely outer Solar System origin on (1) Ceres. *Nature*, 528, 241–244.

De Sanctis, M. C., Ammannito, E., Raponi, A., et al. (2020) Fresh emplacement of hydrated sodium chloride on Ceres from ascending salty fluids. *Nature Astronomy*, 4, 786–793.

De Sanctis, M. C., Coradini, A., Ammannito, E., et al. (2011) The VIR spectrometer. *Space Science Reviews*, 163, 329–369.

De Sanctis, M. C., Frigeri, A., Ammannito, E., et al. (2019) Ac-H-11 Sintana and Ac-H-12 Toharu quadrangles: Assessing the large and small scale heterogeneities of Ceres' surface. *Icarus*, 318, 230–240.

De Sanctis, M. C., Raponi, A., Ammannito, E., et al. (2016) Bright carbonate deposits as evidence of aqueous alteration on (1) Ceres. *Nature*, 536, 1–4.

Dodson-Robinson, S. E., Willacy, K., Bodenheimer, P., Turner, N. J., & Beichman, C. A. (2009) Ice lines, planetesimal composition and solid surface density in the solar nebula. *Icarus*, 200, 672–693.

Ehlmann, B. L., Hodyss, R., Bristow, T. F., et al. (2018) Ambient and cold-temperature infrared spectra and XRD patterns of ammoniated phyllosilicates and carbonaceous chondrite meteorites relevant to Ceres and other Solar System bodies. *Meteoritics & Planetary Science*, 53, 1884–1901.

Farmer, V. C., & Russell, J. D. (1967) Infrared absorption spectrometry in clay studies. *Clays and Clay Minerals*, 15, 121–142.

Ferrari, M., De Angelis, S., De Sanctis, M. C., et al. (2019) Reflectance spectroscopy of ammonium-bearing phyllosilicates. *Icarus*, 321, 522–530.

Fox, V. K., Kupper, R. J., Ehlmann, B. L., et al. (2021) Synthesis and characterization of Fe(III)-Fe(II)-Mg-Al smectite solid solutions and implications for planetary science. *American Mineralogist*, 106, 964–982.

Fu, R. R., Ermakov, A. I., Marchi, S., et al. (2017) The interior structure of Ceres as revealed by surface topography. *Earth and Planetary Science Letters*, 476, 153–164.

Honma, H. (1996) High ammonium contents in the 3800 Ma Isua supracrustal rocks, central West Greenland. *Geochimica et Cosmochimica Acta*, 60, 2173–2178.

Itihara, Y., & Honma, H. (1979) Ammonium in biotite from metamorphic and granitic rocks of Japan. *Geochimica et Cosmochimica Acta*, 43, 503–509.

Jewitt, D. C., & Luu, J. (2004) Crystalline water ice on the Kuiper belt object (50000) Quaoar. *Nature*, 432, 731–733.

Kargel, J. S. (1992) Ammonia-water volcanism on icy satellites: Phase relations at 1 atmosphere. *Icarus*, 100, 556–574.

King, T. V. V., Clark, R. N., Calvin, W. M., Sherman, D. M., & Brown, R. H. (1992) Evidence for ammonium-bearing minerals on Ceres. *Science*, 255, 1551–1553.

Krohn, M. D., & Altaner, S. P. (1987) Near infrared detection of ammonium minerals. *Geophysics*, 52, 924–930.

Kurokawa, H., Ehlmann, B. L., De Sanctis, M. C., (2020) A probabilistic approach to determination of Ceres' average surface composition

from Dawn visible-infrared mapping spectrometer and gamma ray and neutron detector data. *Journal of Geophysical Research: Planets*, 125, e06606.

Lebofsky, L. (1978) Asteroid 1 Ceres: Evidence for water of hydration. *Monthly Notices of the Royal Astronomical Society*, 182, 17P-21P.

Lebofsky, L., Feierberg, M., Tokunaga, A., Larson, H., & Johnson, J. (1981) The 1.7–4.2 μm spectrum of asteroid 1 Ceres: Evidence for structural water in clay minerals. *Icarus*, 48, 453–459.

Li, J.-Y., Reddy, V., Nathues, A., et al. (2016) Surface albedo and spectral variability of Ceres. *The Astrophysical Journal Letters*, 817, L22.

Marchi, S., Ermakov, A. I., Raymond, C. A., et al. (2016) The missing large impact craters on Ceres. *Nature Communications*, 7, 12257.

Matson, D. L., Castillo, J. C., Lunine, J., & Johnson, T. V. (2007) Enceladus' plume: Compositional evidence for a hot interior. *Icarus*, 187, 569–573.

McCord, T., & Sotin, C. (2005) Ceres: Evolution and current state. *Journal of Geophysical Research*, 110, E05009.

McKinnon, W. B. (2008) Could Ceres be a refugee from the Kuiper Belt? *Asteroids, Comets, Meteors Conference*, July 14–18, Baltimore, MD, No 1405, abstract #8389.

Milliken, R. E., & Rivkin, A. S. (2009) Brucite and carbonate assemblages from altered olivine-rich materials on Ceres. *Nature Geoscience*, 2, 258–261.

Mookherjee, M., Redfern, S. A. T., Zhang, M., & Harlov, D. E. (2002) Orientational order–disorder of N(D,H)4+ in tobelite. *American Mineralogist*, 87, 1686–1691.

Neesemann, A., van Gasselt, S., Schmedemann, N., et al. (2019) The various ages of Occator crater, Ceres: Results of a comprehensive synthesis approach. *Icarus*, 320, 60–82.

Papineau, D., Mojzsis, S. J., Karhu, J. A., & Marty, B. (2005) Nitrogen isotopic composition of ammoniated phyllosilicates: Case studies from precambrian metamorphosed sedimentary rocks. *Chemical Geology*, 216, 37–58.

Pizzarello, S., & Williams, L. B. (2012) Ammonia in the early Solar System: An account from carbonaceous meteorites. *The Astrophysical Journal*, 749, 161.

Prettyman, T. H., Yamashita, N., Toplis, M. J., et al. (2017) Extensive water ice within Ceres' aqueously altered regolith: Evidence from nuclear spectroscopy. *Science*, 355, 55–59.

Raponi, A., De Sanctis, M. C., Carrozzo, F. G., et al. (2019) Mineralogy of Occator crater on Ceres and insight into its evolution from the properties of carbonates, phyllosilicates, and chlorides. *Icarus*, 320, 83–96.

Raymond, C. A., Ermakov, A. I., Castillo-Rogez, J. C., et al. (2020) Impact-driven mobilization of deep crustal brines on dwarf planet Ceres. *Nature Astronomy*, 4, 741–747.

Reddy, V., Li, J.-Y., Gary, B. L., et al. (2015) Photometric properties of Ceres from telescopic observations using Dawn Framing Camera color filters. *Icarus*, 260, 332–345.

Rivkin, A. S., Volquardsen, E. L., & Clark, B. E. (2006) The surface composition of Ceres: Discovery of carbonates and iron-rich clays. *Icarus*, 185, 563–567.

Rivkin, A. S., Howell, E. S., & Emery, J. P. (2019) Infrared spectroscopy of large, low-albedo asteroids: Are Ceres and Themis archetypes or outliers? *Journal of Geophysical Research: Planets*, 124, 1393–1409.

Robinson, J. W. (1974) *CRC Handbook of Spectroscopy*. Boca Raton, FL: CRC Press.

Rognini, E., Capria, M. T., Tosi, F., et al. (2020) High thermal inertia zones on Ceres from Dawn data. *Journal of Geophysical Research: Planets*, 125, e05733.

Russell, C. T., & Raymond, C. A. (2011) The Dawn mission to Vesta and Ceres. *Space Science Reviews* 163, 3–23.

Schenk, P., Scully, J., Buczkowski, D., et al. (2020) Impact heat driven volatile redistribution at Occator crater on Ceres as a comparative planetary process. *Nature Communications*, 11, 3679.

Scully, J. E. C., Schenk, P. M., Castillo-Rogez, J. C., et al. (2020) The varied sources of faculae-forming brines in Ceres' Occator crater emplaced via hydrothermal brine effusion. *Nature Communications*, 11, 3680.

Silverstein, R. M., Bassler, G. C., & Morrill, T. C. (1991) *Spectrometric Identification of Organic Compounds*, 5th ed. New York: Wiley.

Srasra, E., Bergaya, F., & Fripiat, J. J. (1994) Infrared spectroscopy study of tetrahedral and octahedral substitutions in an interstratified illite-smectite clay. *Clays and Clay Minerals*, 42, 237–241.

Stein, N. T., Ehlmann, B. L., Palomba, E., et al. (2019) The formation and evolution of bright spots on Ceres. *Icarus*, 320, 188–201.

Stephan, K., Jaumann, R., Zambon, F., et al. (2019) Ceres' impact craters – Relationships between surface composition and geology. *Icarus*, 318, 56–74.

Takir, D., & Emery, J. P. (2012) Outer Main Belt asteroids: Identification and distribution of four 3-mm spectral groups. *Icarus*, 219, 641–654.

Thomas, E. C., Vu, T. H., Hodyss, R., et al. (2019) Kinetic effect on the freezing of ammonium-sodium-carbonate-chloride brines and implications for the origin of Ceres' bright spots. *Icarus*, 320, 150–158.

Usui, F., Hasegawa, S., Ootsubo, T., & Onaka, T. (2019) AKARI/IRC near-infrared asteroid spectroscopic survey: AcuA-spec. *Publications: Astronomical Society of Japan*, 71.

Vedder, T. V. (1965) Ammonium in muscovite. *Geochimica et Cosmochimica Acta*, 29, 221–228.

Vernazza, P., Mothé-Diniz, T., Barucci, M. A., et al. (2005) Analysis of near-IR spectra of 1 Ceres and 4 Vesta, targets of the Dawn mission. *Astronomy and Astrophysics*, 436, 1113–1121.

Vu, T. H., Hodyss, R., Johnson, P. V., & Choukroun, M. (2017) Preferential formation of sodium salts from frozen sodium-ammonium-chloride-carbonate brines – Implications for Ceres' bright spots. *Planetary and Space Science*, 141, 73–77.

Waite, J. H., Jr., Lewis, W. S., Magee, B. A., et al. (2009) Liquid water on Enceladus from observations of ammonia and 40Ar in the plume. *Nature*, 460, 487–490.

Zolotov, M. Y. (2017) Aqueous origins of bright salt deposits on Ceres. *Icarus*, 296, 289–304.

Geomorphology of Ceres

DAVID A. WILLIAMS, ANDREAS NATHUES, AND JENNIFER E. C. SCULLY

10.1 INTRODUCTION

NASA's Dawn mission (Russell & Raymond, 2011; Russell et al., 2016) enabled dwarf planet Ceres to join the group of Solar System bodies that are studied as geological objects instead of only as astronomical objects. Dawn's Framing Camera (FC, Sierks et al., 2011) globally imaged Ceres using a panchromatic (clear) filter on approach (from 10 to 1 km/pixel), and during three orbital phases of the primary mission: Survey (∼415 m/pixel), High-Altitude Mapping Orbit (HAMO: ∼140 m/pixel), and Low-Altitude Mapping Orbit (LAMO: ∼35 m/pixel). In addition, global color filter imagery in seven wavelength bands was obtained during Approach, Survey and HAMO phases, as well as of selected sites from LAMO. When in Extended Mission Phases 1 (XM1) and 2 (XM2), several surface features of high interest were re-visited in all colors and the clear filter at pixel scales better than 35 m. In particular, the Urvara and Occator crater regions were subject to detailed imaging in XM2. A highly elliptical orbit (see Chapter 2) with a pericenter height of about 35 km above the surface led to very high spatial resolution (>3 m/pixel) observations of these craters (Nathues et al., 2020). Dawn's FC provided panchromatic and color band (14-bit grayscale) images of the surface, from which, for example, topographic information via stereo photoclinometry (SPC, Park et al., 2019) and stereo photogrammetry (SPG, Preusker et al., 2016; Jaumann et al., 2017) was derived. The HAMO global RGB color mosaic and the LAMO global clear filter mosaic, in which major geologic features are named, and the HAMO SPG global digital terrain model (DTM) that shows topography, are presented in Figure 10.1. During the Primary Mission, the Dawn Science Team produced global geologic maps during each orbital phase (Buczkowski et al., 2016; Mest et al., 2018: Williams et al., 2018a), and the final LAMO geologic map is presented in Figure 10.2.

In this chapter we discuss the variety of geomorphological features observed on Ceres' surface from Dawn. We begin with a broad overview of the geomorphology, introducing the feature types related to the geologic processes that produced them: Impact, gradation (weathering–erosion–deposition), tectonic, and cryovolcanic/hydrothermal, followed by more detailed descriptions. We conclude with a comparison to asteroid (4) Vesta, in which the differences in geologic features inform us on the

The authors thank Debra Buczkowski and Michelle Kirchoff for very helpful reviews of the chapter. We acknowledge Mr. David Nelson, GIS Specialist and Data Manager of the Ronald Greeley Center, the NASA Regional Planetary Information Facility (RPIF) at Arizona State University, for GIS processing of the global images and geologic map included in this chapter.

different formations and evolutions of the two objects, and a brief discussion of outstanding questions for future Ceres exploration.

10.2 BACKGROUND: CERES GEOMORPHOLOGY FROM PRE-DAWN AND MODELING STUDIES

In the more than 200 years between Ceres' discovery and the arrival of the Dawn spacecraft, much was learned about the dwarf planet from telescopic observations and modeling studies. Telescopic light curve analysis returned a rotation period of ∼9 hours (Chamberlain et al., 2007). Hubble Space Telescope (HST) observations obtained during 2003 and 2004 revealed Ceres' shape to be an oblate spheroid with a mean radius of ∼476 km (Thomas et al., 2005). Combining this size with mass estimates (Michalak, 2000; Viateau & Rapport, 2001) led to a density estimate of ∼2,077 kg/m^3, which is consistent with CM carbonaceous chondrite meteorites (Thomas et al., 2005). A later mass estimate by Pitjeva and Standish (2009) refined the density estimate to be ∼2,075 kg/m^3. Ceres' shape was interpreted to be controlled by hydrostatic equilibrium and was found to be consistent with a central mass concentration suggesting differentiation (Thomas et al., 2005). Moreover, Thomas et al. (2005) found that the surface exhibited no more than ∼10 km of topographic relief (consistent with Millis et al., 1987), and inferred that Ceres' smoother shape in comparison to Vesta was caused by a rheology that was less able to support topography.

Variations in brightness across the surface were also observed in 2003/2004 HST images (with a pixel scale of ∼30 km): 11 surface albedo features from ∼40 km to ∼350 km were identified by Li et al. (2006). Two of these large bright spots are now known to correspond to the impact craters Occator and Haulani (Li et al., 2016). The overall geometric V-Band albedo of the cerean surface is relatively low (0.09) and, while albedo features were observed, the overall albedo range was found to be small in comparison to other small bodies, which was interpreted to be caused by widespread resurfacing by melted water/ice (Li et al., 2006). Radar observations indicated a very loosely packed regolith (Ostro et al., 1979), with a near-surface density of ∼1.24 g/cm^3 (Mitchell et al., 1996) and a low thermal inertia of ∼15 J/m^2/s$^{1/2}$/K. This much lower thermal inertia in comparison to the Moon (∼50 J/m^2/s$^{1/2}$/K) could be because of Ceres' different thermal environment, rather than intrinsically finer particle sizes on Ceres (Spencer et al., 1989; Rivkin et al., 2011).

Lebofsky et al. (1986) first introduced the idea that Ceres is differentiated, after the observation of water-like absorptions in

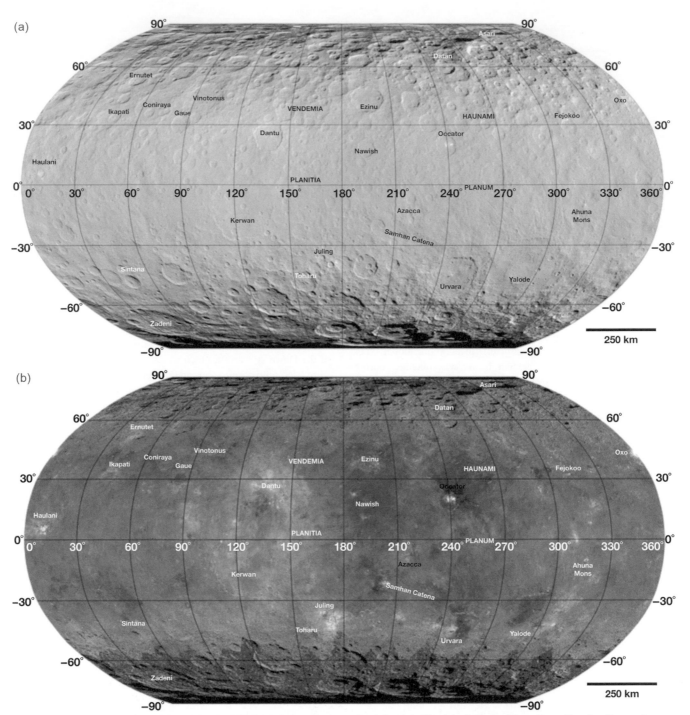

Figure 10.1 (a) Low-Altitude Mapping Orbit (LAMO) global clear filter mosaic of Ceres (∼35 m/pixel), derived from Dawn Framing Camera images. (b) High-Altitude Mapping Orbit (HAMO) global RGB color mosaic of Ceres (∼140 m/pixel), derived from Dawn Framing Camera images (R = 0.96 μm, G = 0.75 μm, B = 0.44 μm). (c) HAMO SPG global digital terrain model (DTM) of Ceres, color-coded for surface elevation, derived from Dawn Framing Camera images. Mosaics are centered on 0°, 180° in Robinson projection, with a 30° latitude–longitude grid. Major named surface features are indicated. A black and white version of this figure will appear in some formats. For the color version, refer to the plate section

Ceres' spectrum. Modeling of Ceres' internal evolution predicted that Ceres differentiated into a water mantle and silicate core, after initial accretion of a mix of ice and rock a few millions of years after CAIs (McCord & Sotin, 2005; Castillo-Rogez & McCord, 2010). Differentiation, and the resulting mineral hydration/dehydration reactions, could have led to heat pulses and dimensional changes causing a dramatic landscape shaped by tectonic deformation,

eruptions, and venting (McCord & Sotin, 2005). Cryovolcanism, from a warm layer at the base of the hydrosphere (Castillo-Rogez & McCord, 2010), was suggested, along with the possibility of a present-day interior liquid water layer (McCord et al., 2011). However, it was also possible that widespread viscous relaxation of the icy surface could have erased many surface features (Bland, 2013). An alternative evolution model proposed that Ceres formed

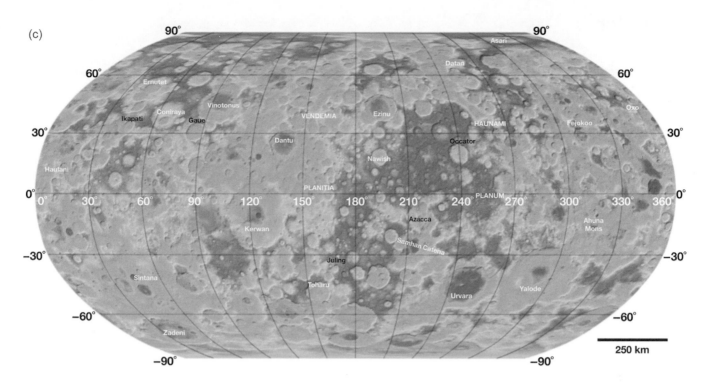

Figure 10.1 (*cont.*)

later, from other planetesimals, and that the low density was due to a porous hydrated silicate composition (Zolotov, 2009).

10.3 OVERVIEW

Like most airless bodies in the Solar System, Ceres as imaged by Dawn displays a heavily cratered surface (Hiesinger et al., 2016; Marchi et al., 2016). This was a surprise, as the inferred ice-rich crust of Ceres (based on telescopic observations) suggested that impact craters should have relaxed and disappeared (Bland, 2013), or become reduced to palimpsests, as observed on the outer planet icy satellites (e.g., Jones et al., 2003). In contrast, Dawn discovered that Ceres has craters preserved at size ranges from 300 km diameter down to the limit of image resolution (Hiesinger et al., 2016). Most established impact crater morphologies are present on Ceres, from simple, bowl-shaped craters to complex craters with central peaks and terraces, polygonal craters, floor-fractured craters, and craters with smooth or lobate floors. However, there are no multi-ring basins. Ceres lacks confirmed impact basins larger than 300 km in diameter, although gravity and topographic data show two-to-three large depressions that could be remnants of ancient impact basins (Marchi et al., 2016). Globally, Ceres' shape is relaxed (see Chapter 12 on Ceres' interior), and this has been interpreted to suggest some process(es) resurfaced Ceres early in its history and erased its largest basins (Marchi et al., 2016).

Ejecta blankets are preserved within and around impact craters. The youngest craters display outstanding bright bluish ejecta fields and rays in color images (Nathues et al., 2016; Stephan et al., 2017). Crater interiors display a variety of features, including terraces and central peaks for larger craters. Lobate materials are found in some crater floors, and have been interpreted as impact melts/slurries (Schenk et al., 2019; Scully et al., 2019a, 2019b); ice-rich landslides and ejecta (Schmidt et al., 2017; Chilton et al., 2019; Duarte et al., 2019; Hughson et al., 2019), and/or cryovolcanic flows (Krohn et al., 2016). Pitted terrain is identified in five-to-seven cerean craters (Sizemore et al., 2017), and is morphologically similar to pitted terrains observed on Vesta and Mars, which are interpreted to result from impact into ice-bearing target materials (Boyce et al., 2012; Denevi et al., 2012; Tornabene et al., 2012).

Gradational features observed on Ceres include the aforementioned lobate materials and ice-rich landslides and ejecta. The ice-rich landslides and ejecta have been categorized into three types (thick and thin landslides, fluidized ejecta – see Section 10.4), with an additional intermediate category (Schmidt et al., 2017; Chilton et al., 2019; Duarte et al., 2019; Hughson et al., 2019). A large landslide in Oxo crater was inferred to have exposed one of the few detectable water ice deposits identified by the Dawn Visible and Infrared Spectrometer (VIR: Combe et al., 2016; Nathues et al., 2017a).

Tectonic features observed on Ceres include impact-induced and non-impact-related feature types. The dominant type is secondary crater chains, which are formed when material ejected during the formation of the primary impact scours the surface in radial patterns. In contrast, pit chains and fractures are non-impact-related tectonic features. The largest pit chains, named the Samhain Catenae, are north of the major impact basins Urvara and Yalode, and are cross-cut by secondary crater chains originating from Urvara and Yalode (Scully et al., 2017). On a smaller scale, several of Ceres' impact craters have heavily fractured floors, and are analogous to lunar floor fractured craters (Buczkowski et al., 2019).

In terms of volcanic features, Ceres distinctive mountain, Ahuna Mons (21 km × 13 km × 4 km high), appears to be a cryovolcanic

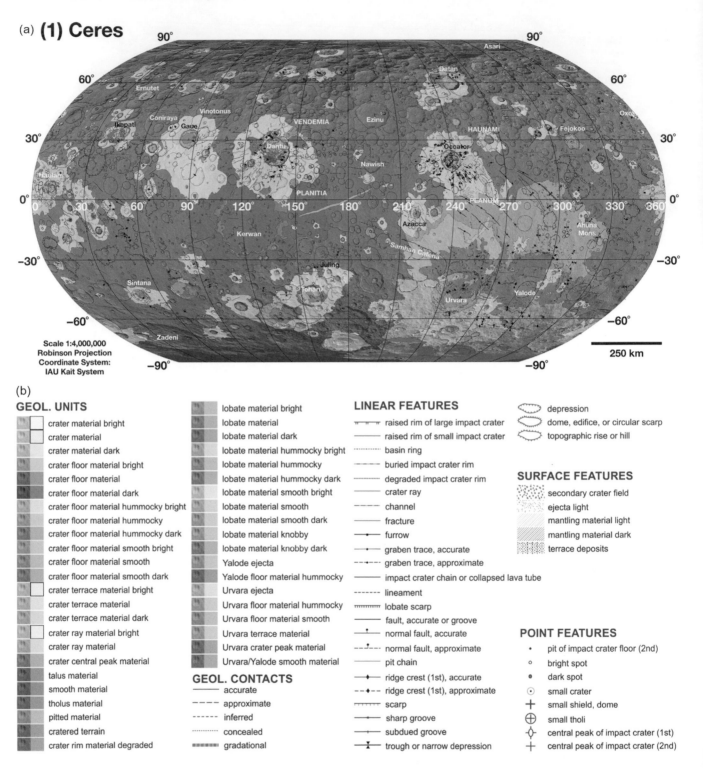

Figure 10.2 (a) Global geologic map of Ceres produced during the Dawn Primary Mission, using Low-Altitude Mapping Orbit (LAMO) images (~35 m/pixel) as the base map. This map is a unification of 15 quadrangle maps, published in a series of papers in *Icarus*, volume 316, December 2018. GIS processing by Andrea Nass, German Aerospace Center (DLR) and David M. Nelson (ASU). (b) Legend for the geological map in (a). GIS processing by Andrea Nass, German Aerospace Center (DLR) and David M. Nelson (ASU).

A black and white version of this figure will appear in some formats. For the color version, refer to the plate section

edifice, formed by inflation or extrusion of a viscous, salt-rich, carbonate-bearing dome, similar in morphology to terrestrial silicic domes (Ruesch et al., 2016). The material making up the bright spots, Cerealia and Vinalia Faculae within Occator crater, including the central Cerealia Tholus dome, are salt-rich liquids containing carbonates, and were emplaced by a combination of cryovolcanic (i.e., extrusion of deep brines) and hydrothermal (i.e., extrusion of slurries from an impact melt chamber) processes

(Nathues et al., 2017b, 2019, 2020; Scully et al., 2019a, 2020; Schmidt et al., 2020).

In Sections 10.4–10.7 we expand our descriptions of these features in greater detail, summarizing the results from the analysis of Dawn images.

10.4 IMPACT FEATURES

Impact craters on Ceres (Figure 10.3) have similar morphologies to those found elsewhere in the Solar System, but also have some differences best interpreted as impacting into a target that is transitional between ice and rock. For example, the morphologies of craters in the <30–40 km diameter range are similar to those on icy satellites of Saturn, such as Tethys and Dione (Hiesinger et al., 2016), except for floor-filling smooth materials that are not found on the icy satellites. These materials most likely represent impact melts or slurries derived from Ceres crust (Schenk et al., 2020), which is rich in hydrothermally-altered phyllosilicates, carbonates, salts (Ammannito et al., 2016; De Sanctis et al., 2016; Raponi et al., 2019), and <30% ice (Bland et al., 2016). Craters >40 km in diameter tend to display pitted floors, floor fractures, and rarely

central pits, indicative of a target weaker than silicate objects like the Moon or Vesta (Hiesinger et al., 2016).

10.4.1 Putative Large Basins

Asteroid collisional models of the primordial Main Belt and extrapolation of the number of large impact craters on Vesta revealed that about 10–15 craters >400 km should have been formed on Ceres (Marchi et al., 2016). Even if Ceres was implanted late in its current position in the Main Belt (De Sanctis et al., 2015), about six-to-seven larger impact basins would have been expected. Instead, Ceres shows no obvious larger impact basins than the 284-km diameter Kerwan. The lack of large basins is likely a consequence of their obliteration facilitated by Ceres' specific (ice-rich) composition and internal evolution. However, Marchi et al. (2016) identified three potential old heavily eroded basins on Ceres, in the initial Dawn topography data. The most notable is a ~800 km diameter depression termed Vendemia Planitia (center ~20°N, 135°E), whose potential rim and floor are studded by many large (>50 km diameter) craters. Interestingly, Vendemia Planitia is also spectrally distinct from the cerean background material, which strengthens the hypothesis of an old, eroded, and relaxed impact basin in which deeper material has been excavated (Marchi et al., 2016).

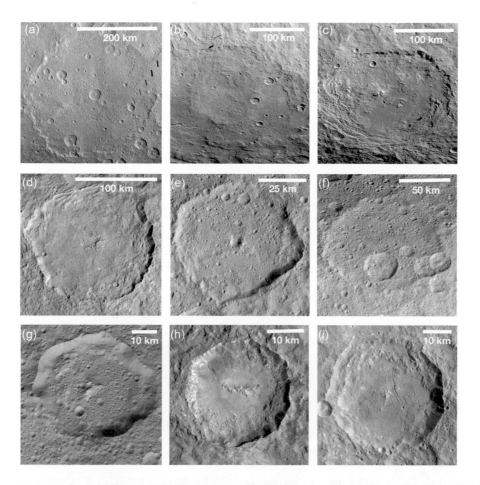

Figure 10.3 Major impact craters on Ceres. (a) Kerwan, a 284-km partly relaxed basin with a central depression; (b) Yalode, a ~260-km basin; (c) Urvara, a ~170-km basin; (d) Dantu, a 126-km complex crater with a possible peak ring–central pit structure; (e) Nawish, a 72-km polygonal crater with a central pit; (f) Coniraya, a 135-km crater inferred to be viscously relaxed (Bland et al., 2016); (g) Ernutet crater, a 52-km crater that contains organic materials (red in image; De Sanctis et al., 2017); (h) Haulani crater, a 34-km crater containing bluish materials and bright rays, possibly the youngest rayed crater on Ceres (~2 Ma (LDM), Krohn et al., 2018); (i) Azacca, ~50-km crater, possibly the oldest rayed crater on Ceres (~76 Ma (LDM), Schmedemann et al., 2016).

10.4.2 Large Craters

Ceres exhibits a few impact basins up to ~300 km diameter. The largest and oldest of these clearly discernable basins is Kerwan (Figures 10.2 and 10.3(a), Williams et al., 2018b), whose center is located slightly south of the cerean equator. Kerwan has a highly discontinuous, polygonal, degraded rim, and shows an extended "smooth" unit, filling its floor as well as areas outside the rim (Williams et al., 2018b). The smooth unit is likely the result of a water–ice rich crust that was liquified by the impact (Williams et al., 2018b). The second and third largest basins are Yalode (diameter ~260 km) and Urvara (diameter ~170 km) (Figure 10.3(b) and (c)), both located on the southern hemisphere in spatial proximity. Cross-cutting relationships and morphologic signatures show that the Urvara impact followed the Yalode impact (Crown et al., 2018), a finding which is supported by crater densities and absolute model ages (Schmedemann et al., 2016; Crown et al., 2018). Both basins resemble complex craters, showing broad relatively flat floors and terraced walls. Because of its younger age, Urvara has a better-preserved morphological expression than Yalode, and exhibits a central peak of 26 km length trending from southwest to northeast. Urvara shows a peculiar morphological dichotomy: extensive smooth materials to the north and east and vast terraced material to the west and south. Prominent ridges and grooves are seen east of the central peak. A few small graben are present inside both basins. Yalode and Urvara ejecta extend far out, for more than two crater diameters, and exhibit troughs, ridges, and scarps. Interestingly, neither Urvara nor Yalode have extensive smooth units similar to Kerwan's, perhaps indicative of crustal heterogeneity between Kerwan and Urvara–Yalode, or alternatively some other mechanism was involved in formation of the Kerwan smooth unit.

10.4.3 Craters

Ceres is depleted in impact craters with diameters greater than 100–150 km (Marchi et al., 2016). However, a number of preserved large craters are present. One of the most prominent is Dantu (126 km diameter). Dantu (Figure 10.3(d)) is located in the northern hemisphere near the center of Vendimia Planitia, and Dantu's floor and ejecta materials show spectral indications for excavated deep crustal material (Stephan et al., 2018, 2019). Dantu exhibits several characteristics, which differ between the northern and southern part of the crater: fracture systems, bright spots, spectral properties, topography, and an asymmetrical ejecta blanket. The central floor shows remnants of a central peak with a potential pit to its west. More obvious central pits are found on the central floors of craters Nawish and Occator. Nawish (Figure 10.3(e)) is a 72 km diameter old impact crater located at northern equatorial latitudes (Frigeri et al., 2018). Nawish reveals a well-defined central depression of about 10 km diameter (Nathues et al., 2020), but lacks the bright material found at Occator (discussed in Section 10.7). Another interesting large impact crater is Coniraya. This 135 km diameter crater (Figure 10.3(f)) is located in the northern hemisphere of Ceres. Based on morphological measurements (low depth to diameter ratio), Bland et al. (2016) argued that Coniraya might be viscously relaxed. The crater is heavily degraded and its ejecta blanket is absent (Pasckert et al., 2018), and is geologically younger than the lesser eroded, 52-km diameter Ernutet crater (Figure 10.3(g)) located in its vicinity. This discrepancy in the

erosion state, the younger Coniraya being more eroded than the older crater Ernutet, led to the hypothesis that Coniraya underwent viscous relaxation caused by a higher volatile content of the subsurface (Pasckert et al., 2018), or conversely it has an altered size–frequency distribution. Ernutet crater is of interest because FC color data show bright deposits that correspond with VIR spectral signatures of aliphatic hydrocarbons (De Sanctis et al., 2017), the strongest concentration of organic materials on Ceres.

10.4.4 Rayed Craters

Particularly striking on the surface of Ceres are several rayed craters, which exhibit bright bluish ejecta blankets (cp. Figure 10.1(b)) in RGB colors (960/740/440 nm). Bluish materials belong to the youngest impact craters on Ceres, and an optical maturation process seems to be responsible for a decreasing reflectance in the 440 nm filter with age (Schmedemann et al., 2016). Prominent examples of rayed craters are Occator, Haulani (Figure 10.3(h)), and Azacca (Figure 10.3(i)). The 92 km diameter complex Occator (Figure 10.4) crater is located on the old elevated Hanami Planum region and shows a unique central depression (Figure 10.4(a)), covered with the brightest material on Ceres (Scully et al., 2020; Nathues et al., 2020, and references therein). The depression exhibits a central dome with unique morphology (Figure 10.4(b)). West of the central depression, a number of less bright faculae (Figure 10.4(c), see Section 10.5) cover the hummocky lobate material (Scully et al., 2019b). Wall collapse and subsequent terrace formation as well as avalanche deposition is evident at Occator. Portions of terrace material along crater walls display isolated sub-parallel fracture systems, mostly oriented parallel to crater wall segments. The largest fracture system is located at the SW portion of Occator, primarily within the terrace material where two fracture systems intersect (Buczkowski et al., 2019). The age of Occator was determined by CSFD measurements to ~22 Ma (Lunar-derived Model – LDM, Neesemann et al., 2019).

Even younger than Occator is the 34 km diameter Haulani crater. CSFD measurements revealed an age of only ~2 Ma (LDM, Krohn et al., 2018). The crater also shows a complex morphology with a variety of lobate flows and tectonic features (Krohn et al., 2018). Its bright ejecta is widespread over the cerean surface, preferentially to the west; some rays extend up to 490 km. The crater shows a sharp steep rim, except for the western segment that collapsed. As similar to other rayed craters, Haulani shows several flow features and pitted terrain, which are indications of a volatile-rich subsurface (Krohn et al., 2018). A further example of a rayed crater is Azacca, a ~50 km diameter crater exhibiting, like Haulani, a central peak. This older crater also shows floor pits and fractures (Buczkowski et al., 2017), as well as a number of faculae that are less bright than in Occator (Stein et al., 2019).

10.5 GRADATIONAL FEATURES

We use the term "gradational" (Greeley, 2013) to include geologic features formed both by degradation (i.e., erosion) and by aggradation (i.e., material transport and deposition). On Ceres, these include mass wasting deposits, that is, various types of landslides: three different types of flow features were first detailed by

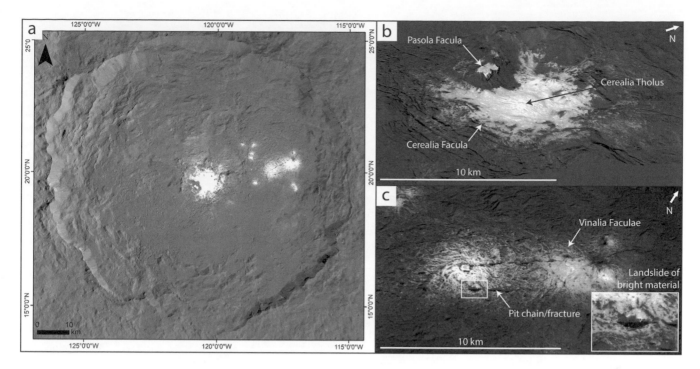

Figure 10.4 Faculae (bright spots) in Occator crater on Ceres. (a) The 92-km diameter Occator crater hosts the two brightest faculae on Ceres. (b) Cerealia Facula and Pasola Facula are located in the center of Occator. Cerealia Facula occurs in and around a central pit, which contains the mound Cerealia Tholus at its base. Pasola Facula occurs as a bright layer atop a ledge. (c) Vinalia Faculae are composed of multiple small bright spots in and around pit chain fracture systems.
Figure adapted from Scully et al. (2020)

Buczkowski et al. (2016) and later more fully characterized by Schmidt et al. (2017) (Figure 10.5).

10.5.1 Type 1 and Type 2 Flow Features (Landslides)

Type 1 flows (Figure 10.5(a)) are thick, tongue-shaped, furrowed flows that often occur on steep slopes, while type 2 (Figure 10.5(b)) flows are thin, spatulate flows with long runouts that are found on shallow slopes (Schmidt et al., 2017). Both types of flows originate at crater rims, flowing into or out of the host crater. However, they are morphologically distinct from dry, ballistically emplaced ejecta deposits seen elsewhere on Ceres. Schmidt et al. (2017) proposed that many type 1 and type 2 flows were triggered by the impact, but then continued to flow and develop post-impact. The morphology and geometry of the type 1 and 2 flows are analogous to ice-rich flows across the Solar System, leading to the hypothesis that the flows had landslide-like behavior post-impact, and are indicative of substantial amounts of subsurface ice, in particular closer to the poles (Schmidt et al., 2017). More recent work by Chilton et al. (2019) has described the characteristics of type 1 and type 2 features in greater detail, and confirmed that a latitude of ~70° separates the two landslide categories: type 1 are mostly located poleward of ~70°, while type 2 are located equatorward. Moreover, Chilton et al. (2019) find that type 1 landslides appear to fail at greater depths in the subsurface, while type 2 have more depth-limited failure. Combe et al. (2016, 2019) identified surface ice in association with type 1 and 2 features in VIR spectral data, consistent with the aforementioned morphological-based evidence for high ice content (Sizemore et al., 2019).

10.5.2 Type 3 Flow Features (Fluidized-Appearing Ejecta)

Type 3 features (Figure 10.5(e) and (f)) are cuspate, sheeted flows that appear to be fluidized. Schmidt et al. (2017) found that type 3 flows, in contrast to types 1 and 2, are strongly coupled to impact timing or size. Hughson et al. (2019) built upon this work by interpreting the type 3 flows as fluidized-appearing ejecta, which have morphological characteristics that are distinct from dry, ballistically emplaced ejecta seen elsewhere on Ceres. The type 3 flows often share analogous morphologies with fluidized ejecta elsewhere in the Solar System but are distinct from rampart ejecta found on Mars and some icy satellites. Hughson et al. (2019) hypothesize that the type 3 flows form when an impact occurs into a low cohesion, ice/silicate mix target material, and then sliding occurs on a low friction, partially icy substrate. While type 1 and 2 flows are interpreted as impact-triggered landslides with significant post-impact development, and type 3 flows are interpreted as fluidized-appearing ejecta, there is no clear cut-off between the three types. Duarte et al. (2019) find that many features have intermediate morphologies in between type 1/2 (Figure 10.5(c)) or type 2/3 flows, indicating that the three types exist along a continuum.

10.5.3 Juling Crater Landslide Deposit

Juling crater, a 20 km diameter crater located at 35°S, 168°E, contains prominent flows on its floor (Figure 10.5(d)), interpreted to be intermediate in type (Duarte et al., 2019). A broad landslide originates from the shadowed, steeply sloping northern wall: it has many morphological characteristics typical of a type 1 flow, but

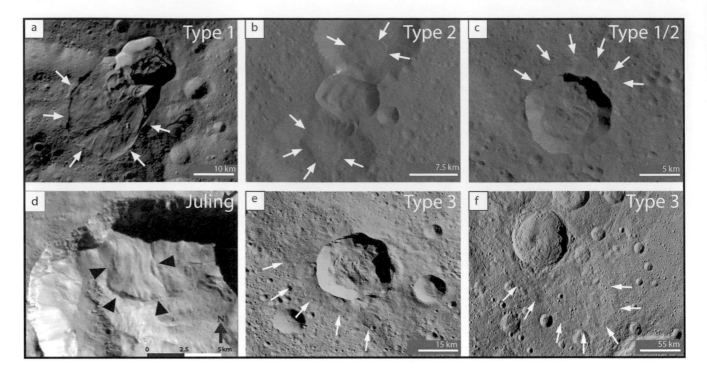

Figure 10.5 Different types of ice-rich flow features observed on Ceres. White or black arrows highlight the edges of the flows. (a, b, c) Type 1, type 2, and intermediate type 1/2 landslides. (d) Type 1/2 intermediate landslide that originated from the shadowed northern wall of Juling crater. (e, f) Type 3 flows, which are interpreted as fluidized-appearing ejecta.
Figure adapted from Sizemore et al. (2019) and Duarte et al. (2019)

other characteristics are more typical of type 2 flows. The VIR spectrometer detected absorption bands characteristic of water ice in the northern wall from which the type 1/2 landslide originates (Raponi et al., 2018), which is consistent with the hypothesis of Schmidt et al. (2017) that these landslides are relatively rich in water ice. Unexpectedly, the water ice signature was observed to change over a six-month period. Model results indicate that the water ice on the shadowed northern wall increased by ∼2 km² during this period, possibly due to a seasonal cyclical process (i.e., water sublimation and condensation) correlated with the solar flux (Raponi et al., 2018).

10.5.4 Dry Mass Wasting

In addition to the relatively ice-rich flows, landslides, and ejecta, there are also multiple occurrences of dry material transport and deposition across Ceres. Talus material is found along the walls of many impact craters. Particularly distinctive talus occurs around the steeply sloping walls (∼30–40°) of Occator crater (Buczkowski et al., 2018; Scully et al., 2019b). These lobes of talus cascade down the crater walls, often from outcrops along the crater rim, forming deposits that appear smooth in the FC images (spatial resolution of a few meters to a few tens of meters per pixel). On the scales of tens to hundreds of meters, the brightness of different talus lobes can vary dramatically, suggesting that materials of contrasting brightness are located next to one another on relatively small spatial scales in the subsurface (Scully et al., 2019b). Dry, ballistically emplaced ejecta blankets also surround impact craters, the youngest of which have distinctive rays and a blue appearance in FC color data (see Section 3.4; Nathues et al., 2016; Stephan et al., 2017).

10.6 TECTONIC FEATURES

Analysis of Dawn images shows that Ceres lacks both globally extensive tectonic features (such as Europa's ridged plains and Vesta's ridge-and-trough terrains; see Chapter 5) and globally extensive compressional faults (such as those on Mercury). Rather, Ceres' more localized tectonic features are subdivided into two categories: (a) larger pit chains (e.g., Samhain Catenae, Nabanna Fossa), and (b) smaller features found in the floor-fractured craters (Figure 10.6). One study interpreted linear features found in the polar regions to be due to thrust faulting (Ruiz et al., 2019), but they are limited in spatial extent, and their interpretation as contractional features has not been widely accepted by the community.

10.6.1 Pit Chains

The Samhain Catenae (Figure 10.6(a) and (b)) are the only set of >1 km wide pit chains identified on Ceres, with a maximum width of ∼11 km, average depth of ∼1 km, and average length of ∼200 km (Buczkowski et al., 2016). They are regional in scale and are found between Occator and Urvara–Yalode craters at an orientation of ∼NW/SE. They are morphologically distinct from the secondary crater chains that originate from Urvara and Yalode craters, which cross-cut the Samhain Catenae in some locations (Scully et al., 2017). Pit chains have more poorly defined rims and more irregular shapes than chains of secondary craters, which have more clearly defined rims and more regular shapes than pits (Buczkowski et al., 2016; Scully et al., 2017). The extensional event that formed the Samhain Catenae is currently hypothesized to be the result of a region of upwelling material arising from convection/diapirism, however a basin-forming impact or freezing of a global/regional

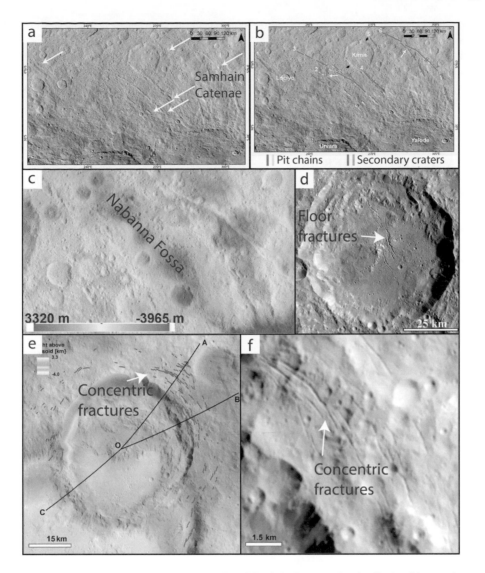

Figure 10.6 Examples of Ceres tectonic features. (a and b) The pit chains of Samhain Catena (red and yellow), with secondary crater chains (pink and magenta), in mapped and unmapped versions. (c) Nabanna Fossa in HAMO topographic data (highest and lowest elevations given in color bar). (d) Floor fractures within Azacca crater. (e and f) Concentric fractures surrounding Ikapati crater in mapped and unmapped versions. Figure adapted from Scully et al. (2017), Buczkowski et al. (2018, 2019), and Otto et al. (2019).
A black-and-white version of this figure will appear in some formats. For the color version, please refer to the plate section

subsurface ocean are also possibilities (Scully et al., 2017). Nabanna Fossa is located between two Samhain Catenae pit chains, and the 2.5 km deep degraded trough is most clearly visible in the HAMO topographic data (Figure 10.6(c)) (Buczkowski et al., 2018). The similar orientation to the Samhain Catenae suggests that Nabanna Fossa was formed in the same extensional event as the Samhain Catenae, but with the Samhain Catenae taking the form of pit chains while Nabanna Fossa took the form of a trough (Buczkowski et al., 2018). The majority of the smaller scale pit chains (generally <1 km wide) are located inside or in the vicinity of impact craters (see Section 10.6.2). These smaller scale pit chains share the poorly defined rims and irregular shapes that are also characteristic of the aforementioned larger pit chains.

10.6.2 Floor Fractured Craters

Twenty-one impact craters on Ceres contain complex patterns of fractures, and have anomalously shallow floors, making them analogous to lunar floor fractured craters (Buczkowski et al., 2016). Some of the fractures within crater floors grade into pit chains, but the dominant morphological expression of extension in these crater floors are fractures (Figure 10.6(d)). Ceres' floor fractured craters range in diameter from 26 km (Haulani crater) to 126 km (Dantu crater: Buczkowski et al., 2016), and include the well-known Occator crater (Buczkowski et al., 2019). The largest of Ceres' floor fractured craters (>50 km) are analogous to class 1 lunar floor fractured craters, because of their radial and/or concentric floor fractures and central peak/pit complexes. The smaller floor fractured craters have v-shaped moats separating the wall from the floor and are less intensely fractured. The formation of the floor fractures has been attributed to cryomagmatic intrusions below the crater floors and/or the intrusion of low viscosity and low density material into the crater walls via solid-state flow (Buczkowski et al., 2016, 2018, 2019, Nathues et al., 2017b; Bland et al., 2018). Surrounding the rims of many of the floor fractured craters are concentric fractures that are a few meters to few kilometers in length and <300 m wide

(Figure 10.6(e) and (f)). They occur within one crater radius of the rim and surround craters with diameters from 20 to 270 km (Otto et al., 2019). The formation of the concentric fractures is attributed to the presence of a shallow, low viscosity subsurface layer, which, because of the association with floor fractures and pitted terrain, may be rich in water ice.

10.7 CRYOVOLCANIC/HYDROTHERMAL/PERIGLACIAL FEATURES

In this section we examine each of Ceres' major endogenic features and the evidence that supports the various hypotheses.

10.7.1 Large Domes

Ceres has between 6 and 12 positive relief topographic features that were interpreted as domes and named "mons" (plural, montes) or "tholus" (plural, tholi), as revealed by regional geologic mapping (Hughson et al., 2018; Krohn et al., 2018; Platz et al., 2018). Ahuna Mons (Figure 10.7(a)) appears to represent the youngest and best preserved of this class of features (Sori et al., 2017), and has

been the most extensively studied. Ahuna Mons is an elliptical topographic feature (21 km × 13 km × 4 km high) on a broad ~2 km topographic rise with a nearby 40 km diameter impact crater cutting into the rise (Ruesch et al., 2016). It has steep (30° to 40°) slopes of talus with a ~300 m depressed summit unit that displays ≤1 km troughs and ridges as well as hummocky material, interpreted as disruption of a brittle layer by extensional forces, and with recent activity at <210 ± 30 Ma (Ruesch et al., 2016). Based on Dawn image analysis (Ruesch et al., 2016) and various modeling studies, the leading interpretation is that Ahuna Mons is a cryovolcanic dome formed by endogenous growth of a salt-rich, carbonate-bearing viscous material (Ruesch et al., 2016). Other domical topographic features, such as Liberalia Mons, and Kaanzaa, Hosil, Mikeli, Cosecha, Dalien, and Wangala Tholi (Figure 10.7(b)) that occur in the same hemisphere (most are in the same quadrant) as Ahuna Mons, have been interpreted as older, more degraded, and topographically relaxed versions of cryovolcanic domes akin to Ahuna Mons (Sori et al., 2017).

10.7.2 Cerealia Tholus

Cerealia Tholus (Figures 10.4(b) and 10.8(a)) is located in the base of Occator's central pit, both of which are covered by the bright

(a)

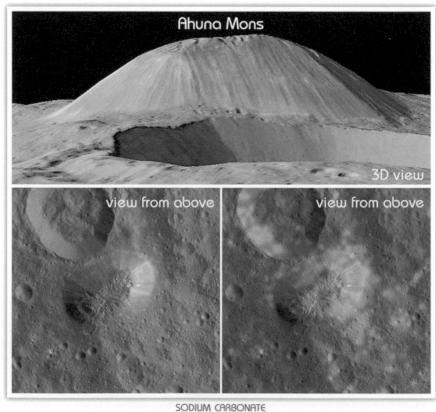

Figure 10.7 (a) Ahuna Mons on Ceres. Top: Perspective view derived from HAMO DTM. Bottom left: LAMO nadir image. Bottom right: LAMO image with overlay of sodium carbonate spectral signature, derived from VIR data. From NASA Photojournal PIA21919. (b) Topographic map of Ceres showing major named domes (tholi) and mountains (mons). Map credit: Preusker et al. (2016).
A black-and-white version of this figure will appear in some formats. For the color version, please refer to the plate section

(b)

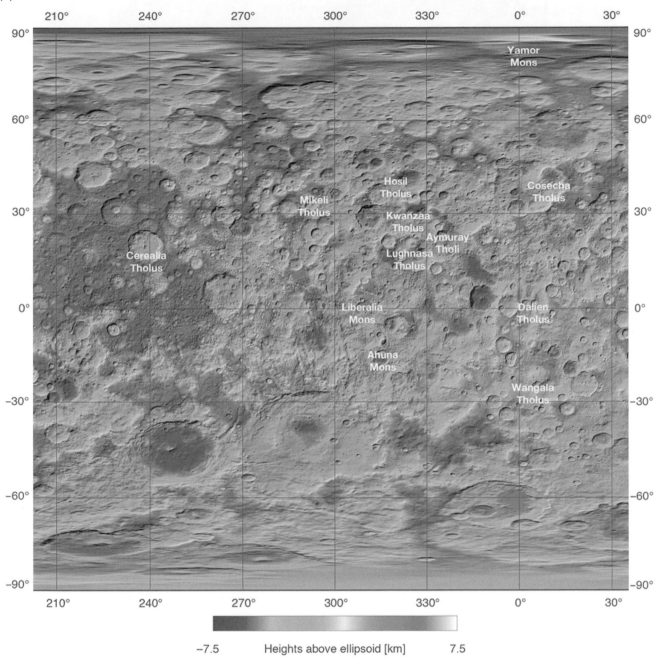

Figure 10.7 (*cont.*)

Cerealia Facula material. A set of fractures with unique (reddish) spectral characteristics in visible to near-infrared wavelengths (Nathues et al., 2019, 2020) cut through the bright material, indicating that the formation of Cerealia Tholus occurred relatively late in the evolution of Occator (Scully et al., 2019a, and references therein) and likely due to an extrusive formation process(es) (Nathues et al., 2017b). Ruesch et al. (2019) and Quick et al. (2019) hypothesize that the extrusion of brines with an increased viscosity due to cooling during ascent constructed the tholus. However, the lack of a large termination scarp around the base of Cerealia Tholus may be inconsistent with a high viscosity brine

(Scully et al., 2020). Schenk et al. (2019) interpret that Cerealia Tholus was formed by laccolithic intrusion and/or volume expansion from freezing of a subsurface volatile reservoir. Conversely, Cerealia Tholus has been interpreted as a candidate pingo on Ceres, because of its hydrologic setting (i.e., bottom of the central pit) and the widespread availably of liquid water/brines that were melted by the impact-generated heat (Schmidt et al., 2020). This scenario requires that the Tholus formed closer in time to the impact, which is not supported by current age estimates of the Tholus (Nathues et al., 2020). The youth of Cerealia Tholus is confirmed by the recent spectral detection of hydrated sodium

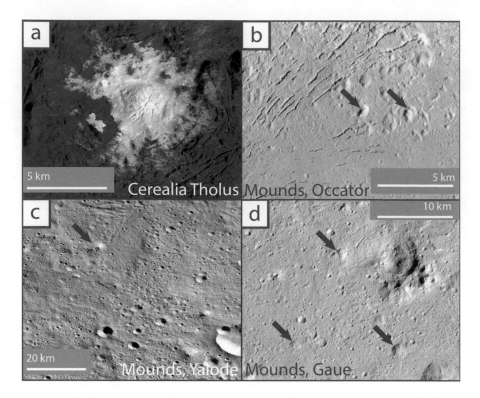

Figure 10.8 Cerealia Tholus and various small mounds on Ceres. (a) Cerealia Tholus, a mound that occurs within Occator crater's central pit and is covered or consists entirely of bright material. (b–d) Examples of small mounds within the lobate material of the floor of Occator crater (b), within the floor of the Yalode impact basin (c), and the floor of Gaue crater (d).
Figure adapted from Sizemore et al. (2019)

chloride on its summit, which is indicative of a recent or ongoing brine extrusion (De Sanctis et al., 2020).

10.7.3 Small Mounds

This class of feature has topographic reliefs of up to hundreds of meters and lateral extents of <10 km, down to hills that are tens to a few hundreds of meters wide (Figure 10.8(b–d)) (Sizemore et al., 2019). The small mounds in Occator's floor are predominantly located in the once liquid-water-rich lobate material, and have been interpreted as impact debris from the Occator crater-forming event (Sizemore et al., 2019). Conversely, Schmidt et al. (2020) suggested that they may also be pingos, ice-cored mounds resulting from periglacial processes. However, their small size makes this interpretation more ambiguous. The small mounds with a more blocky morphology are more likely to be impact debris trapped in the lobate material (Schenk et al., 2019; Sizemore et al., 2019; Schmidt et al., 2020).

10.7.4 Faculae

Occator crater contains well known bright regions, in and surrounding its central pit (named Cerealia Facula: Figures 10.4(b) and 10.8(a)), on a ledge above the central pit (named Pasola Facula: Figure 10.8(a)) and in its eastern crater floor (named Vinalia Faculae) (Figure 10.4(c)). The faculae are mostly composed of sodium carbonate and ammonium chloride, and are up to six times brighter than Ceres' average surface (De Sanctis et al., 2016; Schröder et al., 2017; Raponi et al., 2019). The faculae are widely

accepted to be the remnants of brines exposed on Ceres' surface, which likely reached the surface through impact-induced fractures in the crater floor (Buczkowski et al., 2016) and subsurface (Buczkowski et al., 2019; Scully et al., 2019a, and references therein; Scully et al., 2020; Raymond et al., 2020). The mechanism of faculae emplacement has been interpreted to be either cryovolcanic or hydrothermal.

The cryovolcanic interpretation was first proposed by Nathues et al. (2017b), and primarily stems from the geologically young model ages estimated for the faculae. Nathues et al. (2020) conclude that the majority of Cerealia Facula and Pasola Facula formed ~8 million years ago, with even more recent resurfacing ~1 million year ago, and that Vinalia Faculae formed ~3 million years ago. Occator crater formed ~22 million years ago (Neesemann et al., 2019), resulting in a ~14 to ~21 million year period between crater and faculae formation. When compared to the relatively short duration of an impact-induced melt chamber at Occator (no more than 4 Myr, Bowling et al., 2019), Nathues et al. (2020, and references therein) conclude that an endogenic source is required to form the faculae ~14–21 Myr after the formation of the crater. The aforementioned model ages are based on the lunar derived chronology system for Ceres, but do not differ significantly from ages derived by the asteroid-flux derived chronology (ADM) system (both outlined in Hiesinger et al., 2016).

The hydrothermal interpretation for Occator's faculae was first proposed by Bowling et al. (2019), Schenk et al. (2019), and Scully et al. (2019a). It is based on the morphological similarities between Occator's faculae and hydrothermal deposits elsewhere in the Solar System, in addition to the relative ease of forming a hydrothermal system in a crater of Occator's size in an ice-rich target such as Ceres. The hydrothermal interpretation requires that either the model ages for the formation of Occator and the faculae are significantly closer together, and/or that the duration of the impact-induced melt chamber be significantly extended. Hesse and Castillo-Rogez (2018) modeled the duration of the impact-induced melt chamber using parameters more representative of Ceres than Bowling et al. (2019), and found a melt chamber lifetime of ~12 Myr, which is closer to the aforementioned lifetime required to form faculae (~14–21 Myr). They also found that thermal merging would likely occur between the impact-induced melt chamber and a deep, pre-existing brine reservoir, located at the base of the crust. This deep brine reservoir could potentially provide brines to form faculae as recently as ~1 million years ago. Scully et al. (2020) conclude that the Cerealia-Facula-forming and Pasola-Facula-forming brines were sourced from the thermally merged, impact-induced melt chamber and the deep brine reservoir, while the Vinalia-Faculae-forming brines were only sourced in the deep brine reservoir.

Somewhat less bright faculae are also found elsewhere on Ceres, usually in the floors and walls of impact craters. Stein et al. (2019) hypothesize that faculae similar to Occator's formed in impact craters on Ceres throughout its history. They were then buried, and re-excavated by later impacts, explaining the strong association between faculae and crater floors and walls.

Based on the current model ages for the faculae and model-derived durations for the impact-induced melt chamber, a contribution from a deep brine reservoir is required to form the faculae, which is in keeping with a cryovolcanic origin. However, the formation of the faculae is intimately linked with their location in an impact crater: it seems unlikely that the faculae would have formed outside of an impact crater without the concentration of brines at depth. The conclusion at the end of the Dawn mission (Raymond et al., 2020) is that both cryovolcanic and hydrothermal processes had a role in faculae formation.

10.8 COMPARING THE SURFACES OF CERES AND VESTA

The goal of the Dawn mission was to characterize the two most massive objects in the Main Asteroid Belt: asteroid (4) Vesta and dwarf planet Ceres (Russell & Raymond, 2011). Vesta, visited first by Dawn from July 2011 to September 2012 because of its closer position to Earth, was determined to be an irregular silicate body, consistent with predictions based on Hubble telescopic observations (Thomas et al., 1997a, 1997b). Vesta's irregular shape (see also Chapter 5) was caused by two large impacts at its south pole (Jaumann et al., 2012), forming the older, ~400 km diameter Veneneia basin superposed by the younger, 505 km diameter Rheasilvia basin (Marchi et al., 2012; Russell et al., 2012; Schenk et al., 2012). The general consensus is that these two impacts excavated the bulk of the material making up the Vesta asteroid family, including howardite–eucrite–diogenite (HED) basaltic achondrite meteoroids, the source of the many of the found HED meteorites on Earth, whose common origin is based on the spectral and compositional similarities between Vesta and the HED meteorites (McCord et al., 1970; De Sanctis et al., 2012). Gravity data (see

Chapter 12) were found to be consistent with a Vesta that differentiated into a crust, mantle, and core (Russell et al., 2012), demonstrating that silicate bodies at least as small as ~525 km (mean diameter) underwent differentiation and planet-like interior melting processes in the early Solar System.

The surface of Vesta is dominated by impact craters of all sizes and preservation states, including those with terraces and steep cliffs at basin margins, as well as atypical "asymmetric craters" with sharp rims on upslope sides and subdued rims on downslope sides, resulting from impacts on steeply sloped terrain. Vesta has a high topography to radius ratio (Jaumann et al., 2012) of ~30% (cf., Mars and the Moon ~1%), indicative of a surface with lots of steep slopes resulting from impact bombardment and degradation. Vesta is also dominated by ridge-and-trough terrains hundreds of kilometers long in the equatorial region (Divalia Fossae) and northern latitudes (Saturnalia Fossae), interpreted as graben formed as a tectonic response to the Veneneia and Rheasilvia impact events (Jaumann et al., 2012).

In contrast, the surface of Ceres, while heavily cratered, does not display either impact basins larger than ~300 km diameter, or extensive asymmetric craters, or ridge-and-trough systems. While Vesta's surface preserves the results of impact cratering into hard silicate rock throughout its history, the lack of large impact basins on Ceres suggests a surface material more conducive to resurfacing and alteration: resurfacing to remove the largest basins, alteration to produce the spectral signatures of ammoniated phyllosilicates, carbonates, and salts as recorded by VIR (Ammannito et al., 2016; De Sanctis et al., 2016; Raponi et al., 2019; see also Chapter 7). This and other evidence, such as the interpretation of the Kerwan impact crater's smooth material as an impact-induced melt sheet from a target rich in water ice (Williams et al., 2018b), suggests that Ceres surface was compositional and physically heterogeneous and different than Vesta's. Ceres surface is dominated by weaker silicates, other minerals such as carbonates and salts, and extensive water ice (Chapter 7). Rare exposures of water ice (Combe et al., 2016, 2019), lobate flows, and landslides with morphologies consistent with terrestrial rock glaciers or otherwise ice-mobilized flows (Schmidt et al., 2017; Chilton et al., 2019; Duarte et al., 2019; Hughson et al., 2019), carbonate- and salt-bearing materials in Ceres' bright spots (De Sanctis et al., 2016; Raponi et al., 2019), among other evidence, all support a planetary surface for Ceres in which water/ice had a central role in its formation and evolution, in contrast to Vesta, whose surface is composed of "harder" silicates with a very limited role for water/ice in its evolution. It has been proposed that, because of the presence of the ammoniated phyllosilicates (Ammannito et al., 2016; Chapter 9), Ceres may have formed in a different part of the Solar System than Vesta (De Sanctis et al., 2018). If so, then the differences in the two most massive objects in the Main Asteroid belt as revealed by the NASA Dawn Mission may have revealed more evidence of a dynamic early Solar System (see Part III).

10.9 SUMMARY AND REMAINING QUESTIONS

The geomorphology of Ceres displays surface features indicative of all the major types of planetary geologic processes: impact cratering, gradation, tectonism, and volcanism. Impact cratering was the dominant surface-modifying process, as exemplified by the large diversity of impact crater morphologies. The variation in crater degradation suggests a heterogeneous surface with variations in ice content that would have affected crater relaxation. The variation in landslide morphology also suggests variations in ice content of Ceres crust. The limited variety and extent of tectonic features favors local rather than global events, such as brine intrusions or disruption by impacts. The presence of geologically young features such as Ahuna Mons and the Occator faculae suggest ongoing cryovolcanic and hydrothermal activity from a brine-containing interior.

The successful exploration of Ceres by the Dawn mission and identification of its surface features has demonstrated that Ceres is a relict ocean world (Castillo-Rogez et al., 2020), such that it has been included in the NASA Roadmap for Ocean Worlds (Hendrix et al., 2019) as a world whose past and present habitability should be assessed. Key questions that should be addressed by a future mission include: (1) Do liquid water/brines still exist inside Ceres, and, if so, how extensive and how deep are they? (2) What are the specific compositions of Ceres's bright materials, including salts, carbonates, and organics? (3) How much heat remains in Ceres interior? (4) Could extinct or extant microbial life exist within Ceres? As a consequence of Dawn's discoveries at Ceres, a Mission Concept Study was funded by NASA in 2019–2020 to assess the scientific merit and architecture for a new mission to Ceres to assess its habitability. This study looked at not only a new orbiter, but also *in situ* exploration using lander, rover/hopper, and possible sample return architectures. The final report of this study was delivered to NASA in August 2020 and emphasizes that a sample return mission that brings back samples of both dark and bright materials from Vinalia Facula would provide the most science per dollar to answer these questions about Ceres habitability.

REFERENCES

Ammannito, E., De Sanctis, M. C., Ciarniello, M., et al. (2016) Distribution of phyllosilicates on the surface of Ceres. *Science*, 353, aaf4279.

Bland, M. T. (2013) Predicted crater morphologies on Ceres: Probing internal structure and evolution. *Icarus*, 226, 510–521.

Bland, M. T., Raymond, C. A., Schenk, P. M., et al. (2016) Composition and structure of the shallow subsurface of Ceres revealed by crater morphology. *Nature Geoscience*, 9, 538–542.

Bland, M. T., Sizemore, H. G., Buczkowski, D. L., et al. (2018) Why is Ceres lumpy? Surface deformation induced by solid-state subsurface flow. *49th Lunar and Planetary Science Conference*, March, Houston, TX, Abstract #1627.

Bowling, T. J., Ciesla, F. J., Davison, T. M., et al. (2019) Post-impact thermal structure and cooling timescales of Occator crater on asteroid 1 Ceres. *Icarus*, 320, 110–118.

Boyce, J., Wilson, L., Mouginis-Mark, P. J., Hamilton, C. W., & Tornabene, L. L. (2012) Origin of small pits in martian impact craters. *Icarus*, 221, 262–275.

Buczkowski, D. L., Schmidt, B. E., Williams, D. A., et al. (2016) The geomorphology of Ceres. *Science*, 353.

Buczkowski, D. L., Scully, J. E. C., Quick, L., et al. (2019) Tectonic analysis of fracturing associated with Occator crater. *Icarus*, 320, 49–59.

Buczkowski, D. L., Sizemore, H. G., Bland, M. T., et al. (2018) Floor-fractured craters on Ceres and implications for interior processes. *Journal of Geophysical Research*, 123, 3188–3204.

Buczkowski, D. L., Williams, D. A., Scully, J. E. C., et al. (2017) The geology of the Occator quadrangle of dwarf planet Ceres: Floor-fractured craters and other geomorphic evidence of cryomagmatism, *Icarus*, 316, 128–139.

Castillo-Rogez, J. C., & McCord, T. B. (2010) Ceres' evolution and present state constrained by shape data. *Icarus*, 205, 443–459.

Castillo-Rogez, J. C., Neveu, M., Scully, J. E. C., et al. (2020) Ceres: Astrobiological target and possible ocean world. *Astrobiology*, 20, 269–291.

Chamberlain, M. A., Sykes, M. V., & Esquerdo, G. A. (2007) Ceres light-curve analysis – Period determination. *Icarus*, 188, 451–456.

Chilton, H. T., Schmidt, B. E., Duarte, K., et al. (2019) Landslides on Ceres: Inferences into ice content and layering in the upper crust. *Journal of Geophysical Research*, 124, 1512–1524.

Combe, J.-Ph., McCord, T. B., Tosi, F., et al. (2016) Detection of local H2O exposed at the surface of Ceres. *Science*, 353, aaf3010.

Combe, J.-Ph., Raponi, A., Zambon, F., et al. (2019) Exposed H_2O-rich areas detected on Ceres with the Dawn visible and infrared mapping spectrometer. *Icarus*, 318, 22–41.

Crown, D. A., Sizemore, H. G., Yingst, R. A., et al. (2018) Geologic mapping of the Urvara and Yalode Qudrangles of Ceres. *Icarus*, 316, 167–190.

De Sanctis, M. C., Ammannito, E., Capria, M. T., et al. (2012) Spectroscopic characterization of mineralogy and its diversity across Vesta. *Science*, 336, 697–700.

De Sanctis, M. C., Ammannito, E., Carrozzo, F. G., et al. (2018) Ceres' global and localized mineralogical composition determined by Dawn's Visible and Infrared Spectrometer (VIR). *Meteoritic & Planetary Science*, 53, 1844–1865.

De Sanctis, M. C., Ammannito, E., McSween, H., et al. (2017) Localized aliphatic organic material on the Surface of Ceres. *Science*, 355, 719–722.

De Sanctis, M. C., Ammannito, E., Raponi, A., et al. (2015) Ammoniated phyllosilicates with a likely outer Solar System origin on (1) Ceres. *Nature*, 528, 241–244.

De Sanctis, M. C., Ammannito, E., Raponi, A., et al. (2020) Fresh emplacement of hydrated sodium chloride on Ceres from ascending salty fluids. *Nature Astronomy*, 4, 786–793.

De Sanctis, M. C., Raponi, A., Ammannito, E., et al. (2016) Bright carbonate deposits as evidence of aqueous alteration on (1) Ceres. *Nature*, 536, 54–57.

Denevi, B. W., Blewett, D. T., Buczkowski, D. L., et al. (2012) Pitted terrain on Vesta and implications for the presence of volatiles. *Science*, 338, 246–249.

Duarte, K., Schmidt, B. E., Chilton, H., et al. (2019) Landslides on Ceres: Diversity and geologic context. *Journal of Geophysical Research*, 124, 3329–3343.

Frigeri, A., Schmedemann, N., Williams, D. A., et al. (2018) The geology of the Nawish quadrangle of Ceres: The rim of an ancient basin. *Icarus*, 316, 114–127.

Greeley, R. (2013) *Introduction to Planetary Geomorphology*. New York: Cambridge University Press.

Hendrix, A. R., Hurford, T. A., Barge, L. M., et al. (2019) The NASA roadmap to ocean worlds. *Astrobiology*, 19, 1.

Hesse, M. A., & Castillo-Rogez, J. C. (2018) Thermal evolution of the impact-induced cryomagma chamber beneath Occator crater on Ceres. *Geophysical Research Letters*, 46, 1213–1221.

Hiesinger, H., Marchi, S., Schmedemann, N., et al. (2016) Cratering on Ceres: Implications for its crust and evolution. *Science*, 353, 4759.

Hughson, K., Russell, C. T., Schmidt, B. E., et al. (2019) Fluidized appearing ejecta on Ceres: Implications for the mechanical properties, frictional properties, and composition of its shallow subsurface. *Journal of Geophysical Research*, 124, 1819–1839.

Hughson, K., Russell, C. T., Williams, D. A., et al. (2018) The Ac-H-5 (Fejokoo) quadrangle of Ceres: Geologic map and geomorphological evidence for ground ice mediated surface processes. *Icarus*, 316, 63–83.

Jaumann, R., Williams, D. A., Buczkowski, D. L., et al. (2012) Vesta's shape and morphology. *Science*, 336, 687–690.

Jaumann, R., Preusker, F., Krohn, K., et al. (2017) Topography and geomorphology of the interior of Occator crater on Ceres. *48th Lunar and Planetary Science Conference*, March, Houston, TX, Abstract #1440.

Jones, K.B., Head, J., Pappalardo, R. T., & Moore, J. M. (2003) Morphology and origin of palimpsests on Ganymede from Galileo observations. *Icarus*, 164, 197–212.

Krohn, K., Jaumann, R., Otto, K. A., et al. (2018) The unique geomorphology and structural geology of the Haulani crater of dwarf planet Ceres as revealed by geological mapping of equatorial quadrangle Ac-6 Haulani. *Icarus*, 316, 84–98.

Krohn, K., Jaumann, R., Stephan, K., et al. (2016) Cryogenic flow features on Ceres: Implications for crater-related cryovolcanism on dwarf planet Ceres. *Geophysical Research Letters*, 43, 11994–12003.

Lebofsky, L. A., Sykes, M. V., Tedesco, E. F., et al. (1986) A refined 'standard' thermal model for asteroids based on observations of 1 Ceres and 2 Pallas. *Icarus*, 68, 239–251.

Li, J.-Y., Mcfadden, L. A., Parker, J. W., et al. (2006) Photometric analysis of 1 Ceres and surface mapping from HST observations. *Icarus*, 182, 143–160.

Marchi, S., Ermakov, A., Raymond, C. A., et al. (2016) The missing large impact craters on Ceres. *Nature Communications*, 7, 12257.

Marchi, S., McSween, H. Y., O'Brien, D. P., et al. (2012) The violent collisional history of asteroid 4 Vesta. *Science*, 336, 690–693.

McCord, T. B., Adams, J. B., & Johnson, T. V. (1970) Asteroid Vesta: Spectral reflectivity and compositional implications. *Science*, 168, 1445–1447.

McCord, T. B., Castillo-Rogez, J. C., & Rivkin, A. (2011) Ceres: Its origin, evolution and structure and Dawn's potential contribution. *Space Science Reviews*, 163, 63–76.

McCord, T. B., & Sotin, C. (2005) Ceres: Evolution and current state. *Journal of Geophysical Research*, 110, E05009.

Mest, S. C., Crown, D. A., Berman, D. C., et al. (2018) The HAMO-based global geologic map and chronostratigraphy of Ceres. *49th Lunar and Planetary Science Conference*, March, Houston, TX, Abstract #2730.

Michalak, G. (2000) Determination of asteroid masses – I. (1) Ceres, (2) Pallas and (4) Vesta. *Astronomy & Astrophysics*, 360, 363–374.

Millis, R. L., Wasserman, L. H., Franz, O. G., et al. (1987) The size, shape, density, and albedo of Ceres from its occultation of BD+8°471. *Icarus*, 72, 507–518.

Mitchell, D. L., Ostro, S. J., Hudson, R. S., et al. (1996) Radar observations of asteroids 1 Ceres, 2 Pallas, and 4 Vesta. *Icarus*, 124, 113–133.

Nathues, A., Hoffmann, M., Platz, T., et al. (2016) FC color images of dwarf planet Ceres reveal a complicated geological history. *Planetary and Space Science*, 134, 122–127.

Nathues, A., Platz, T., Hoffmann, M., et al. (2017a) Oxo crater on (1) Ceres: Geological history and the role of water-ice. *Astronomical Journal*, 154, 84–96.

Nathues, A., Platz, T., Thangjam, G., et al. (2017b) Evolution of Occator crater on (1) Ceres. *Astronomical Journal*, 153,112–123.

Nathues, A., Platz, T., Thangjam, G., et al. (2019) Occator crater in color at highest spatial resolution. *Icarus*, 320, 24–38.

Nathues, A., Schmedemann, N., Thangjam, G., et al. (2020) Recent cryovolcanic activity at Occator crater on Ceres. *Nature Astronomy*, 4, 794–801.

Neesemann, A., van Gesselt, S., Schmedemann, N., et al. (2019) The various ages of Occator crater, Ceres: Results of a comprehensive synthesis approach. *Icarus*, 320, 60–82.

Ostro, S. J., Pettengill, G. H., Shapiro, I. I., Campbell, D. B., & Green, R. R. (1979) Radar observations of asteroid 1 Ceres. *Icarus*, 40, 355–358.

Otto, K. A., Marchi, S., Trowbridge, A., Melosh, H. J., & Sizemore, H. G. (2019) Ceres crater degradation inferred from concentric fracturing. *Journal of Geophysical Research: Planets*, 124, 1188–1203.

Park, R. S., Vaughan, A. T., Konopliv, A. S., et al. (2019) High-resolution shape model of Ceres from stereophotoclinometry using Dawn imaging data. *Icarus*, 319, 812–827.

Pasckert, J. H., Hiesinger, H., Ruesch, O., et al. (2018) Geologic mapping of the Ac-2 Coniraya Quadrangle of Ceres from NASA's Dawn Mission: Implications for a heterogeneously composed crust. *Icarus*, 316, 28–45.

Pitjeva, E. V., & Standish, E. M. (2009) Proposals for the masses of the three largest asteroids, the Moon–Earth mass ratio and the astronomical unit. *Celestial Mechanics & Dynamical Astronomy*, 103, 365–372.

Platz, T., Natheus, A., Sizemore, H. G., et al. (2018) Geological mapping of the Ac-10 Rongo Quadrangle of Ceres. *Icarus*, 316, 140–153.

Preusker, F., Scholten, F., Matz, K.-D., et al. (2016) Dawn at Ceres – Shape model and rotational state. *47th Lunar and Planetary Science Conference*, March, Houston, TX, Abstract #1954.

Quick, L. C., Buczkowski, D. L., Ruesch, O., et al. (2019) A possible brine reservoir beneath Occator crater: Thermal and compositional evolution and formation of the Cerealia dome and Vinalia Faculae. *Icarus*, 320, 119–135.

Raponi, A., De Sanctis, M. C., Carrozzo, F. G., et al. (2019) Mineralogy of Occator crater on Ceres and insight into evolution from the properties carbonates, phyllosilicates, and chlorides. *Icarus*, 320, 83–96.

Raponi, A., De Sanctis, M. C., Frigeri, A., et al. (2018) Variations in the amount of water ice on Ceres' surface suggest a seasonal water cycle. *Science Advances*, 4, eaao3757.

Raymond, C. A., Ermakov, A. I., Castillo-Rogez, J. C., et al. (2020) Impact-driven mobilization of deep crustal brines on dwarf planet Ceres. *Nature Astronomy*, 4, 741–747,

Rivkin, A. S., Li, J.-Y., Milliken, R. E., et al. (2011) The surface composition of Ceres. *Space Science Reviews*, 163, 95–116.

Ruesch, O., Platz, T., Schenk, P., et al. (2016) Cryovolcanism on Ceres. *Science*, Volume 353.

Ruesch, O., Quick, L., Landis, M. E., et al. (2019) Bright carbonate surfaces on Ceres as remnants of salt-rich water fountains. *Icarus*, 320, 39–48.

Ruiz, J., Jiménez-Díaz, A., Mansilla, F., et al. (2019) Evidence of thrust faulting and widespread contraction of Ceres. *Nature Astronomy*, 3, 916–921.

Russell, C. T., & Raymond, C. A. (2011) The Dawn mission to Vesta and Ceres. *Space Science Reviews*, 163, 3–23.

Russell, C. T., Raymond, C. A., Ammannito, C. A., et al. (2016) Dawn arrives at Ceres: Exploration of a small, volatile-rich world. *Science*, 353, 1008–1010.

Russell, C. T., Raymond, C. A., Coradini, A., et al. (2012) Dawn at Vesta: Testing the protoplanet paradigm. *Science*, 336, 684–686.

Schenk, P., O'Brien, D. P., Marchi, S., et al. (2012) The geologically recent giant impact basins at Vesta's south pole. *Science*, 336, 694–697.

Schenk, P., Scully, J., Buczkowski, D., et al. (2020) Impact-driven brine-melt, volatile distribution, and brine effusion in crater floor deposits on a transitional ice-salt-silicate-rich dwarf planet at Occator crater, Ceres. *Nature Commications*, 11, 3679.

Schenk, P., Sizemore, H. G., Schmidt, B., et al. (2019) The central pit and dome at Cerealia Facula bright deposit and floor deposits in Occator crater, Ceres: Morphology, comparisons and formation. *Icarus*, 320, 159–187.

Schmedemann, N., Kneissl, T., Neesemann, A., et al. (2016) Timing of optical maturation of recently exposed material on Ceres. *Geophysical Research Letters*, 43, 11987–11993.

Schmidt, B. E., Hughson, K. H. G., Chilton, H. T., et al. (2017) Geomorphological evidence for ground ice on dwarf planet Ceres. *Nature Geoscience*, 10, 338–343.

Schmidt, B. E., Sizemore, H. G., Hughson, K. H. G., et al. (2020) Post-impact cryo-hydrologic formation of small mounds and hills in Ceres' Occator crater. *Nature Geoscience*, 13, 605–610.

Schröder, S. E., Mottola, S., Carsenty, U., et al. (2017) Resolved spectro-photometric properties of the Ceres surface from Dawn Framing Camera images. *Icarus*, 288, 201–225.

Scully, J. E. C., Bowling, T., Bu, C., et al. (2019a) Synthesis of the special issue: The formation and evolution of Occator crater. *Icarus*, 320, 213–225.

Scully, J. E. C., Buczkowski, D. L., Raymond, C. A., et al. (2019b) Ceres' Occator crater and its faculae explored through geologic mapping. *Icarus*, 320, 7–23.

Scully, J. E. C., Buczkowski, D. L., Schmedemann, N., et al. (2017) Evidence for the interior evolution of Ceres from geologic analysis of fractures. *Geophysical Research Letters*, 44, 9564–9572.

Scully, J. E. C., Schenk, P. M., Castillo-Rogez, J. C., et al. (2020) The varied sources of faculae-forming brines in Ceres' Occator crater, emplaced via hydrothermal brine effusion. *Nature Communications*, 11, 3680.

Sierks, H., Keller, H. U., Jaumann, R., et al. (2011) The Dawn Framing Camera. *Space Science Reviews*, 163, 263–328.

Sizemore, H. G., Platz, T., Prettyman, T. H., et al. (2017) Pitted terrain on dwarf planet Ceres: Morphological evidence for shallow volatiles at low and mid latitudes. *Geophysical Research Letters*, 44, 6570–6578.

Sizemore, H. G., Schmidt, B. E., Buczkowski, D. A., et al. (2019) A global inventory of ice-related morphological features on dwarf planet Ceres: Implications for the evolution and current state of the cryosphere. *Journal of Geophysical Research*, 124, 1650–1689.

Sori, M. M., Byrne, S., Bland, M. T., et al. (2017) The vanishing cryovolcanoes of Ceres. *Geophysical Research Letters*, 44, 1243–1250,

Spencer, J. R., Lebofsky, L. A., & Sykes, M. V. (1989) Systematic biases in radiometric diameter determinations. *Icarus*, 78, 337–354.

Stein, N. T., Ehlmann, B. L., Palomba, E., et al. (2019) The formation and evolution of bright spots on Ceres. *Icarus*, 320, 188–201.

Stephan, K., Jaumann, R., Krohn, K., et al. (2017) An investigation of the bluish material on Ceres. *Geophysical Research Letters*, 44, 1660–1668.

Stephan, K., Jaumann, R., Wagner, R., et al. (2018) Dantu's mineralogical properties – A view into the composition of Ceres' crust. *Meteoritics & Planetary Science*, 53, 1866–1883.

Stephan, K., Jaumann, R., Zambon, F., et al. (2019) Ceres' craters – relationships between surface composition and geology. *Icarus*, 318, 56–74.

Thomas, P. C., Binzel, R. P., Gaffey, M. J., et al. (1997a) Impact excavation on asteroid 4 Vesta: Hubble Space Telescope results. *Science*, 277, 1492–1495.

Thomas, P. C., Binzel, R. P., Gaffey, M. J., et al. (1997b) Vesta: Spin pole, size and shape from HST images, *Icarus*, 128, 88–94.

Thomas, P. C., Parker, J. W., McFadden, L. A., et al. (2005) Differentiation of the asteroid Ceres as revealed by its shape. *Nature*, 437, 224–226.

Tornabene, L., Osinski, G. R., McEwen, A. S., et al. (2012) Widespread crater-related pitted materials on Mars: Further evidence for the role of target volatiles during the impact process. *Icarus*, 220, 348–368.

Viateau, B., & Rapport, N. (2001) Mass and density of asteroids (4) Vesta and (11) Parthenope. *Astronomy & Astrophysics*, 370, 602–609.

Williams, D. A., Buczkowski, D. L., Mest, S. C., et al. (2018a) Introduction: The geological mapping of Ceres. *Icarus*, 316, 1–13.

Williams, D. A., Kneissl, T., Neesemann, A., et al. (2018b) The geology of the Kerwan quadrangle of dwarf planet Ceres: Investigating Ceres' oldest impact basin. *Icarus*, 316, 99–113.

Zolotov, M. Y. (2009) On the composition and differentiation of Ceres. *Icarus*, 204, 183–193.

Ceres' Internal Evolution

JULIE CASTILLO-ROGEZ AND PHILIP BLAND

11.1 INTRODUCTION

With a diameter of 940 km and a density of 2,162 kg/m^3, Ceres is the largest and most water-rich body in the inner Solar System, in terms of relative water to rock fraction ($\sim$1:1). Its maximum internal pressure is about 170 MPa, which implies it might have retained some porosity in its shallow interior. The Dawn mission has returned detailed information on the geophysical state of Ceres and its composition, the first of its kind for a mid-sized icy body. These observations add to the knowledge of icy body evolution gathered by previous missions to broaden our understanding of the mechanisms driving the internal differentiation of mid-sized, that is, low-gravity, volatile-rich bodies. Conversely, indirect information on Ceres' evolution can be inferred from observations obtained at other bodies that share similar physical properties and that have been observed by spacecraft (Figure 11.1). This area of research is important because internal evolution determines the formation and preservation of habitable niches, as has been suggested for Ceres (Castillo-Rogez et al., 2020). Many of these objects are believed to have hosted a deep ocean, at least in their early history, resulting from the melting of accreted ice phases. Such melting was driven by short-lived radioisotopes and/or accretional and gravitational energy in larger bodies, such as dwarf planets and large icy moons. However, constraints on the interior evolution of these objects are scarce, and the behavior of material in their low gravity is only partially understood.

This chapter reviews knowledge gained from geophysical data of Ceres returned by the Dawn mission, and the implications for understanding the evolution of similar bodies. Section 11.2 summarizes observational constraints on the internal and chemical evolution of Ceres, carbonaceous chondrite parent bodies, and icy moons. Then, we attempt to fill in the gaps in our understanding of Ceres by addressing the drivers of differentiation and co-dependencies between chemical and physical processes (Section 11.3). Specific interior and evolution models of Ceres that aim to explain the Dawn observations are summarized in Section 11.4. In the conclusion, we highlight specific directions for future work in order to progress in our understanding of differentiation in icy bodies.

Part of this research was carried out at the Jet Propulsion Laboratory, California Institute of Technology, under a contract with the National Aeronautics and Space Administration (80NM0018D004). © 2020. All rights reserved. The authors are thankful to Anton Ermakov for his review of this chapter and for his help addressing some of the reviewer's comments. The authors also acknowledge input from Thomas Prettyman, whose detailed review and useful suggestions helped improved the quality of this manuscript. Lastly, the authors thank the editors of this book for their guidance.

11.2 OBSERVATIONAL CONSTRAINTS ON CERES AND ICY BODY INTERIORS

11.2.1 Summary of Observational Constraints on Ceres' Internal Evolution and Current State

Dawn's extensive mapping of Ceres at multiple spectral and spatial wavelengths has yielded key information on its interior structure (see Chapter 12) and chemical evolution (see Chapter 7). The general consensus is that Ceres' volatiles went through a phase of global melting, pervasive aqueous alteration, and partial differentiation (i.e., separation of the rock phase and the volatile phase) into a hydrated rocky mantle and a volatile-rich crust, the latter being about 40 km thick (Ermakov et al., 2017). However, it has also been suggested that Ceres directly accreted hydrated materials from smaller planetesimals, such as CI/CM parent bodies (Zolotov, 2020) and little or no ice (Figure 11.2). We focus the bulk of this chapter on the model of Ceres' interior that is stratified in an ice-rich crust and rocky mantle based on the preponderance of evidence for ice in Ceres' shallow subsurface (Sizemore et al., 2019). Alternative models are addressed in Section 11.4.

A slight departure from nonhydrostaticity results in a large uncertainty in the density of the rocky mantle, between 2,430 and 2,950 kg/m^3 (Ermakov et al., 2017; Mao & McKinnon, 2018). Even though that estimate matches the bulk density of CI chondrites (Macke et al., 2011), in practice the volatile component of CI chondrite material, the salts, and part of the organic matter should have separated from the rock phase. A high abundance of carbon inferred from the Dawn gamma ray and neutron detector (GRaND) instrument (Prettyman et al., 2017, 2018; see review in Chapter 8) and interpreted as amorphous carbon (Marchi et al., 2019), as well as the abundance of salts found on the surface, supports this hypothesis. Hence, the density of Ceres' solid mantle material is likely denser than the average CI chondrite material, suggesting presence of porosity. However, the uncertainty on mantle density corresponds to a porosity that may range between 0 and >30%. Since many porosity-filling materials (e.g., organic matter, salts, water) could be in liquid form in Ceres' internal conditions, understanding the conditions in which a rocky mantle separates from the volatile phase is a key to understanding its evolution.

Modeling of topographic relaxation by Fu et al. (2017) suggests the presence of a low viscosity layer below the crust on a global scale, and infers a viscosity in that region lower than 10^{21} Pa s. This loose constraint does not allow quantification of the amount of liquid in the upper mantle. Additional indication of deep liquid comes from the detection at Cerealia Facula of hydrohalite, a hydrated salt (NaCl·2H$_2$O), whose stability at Ceres' surface temperature is expected to be limited to a few thousand years at most (De Sanctis et al., 2020). This discovery suggested brine exposure is recent and potentially ongoing. Combined with geological and geophysical observations (e.g., Park et al., 2020; Scully et al., 2020), it indicates that the Occator faculae feed from a brine source below the crust (Raymond et al., 2020). Based on gravity data, explanations for the emplacement of Ahuna Mons also involve brines sources from the upper mantle (Ruesch et al., 2016, 2019).

Information derived from topography and gravity analyses cannot probe below 100 km depth (Fu et al., 2017) and the presence of a metallic core cannot be ruled out based on available data (King et al., 2018). However, the density range inferred for the mantle suggests Ceres' rock never reached temperatures leading to full silicate dehydration (~850 K), which sets important constraints on the modalities of its internal evolution.

A key question opened by observations of carbonaceous chondrites and a few icy bodies is how fast the rock settled to form a compact mantle, if it settled at all (Travis et al., 2018). The long-term suspension of particles tens of microns large could extend the interaction of rock particles with water and thus affect the extent of leaching of major elements from the rock, which form the basis for brines and salts. In such a scenario, the thermophysical properties and stability of mixtures prior to and after aqueous alteration would vary significantly with major implications for the long-term preservation of liquid in bodies of Ceres' size and above (Section 11.3). Observations of Ceres' surface mineralogy suggest alteration was advanced because of the dominance of magnesium serpentine (De Sanctis et al., 2015) (see also Figure 11.4). Indeed, iron-rich minerals are unstable in hydrothermal conditions (e.g., Scott et al., 2002). However, the presence of a small fraction (up to 7 vol.%) of cronstedtite, an iron-rich form of serpentine, cannot be ruled out from the fitting of Ceres' global surface spectrum (De Sanctis et al., 2015). Hence, it is possible aqueous alteration in Ceres did not proceed until equilibrium.

Ceres volatile to rock mass fractions of about 47:53 leads to a salinity in the early global ocean of about 5 wt.% at equilibrium (Castillo-Rogez et al., 2018). Hence, an additional constraint on the extent of aqueous alteration may be inferred from salinity estimated based on surface composition. Regarding salt characterization, Dawn data can only provide a robust detection of carbonates and certain chlorides, and the relative positions of the 3.4- and 4-μm bands are reliable markers of the type of cation associated with the carbonate or bicarbonate ions (see Palomba et al., 2019). Hydrated carbonates have been found in a few sites, such as Haulani crater (Tosi et al., 2018), Juling crater, and Oxo crater (Carrozzo et al., 2018). Chlorides in general do not have vibration bands in the infrared but they can be, and have been detected if bound with ammonium (Raponi et al., 2019) or if they are in hydrate form (De Sanctis et al., 2020; Chapter 7). These pieces of information point to a high abundance of salt in Ceres' early ocean, a likely signature of advanced aqueous alteration. Evidence for abundant salts is found across Ceres' surface in the form of hundreds of "bright spots" (Stein et al., 2017). These are believed to have several origins: excavation of shallow salt lenses via small impacts,

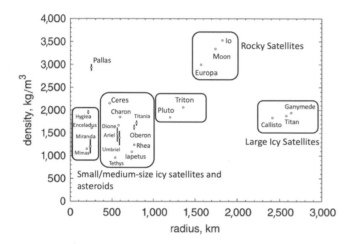

Figure 11.1 Ceres is a mid-sized icy body, like many other objects in the outer Solar System, and at least one large Main Belt asteroid (Hygiea). Knowledge gained from observations of these bodies can help derive a coherent picture of their evolution.

After Hussmann et al. (2006)

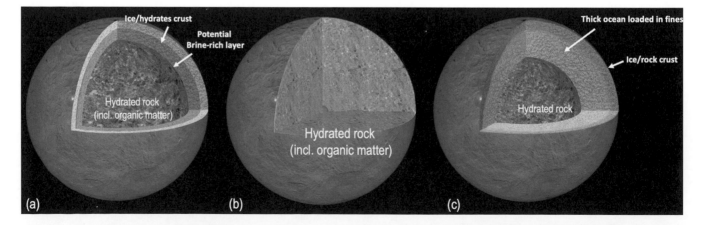

Figure 11.2 The three main interior models suggested for Ceres. (a) Three-layer interior with a mantle of hydrated rock, crust of ice and hydrates, and 10s-km thick brine-rich layer at the interface between the two (e.g., Ermakov et al., 2017; Fu et al., 2017; Castillo-Rogez et al., 2019). (b) Undifferentiated mixture of silicates and organic matter with little or no ice (Zolotov, 2020). (c) Rock-rich crust above a 100s-km thick ocean and solid rocky core (Travis et al., 2018).
Credit: NASA/JPL-Caltech/UCLA/MPS/ DLR/IDA/Francois Rogez

concentration of salts in impact melt chambers, as is the case of Occator crater's faculae (Raymond et al., 2020), or extrusion of brines from a deep reservoir to the surface by mechanisms such as mud-based volcanism in the specific case of Ahuna Mons (Ruesch et al., 2019). Additional information on Ceres' geology can be found in Chapter 10.

11.2.2 Constraints on Differentiation Processes in Carbonaceous Chondrite Parent Bodies

Primordial asteroids accreted as a mixture of nebular fines, ice, and coarse spherules (chondrules). Carbonaceous chondrite (CC) meteorites are samples of these asteroids. A subset of these meteorites show evidence of significant hydrothermal alteration that occurred within their asteroid parent bodies (e.g., Brearley, 2003). Counterintuitively, the most chemically pristine (CI and CM chondrites, which closely match the solar photosphere) have experienced the most pervasive aqueous alteration. In those specimens, the mineralogical products of aqueous alteration are ubiquitous. In the CMs, there is a surprisingly narrow range in the overall modal volume (vol.%) of the most abundant phases (Mg-serpentine (25–33%) and Fe-cronstedtite (43–50%)) (Howard et al., 2009). The range in modal total phyllosilicate is 73–79% (Howard et al., 2009). Altogether, about 78 products of aqueous alteration have been reported in CC meteorites (e.g., Brearley, 2006), including serpentines, a variety of clays, sulfides, carbonates, oxides, hydroxides, and chlorides, as well as 1,000s of organic compounds (Alexander et al., 2015). Furthermore, two clasts of halite believed to stem from volcanically active C-type asteroids have been found in H-type breccias (Zag and Monahans, Rubin et al., 2002). These clasts trapped a large number of compounds from the liquid medium where the clasts precipitated, including organic compounds, pyroxene and olivine, and a variety of hydrated minerals (e.g., Chan et al., 2018).

CI and CM meteorites have chemical compositions that are a close match to the solar photosphere – in the case of CI chondrites at least 40 elements within 10% of the photosphere value (Lodders et al., 2009; Maiorca et al., 2014). The solar/meteoritic ratio is independent of volatility, or the affinity of elements for certain environments. Importantly, when considering the hydrothermal history of a geological body, it is also independent of solubility. Highly soluble elements such as Na, K, and F are also within 10% of solar (Lodders et al., 2009; Maiorca et al., 2014). CM chondrites show a modest, monotonic depletion relative to solar in elements with condensation temperatures <1,350 K. However, as with CI, there is no evidence of solubility-related fractionation. The unusual chemistry of CI and CM chondrites, abundant water, and mix of complex organics, have led many to identify these meteorites – and their primordial asteroid parent bodies – as the precursors from which many other Solar System bodies accreted. In many cases, the objects that we currently observe in the Solar System – icy moons, C- and D-type asteroids, or Ceres – may be the geophysically evolved products of material with CM/CI bulk chemistry (Desch et al., 2018). This is discussed in more detail in Section 11.3.1.

11.2.2.1 Porosity

Although they are now lithified, primordial CC precursors (and their asteroid parent bodies) were initially highly porous. Studies of experimentally synthesized fine-grained material (Blum, 2004; Blum & Schrapler, 2004), and modeled accreted aggregates (Ormel et al., 2008) indicate that primordial matrix porosities were in the 70–80% range. Bland et al. (2011) used fabric analysis to show that the primary matrix porosity of the Allende CV chondrite, pre-compaction, was of order 70–80%. A review of chondrite porosity data (Sasso et al., 2009; Macke et al., 2011) by Bland et al. (2014) supports this estimate, indicating that some extant meteorites still have high porosity: current matrix porosity in the most porous OCs must be 60–70%; 60% in the most porous CVs and 66–70% in the most porous COs (Bland et al., 2014).

11.2.2.2 Permeability

A rock can exhibit permeability at the grain scale, or in fractures, or both. Large scale fracture networks – and associated zones of alteration around them – are not observed in aqueously altered meteorites. Rather, in CM and CI chondrites, pervasive "homogeneous" alteration at the grain-scale indicates that water must have been present along grain boundaries, so grain-scale permeability may have dominated. It is well known that permeability varies in proportion to pore-size, which in turn is related to grain size, sorting, grain shape, grain packing, and the degree of cementation (Dullien, 1992). Qualitatively, even at very high porosity values, the permeability of a porous medium will be low if the grain size is small. This is well known in the terrestrial environment. Siltstones and claystones, which are terrestrial analogues with comparable pore-sizes to meteorite samples, act as barriers to flow over millions of years. Although chondrite precursors were highly porous, it does not follow that they were highly permeable.

Bland et al. (2009) showed that earlier asteroid geophysical models may have used permeability estimates around six orders-of-magnitude too high. For the lower permeability estimates, liquid water flow would be reduced to millimeter-scale distances in asteroids with diameters of 100–200 km – consistent with observations of unfractionated primordial chemistry in chondrites. But in much larger Ceres-class asteroids this restriction would not apply. Even at low permeability, flow length scales may be sufficient to allow convection. Under these circumstances, we might predict that samples derived from Ceres would show alteration-related chemical fractionation. Interestingly, elemental observations of Ceres' surface point to a deficiency in iron in comparison to CI chondrite material (Prettyman et al., 2017).

11.2.2.3 Isotopic and Chemical Heterogeneity

Heterogeneity in carbonate oxygen isotopes is observed in carbonaceous chondrites. This requires formation at varying temperatures and water-to-rock ratios, previously interpreted as consistent with large-scale open-system circulation of fluids through the body (Young et al., 1999; Figure 11.3). Carbonate heterogeneity is not limited to O-isotopes. "Clumped" isotope and C-isotope data (Guo & Eiler, 2007; Alexander et al., 2015; Fujiya et al., 2015), and chemical data (Fujiya et al., 2015), also indicate formation over a range of temperatures, even in the same meteorite (Guo & Eiler, 2007) and under different physicochemical conditions (e.g., redox states). This heterogeneity is observed at fine scales (100s μm; Fujiya et al., 2015). Grain-scale heterogeneity and millimeter- to

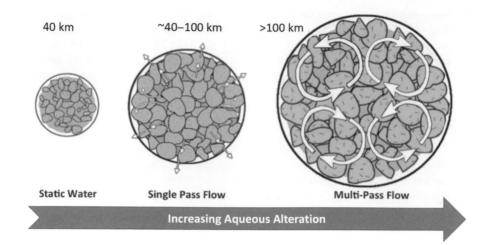

Figure 11.3 The regime of hydrothermal circulation in water-rich bodies, dependent on object size, as suggested by Young et al. (2003). In bodies only tens of kilometers in diameter, fluid convection triggered by ^{26}Al decay heat is expected to be non-existent (Left) or weak, possibly limited to a single episode of convective upwelling (Center). Objects a few hundreds of kilometers in diameter could be subject to sustained convective flow (Right), increasing the effective water-to-rock mass ratio. (After Young et al., 2003; figure permission from Castillo-Rogez & Young, 2016.). Bland and Travis (2017) have suggested an alternative model (the so-called mudball model), where fine particles remain entrained with fluid, rather than water circulating through static rock.

centimeter-scale homogeneity are predictions of the mudball model outlined by Bland and Travis (2017).

11.2.2.4 Alteration Temperatures

Inferred alteration temperatures from isotopic analyses of CM carbonates are 20–71°C (Guo & Eiler, 2007) and 0–130°C (Alexander et al., 2015). The presence of dolomite indicates a (relatively) high initial parent body temperature, in excess of 120°C (Lee et al., 2014). Reaction modeling of phyllosilicate formation (Dyl et al., 2010; Zolotov, 2014) predicts temperatures at the lower end of this range. Considering CM parent bodies, these observations suggest an object where internal temperature was moderated, and that did not have a steep thermal gradient, that is, not an onion shell.

11.2.2.5 Water/Rock Ratio

Estimates of primordial water/rock (W/R) ratio in CCs vary widely. Here we define "primordial" as the ratio at time of accretion, prior to ice melting and onset of alteration. We need to consider how ice may have been accreted into the parent asteroid: whether as ice "rinds" around matrix fines (in which case the W/R ratio should track matrix abundance), or as discrete clasts of varying size. Attempts to constrain the W/R ratio are based on thermodynamic modeling (e.g., Zolensky et al., 1989), geophysical modeling (Young et al., 2003; Figure 11.3), and oxygen-isotopic compositions of aqueously formed minerals (e.g., Leshin et al., 1997; Clayton & Mayeda, 1999; Benedix et al., 2003; Guo & Eiler, 2007; Schrader et al., 2011). From these analyses it is inferred that CI and CM chondrites experienced low-temperature aqueous alteration (<150°C) under variable, but relatively high water/rock (W/R) ratios (0.6–1 by volume, or even higher). In objects hundreds of kilometers in diameter, like some of the carbonaceous chondrite parent bodies, hydrothermal convection in a global ocean triggered by ^{26}Al decay was expected to be sustained for tens of

millions years (Young et al., 2003; Bland & Travis, 2017; Figure 11.3), leading to high (≫1) effective water-to-rock ratios. In the case of a 1,000-km large, water-rich (∼47 vol.%) body, sedimentation of millimeter- to centimeter-large rock particles (e.g., chondrules, Neveu & Desch, 2015) proceeded during the differentiation phase so that these particles may be only partially altered (see more detail in Section 11.3.2).

11.2.2.6 Observations of Large C-type Asteroids

Observations with the SPHERE instrument part of the Very Large Telescope (Vernazza et al., 2018) recently brought new constraints on the physical evolution of large water-rich asteroids. They revealed that (10) Hygiea, a ∼430-km-large asteroid that shares similar spectral properties with Ceres (Vernazza et al., 2017; Rivkin et al., 2019), is also water rich with a density of ∼1,940 kg/m^3 (Vernazza et al., 2019; see Chapter 1). Hygiea's smooth, globally relaxed shape is also consistent with the presence of a water rich crust. The detection of at least one Main Belt comet in the Hygiea family (Sheppard & Trujillo, 2015) is additional evidence for the separation of a water-rich crust. Like in the case of CI parent bodies, ice melting and differentiation could be triggered by ^{26}Al, leading to aqueous alteration and differentiation of a water-rich shell.

On the other hand, the "golf-ball" surface of (2) Pallas suggests that asteroid's surface contains little ice despite evidence for pervasive aqueous alteration (Marsset et al., 2020). The albedo of the 510-km-large asteroid is at least twice that of the average C-type asteroid, which Marsset et al. (2020) interpreted as evidence for abundant salts. These authors concluded that Pallas likely developed an icy crust evolved from a global ocean that got stripped as a consequence of intensive bombardment. These observations combined with a rock-like bulk density of about 2,700 kg/m^3 suggest Pallas could represent the rocky mantle of the proto-Pallas.

Additional constraints on large asteroid interior structure may be found at asteroid families. For example, the Themis family contains ∼10,000 members and is believed to have formed from the catastrophic disruption from a precursor that was about 390–450 km in size (Marzari et al., 1995). Variations in spectral properties found across the family suggest partially differentiation (Ziffer et al., 2011), although space weathering may be a confounding parameter (Marsset et al., 2016). Numerical simulations suggest that, for a time of formation of 3.5 My, inferred for CM parent bodies (Jogo et al., 2017), Themis should be partially differentiated in a rocky mantle, water-rich shell, and a relatively pristine crust (Castillo-Rogez & Schmidt, 2010). Stripping of the crust by a large impact could explain the presence of ice and organic matter at shallow depth on (24) Themis, that may be exposed by impacts (Rivkin & Emery, 2010) and the formation of Main Belt comets among the family members (Hsieh, 2012).

A large number of large C-type asteroids do not show evidence for water of hydration on the surface, which Neveu and Vernazza (2019) interpreted as evidence for formation in the absence of ^{26}Al, potentially farther in the outer Solar System than the region source of asteroids displaying evidence for water of hydration on their surface.

In summary, large C-type asteroids underwent various evolution pathways depending on their time of formation and size. However, the set of bodies for which geophysical constraints (e.g., density, geology) are available is limited to a handful, which prevents population-scale inferences.

11.2.3 Icy Moons

Additional clues on Ceres' interior may be gained from comparative studies with other water-rich bodies in the outer Solar System. Indeed, geophysical observations for several relevant objects have been obtained by past space missions, in particular Cassini at Enceladus and New Horizons at Pluto. Searching for commonalities at these bodies is warranted considering the presence of super volatiles on Ceres and its likely origin in the outer Solar System (see Chapter 9).

Space missions to mid-sized and large icy moons have revealed these bodies to be differentiated to some extent. However, interpretation of the gravity data is not sufficiently unique to reveal robust constraints on their deeper interior or identify the presence of metallic cores. Non-hydrostatic anomalies represent an additional source of confusion depending on the object size and rotation period (Gao & Stevenson, 2013). This is likely an issue at Callisto (Mueller & McKinnon, 1989). This Galilean moon appears to be mostly undifferentiated based on the moment of inertia derived from the Galileo gravity science data (Anderson et al., 2001). On the other hand, the signature of an induced magnetic field in the Galileo magnetometer data suggests the presence of a deep ocean (Zimmer et al., 2000). Further observations are needed to fully understand Callisto's interior structure. Complementary information comes from magnetic measurements that can detect induced fields in thick oceans and the dynamo signature of a magnetic field generated by convection in a metallic core, as was found at Jupiter's moon Ganymede (Kivelson et al., 2002).

For objects that are less evolved, at least as indicated by a large moment of inertia, it is possible to gain insights into mantle properties. Titan and Enceladus both appear to have low density mantles, at about 2,400–2,600 kg/m^3, like Ceres (Iess et al., 2014) (Figure 11.4), if it is assumed that departure from hydrostaticity is not acting as a major confounding parameter. Enceladus' low

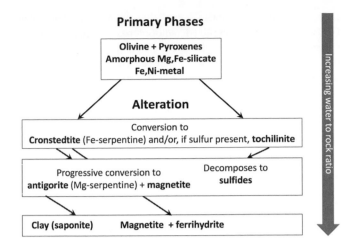

Figure 11.4 Progression of mineralogy formed as a consequence of aqueous alteration in carbonaceous chondrites (after McSween et al., 2018). Note that sulfates are not expected to form in Ceres (Castillo-Rogez et al., 2018). Those formed in carbonaceous chondrites are likely of terrestrial origin (Gounelle & Zolensky, 2001).

density mantle is interpreted as evidence for the presence of porosity (Choblet et al., 2017). Titan's high-pressure mantle on the other hand would not hold any porosity. Instead, geochemical modeling suggests Titan's mantle could be storing a significant amount of organic matter (Neri et al., 2019). This possibility has also been suggested by Zolotov (2020) for Ceres.

The study of icy moons also yields constraints on rock–water differentiation using salinity estimates from geophysical observations. For example, magnetometry can reveal a magnetic field induced in a deep ocean (e.g., Zimmer et al., 2000), although this measurement is a height integrated conductance, i.e., it is a function of both the ocean salinity and its thickness. In the case of Europa, a thin crust, and thus thick ocean, inferred from geological constraints (e.g., Schenk, 2002) would correspond to a salinity $<\sim 1$ wt.%. This is consistent with low salinities determined by theoretical modeling (e.g., ~ 1.2 wt.%, Zolotov & Shock, 2001). As another example, a salinity of 2.5 wt.% has been inferred for Enceladus' ocean from ice grains sampled by the Cassini Dust Analyzer (Postberg et al., 2011). These results overall suggest icy moon oceans are brackish and it is possible that, similarly, aqueous alteration did not proceed to chemical equilibrium in Ceres, which would yield a salinity of 5 wt.% in the early ocean (Castillo-Rogez et al., 2018).

11.3 DRIVERS AND CONSEQUENCES OF DIFFERENTIATION IN WATER-RICH BODIES

Key drivers of internal differentiation have been explored in previous reviews (e.g., Castillo-Rogez & Young, 2016). In the case of planetesimals, the time of formation drives the number of short-lived radioisotopes (^{26}Al and ^{60}Fe) accreted. Short-lived radioisotope decay heat can have a significant impact on the early history of planetesimals, driving melting, aqueous alteration, and differentiation, processes that would otherwise not happen in these small bodies. In larger (>1,000 km diameter) objects, accretional heating may also affect early internal evolution by melting their outer (100s km) regions. Long-lived radioisotopes can play a role in sustaining long-term internal activity and possibly liquid for bodies that are ≥1,000 km. Feedbacks between chemical and physical properties are such that it is not possible to present the different effects in a straightforward manner. First, we review the state of understanding of aqueous alteration in mid-sized icy bodies. Then we address physical and chemical feedbacks and detail a few specific processes of importance to Ceres.

11.3.1 Products of Aqueous Alteration

The size of the object and relative amount of water-to-rock drives the extent of aqueous alteration. A temperature greater than 0°C is critical in getting serpentinization started (Allen & Seyfried, 2004) and completed. Once started, several processes keep serpentinization going. First, based on observations of terrestrial rock, the volume change incurred by silicate grains results in the development of microcracks so that the surface area of rock exposed to water keeps increasing as serpentinization proceeds (Vance et al., 2007). Also, serpentinization is a very exothermic process that helps maintain warm temperatures.

Progress of the aqueous alteration of primary phases condensed in the solar nebula is summarized in Figure 11.4. While carbonaceous chondrites are very relevant analogues to Ceres, the much larger size of the dwarf planet can potentially explain some differences with Ceres' surface mineralogy observed by the Dawn mission. Ammonium–cation exchange in clays is observed at temperatures lower than 50°C (see Neveu et al., 2017 for a review of the experimental literature on this topic; see also Chapter 9). This

sets a constraint on the environment in which aqueous alteration proceeded in Ceres. Temperatures in carbonaceous chondrite parent bodies were inferred from the mineralogy to be higher than 50°C (e.g., Keil, 2000 for a review), which could explain why ammoniated clays have not been found in carbonaceous chondrites.

Another important consequence of object size is the partial pressure that can be reached by various gas species, and especially hydrogen. The partial pressure of hydrogen, pH_2, increases due to oxidation reactions, in particular $3 Fe + 4 H_2O => Fe_3O_4 + 4 H_2$, destabilizing iron-rich phases in serpentine and other minerals (e.g., Sleep et al., 2004). In large water-rich bodies that can sustain a high partial pressure of hydrogen, this process leads to a highly reducing environment (e.g., McKinnon & Zolensky, 2003). The maximum pH_2 is determined by lithostatic pressure; in Ceres $\log_{10} p(H_2)$ up to 2 (in MPa) (Figure 11.5). Furthermore, the consumption of protons and relative increase of free OH^- results in pH increase, typically >9, creating an alkaline environment.

Additional drivers of environmental conditions are the original volatile composition. Objects like Ceres and CC parent bodies likely accreted CO_2 and CO, as indicated by the large abundance of carbonates found in these bodies (De Sanctis et al., 2015). These species tend to increase the oxidizing potential of the ocean. They can remove H_2 from the system via Fischer-Tropsch type reactions, assuming catalysts and/or kinetically favorable conditions are present (see Section 11.3.2).

Many elements are very mobile in hydrothermal environments, especially alkali and alkaline earth metals. The redistribution of elements from the rock to liquid phase during aqueous alteration in water-rich bodies has been extensively modeled (e.g., Zolotov & Shock, 2001; Neveu et al., 2017, and references therein).

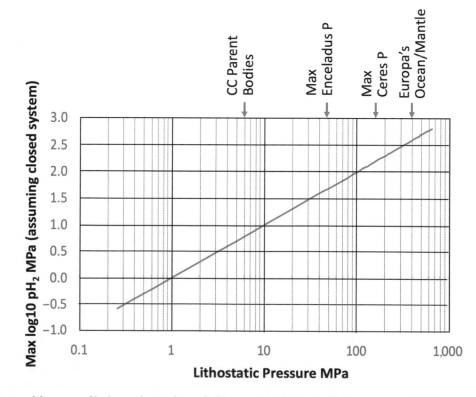

Figure 11.5 Maximum partial pressure of hydrogen that can be reached in various bodies. An object of Ceres' size could keep >100 MPa of hydrogen and other gas species under pressure, whereas redox conditions in the carbonaceous chondrite parent bodies were more oxidizing.

11.3.2 Chemical and Physical Feedbacks

Some of the key feedbacks between chemistry and physical properties resulting from aqueous alteration are summarized here.

11.3.2.1 Physical Evolution and Aqueous Alteration

Physical differentiation impacts the extent of aqueous alteration. Specifically, settling of the rock phase determines the timescale during which water interacts with rock. The terminal velocity (Stokes velocity) v of a particle of size a and density ρ sinking through a liquid layer of density ρ' is described as:

$$v = \frac{\frac{4}{3}\pi a^3(\rho' - \rho)g}{6\pi\eta a} \tag{11.1}$$

The velocity is a function of gravity g (0.027 m/s^2) and the dynamic viscosity of the liquid layer η. Taking Earth's ocean dynamic viscosity as a reference, for example, 0.00189 at 0°C (El-Dessouky & Ettouney, 2002), and particle density of 3,000 kg/m^3, the settling timescale inside Ceres is very fast: between about one year for particles of >0.1 mm and 1 My for micron-sized particles. Particles smaller than 1 μm might never settle in a perennial ocean (Travis et al., 2018). Neveu and Desch (2015) suggested that if Ceres accreted from chondritic material, which contains a mixture of mm-sized chondrules and very fine, submicron sized particles, then the chondrules would settle rapidly and form an inner mantle that may be ∼350 km in radius, whereas fines (particles 10s μm in size) could remain suspended for an extended periods of time. A similar early structure is assumed by Travis et al. (2018). Even in the most pessimistic case from a reaction kinetics standpoint, the progression rate of serpentinization across a rocky layer is at least 50 μm/year (Vance & Melwani Daswani, 2020). Hence, chondrules may be only partially serpentinized in the course of their journey toward the formation of a rocky mantle. On the other hand, suspended fines may be fully aqueously altered.

The extent of aqueous alteration determines the thermophysical properties of the mixture. Indeed, the thermal conductivity of hydrated rock (e.g., clays, serpentine, carbonates) is a factor 2–3 lower than that of anhydrous silicates (e.g., olivine, pyroxene) (Clauser & Huenges, 1995). Furthermore, iron metal-rich compounds may be redistributed as a consequence of aqueous alteration. For example, the iron-rich phases of olivine and pyroxene turn to magnetite and sulfides. These may remain embedded in the rock but particles richer in metal may separate from the rock by sinking faster (Scott et al., 2002). Elemental observations of Ceres' surface point to a deficiency in iron in comparison to CI chondrite material (Prettyman et al., 2017). This has been interpreted as evidence for iron fractionation during a global scale differentiation event as a result of the faster sinking of metal-rich, i.e., denser particles (Prettyman et al., 2017). A possible alternative or additional fractionation mechanism may be via circulation through a static but porous and permeable rocky mantle (e.g., Neveu et al., 2015).

This is consistent with Scott et al. (2002), who suggested that upon differentiation in a large ocean, a rocky mantle may be stratified or present a gradient in metal-rich species with depth. The analysis by King et al. (2018) using geophysical constraints returned by Dawn shows that this possibility cannot be discarded at

Ceres. This configuration, combined with the increased compressibility of hydrated silicates versus their anhydrous counterparts and the potential precipitation of salts in pores (Neveu et al., 2015) likely impeded convection onset in the mantle, as shown previously for Titan (Castillo-Rogez & Lunine, 2010). In summary, rocky mantles in mid-sized and large icy bodies could show both increasing density and thermal conductivity (due to increased metal content) with depth. The thermal implications of this situation have not been explored yet.

11.3.2.2 Redistribution of Radioisotopes

Uranium and thorium form oxides that remain associated with the rock phase. Experimental research on ammonium suggests however that NH_4^+ can substitute to K+ and Ca+ (to a lower extent) at temperatures less than 50°C (see Neveu et al., 2017). This increases the mobility of K+, one of the major radioisotopes, in hydrothermal environments so that leaching of K+ from the rock could be extensive. The thermal implications of this redistribution is twofold: (a) the loss of ^{40}K from the rock can significantly decrease the maximum temperature reached in Ceres' core at present, by up to 200 degrees, all other parameters being equal; (b) the concentration of potassium-rich brines at the base of the crust can locally increase the deep brine layer up to ∼7 degrees at present (Castillo-Rogez et al., 2019).

11.3.2.3 Fate of Major Volatiles

Ammonium observed across Ceres' surface likely comes from the accretion of ammonia hydrates. An alternative origin from the desorption of organic matter hypothesized by McSween et al. (2018) is not supported by internal evolution models: the required temperatures are never met (Travis et al., 2018) or met too late in Ceres' history (Castillo-Rogez et al., 2019), when the ocean is mostly frozen, to explain the global and homogeneous distribution of that material on Ceres' surface. The majority of ammonia is removed in the form of ammonium and stored in phyllosilicates and in the form of salts. In this condition, it loses its antifreeze properties. Geochemical modeling shows that the majority of the ammonia may be consumed and the rest concentrates in residual liquid (Castillo-Rogez et al., 2018).

The large abundance of carbonates found on Ceres' surface (De Sanctis et al., 2015; Carrozzo et al., 2018) suggests Ceres accreted a lot of CO and CO_2. Once in solution, these compounds indeed could form carbonate and bicarbonate ions at pressures <∼45 MPa. At higher pressure, CO_2 could instead be trapped in the form of gas hydrates (or "clathrates") (Choukroun et al., 2013). Although it is suggested that CO_2 could eventually be converted into methane in many planetary environments by Fischer-Tropsch type synthesis (Holm et al., 2015), that reaction is known to be sluggish (Glein et al., 2009). It may be accelerated in the presence of catalysts (e.g., Sleep et al., 2004). If proceeding on a global scale, removal of CO_2 impacts the pH of the solution and drives the mineralogy produced during the freezing of the ocean. Depending on the nature of the enclosed gas species, the clathrate hydrates may sink (e.g., CO_2 hydrates) or accumulate at the base of growing ice shell (e.g., CH_4 hydrates). The thermal conductivities of hydrates are significantly lower in comparison to water ice for a given temperature, by a factor 5–10 (e.g., Durham et al., 2010). On the other hand, the strength of gas hydrates can be two-to-three orders of magnitude greater than ice. These two properties act in impeding heat transfer and keeping the interior warm (see Section 11.4.2).

Another possible mechanism shifting the partial pressures of key gas species is the opening of the system by exposure to vacuum by large impacts. Ceres was likely subject to basin-forming impacts during its early history (Marchi et al., 2016) and it is unlikely that Ceres' ocean remained a closed system during that period. This is supported by the surface mineralogy – assuming it is representative of Ceres' deep interior – that suggests formation under a partial pressure of hydrogen that is several orders of magnitude less than the theoretical partial pressure expected in Ceres' interior (see Figure 11.5). Loss of H_2 and CO_2 could thus shift the pH in opposite directions.

11.4 EVOLUTION PATHWAYS FOR CERES

Three main evolution pathways have been suggested for Ceres. The model by Zolotov (2020) explains the observed moment of inertia by modeling porosity distribution in an undifferentiated mixture of organic compounds and hydrated silicates. While novel, this model does not track the long-term thermal evolution of the deep interior. Organic compounds and hydrated silicates have thermal conductivities a factor three-to-five lower than ice, and the addition of a porous crust further acts as an insulating layer. Previous work showed that a model of this type would not be thermally stable. Instead, the rock would dehydrate on a timescale of hundreds of million years (Castillo-Rogez, 2011; Neumann et al., 2020), leading to the formation of a volatile-rich crust and dry silicate mantle. This model also does not provide obvious mechanisms for the recent activity observed at Occator crater and Ahuna Mons. In the rest of this section, we focus on partially differentiated interior models, in which a volatile-rich crust has separated from the rock following melting on a global scale.

11.4.1 Mantle Evolution

Two end state models for Ceres' rocky mantle have been presented in the literature. Pre-Dawn models have assumed rapid settling of the mantle accompanied by rapid freezing of an icy crust so that Ceres' interior is mostly frozen a few 100 My after formation (McCord & Sotin, 2005; Castillo-Rogez & McCord, 2010). Precipitation of salts in veins or pores results in further lithification of the mantle (Neveu et al., 2015). Gradual sedimentation of particles of different sizes and metal contents as well as cementation by salt could lead to a mantle structure stable against solid-state convection. In this case, the mantle temperatures could reach the dehydration temperature of phyllosilicates, around 850K (Castillo-Rogez et al., 2019). However, the behavior of organic matter, which can be mobile – low viscosity or even liquid – at mantle temperatures (100s °C) has not been investigated in that type of planetary environment. It could potentially separate from the rock and contribute to convecting heat upward.

A more recent generation of models suggests that if not all the silicates settled early on, then slow convection in a "mudball," that is, ocean loaded in fines, could lead to the long-lived persistence of liquid in Ceres and other mid-sized bodies (Neveu & Desch, 2015;

Travis & Schubert, 2015; Bland & Travis, 2017; Travis et al., 2018). This type of structure has been proposed as a way to enhance tidal heat generation in Enceladus' porous core (Roberts, 2015; Travis & Schubert, 2015), a model that has been adopted by the community following some refinements (Choblet et al., 2017). This model yielded predictions that appear consistent with the Dawn observations, and in particular the presence of large-scale gravity anomalies in the mantle (Raymond et al., 2020) expressing lateral variations in brine versus silicate content, the preservation of high porosity in the mantle, and limited heating of the rock. On the other hand, the density profile predicted in Travis et al. (2018) for a reference model is significantly different from the Ermakov et al. (2017) and Mao and McKinnon (2018) results. In particular, the predicted upper mantle density is 1,900 kg/m^3 instead of >2,400 kg/m^3. The density of the mantle is determined by the physics of fines. Although the Travis et al. (2018) application of hydrothermal circulation to Ceres is elaborate and builds on multiple studies, it needs additional refinements in order to fully address the behavior of particles in a thick ocean loaded in salts and organic matter. However, the abundance and nature of solutes as well as environmental conditions (e.g., pH) play an important role in the fate of fine silicate particles. They may drive particle dissolution and flocculation, that is, the aggregation of particles due, for example, to electrical charging, a phenomenon well studied on Earth (e.g., Kranck, 1973). Sirono (2013) showed that in the absence of salts driving coagulation, fines might never settle.

11.4.2 Formation and Evolution of the Crust

A key open question until recently is the extent to which an undifferentiated ice-rock upper crust could remain on Ceres over the long-term. This prospect is unlikely if Ceres' formation involved sufficient accretional heat to warm up, break, and lead the crust to founder (Castillo-Rogez & McCord, 2010) (pathway #2 in Figure 11.6). However, formation in the Kuiper belt, as suggested by Neumann et al. (2020), likely involved little accretional heat, since accretion could span tens of million years (pathway #1 in Figure 11.6). Hence, if Ceres had an undifferentiated crust early on, it most likely foundered early in Ceres' history. Admittance analysis by Ermakov et al. (2017) supports the absence of a thick undifferentiated crust on Ceres (see Chapter 12).

Top-down freezing of the crust is expected to lead to the increasing accretion of impurities (e.g., rock particles and salts) with depth, based on terrestrial analogues (e.g., frozen lakes). The Dawn observations suggest this gradient exists in Ceres' crust (Park et al., 2020). As the crust was growing, a small fraction of soluble impurities (salts, gas, and organic matter) could become trapped at the ice grain boundaries (e.g., Santibáñez et al., 2019), but the majority of soluble species though concentrate in the residual liquid, potentially leading to saturated brine solutions (Castillo-Rogez et al., 2018, 2019). The freezing sequence modeled by Castillo-Rogez et al. (2018) (Figure 11.7) reproduces salts detected by the Dawn mission, namely sodium carbonate, ammonium chlorides, and hydrohalite (see Chapter 7). That modeling also shows that clathrate hydrates are the dominant form of water under a few megapascal of pressure.

The fate of the salts is not understood. With their densities being greater than the ocean density, it is expected that the salts would sink in the shrinking ocean. Yet, Ceres' surface displays hundreds

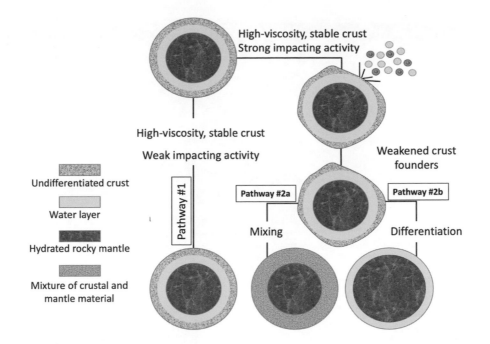

Figure 11.6 Possible evolution pathways for Ceres' crust in the first 100s of million years after formation (based after Neumann et al., 2020). Depending on the duration of accretion and amount of accretional heat involved, an undifferentiated upper crust likely foundered early in Ceres' history. An extreme scenario (pathway #1) assumes that a thick undifferentiated layer could remain until present if Ceres accreted late, that is, without 26-aluminum decay heat and weak impacting history. However, this scenario is not consistent with the Dawn geophysical data. A more likely evolution pathway (#2) assumes that an undifferentiated crust foundered early on as a consequence of weaknesses introduced by impacting.

of small bright areas characterized or interpreted as salt deposits (Stein et al., 2017; Carrozzo et al., 2018). Stein et al. (2019) pointed out that salt lenses present in the shallow subsurface are expected to be broken down and scattered on a timescale of 0.5–1 Gy. Hence the salt deposits found across Ceres' surface must be recently sourced from the deeper subsurface.

The Castillo-Rogez et al. (2019) model tests the prospect for the long-term preservation of liquid below a water-rich crust, using a model that combines compositional and physical constraints derived from the Dawn observations: the reference structure is the two-layer density profile inferred by Ermakov et al. (2017) with the rheological constraints from Fu et al. (2017), that includes abundant clathrates in the crust and a ∼60-km layer of brines and silicates below the crust. Castillo-Rogez et al. (2019) and Formisano et al. (2020) show that a high abundance of hydrates (up to 70 vol.%) should stabilize the crust against solid-state convection. This effect combined with the hydrates' low thermal conductivities could explain the preservation of deep fluid at present (Castillo-Rogez et al., 2019), especially below Hanami Planum's thick crust (∼55 km) (Raymond et al., 2020). Hydrates have also been suggested as responsible for the preservation of liquid in Pluto and could be a common feature of mid-sized and large icy bodies (Kamata et al., 2019).

Several mechanisms for supplying brines to the shallow subsurface of Ceres over the long-term are emerging. One is that Ahuna Mons in particular could have formed from slurry extrusion (Ruesch et al., 2019) stemming from a long-lived convective muddy mantle. The 5–10 vol.% sodium carbonate found on the flank of this mons (Zambon et al., 2017) suggests a source temperature of ∼255 K (Castillo-Rogez et al., 2019, Figure 11.4). Such a warm temperature may be explained by the Travis et al. (2018) evolution model but is not explained by the "insulating," clathrate-rich crust model of Castillo-Rogez et al. (2019).

The various models for Ceres' evolution currently published are summarized in Table 11.1. Further exploration of these models should aim to further test their viability against the various geophysical, geological, and compositional constraints inferred from the Dawn data.

11.5 CONCLUSION AND FUTURE PERSPECTIVES

By observing dwarf planet Ceres in great detail, the Dawn mission has brought unique datasets whose analysis has led to a better understanding of the evolution of Ceres and can serve as a reference for predicting the evolution and current states of other mid-sized, ice-rich bodies (icy moons, other large icy asteroids like 10 Hygiea, and dwarf planets). Mineralogical observations indicate a global scale event of aqueous alteration. Gravity and topography data show partial differentiation that is best explained so far by the separation of a water-rich crust. Elemental measurements further point to differential settling of metal-rich particles during the differentiation phase, likely when Ceres held a deep ocean. The level of detail returned by Dawn revealed that Ceres is a complex body, in particular from a geochemical standpoint, where salts and gas species are major players. Ceres may be representative of objects small enough (≲1,000 km across) to host internal mudballs for extended periods of time (Bland & Travis, 2017), resulting in long-lived water–rock interaction. However, constraints on mantle density and rheology from the Dawn data are ambiguous. Three major unknowns in our understanding of Ceres' deep interior and evolution remain at this time: (a) the state of the rocky mantle: coherent/consolidated, in the form of a convecting mud

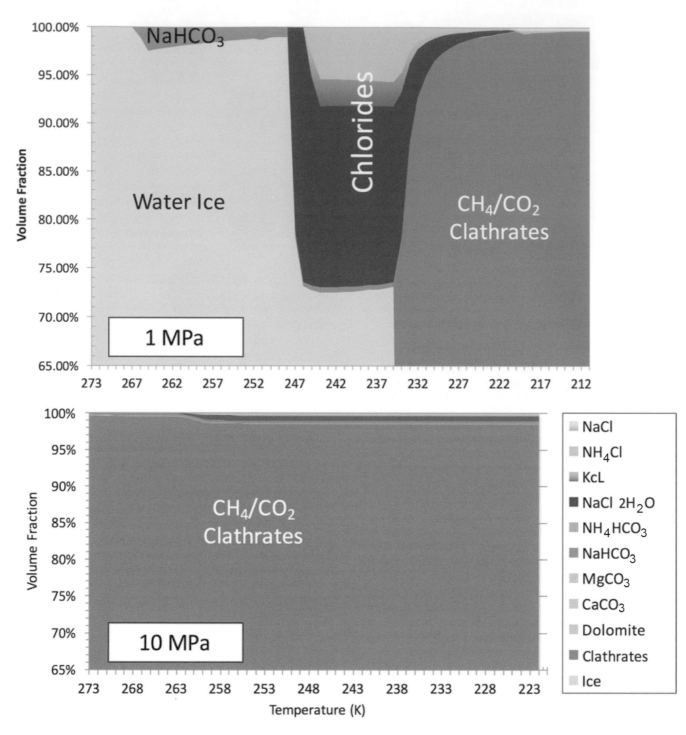

Figure 11.7 Precipitation sequence of solids in Ceres' ocean upon freezing. The top chart shows that at low pressures (a few MPa) the bulk of water freezes in the form of ice with a small fraction (20 vol.%) of salts. The bottom chart shows that under higher pressure, water freezes primarily in the form of clathrate hydrates, and salts are of low abundance (Castillo-Rogez et al., 2018).

A black and white version of this figure will appear in some formats. For the color version, refer to the plate section

(Travis et al., 2018), or a weakly consolidated but static rocky mantle with potentially significant porosity (Neveu et al., 2015), as has been suggested for Enceladus (Choblet et al., 2017); (b) the extent of liquid in Ceres, in the form of a 10s-km-thick residual liquid layer with pore liquid (Fu et al., 2017), localized seas, hundreds of kilometers in extent and maybe tens of kilometers deep (Castillo-Rogez et al., 2019; Raymond et al., 2020), and/or

local brine pockets; (c) the extent of aqueous alteration during differentiation with implications on the amount of salts produced during that phase and the fate of these salts, as well as of organic matter. Whether (near) chemical equilibrium was reached in Ceres' early ocean will remain uncertain until refined mineralogical characterization of Ceres' surface by an in-situ mission (or sample return) becomes available.

Table 11.1 *Summary of the four main models proposed for Ceres in terms of their ability to reproduce the Dawn observations.*

See Figure 11.2 for illustration. Italic text is used when describing inconsistencies with observations. Cells in light grey indicate areas whose consistency with the Dawn observations requires further analyses. Cells in dark grey indicate aspects of the model that appear intrinsically inconsistent with the Dawn observations.

Model	Main Characteristics	Consistency with Dawn Observations?		
		Chemistry	Geophysics	Geology
Zolotov (2020)	Undifferentiated; no ice; low density from porosity and organic matter; no liquid	Explains high carbon content inferred from VIR and GRaND; *does not address the occurrence of sodium carbonate and other salts*	Matches density profile; *topography relaxation not verified; long-term thermal evolution of interior not modeled*	*Inconsistent with evidence for ice in Ceres' shallow subsurface; does not provide a framework for the activity observed at Occator and Ahuna Mons*
Neumann et al. (2020)	Late accretion in the Kuiper belt yields a partially differentiated crust that eventually founders	Explains hydrated materials accreted from planetesimals and large fraction of organic matter; preserves deep liquid layer until present; *does not address the occurrence of sodium carbonate and other salts*	Consistent with Dawn-derived density profile; *predicted crustal properties (strength, density), not verified against the Dawn observations*	Consistent with Dawn observations; Slurry-based upper mantle origin for Ahuna Mons formation matches gravity data
Neveu et al. (2015); Castillo-Rogez et al. (2018; 2019)	Solid mantle, mud between mantle and crust 10s km thick	Explains pervasive rock hydration in global early ocean; includes clathrates in crust; forms sodium carbonate; *does not address the fate of organics*	Consistent with Dawn-derived density profile. Explains long-term retention of liquid	Model uses crustal composition consistent with geological observations
Travis et al. (2018); Neveu and Desch (2015)	Rocky core, thick (100s km) ocean loaded in fines	Explains pervasive rock hydration in global early ocean; *does not address the fate of organics*	Explains long-term retention of liquid, recent activity, and (potentially) gravity anomalies in deep mantle; *density profile is not consistent with admittance-derived constraints*	*Crust model does not explain high crustal strength derived from geological observations*

Ceres falls under the category of ocean worlds (Hendrix et al., 2019), that is, ice-rich bodies that have preserved a deep ocean until present. Inferences so far for Ceres' interior raise important questions. First, the prospect for a low porosity mantle filled with brines for extended period of time could offer a medium propitious for advanced geochemistry involving organic matter (Travis et al., 2018). This geophysical context is also relevant to Enceladus (Choblet et al., 2017). Several geological landmarks suggest local connections between the deep interior and the surface in the form of brine- or mud-driven volcanism (Ahuna Mons, Ruesch et al., 2019). Large impacts may play a role in creating pathways for material transfer in the form of deep impact melt chambers and fractures (Bowling et al., 2019; Raymond et al., 2020), providing an opportunity to glimpse Ceres' internal composition. An assessment of Ceres' astrobiological significance based on the Dawn observations can be found in Castillo-Rogez et al. (2020).

Differentiation is a difficult problem to model and simulate in the laboratory because settling of particles against a convecting hydrothermal background in low gravity depends on a number of processes, some of which are poorly constrained. In particular,

chemistry plays a major role in driving physical differentiation per the role of particle ionization. Experimental and theoretical research needed to improve our understanding of ocean worlds has been summarized in previous work (especially Hendrix et al., 2019; Castillo-Rogez et al., 2020). The most critical aspect pertaining to mid-sized bodies like Ceres is about understanding the physics of mixtures involving brines, silicates, and organic matter in low-gravity conditions. In particular, the fractionation of organic matter between the rocky mantle, icy shell, and residual ocean remains to be investigated in detail. In carbonaceous chondrites, soluble organic matter has been found in association with the silicate matrix (Le Guillou et al., 2014). Hence, part of the organic matter accreted and produced in Ceres is expected to be trapped in the rocky mantle where it can mature as a result of slow heating under pressure.

The Dawn observations and recent high-resolution observation of large asteroid surfaces provide new leads for exploring differentiation processes. They also raise new questions about the modalities of heat transfer in a geochemically complex body, with implications for all mid-sized and large water-rich bodies across the Solar System. The prospect, suggested by multiple observations, that Ceres could host liquid at present, either in local ponds or on a global scale, points to the role of hydrates in Ceres' crust in slowing down heat transfer (Castillo-Rogez et al., 2019). A major caveat though is that non-hydrostatic anomalies are deceiving and hide the actual extent of differentiation (Mao & McKinnon, 2018; Chapter 12). Future observations should attempt to obtain the true moment of inertia of Ceres, for example by directly measuring the precession constant. Knowledge gained at Ceres can be applied to other icy worlds, and in particular heat-starved bodies such as Callisto and Pluto. For example, these objects likely benefit from the occurrence of "exotic" volatile species or forms of water ice, like gas and salt hydrates, that decrease eutectic temperature and thermophysical properties. Furthermore, late activity may be driven by the concentration of brines at the base of the crust. Hence, candidate ocean worlds may extend to the far outer Solar System.

REFERENCES

Alexander, C. M. O'D., Bowden, R., Fogel, M. L., & Howard, K. T. (2015) Carbonate abundances and isotopic compositions in chondrites. *Meteoritics & Planetary Science*, 50, 810–833.

Allen, D. E., & Seyfried Jr., W. E. (2004) Serpentinization and heat generation: Constraints from Lost City and Rainbow hydrothermal systems. *Geochimica et Cosmochimica Acta*, 68, 1347–1354.

Anderson, J. D., Jacobson, R. A., McElrath, T. P., et al. (2001) Shape, mean radius, gravity field and interior structure of Callisto. *Icarus*, 153, 157–161.

Benedix, G. K., Leshin, L. A., Jackson, T., & Thiemens, M. H. (2003) Carbonates in CM2 chondrites: Constraints on alteration conditions from oxygen isotopic compositions and petrographic observations. *Geochimica et Cosmochimica Acta*, 67, 1577–1588.

Bland, P. A., Collins, G. S., Davison, T. M., et al. (2014) Pressure-temperature evolution of primordial Solar System solids during impact-induced compaction. *Nature Communications*, 5, 5451–5451.

Bland, P. A., Howard, L. E., Prior, D. J., et al. (2011) Earliest rock fabric formed in the Solar System preserved in a chondrule rim. *Nature Geoscience*, 4, 244–247.

Bland, P. A., Jackson, M. D., Coker, R. F., et al. (2009) Why aqueous alteration in asteroids was isochemical: High porosity ≠ high permeability. *Earth and Planetary Science Letters*, 287, 559–568.

Bland, P. A., & Travis, B. J. (2017) Giant convecting mud balls of the early Solar System. *Science Advances*, 3, e1602514.

Blum, J. (2004) Grain growth and coagulation. In A. N. Witt, G. C. Clayton, & B. T. Draine (eds.), *Astrophysics of Dust*. San Francisco, CA: Astronomical Society of the Pacific, pp. 369–391.

Blum J., & Schrapler, R. (2004) Structure and mechanical properties of high-porosity macroscopic agglomerates formed by random ballistic deposition. *Physical Review Letters*, 93: 115503-1–115503-4.

Bowling, T. J., Ciesla, F. J., Davison, T. M., et al. (2019) Post-impact thermal structure and cooling timescales of Occator crater on Asteroid 1 Ceres. *Icarus*, 320.

Brearley, A. J. (2003) Nebular versus parent body processing. in A. M. Davis (ed.), *Treatise on Geochemistry*, Vol. 1. Amsterdam: Elsevier, pp. 247–268.

Brearley, A. J. (2006) The action of water. In D. S. Lauretta, & H. Y. McSween, Jr. (eds.), *Meteorites and the Early Solar System II* (pp. 587–624). Tucson: University of Arizona Press.

Carrozzo, F. G., De Sanctis, M. C., Raponi, A., et al. (2018) Nature, formation, and distribution of carbonates on Ceres. *Science Advances*, 4, e1701645.s.

Castillo-Rogez, J. C. (2011) Ceres – Neither a porous nor salty ball. *Icarus*, 215, 599–602.

Castillo-Rogez, J. C., Hesse, M., Formisano, M., et al. (2019) Conditions for the long-term preservation of a deep brine Rreservoir in Ceres. *Geophysical Research Letters*, 46, 1963–1972.

Castillo-Rogez, J. C., & Lunine, J. I. (2010) Evolution of Titan's rocky core constrained by Cassini observations. *Geophysical Research Letters*, 37, L20205.

Castillo-Rogez, J. C., & McCord, T. B. (2010) Ceres' evolution and present state constrained by shape data. *Icarus*, 205, 443–459.

Castillo-Rogez, J. C., Neveu, M., McSween, H. Y., et al. (2018) Insights into Ceres' evolution from surface composition. *Meteoritics & Planetary Science*, 53, 1820–1843.

Castillo-Rogez, J. C., Neveu, M., Scully, J. E. C., et al. (2020) Ceres: Astrobiological target and possible ocean world. *Astrobiology*, 20, 269–291.

Castillo-Rogez, J. C., & Schmidt, B. E. (2010) Geophysical evolution of the Themis family parent body. *Geophysical Research Letters*, 37, L10202.

Castillo-Rogez, J. C., & Young, E. D. (2016) Origin and evolution of volatile-rich planetesimals. In L. Elkins-Tanton, & B. Weiss (eds.), *Planetesimal Differentiation*. Cambridge: Cambridge University Press, pp. 92–114.

Chan, Q. H. S., Zolensky, M. E., Kebukawa, Y., et al. (2018) Organic matter in extraterrestrial water-bearing salt crystals. *Science Advances*, 4, eaao3521.

Choblet, G., Tobie, G., Sotin, C., et al. (2017) Powering prolonged hydrothermal activity inside Enceladus. *Nature Astronomy*, 1, 841–847.

Choukroun, M., Kieffer, S., Lu, X., & Tobie G. (2013) Clathrate hydrates: Implication for exchange processes in the outer Solar System. In S. M. Gudipati, & J. C. Castillo-Rogez (eds.), *Science of Solar System Ices*, 3rd ed. (Astrophysics and Space Science Library, 356, pp. 409–454). New York: Springer.

Clauser, C., & Huenges, E. (1995) Thermal conductivity of rocks and minerals. In T. J. Ahrens (ed.), *Rock Physics & Phase Relations*. Washington, DC: American Geophysical Union.

Clayton, R. N., & Mayeda, T. K. (1999) Oxygen isotope studies of carbonaceous chondrites. *Geochimica et Cosmochimica Acta*, 63, 2089–2104.

De Sanctis, M. C., Ammanito, E., Raponi, E., et al. (2015) Ammoniated phyllosilicates with a likely outer Solar System origin on (1) Ceres. *Nature*, 528, 241–244.

De Sanctis, M. C., Ammannito, E., Raponi, A., et al. (2020) Recent emplacement of hydrated sodium chloride on Ceres from ascending salty fluids. *Nature Astronomy*, 4, 786–793.

Desch, S. J., Kalyaan, A., & Alexander, C. M. O'D. (2018) The effect of Jupiter's formation on the distribution of refractory elements and inclusions in meteorites. *The Astrophysical Journal Supplement Series*, 238, 11.

Dullien, F. A. L. (1992) *Porous Media, Fluid Transport and Pore Structure*, 2nd ed. Cambridge, MA: Academic Press.

Durham, W. B., Prieto-Ballesteros, O., Goldsby, D. L., & Kargel, J. S. (2010) Rheological and thermal properties of icy materials. *Space Science Reviews*, 153, 273–298.

Dyl, K. A., Manning, C. E., & Young, E. D. (2010) The implications of cronstedtite formation in water-rich planetesimals and asteroids. *Astrobiology Science Conference 2010: Evolution and Life: Surviving Catastrophes and Extremes on Earth and Beyond*, April 2010, League City, TX. LPI Contrib. 1538, #5627.

El-Dessouky, H. T., & Ettouney, H. M. (2002) *Fundamentals of Salt Water Desalination*. Amsterdam: Elsevier Science B.V.

Ermakov, A. I., Fu, R. R., Castillo-Rogez, J. C., et al. (2017) Constraints on Ceres' internal structure and evolution from its shape and gravity measured by the Dawn spacecraft. *Journal of Geophysical Research: Planets*, 122, 2267–2293.

Formisano, M., Federico, C., Castillo-Rogez, J., De Sanctis, M. C., & Magni, G. (2020) Thermal convection in the crust of the dwarf planet Ceres. *Monthly Notices of the Royal Astronomical Society*, 494, 5704–5712.

Fu, R., Ermakov, E., Marchi, S., et al. (2017) Interior structure of the dwarf planet Ceres as revealed by surface topography. *Earth and Planetary Science Letters*, 476, 153–164.

Fujiya, W., Sugiura, N., Marrocchi, Y., et al. (2015) Comprehensive study of carbon and oxygen isotopic compositions, trace element abundances, and cathodoluminescence intensities of calcite in the Murchison CM chondrite. *Geochimica et Cosmochimica Acta*, 70, 101–117.

Gao, P., & Stevenson, D. J. (2013) Nonhydrostatic effects and the determination of icy satellites' moment of inertia. *Icarus*, 226, 1185–1191.

Glein, C. R., Desch, S. J., & Shock, E. L. (2009) The absence of endogenic methane on Titan and its implications for the origin of atmospheric nitrogen. *Icarus*, 204, 637–644.

Gounelle, M., & Zolensky, M. E. (2001) A terrestrial origin for sulfate veins in CI1 chondrites. *Meteoritics & Planetary Science*, 36, 1321–1329.

Guo, W., & Eiler, J. M. (2007) Temperatures of aqueous alteration and evidence for methane generation on the parent bodies of the CM chondrites. *Geochimica et Cosmochimica Acta*, 71, 5565–5575.

Hendrix, A. R., Hurford, T. A., Barge, L. M., et al. (2019) The NASA roadmap to ocean worlds. *Astrobiology*, 19.

Holm, N. G., Oze, C., Mousis, O., Waite, J. H., & Guilbert-Lepoutre, A. (2015) Serpentinization and the formation of H_2 and CH_4 on celestial bodies (planets, moons, comets). *Astrobiology*, 15, 587–600.

Howard, K. T., Benedix, G. K., Bland, P. A., & Cressey, G. (2009) Modal mineralogy of CM2 chondrites by X-ray diffraction (PSD-XRD). Part 1: Total phyllosilicate abundance and the degree of aqueous alteration. *Geochimica et Cosmochimica Acta*, 73, 4576–4589.

Hsieh, H. H. (2012) Main-belt comets as tracers of ice in the inner Solar System. *Proceedings of the International Astronomical Union 8*, Issue S293 (Formation, Detection, and Characterization of Extrasolar Habitable Planets), 212–218.

Hussmann, H., Sohl, F., & Spohn, T. (2006) Subsurface oceans and deep interiors of medium-sized outer planet satellites and large trans-Neptunian objects. *Icarus*, 185, 258–273.

Iess, L., Stevenson, D. J., Parisi, M., et al. (2014) The gravity field and interior structure of Enceladus. *Science*, 344, 78–80.

Jogo, K., Nakamura, T., Ito, M., et al. (2017) Mn-Cr ages and formation conditions of fayalite in CV3 carbonaceous chondrites: Constraints on the accretion ages of chondritic asteroids. *Geochimica et Cosmochimica Acta*, 199, 58–74.

Kamata, S., Nimmo, F., Sekine, Y., et al. (2019) Pluto's ocean is capped and insulated by gas hydrates. *Nature Geoscience*, 12, 407–410.

Keil, K. (2000) Thermal alteration of asteroids: Evidence from meteorites. *Planetary and Space Science*, 48, 887–903.

King, S. D., Castillo-Rogez, J. C., Toplis, M. J., et al. (2018) Ceres internal structure from geophysical constraints. *Meteoritics & Planetary Science*, 53, 1999–2007.

Kivelson, M. G., Khurana, K. K., & Volwerk, M. (2002) The permanent and inductive magnetic moments of Ganymede. *Icarus*, 157, 507–522.

Kranck, K. (1973) Flocculation of suspended sediment in the sea. *Nature*, 246, 348–350.

Le Guillou, C., Bernard, S., Brearley, A. J., & Remusat, L. (2014) Evolution of organic matter in Orgueil, Murchison and Renazzo during parent body aqueous alteration: In situ investigations. *Geochimica et Cosmochimica Acta*, 131, 368–392.

Lee, M. R., Lindgren, P., & Sofe, M. R. (2014) Aragonite, breunnerite, calcite and dolomite in the CM carbonaceous chondrites: High fidelity recorders of progressive parent body aqueous alteration. *Geochimica et Cosmochimica Acta*, 144, 126–156.

Leshin, L. A., Rubin, A. E., & McKeegan, K. D. (1997) The oxygen isotopic composition of olivine and pyroxene from CI chondrites. *Geochimica et Cosmochimica Acta*, 61, 835–845.

Lodders, K., Palme, H., & Gail, H. P. (2009) Abundances of elements in the Solar System, in J. E. Trumper (ed.), *Landolt-Bornstein, New Series, Astronomy and Astrophysics*. Berlin: Springer Verlag.

Macke, R. J., Consolmagno, G. J., & Britt, D. T. (2011) Density, porosity, and magnetic susceptibility of carbonaceous chondrites. *Meteoritics & Planetary Science*, 46, 1842–1862.

Maiorca, E., Uitenbroek, H., Uttenthaler, S., et al. (2014) A new solar fluorine abundance and a fluorine determination in the two open clusters M67 and NGC 6404. *The Astrophysical Journal*, 788, 149.

Mao, X., & McKinnon, W. B. (2018) Faster paleospin and deep-seated uncompensated mass as possible explanations for Ceres' present-day shape and gravity. *Icarus*, 299, 430–442.

Marchi, S., Ermakov, A. I., Raymond, C. A., et al. (2016) The missing large impact craters on Ceres. *Nature Communications*, 7, 12257.

Marchi, S., Raponi, A., Prettyman, T. H., et al. (2019) An aqueously altered carbon-rich Ceres. *Nature Astronomy*, 3, 140–145.

Marsset, M., Brož, M., Vernazza, P., et al. (2020) The violent collisional history of aqueously evolved (2) Pallas. *Nature Astronomy*, 4, 569–576.

Marsset, M., Vernazza, P., Birlan, M., et al. (2016) Compositional characterization of the Themis family. *Astronomy & Astrophysics*, 586, id.A15.

Marzari, F., Davis, D., & Vanzani, V. (1995) Collisional evolution of asteroid families. *Icarus*, 113, 168–187.

McCord, T. B., & Sotin, C. (2005) Ceres: Evolution and current state. *Journal of Geophysical Research*, 110, EO5009–EO5014.

McKinnon, W. B., & Zolensky, M. E. (2003) Sulfate content of Europa's ocean and shell: Evolutionary considerations and some geological and astrobiological implications. *Astrobiology*, 3, 879–897.

McSween, H. Y., Emery, J. P., Rivkin, A. S., et al. (2018) Carbonaceous chondrite analogs for the composition and alteration of Ceres. *Meteoritics & Planetary Science*, 53, 1793–1804.

Mueller, S., & McKinnon, W. B. (1989) Three-layered models of Ganymede and Callisto: Compositions, structures, and aspects of evolution. *Icarus*, 76, 437–464.

Néri, A., Guyot, F., Reynard, B., & Sotin, C. (2019) A carbonaceous chondrite and cometary origin for icy moons of Jupiter and Saturn. *Earth and Planetary Science Letters*, 530, id. 115920.

Neumann, W., Jaumann, R., Castillo-Rogez, J. C., Raymond, C. A., & Russell, C. T. (2020) Ceres' partial differentiation: Undifferentiated crust mixing with a water-rich mantle. *Astronomy and Astrophysics*, 633, A117.

Neveu, M., & Desch, S. J. (2015) Geochemistry, thermal evolution, and cryovolcanism on Ceres with a muddy mantle. *Geophysical Research Letters*, 42, 10,197–10,206.

Neveu, M., Desch, S.J., & Castillo-Rogez, J. C. (2015) Core cracking and hydrothermal circulation can profoundly affect Ceres' geophysical evolution. *Journal of Geophysical Research*, 120, 123–154.

Neveu, M., Desch, S. J., & Castillo-Rogez, J. C. (2017) Aqueous geochemistry in icy world interiors: Fate of antifreezes and radionuclides. *Cosmochimica et Geochimica Acta*, 212, 324–371.

Neveu, M., & Vernazza, P. (2019) IDP-like asteroids formed later than 5 Myr after Ca-Al-rich inclusions. *The Astrophysical Journal*, 875, id. 30.

Ormel, C. W., Cuzzi, J. N., & Tielens, A. G. G. M. (2008) Co-accretion of chondrules and dust in the solar nebula. *The Astrophysical Journal*, 679, 1588–1610.

Palomba, E., Longobardo, A., De Sanctis, M. C., et al. (2019) Compositional differences among Bright Spots on the Ceres surface. *Icarus*, 320, 202–212.

Park, R. S., Konopliv, A. S., Ermakov, A. I., et al. (2020) Evidence of non-uniform crust of Ceres from Dawn's high-resolution gravity data. *Nature Astronomy*, 4, 748–755.

Postberg, F., Schmidt, J., Hillier, J., Kempf, S., & Srama, R. (2011) A saltwater reservoir as a source of compositionally stratified plume on Enceladus. *Nature*, 474, 620–622.

Prettyman, T. H., Yamashita, N., Ammannito, E., et al. (2018) Elemental composition and mineralogy of Vesta and Ceres: Distribution and origins of hydrogen-bearing species. *Icarus*, 318, 42–55.

Prettyman, T. H., Yamashita, N., Toplis, M. J., et al. (2017) Extensive water ice within Ceres' aqueously altered regolith: Evidence from nuclear spectroscopy. *Science*, 355, 55–59.

Raponi, M. C., De Sanctis, F. G., Carrozzo, M., et al. (2019) Mineralogy of Occator crater on Ceres and insight into its evolution from the properties of carbonates, phyllosilicates, and chlorides. *Icarus*, 320, 83–96.

Raymond, C. A., Castillo-Rogez. I., Ermakov, S., et al. (2020) Impact-driven mobilization of deep crustal brines on dwarf planet Ceres. *Nature Astronomy*, 4, 741–747.

Rivkin, A. S., & Emery, J. P. (2010) Detection of ice and organics on an asteroidal surface. *Nature*, 464, 1322–1323.

Rivkin, A. S., Howell, E. S., & Emery, J. P. (2019) Infrared spectroscopy of large, low-albedo asteroids: Are Ceres and Themis archetypes or outliers? *Journal of Geophysical Research*, 124, 1393–1409.

Roberts, J. H. (2015) The fully core of Enceladus. *Icarus*, 258, 54–66.

Rubin, A. E., Zolensky, M. E., & Bodnar, R. J. (2002) The halite-bearing Zag and Monahans (1998) meteorite breccias: Shock metamorphism, thermal metamorphism and aqueous alteration on the H-chondrite parent body. *Meteoritics & Planetary Science*, 37, 125–141.

Ruesch, O., Genova, A., Neumann, W., et al. (2019) Slurry extrusion on Ceres from a convective mud-bearing mantle. *Nature Geoscience*, 12, 505–509.

Ruesch, O., Platz, T., Schenk, P., et al. (2016) Cryovolcanism on Ceres. *Science*, 353, aaf4286.

Santibanez, P. A., Michaud, A. B., Vick-Majors, T. J., et al. (2019) Differential incorporation of bacteria, organic matter, and inorganic ions into lake ice during ice formation. *Journal of Geophysical Research – Biogeosciences*, 124, 585–600.

Sasso, M. R., Macke, R. J., Boesenberg, J. S., et al. (2009) Incompletely compacted equilibrated ordinary chondrites. *Meteoritics & Planetary Science*, 44, 1743–1753.

Schenk, P. M. (2002) Thickness constraints on the icy shells of the galilean satellites from a comparison of crater shapes. *Nature*, 417, 419–421.

Schrader, D. L., Franchi, I. A., Connolly, H. C. Jr., et al. (2011) The formation and alteration of the Renazzo-like carbonaceous chondrites I: Implications of bulk-oxygen isotopic composition. *Geochimica et Cosmochimica Acta*, 75, 308–325.

Scott, H. P., Williams, Q., & Ryerson, F. J. (2002) Experimental constraints on the chemical evolution of icy satellites. *Earth and Planetary Science Letters*, 203, 399–412.

Scully, J. E. C., Schenk, P. M., Buczkowski, D. L., et al. (2020) Formation of the bright faculae in Ceres' Occator crater via long-lived brine effusion in a hydrothermal system. *Nature Communications*, 11, 3680.

Sheppard, S. S., & Trujillo, C. (2015) Discovery and characteristics of the rapidly rotating active asteroid (62412) 2000 SY178 in the Main Belt. *Astronomy Journal*, 149, id. 44.

Sirono, S.-I. (2013) Differentiation of silicates from H2O ice in an icy body induced by ripening. *Earth, Planets and Space*, 65, 1563–1568.

Sizemore, H. G., Schmidt, B. E., Buczkowski, D. A., et al. (2019) A global inventory of ice-related morphological features on dwarf planet Ceres: Implications for the evolution and current state of the cryosphere. *Journal of Geophysical Research*, 124, 1650–1689.

Sleep, N. H., Meibom, A., Fridriksson, Th., Coleman, R. G., & Bird, D. K. (2004) H2-rich fluids from serpentinization: Geochemical and biotic implications. *Proceedings of the National Academy of Sciences (USA)*, 101, 12818–12823.

Stein, N. T., Ehlmann, B. L., Bland, M., Castillo-Rogez, J., & Stevenson, D. (2019) The formation and timing of near-surface Na-carbonate deposits on Ceres. *Europlanet Science Congress*, 13, EPSC-DPS2019–1194-1.

Stein, N. T., Ehlmann, B. L., Palomba, E., et al. (2017) The formation and evolution of bright spots on Ceres, *Icarus*, 320, 188–201.

Tosi, F., Carrozzo, F. G., Raponi, A., et al. (2018) Mineralogy and temperature of crater Haulani on Ceres. *Meteoritics & Planetary Science*, 53, 1902–1924.

Travis, B. J., Bland, P. A., Feldman, W. C., & Sykes, M. (2018) Hydrothermal dynamics in a CM-based model of Ceres. *Meteoritics & Planetary Science*, 53, 2008–2032.

Travis, B. J., & Schubert, G. (2015) Keeping Enceladus warm. *Icarus*, 250, 32–42.

Vance, S. E., Harnmeijer, J., Kimura, J., et al. (2007) Hydrothermal systems in small ocean planets. *Astrobiology*, 7, 987–1005.

Vance, S. E., & Melwani Daswani, M. (2020) Serpentinite and the search for life beyond Earth. *Philosophical Transactions of the Royal Society A*, 378, 20180421.

Vernazza, P., Brož, M., Drouard, A., et al. (2018) The impact crater at the origin of the Julia family detected with VLT/SPHERE? *Astronomy & Astrophysics*, 618, id.A154.

Vernazza, P., Castillo-Rogez, J., Beck, P., et al. (2017) Different origins or different evolutions? Decoding the spectral diversity among C-type asteroids. *The Astronomical Journal*, 153, 72.

Vernazza, P., Jorda, L., Ševeček, P., et al. (2019) A basin-free spherical shape as an outcome of a giant impact on asteroid Hygiea. *Nature Astronomy*, 4, 136–141.

Young, E. D., Ash, R. D., England, P. & Rumble, D. (1999) Fluid flow in chondritic parent bodies: Deciphering the composition of planetesimals. *Science*, 286, 1331–1335.

Young, E. D., Zhang, K. K., & Schubert, G. (2003) Conditions for water pore convection within carbonaceous chondrite parent bodies – Implications for planetesimal size and heat production. *Earth and Planetary Science Letters*, 213, 249–259.

Zambon, F., Raponi, A., Tosi, F., et al. (2017) Spectral analysis of Ahuna Mons mission's visible-infrared spectrometer. *Geophysical Research Letters*, 44, 97–104.

Ziffer, J., Campins, H., Licandro, J., et al. (2011) Near-infrared spectroscopy of primitive asteroid families. *Icarus*, 213, 538–546.

Zimmer, C., Khurana, K. K., & Kivelson, M. G. (2000) Subsurface oceans on Europa and Callisto: Constraints from Galileo magnetometer observations. *Icarus*, 147, 329–347.

Zolensky, M. E., Bourcier, W. L., & Gooding, J. L. (1989) Aqueous alteration on the hydrous asteroids – Results of EQ3/6 computer simulations. *Icarus*, 78, 411–425.

Zolotov, M. Y. (2014) Formation of brucite and cronstedtite-bearing mineral assemblages on Ceres. *Icarus*, 228, 13–26.

Zolotov, M. Y. (2020) The composition and structure of Ceres' interior. *Icarus*, 335, 113404.

Zolotov, M. Y., & Shock, E. L. (2001) Composition and stability of salts on the surface of Europa and their oceanic origin. *Journal of Geophysical Research*, 106, 32815–32827.

12

Geophysics of Vesta and Ceres

ANTON I. ERMAKOV AND CAROL A. RAYMOND

12.1 INTRODUCTION

Asteroid (4) Vesta and dwarf planet (1) Ceres are examples of a population of early-forming protoplanets that coalesced to form the terrestrial planets. Despite their relatively small sizes, their internal structures reflect complex evolutionary paths from their formation to the present day. These two objects reveal divergent geophysical evolutions that likely represent endmembers of this early population of planetesimals. As such, they provide constraints on the chemical gradients in the protoplanetary disk, as well as on the dynamical evolution during planet growth and migration.

The goal of this chapter is to review the improved constraints on the internal structures of Vesta and Ceres provided by the data from the Dawn mission. These constraints can be used to better understand the conditions within the protoplanetary disk specific to the locations and times at which Vesta and Ceres formed, and what internal processes drove their evolution to the observed states. The evolution of the internal structure of these bodies critically depends on: (1) the abundance of volatiles in the interior, which controls the internal dynamics and heat transport; (2) the time of accretion, which determines the amount of short-lived radioisotopes (most importantly ^{26}Al and ^{60}Fe) incorporated into the body, which, in turn, sets the peak interior temperature enabling physical and chemical differentiation; (3) the amount of the long-lived radioisotopes (^{40}K, ^{232}Th, ^{235}U, ^{238}U) that provide a continuous heat source; and (4) surface temperature that sets the surface boundary condition (McCord & Sotin, 2005; Castillo-Rogez & McCord, 2010; Neumann et al., 2014, 2020; Neveu & Desch, 2015). As we will demonstrate, the interplay between these factors determines the divergent geophysical evolutions of Vesta and Ceres.

We study the internal structure by modeling the body's response to various types of forcings and comparing this modeled response to the observed one. The predominant forcing on Vesta and Ceres is their fast rotation, but large impacts have also played a role. The body's hydrostatic response to rotation (see Section 12.2.4) leads to its oblateness. Vesta currently rotates with a period of 5.3 hours (Konopliv et al., 2014) and Ceres rotates with a period of 9.1 hours (Konopliv et al., 2018). For a given rotation rate, a body in hydrostatic equilibrium with a higher density in the center will be less oblate. Dermott (1979) and Tricarico (2014) used a theory of figures for a multi-layer body to show how the hydrostatic shapes of rapidly rotating asteroids can be used to infer their internal structures. Once hydrostatic flattening is accounted for, the amplitude of remaining non-hydrostatic topography is ultimately controlled by material strength (Johnson & McGetchin, 1973; Melosh, 2011), which reflects the resistance of the interiors to deformation on geological time scales. Thus, the observation of non-hydrostatic topography can yield constraints on the mechanical properties of

the interior, such as viscosity, which control the relaxation of non-hydrostatic topography over time.

Vesta is an archetype of a large, mostly intact, differentiated protoplanet (Zuber et al., 2011). Vesta's differentiated state was postulated based on the nearly perfect match between Vesta's telescopic reflectance spectrum and the spectra of the plentiful howardite–eucrite–diogenite (HED) meteorites, first recognized by McCord et al. (1970) and reviewed in Chapter 3. Based on this match, geochemical constraints from the HED meteorites were used to model Vesta's interior structure, which indicated a basaltic crust on top of a peridotite mantle, overlying an iron core (e.g., Ruzicka et al., 1997). The inferred state of differentiation of this 530-km diameter protoplanet points to rapid accretion within 2 My of the condensation of the Calcium–Aluminum Inclusions (CAIs, Neumann et al., 2014), resulting in pervasive melting due to the heat released from the decay of ^{26}Al (Ghosh & McSween, 1998, also see Chapter 4), which may have produced a partial magma ocean (Mandler & Elkins-Tanton, 2013). Vesta's fast rotation makes its shape highly sensitive to its differentiation state, and the early melting episode allowed its shape to relax to hydrostatic equilibrium (Fu et al., 2014). As we will show, Dawn's observations allow us to disentangle Vesta's fossil shape from the effects of its violent geological evolution, which was marked by two large-scale impact events.

Understanding of Ceres' internal structure prior to Dawn was primarily based on telescopic observations of its shape (e.g., Keck II Observatory, Carry et al. (2008); Hubble Space Telescope (HST), Thomas et al., 2005) and on its reflectance spectrum. Ceres' gravitational effect on the orbits of other asteroids and Mars allowed determination of Ceres' mass (e.g., Kovačević & Kuzmanoski, 2007; Konopliv et al., 2011b). This mass estimate combined with the mean radius of about 470 km from the shape models led to a mean density of $\approx 2,000 \text{ kg/m}^3$ – indicating a significant fraction of low-density material consistent with ice. The inferred low density and mineral hydration features in the reflectance spectrum (Lebofsky, 1978; Lebofsky et al., 1981; Rivkin et al., 2006; Milliken & Rivkin, 2009) led to the conclusion that water played a significant role in Ceres' internal structure (McCord & Sotin, 2005).

Being by far the largest body in the asteroid belt, Ceres more closely approaches hydrostatic equilibrium than other major asteroids. Observations prior to the Dawn mission indicated that Ceres' shape is consistent with an oblate ellipsoid (e.g., Thomas et al., 2005). However, unlike with Vesta, in all pre-Dawn determinations of Ceres' shape, the error of the ellipsoidal fits was dominated by the noise in the measurements as opposed to non-hydrostatic effects in the shape. The polar flattening was used to estimate the degree of differentiation of Ceres. However, discrepancies in polar flattening factors amongst the pre-Dawn shape models led to significantly different conclusions about Ceres' interior. Thomas et al. (2005) concluded that Ceres is a differentiated body using the HST observations, whereas Zolotov (2009) concluded Ceres is chemically undifferentiated using the more oblate shape derived with the ground-based adaptive optics observations (Carry et al., 2008). Even though the uncertainty in Ceres' shape was large, a fully differentiated interior with a dense, iron core was excluded by Ceres' shape observations prior to Dawn (Castillo-Rogez & McCord, 2010). For more details on the differentiation of Ceres see Chapter 11.

The Dawn mission performed a comprehensive mapping of the gravity fields and shapes of Vesta and Ceres from a series of polar orbits at different altitudes (see Chapter 2). Images from the Dawn Framing Camera (FC) have been used to construct the shape models of Vesta and Ceres. The shape models were developed by two different techniques: stereophotogrammetry (SPG) and stereophotoclinometry (SPC) (Raymond et al., 2011). In the SPG technique, the shape is reconstructed by correlating images of the same part of the surface viewed from different angles with subsequent bundle adjustment. Image data acquisition was designed to obtain at least three different viewing geometries under similar illumination conditions suitable for stereophotogrammetric analysis (Jaumann et al., 2012). Additional image data were collected to optimize for the SPC technique, which uses surface brightness in addition to the geometric information to reconstruct surface slopes by modeling the surface reflectance function and, also, independently solving for the surface albedo. Typically, a single globally optimized reflectance function is used in SPC.

The SPG models of Vesta and Ceres were reported in Jaumann et al. (2012) and Preusker et al. (2016), respectively. The SPC models were produced by Gaskell (2012) for Vesta and by Park et al. (2019) for Ceres. The systematic difference between the SPG and SPC models for both Vesta and Ceres can be characterized by a long-wavelength, primarily zonal pattern with an amplitude on the order of 100 meters (Ermakov, 2017), which is small enough to not affect the geophysical analyses. Table 12.1 summarizes the available global shape models of Vesta and Ceres.

The gravity models are developed by analyzing the spacecraft radio-tracking data. The radio-tracking data consist of the measurements of the range to the spacecraft in a series of polar orbits at different altitudes, as well as the range-rate (e.g., Konopliv et al., 2011a). The range-rate measurement is primarily used for a gravity field determination, whereas the range measurement is mostly sensitive to the body's ephemeris. The accuracy of the gravity model is set by the radio-tracking data from the lowest altitude circular orbit with complete coverage (Bills & Ermakov, 2019). The science orbits were circular and polar with one exception – Dawn's second extended mission (XM2) at Ceres that had a high eccentricity and low pericenter altitude (35 km). Table 12.2 summarizes gravity models produced from the Dawn data.

The rest of the chapter is structured as follows. In Section 12.2, we describe the methodology of using gravity and shape data to constrain internal structure. In Section 12.3, we discuss the results of Dawn's geophysical investigation of Vesta. In Section 12.4, we describe Dawn's geophysical results at Ceres. The synthesis of the results and comparison of Vesta's and Ceres' internal structures and evolution is provided in Section 12.5. Open questions are summarized in Section 12.6.

Table 12.1 *Summary of global shape models for Vesta and Ceres.*

The resolution of these models with the exception of the Ceres SPG is defined by the lowest altitude circular orbit of the spacecraft, from which the images used in the shape reconstruction were collected. These orbits were named Low-Altitude Mapping Orbits and had mean altitudes of 210 km and 365 km at Vesta and Ceres, respectively. The SPG model for Ceres is based on the High-Altitude Mapping Orbit data (altitude of 1,475 km).

Model	Smallest image pixel footprint size (m)	Model resolution (m)	Reference to the dataset
Vesta SPG	20	70	Jaumann et al. (2012)
Vesta SPC	20	20–50	Gaskell (2012)
Ceres SPG	140	140	Preusker et al. (2016)
Ceres SPC	35	100	Park et al. (2019)

Table 12.2 *Summary of gravity models for Vesta and Ceres produced from the Dawn mission.*

The gravity model resolution is defined as the half-wavelength of the spherical harmonic degree equal to the degree strength (see Section 12.2.1). The degree strength is variable, reflecting the variable effective resolution of the gravity models.

Model	Minimum degree strength	Maximum degree strength	Worst spatial resolution (km)	Best spatial resolution (km)	Reference to the dataset
Vesta, spherical harmonics	18	20	46	41	Konopliv et al. (2014)
Vesta, ellipsoidal harmonics	≈ 15		≈ 54		Park et al. (2014)
Ceres, spherical harmonics	18	59	82	25	Park et al. (2020)

12.2 METHODS

12.2.1 Shape and Gravity Modeling Using Spherical Harmonics

The shape models can be represented in the form of a spherical harmonic expansion:

$$r(\lambda, \phi) = R\sum_{n=0}^{\infty}\sum_{m=0}^{n} \bar{P}_{nm}(sin\phi)[\bar{A}_{nm}cos(m\lambda) + \bar{B}_{nm}sin(m\lambda)],$$
(12.1)

where $\bar{A}_{nm}$ and $\bar{B}_{nm}$ are the normalized shape spherical harmonic coefficients, λ is longitude, ϕ is latitude, n is the spherical harmonic degree, m is order, R is the mean radius, and $\bar{P}_{nm}$ are the normalized associated Legendre functions. The normalized shape spherical harmonic coefficients are related to the unnormalized shape spherical harmonic coefficients as $(\bar{A}_{nm}, \bar{B}_{nm})N_{nm} = (A_{nm}, B_{nm})$. The normalization factor N_{nm} is defined as:

$$N_{nm} = \sqrt{(n-m)!(2-\delta_{0m})(2n+1)/(n+m)!},$$
(12.2)

where δ_{0m} is the Kronecker delta. Gravity models (Konopliv et al., 2014, 2018; Park et al., 2020) are also represented in spherical harmonics:

$$U(r,\lambda,\phi) = \frac{GM}{r}\sum_{n=2}^{\infty}\sum_{m=0}^{n}\left(\frac{R_0}{r}\right)^{n}\bar{P}_{nm}(sin\phi)[\bar{C}_{nm}cos(m\lambda)$$
$$+ \bar{S}_{nm}sin(m\lambda)],$$
(12.3)

where U is the gravitational potential, G is the universal gravitational constant, M is the mass of the body, R_0 is the reference radius, and $\bar{C}_{nm}$ and $\bar{S}_{nm}$ are the normalized spherical harmonic coefficients. The normalization is done in a similar way as for the shape (Eq. 12.2). The gravitational acceleration is given by the gradient of the potential. The zonal coefficients ($m = 0$) are

often written as $\bar{J}_n = -\bar{C}_{n0}$. The summation in Eq. (12.3) starts from $n = 2$ as the $n = 1$ terms are zero since the coordinate origin is placed at the center of mass. The center of the figure is defined as the center of mass for the shape assuming a homogeneous interior. The vector from the center of the figure to the center of mass is defined as the center-of-mass–center-of-figure (COM-COF) offset. The degree-2 gravity, shape, and gravity from the shape coefficients for Vesta and Ceres are provided in Table 12.3.

The spherical harmonic series converges outside of the minimum radius sphere circumscribing the body (a.k.a. the Brillouin sphere). Therefore, mapping gravity anomalies directly on the surface of an irregularly shaped body might lead to unrealistically large anomalies due to the divergence of the spherical harmonic series. As an alternative to the spherical harmonic series, ellipsoidal harmonics can be used that have the convergence region outside of the circumscribing ellipsoid that comes closer to the surface than the circumscribing sphere. The gravity models of Vesta were developed using both spherical harmonic functions (Konopliv et al., 2014) and ellipsoidal harmonic functions (Park et al., 2014). Currently, there is no ellipsoidal harmonic-based model for the gravity of Ceres. Mapping the gravity field of Ceres on the best-fit ellipsoid of the shape leads to unrealistically large anomalies in the polar regions with high gravity resolution dominated by the XM2 data. The higher local degree of expansion causes larger divergence, which is more evident in the polar regions since these regions lie further below the circumscribing sphere. Therefore, gravity anomalies of Ceres in this chapter (Figure 12.5) are mapped on a 482-km radius sphere.

Having both shape and gravity expressed in spherical harmonics allows for conducting spectral analysis of the two datasets. To study how the amplitude of gravity or topography changes with spatial scale, we define the variance spectrum of the gravity (denoted by the superscript gg) and topography (denoted by the superscript tt) and cross-spectrum (denoted by the superscript gt).

Table 12.3 *Normalized second-degree shape, gravity, and gravity-from-shape coefficients of Vesta and Ceres.*

The reference radii for the observed gravity fields and gravity-from-shape of Vesta and Ceres are 265.0 and 470.0 km, respectively. The mean radius of Vesta is 260.359 and 260.267 km for the SPG and SPC models, respectively. The mean radius of Ceres is 469.470 and 469.480 km for the SPG and SPC models, respectively.

Model	$\bar{C}_{20} \times 10^5$ or $\bar{A}_{20} \times 10^5$	$\bar{C}_{21} \times 10^5$ or $\bar{A}_{21} \times 10^5$	$\bar{S}_{21} \times 10^5$ or $\bar{B}_{21} \times 10^5$	$\bar{C}_{22} \times 10^5$ or $\bar{A}_{22} \times 10^5$	$\bar{S}_{22} \times 10^5$ or $\bar{B}_{22} \times 10^5$	Reference to the dataset
Vesta's shape	−6,525.60	164.17	76.05	604.20	271.24	Gaskell (2012)
	−6,533.97	167.96	68.00	622.23	223.21	Jaumann et al. (2012)
Vesta's gravity from shape	−3,519.27	48.66	33.98	422.49	180.14	Gaskell (2012)
	−3,525.39	50.35	31.17	434.87	146.73	Jaumann et al. (2012)
Vesta's gravity	−3,177.94	0.00	0.00	418.50	124.54	Konopliv et al. (2014)
Ceres' shape	−2,336.15	34.77	−38.70	−3.65	117.47	Park et al. (2019)
	−2,332.78	34.11	−39.03	−3.65	118.05	Preusker et al. (2016)
Ceres' gravity from shape	−1,362.28	20.14	−24.14	−2.37	76.22	Park et al. (2019)
	−1,360.15	19.76	−24.32	−2.34	76.60	Preusker et al. (2016)
Ceres' gravity	−1,185.10	0.00	0.00	24.70	−27.44	Park et al. (2020)

These spectra are found using the normalized spherical harmonic coefficients as follows:

$$V_n^{gg} = \sum_{m=0}^{n} \bar{C}_{nm}^2 + \bar{S}_{nm}^2; V_n^{tt} = \sum_{m=0}^{n} \bar{A}_{nm}^2 + \bar{B}_{nm}^2;$$
$$V_n^{gt} = \sum_{m=0}^{n} \bar{C}_{nm}\bar{A}_{nm} + \bar{S}_{nm}\bar{B}_{nm}. \tag{12.4}$$

The admittance Z_n and correlation spectra R_n are found using the variance spectra (Wieczorek, 2015; Ermakov et al., 2017a):

$$Z_n = \frac{GM}{R^3}(n+1)\frac{V_n^{gt}}{V_n^{tt}}; R_n = \frac{V_n^{gt}}{\sqrt{V_n^{gg}V_n^{tt}}}. \tag{12.5}$$

The optional pre-factor $\frac{GM}{R^3}(n+1)$ is used here to find the admittance between radial gravitational acceleration (in units of mGal $= 10^{-5}$ m/s^2) and surface topography (in units of km). Correspondingly, admittance has units of mGal/km. Admittance can be thought of as a transfer function between gravity and topography and can be used to evaluate the mechanism that supports the topography. The correlation is unitless and is used as a measure of the phase relationship between gravity and topography.

The Root-Mean-Square (RMS) spectra of either gravity or topography are found as:

$$M_n^{gg} = \left(\frac{V_n^{gg}}{2n+1}\right)^{\frac{1}{2}}; M_n^{tt} = \left(\frac{V_n^{tt}}{2n+1}\right)^{\frac{1}{2}}. \tag{12.6}$$

The RMS spectrum has the same dimension as the spherical harmonic coefficients. The RMS spectrum computed using the gravity coefficients (from Eq. 12.3) represents the gravity power. The RMS spectrum computed using the errors of the gravity coefficients represents the error spectrum. Figure 12.1 shows the RMS spectra of the gravity signal and error (or noise) for Vesta and Ceres. The global degree strength is defined as the spherical harmonic degree at which the RMS of the signal is equal to the RMS of the error. It follows from Figure 12.1 that the gravity models both for Vesta and Ceres globally have SNR > 1 for $n \lesssim 18$, thus global degree strength is ≈ 18 for Vesta and Ceres gravity models.

12.2.2 Isostasy

Isostasy is defined by the state in which there are no lateral gradients in hydrostatic pressure that could drive lateral flow below a certain depth called the depth of compensation (Watts, 2001; also see Beuthe et al., 2016; Cadek et al., 2019). If a planetary body's topography is in the state of (Airy) isostasy (a.k.a. isostatically compensated), there exists a relationship between the surface topography and topography of the bottom interface of the compensated layer. A floating iceberg is a good example: knowing the height of the top of the iceberg above the water surface, we get that the depth to its bottom part is about $\left|\frac{\rho_{ice}}{\rho_{ice}-\rho_{water}}\right| \approx \frac{1000 \text{ kg/m}^3}{100 \text{ kg/m}^3} \approx 10$ times larger than the surface height. To calculate the amplitude of the bottom interface, we used a viscous flow model (Ermakov et al., 2017a), where viscous relaxation was computed for an arbitrary topographic load employing a two-layer spherical body with an inviscid inner layer and uniform density and viscosity outer layer. Ermakov et al. (2017a) called this model of isostasy "dynamic isostasy": as the two interfaces continue to relax, the ratio of their amplitudes approaches a degree-dependent constant w_n dubbed "shape ratio" in Cadek et al. (2019). Shape ratio is not to be confused with the topographic ratio defined as the ratio of heights with respect to equipotential surfaces (Cadek et al., 2019). For example, in the Cartesian case, w_n is equal to $\rho_1/\Delta\rho$, where $\Delta\rho$ is the density contrast between mantle and crust and ρ_1 is the crustal density. The gravity–topography admittance of a generic isostatic model can be found in the following way:

$$Z_n = \frac{GM}{R^3} \cdot \frac{3(n+1)}{2n+1} \cdot \left(\frac{\rho_1}{\bar{\rho}} - \frac{\Delta\rho}{\bar{\rho}}\left(\frac{R-h}{R}\right)^{n+2} w_n\right), \tag{12.7}$$

where w_n is the absolute value of the ratio of the crust–mantle topography to the outer topography, h is the crustal thickness (assumed to be equal to the depth of compensation), and $\bar{\rho}$ is the mean density. Also, see the appendix of Ermakov et al. (2017a) for more details on computing w_n.

12.2.3 Gravity Anomalies

All gravity anomalies are defined as the difference between the observed gravitational acceleration and gravitational acceleration computed assuming some kind of model. The anomalies are typically computed on a reference surface, such as an ellipsoid that best approximates the equipotential surface. Also, typically only the acceleration component normal to the reference surface is presented (e.g., in Figures 12.3(b), 12.5(b), and 12.5(c)). In this chapter, we adopt the following definitions. For the free-air anomaly, the modeled gravity is that of a body in hydrostatic equilibrium. For the Bouguer anomaly, the modeled gravity is that of the body's observed shape assuming some internal structure (e.g., constant density crust on top of hydrostatic, constant density mantle). For the isostatic anomaly, the modeled gravity is that of a body, whose topography is isostatically compensated. Knowing the surface topography, we can use w_n from Eq. (12.7) to compute the topography of the crust–mantle interface. After doing so, we compute the model gravity of the isostatic body with the observed outer shape and computed isostatic crust–mantle interface. An isostatic anomaly is defined as the difference between the observed gravity and gravity of the isostatic body (Eq. 16 in Ermakov et al., 2017a).

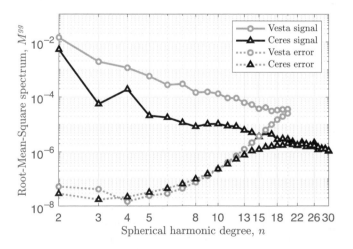

Figure 12.1 Gravity signal and error RMS spectra for Vesta (gray curves, using VESTA20G model, Konopliv et al., 2014) and Ceres (black curves, using CERES70E model, Park et al., 2020). The globally averaged degree strength is defined by the intersection of the gravity signal and gravity error spectra.

Even though the gravity anomalies are often thought of as gravity observations, their computation depends on the model assumptions. The interior density structure needs to be assumed for the free-air and Bouguer anomaly. Besides, a specific mechanism of compensation (e.g., Airy versus Pratt) needs to be assumed for computing the isostatic anomaly.

12.2.4 Interior Structure Modeling

The interior models are constructed by satisfying the observational constraints from the gravity and shape as well as any other a priori information such as compositional constraints from remote sensing data. The basic constraints are the total mass and volume of the body. Hydrostatic equilibrium provides an additional constraint that connects the internal structure to the observed oblateness of the gravity or of the shape (e.g., Dermott, 1979; Tricarico, 2014). Hydrostatic equilibrium is a state in which the force of gravity is balanced by the sum of the pressure gradient and centrifugal force. The gravity oblateness is typically expressed in terms of zonal degree-2 gravity coefficient, or $\bar{J}_2$. The shape oblateness is typically expressed as the polar flattening of the best-fit ellipsoid: $f_p = (a - c)/a$, where a and c are equatorial and polar ellipsoid axes, but can also be expressed through $\bar{J}_2^{shape} = -\bar{A}_{20}$. The hydrostatic equilibrium constraint requires careful thought as the non-hydrostatic effects could be significant both in the shape and gravity, especially for such small, low-surface gravity bodies as Vesta and Ceres.

The simplest internal structure model of a differentiated asteroid is a model with two uniform density layers (Toplis et al., 2013; Ermakov et al., 2014, 2017a). Such an internal structure in hydrostatic equilibrium is fully characterized by five parameters: two radii, two densities, and the rotation period that affects the oblateness of the two interfaces. The volume-equivalent outer radius, rotation period, and total mass are the three fixed quantities as the uncertainties in these parameters are typically small. This leaves only two unknown parameters. A family of solutions for the remaining two unknowns is found by satisfying the observed gravity or shape oblateness (see figure 8 in Ermakov et al., 2017a).

12.3 GEOPHYSICAL RESULTS AT VESTA

12.3.1 Vesta's Shape

Vesta's shape is dominated by its oblateness $(a - c = 56 \text{ km})$ (Jaumann et al., 2012; Chapter 5). The shape reconstruction of Vesta using FC images revealed significant relief with respect to the best-fit ellipsoid with ellipsoidal heights ranging from -22.3 km to $+19.1 \text{ km}$ (Jaumann et al., 2012). Dawn observed not one but two giant impact basins in the southern hemisphere of Vesta: a larger one named Rheasilvia and a smaller and older one named Veneneia. The two large basins make the apparent shape of Vesta more oblate and likely contribute to the COM–COF offset. As a consequence of the two giant impacts, Vesta's global shape significantly deviates from hydrostatic equilibrium (Bills et al., 2014; Ermakov et al., 2014; Konopliv et al., 2014; Park et al., 2014). If the entire shape of Vesta is fitted with a biaxial ellipsoid, the polar flattening factor f_p is even larger than expected for a

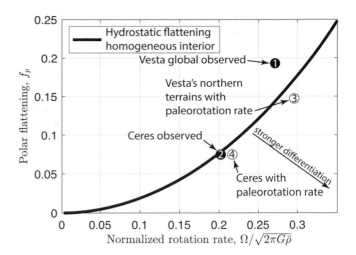

Figure 12.2 Polar flattening factors for Vesta and Ceres. The hydrostatic flattening for a homogeneous body is shown as a solid black curve. Points 1 and 2 denote the present-day global flattening factors at the present-day rotation rates of Vesta and Ceres, respectively. Point 3 denotes the inferred paleorotation rate and flattening for the northern shape of Vesta (see Section 12.3.3). Point 4 denotes the inferred paleo-rotation rate for Ceres (see Section 12.4.3). Note that the rotation rate is normalized in order to enable comparisons between homogeneous bodies of different densities.

homogeneous body in hydrostatic equilibrium (Figure 12.2). In other words, Vesta's global flattening factor corresponds to a much shorter rotation period of a homogeneous body: 4.3 hours compared to the present period of 5.3 hours.

Hydrocode modeling of the two impact basins (Ivanov & Melosh, 2013) showed that the portions of Vesta northwards of the belt of thickened crust surrounding the southern impact basins have not been significantly deformed by impacts and, consequently, are a reliable representation of Vesta's fossil shape prior to the impacts. Ermakov et al. (2014) computed an ellipsoidal fit to the presumed undeformed northern cap of Vesta and found that its flattening factor is significantly less than that of the global shape but still corresponds to a shorter-than-present rotation period of 4.9 hours. Thus, we are presented with a puzzle: from the shape alone, the interior of Vesta appears nearly homogenous or even "anti-differentiated," but the geochemical models based on the HEDs (e.g., Ruzicka et al., 1997) yield a differentiated Vesta. In order to resolve this puzzle, we need to include information from the gravity field data.

12.3.2 Vesta's Gravity

Vesta's gravity model yields a COM-COF offset of 1.4 km (Ermakov et al., 2014; Konopliv et al., 2014). Surprisingly, the offset is the smallest in the z-direction. The COM-COF offset could indicate either that the core is not centered at the center of the figure or it could represent a degree-1 crustal thickness or density variation with thinner (denser) crust in the eastern hemisphere.

The observed $\bar{J}_2$ gravity coefficient of Vesta is lower than the value expected from a homogeneous density distribution. This implies a density concentration toward the center of Vesta (Ermakov et al., 2014). In addition, the gravity-topography admittance decreases with the spherical harmonic degree (see Figure 12.6(b)), and since higher harmonic degrees are sensitive to shallower structure, the decrease of the admittance with degree also indicates an increase of density with depth.

Ermakov et al. (2014) computed the Bouguer gravity anomalies for Vesta for a range of internal structures. The strongest positive anomaly is observed in the highest-standing terrain named Vestalia Terra at $\approx 15°$ E and $30°$ S (Figure 12.3(a)). The central peak of the Rheasilvia basin ($\approx 90°$ E and $75°$ S) is associated with a positive Bouguer gravity anomaly of $+100$ mGal. Negative Bouguer anomalies occur at the older set of impact-induced troughs called Saturnalia Fossae that were created by the Veneneia impact (Scully et al., 2014) and within the ejecta of the giant impact basins. Finally, it is worth noting that the Bouguer gravity anomalies poorly correlate with the geologic or mineralogic units, which indicates that surface composition is likely not representative of the subsurface structure. Potential sources of these anomalies are discussed in Section 12.3.4.

12.3.3 Vesta's Internal Structure Modeling

Due to the non-uniqueness of gravity inversions, it is not possible to identify a single density structure that satisfies the gravity and shape data. Fortunately, the Dawn mission confirmed a high-fidelity match between Vesta's surface composition and the HED meteorites, thus providing a powerful constraint on Vesta's thermochemical evolution. While the Dawn gravity data alone do not require a specific core density or radius (Bills et al., 2014), Russell et al. (2012) used the HED-derived constraints, coupled with the measured $\bar{J}_2$ gravity coefficient, to argue for a high-density iron core at Vesta's center, since both are consistent with, and linked in prior geochemical evolution models based on the HEDs (e.g., Ruzicka et al., 1997).

The integral constraints on the internal structure from Dawn gravity and shape data were cross-checked against the geochemical evolutionary models under the assumption that Vesta's bulk composition started chondritic and was not altered over Vesta's evolution. Using such analysis, Toplis et al. (2013) modeled the thermodynamic evolution of Vesta starting from different bulk chondritic compositions. The solution that best matched the geochemical and geophysical constraints yielded a core mass fraction of 15% for a bulk composition dominated by material affine to volatile-depleted H-chondrites, with a smaller fraction (25%) of CM-chondritic-like material. This core mass fraction and implied crustal thickness are consistent with those implied by Vesta's gravity field (Ermakov et al., 2014); see Chapter 4 for further details. Finally, the iron core interpretation is supported by the observed remanent magnetization of a eucritic meteorite (Allan Hills A81001) that implies an ancient geodynamo sustained in Vesta's once-molten iron core (Fu et al., 2012).

Fu et al. (2014) studied the relaxation of Vesta's topography using finite elements (Bangerth et al., 2007) coupled with a thermal evolution model. They found that the deep interior of Vesta could reach high temperatures corresponding to viscosities $<10^{19}$ Pa s both for olivine-rich silicates and for sulfur-rich metal (Hirth & Kohlstedt, 1996; Ghosh & McSween, 1998; Dobson et al., 2000). Therefore, Vesta's core is expected to have closely approached the state of hydrostatic equilibrium. Since the interior cooled more slowly than the outer parts of Vesta, the core's shape was frozen in later than those of the crust and the mantle.

Tkalcec et al. (2013) found evidence of convection in Vesta's mantle within the first 50 million years after the formation of CAIs from the crystallographic lattice orientation of olivine grains in a diogenitic meteorite. Modeling results indicate that convective shutdown likely occurred within a few My of Vesta's formation for the stagnant lid case (Sterenborg & Crowley, 2013) or, perhaps, even earlier in the case of heat pipe cooling regime dominated by advection of melt to the surface (Moore & Webb, 2013), which could apply to Vesta due to the believed efficiency of melt extraction on similarly sized asteroids (Wilson & Keil, 2012).

The end of the relaxation window is sensitive to the time of the convective shutdown. Fu et al. (2014) concluded that the present-day non-hydrostaticity would have likely quickly relaxed within 40 My after shutdown of mantle convection. If an insulating layer of megaregolith with a 5-km thickness is added on the surface, the relaxation window is prolonged to 200 My after the convective shutdown. Subsequently, the surface of Vesta effectively does not relax and, therefore, could record its equilibrium shape attained at that early epoch at the paleo-rotational period, which could be different from the present-day rotation period.

Ermakov et al. (2014) presented a family of two-layer internal structure models of Vesta by satisfying the following constraints: (1) the observed $\bar{J}_2$ gravity coefficient and (2) the flattening factor of the northern shape undeformed by the giant impacts. The free parameters were the core size, core density, as well as the rotation period at which hydrostatic equilibrium was computed. The core shape was assumed to be hydrostatic for the chosen rotation period. Ermakov et al. (2014) concluded that in order to satisfy the above-mentioned constraints, the rotation period needs to be between 4.83 and 4.93 hours. If the density for the core is assumed to be between 5,000 and 8,000 kg/m^3 (corresponding to Fe-S or Fe-Ni composition, respectively), the core volume-equivalent radius is constrained to be between 110 and 162 km.

The polar moment of inertia of Vesta was measured from the precession of Vesta's spin axis. However, the low accuracy of this measurement $\left(\sigma_{\frac{C}{MR^2}} \approx 0.1 \right)$ did not permit using it to infer the density structure (Alex Konopliv, 2020, personal communication). Further reanalysis of Vesta's landmark data could potentially improve the accuracy of the moment of inertia determination.

12.3.4 Vesta's Crustal Heterogeneity

While Dawn's exploration confirmed Vesta to be differentiated with a core, mantle, and crust, the gravity, topography, and compositional mapping data sets attest to significant variations in crustal properties (see Chapters 3, 4, and 6). Gravity variations can be interpreted as variations in crustal thickness (Figure 12.4(a)), potentially with a vertical density gradient, or lateral variations in density (Figure 12.4(b)). Moreover, the origin of the COM-COF offset adds to the ambiguity as it may or may not be absorbed by offsetting the core. The ambiguity can be mitigated by correlation with geomorphologic and compositional data. It is expected that such crustal thickness and/or density variations would arise from the significant impact processing of Vesta's crust and mantle, but they can also be ascribed to an initially heterogeneous crust/mantle architecture resulting from complex magmatic processes (Raymond et al., 2017).

Vesta's crustal thickness, derived from inversions using Dawn's gravity and shape data (Ermakov et al., 2014) that minimize the residuals to a three-layer model with fixed core and mantle

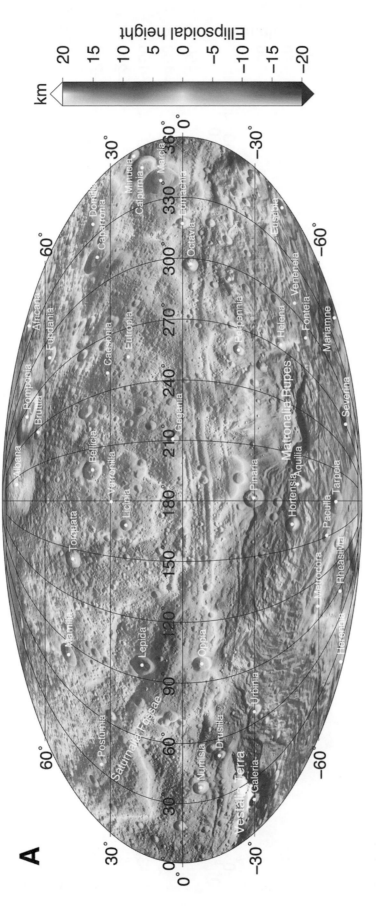

Figure 12.3 (a) Map of Vesta's topography in the Mollweide cartographic projection. (b) Map of Vesta's Bouguer gravity anomaly computed using the VESTA20G gravity model (Konopliv et al., 2014) and SPG shape model (Jaumann et al., 2012). The gravity anomalies are computed on a 293.2 × 266.5 km ellipsoid. The internal structure for computing the Bouguer anomaly was taken from Table 3 in Ermakov et al. (2014). The Bouguer gravity anomaly was expanded up to $n = 16$. Here and elsewhere, longitudes increase eastward from 0° to 360°. A black and white version of this figure will appear in some formats. For the color version, refer to the plate section.

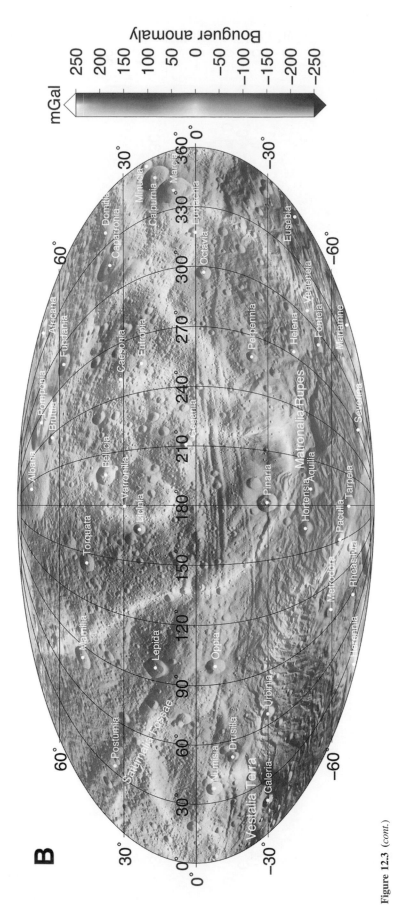

Figure 12.3 *(cont.)*

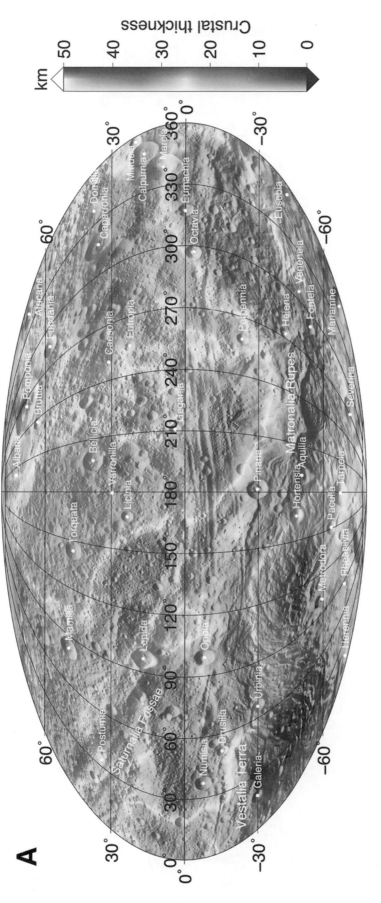
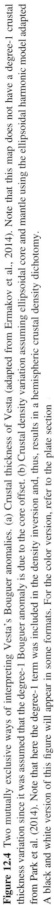

Figure 12.4 Two mutually exclusive ways of interpreting Vesta's Bouguer anomalies. (a) Crustal thickness of Vesta (adapted from Ermakov et al., 2014). Note that this map does not have a degree-1 crustal thickness variation since it was assumed that the degree-1 Bouguer anomaly is due to the core offset. (b) Crustal density variation assuming ellipsoidal core and mantle using the ellipsoidal harmonic model adapted from Park et al. (2014). Note that here the degree-1 term was included in the density inversion and, thus, results in a hemispheric crustal density dichotomy. A black and white version of this figure will appear in some formats. For the color version, refer to the plate section

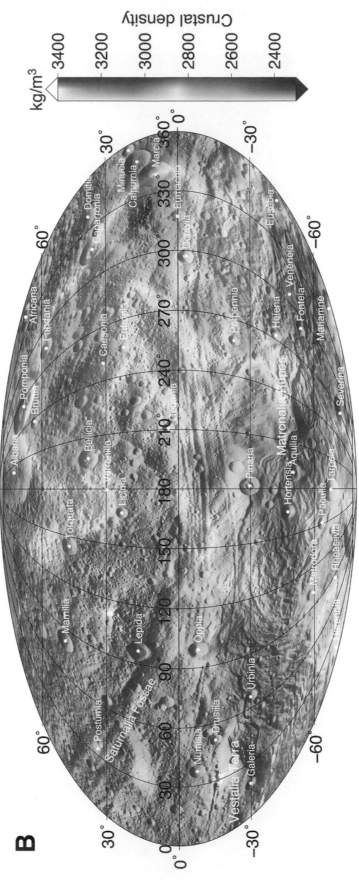

Figure 12.4 (*cont.*)

densities and fixed core size, is shown in Figure 12.4(a). In these models, the crustal density is 2,900 kg/m^3 and the mantle and core densities are 3,200 kg/m^3 and 7,800 kg/m^3, respectively. For such a model, the derived crustal thicknesses range from 0 to >55 km (the lower bound is set to zero). It is not possible to derive a unique map of crustal thickness for a three-layer model given only gravity and shape data. Perturbing the densities from the chosen values affects the absolute values of the crustal thickness, but the overall pattern of variation does not significantly change. Thus, Figure 12.4(a) should be thought of as a map of relative crustal thickness.

Several strong crustal thickness anomalies are seen in the crustal thickness map. The strongest positive Bouguer gravity anomaly in the Vestalia Terra region (see Figure 12.3(b)), results in a local crustal thinning. Vestalia Terra lies in the direction of the COM-COF offset and represents the highest-standing terrain on Vesta. Its high topography appears to predate the Rheasilvia and Veneneia impacts, as those basins carve the edges of Vestalia Terra (see Chapter 5), which supports the hypothesis that the subsurface source of the Vestalia Terra positive anomaly is likely higher-density ultramafic mantle material (Raymond et al., 2013). It is difficult to probe the bedrock in this region because of the mantling by the Rheasilvia ejecta, but several small impact craters indicate a localized presence of diogenitic material (Buczkowski et al., 2014). Thus, Vestalia Terra could be the result of magmatic processes that predate the Rheasilvia and Veneneia basins, or it may be explained as a surviving intact remnant of the original vestan crust and mantle. The largest crustal thicknesses are associated with thick ejecta from the giant impact basins Veneneia and Rheasilvia, and the older Saturnalia Fossae in the northern hemisphere, shown to be a trough system that resulted from the Veneneia impact (Scully et al., 2014). These regions are likely dominated by a highly fractured, porous material that is reflected in the lower overall density (or equivalently larger crustal thickness) of the regions.

The distribution of orthopyroxene and olivine absorption bands and the inferred variations in density (Park et al., 2014) across the vestan surface attest to a complex magmatic history. A broad swath of the vestan surface, extending from the eastern rim of the Rheasilvia basin (near Matronalia Rupes) in the south to high northern latitudes, and from ≈30° W to 150° E longitude, shows enrichment in orthopyroxene-rich diogenite. In a model that assumes that gravity anomalies are due to density variations within the crust (as opposed to variations in crustal thickness), this region also shows higher density material, as illustrated in the density map shown in Figure 12.4(b) (after Park et al., 2014). We note, however, that comparing observed surface mineralogy and gravity suffers from the circumstance that the mineralogical information only applies to a thin, gravitationally negligible layer. Raymond et al. (2017) summarized evidence from gamma ray and neutron data (Prettyman et al., 2013) and IR spectral data (Ammannito et al., 2013) that support the hypothesis that this high-density, diogenite-enriched region represents a concentration of shallow plutons of diogenitic (upper mantle) composition.

Ammannito et al. (2013) reported that olivine is present at high concentration on Vesta's surface only in a few unexpected locations: on crater walls and in ejecta scattered diffusely over a broad area surrounding Bellicia crater in the northern hemisphere, which is at the edge of the region of anomalous density. Ruesch et al. (2014) also identified other possible olivine-enriched areas that are associated with diogenite-like pyroxene. The higher diogenite and olivine concentrations on the surface in this region appear to be associated with a few impacts large enough to distribute the mid-lower-crustal materials onto the surface. These observations constitute strong evidence in favor of a crustal architecture that includes shallow plutons, as predicted by Beck and McSween (2010) and Mandler and Elkins-Tanton (2013).

12.3.5 Vesta's Missing Olivine

A major puzzle arising from Dawn's data regarding Vesta's interior is the lack of evidence for an olivine-rich mantle predicted by the HED-based models (e.g., Ruzicka et al., 1997). Hydrocode impact simulations that assumed an olivine mantle beneath a eucritic crust (Jutzi & Asphaug, 2011; Ivanov & Melosh, 2013; Jutzi et al., 2013; see Chapter 16) predicted excavation of the dense olivine mantle, and a characteristic mascon-like gravity anomaly, positive near the basin's central peak and negative in the basin's floor. However, while Dawn revealed not one, as was inferred earlier by the HST observations (Thomas et al., 1997), but two overlapping giant impact basins, Rheasilvia and Veneneia, located near the south pole, which increased the probability of mantle excavation, the expected olivine exposures were not detected (Ammannito et al., 2013; McSween et al., 2013). The crustal thickness inversion showed that the thinnest crust is located within the floors of the two basins, with a modest crustal thinning beneath the central peak of the Rheasilvia basin, and no mantle uplift was found in association with the Veneneia basin (Ermakov et al., 2014). The lack of substantial high-density material beneath the central peak is consistent with the lack of clear olivine detections on the floors of Rheasilvia and Veneneia basins (McSween et al., 2013), although Li and Milliken (2015) find evidence for small olivine-rich areas around Rheasilvia's central mound. Instead, the floors of these basins appear to be dominated by diogenite and howardite (Ammannito et al., 2013; Prettyman et al., 2013).

The lack of surface olivine could indicate a thicker crust than assumed in the hydrocode simulations. Clenet et al. (2014) conducted a smooth particle hydrodynamics (SPH) modeling and favored a thick (over 80 km) eucritic crust intruded by diogenitic plutons based on the inferred excavation depth in the Rheasilvia and Veneneia basin. However, the existence of a thick crust and large core does not leave enough space within Vesta to accommodate all of the olivine needed to explain the enrichment patterns in the HEDs based on the cosmochemical abundances (Righter & Drake, 1997; Ruzicka et al., 1997; Toplis et al., 2013). To resolve this inconsistency, Consolmagno et al. (2015) suggested either that Vesta formed with a non-chondritic bulk composition or that it underwent a global physical alteration, potentially due to large-scale collisions, that affected its bulk composition and interior structure.

However, non-chondritic bulk composition might not be required if olivine was sequestered in the deep mantle via the two-stage crystallization model (Mandler & Elkins-Tanton, 2013), which is supported by Dawn's compositional and gravity data, as discussed by Raymond et al. (2017). The paucity of olivine detections could indicate that magmatism on Vesta was more complex than the differentiation sequence inferred from the HEDs, and that equilibrium crystallization did not proceed to completion (see discussion in Chapter 4). Instead, the observations are consistent with the hypothesis that Vesta's early melting resulted in a partial magma ocean that crystallized olivine in the deepest mantle, which transitioned to a regime in which residual melts were delivered to the crust and upper mantle, where the melts underwent fractional crystallization (Mandler & Elkins-Tanton, 2013). In addition, such fractional crystallization can account for trace element variations in HEDs (e.g., Barrat et al., 2010) that have been invoked to challenge the notion of a single parent body (e.g., Scott et al., 2009).

12.4 GEOPHYSICAL RESULTS AT CERES

12.4.1 Ceres' Shape

The Dawn images allowed accurate measurement of the polar flattening of Ceres and an estimate of the magnitude of its non-hydrostatic topography (see Chapter 10). It can be seen in Figure 12.2 that Ceres' polar flattening corresponds to a nearly homogeneous internal structure (also see Table 12.3). If fit with a biaxial ellipsoid, the difference between the polar and equatorial axes is 36.2 km. The topographic range is much smaller, with ellipsoidal heights ranging from –7.3 km to 9.5 km (Park et al., 2019; Figure 12.5(a)). The overall smoothness of Ceres' topography compared to that of Vesta is also manifested by a lower topographic RMS spectrum across all observed wavelengths (Ermakov et al., 2017a).

The magnitude of Ceres' non-hydrostaticity can be estimated by fitting Ceres' shape with a triaxial ellipsoid. The fit reveals that Ceres possesses a significant equatorial flattening of approximately 2 km (Ermakov et al., 2017a). Hanami Planum – a region of high-standing topography between 180° E and 270° E – represents the major contribution to the triaxiality of Ceres. Therefore, we note that even though Ceres appears to be more hydrostatic than Vesta, the magnitude of non-hydrostatic effects on Ceres is significant. In practice, non-hydrostaticity limits the applicability of the hydrostatic equilibrium constraint and, therefore, presents a major geophysical source of uncertainty in interpreting the gravity and shape data.

12.4.2 Ceres' Gravity

The first gravity models of Ceres (Park et al., 2016) indicated that Ceres' topography could be isostatically compensated based on relatively low power of gravity anomalies. The major non-hydrostatic term in Ceres' gravity is the sectoral (i.e., $n = m$) degree-2 term (Table 12.3), which is negatively correlated with the sectoral degree-2 term in the shape. The admittance for the degree-2 sectoral term is, therefore, also negative, ≈ -36 mGal/km. Negative admittances are rarely observed. Early gravity and shape models of Titan showed a negative admittance at degree-3. Hemingway et al. (2013) proposed an explanation for Titan's negative degree-3 admittance by freezing and melting at the base of the shell that creates the bottom loading of the shell. However, this was later disproved by newer gravity and shape models (Durante et al., 2019). Beyond degree-2, the observed low gravity-topography admittances for Ceres are well explained by an isostatic model (Ermakov et al., 2017a). This is also manifested by the fact that that isostatic anomalies (Figure 12.5(c)) are a factor ≈ 2 lower than the Bouguer anomalies (Figure 12.5(b)).

The Dawn XM2 mission at Ceres provided high-resolution gravity data in the narrow strip along the migration path of the spacecraft's pericenter. The pericenter of XM2 migrated southward from just north of the Occator crater. The pericenter passed through the center of Occator and continued south passing through Urvara crater and then into the opposite (Eastern) hemisphere. The degree strength in the swath around the XM2 pericenter reaches degree 59, whereas elsewhere the gravity field at Ceres is accurate up to degree 18 (Park et al., 2020).

12.4.3 Ceres' Internal Structure Modeling

There are three types of constraints on the bulk internal structure of Ceres. The first type of constraint comes from the hydrostatic flattening of Ceres' shape or gravity. The second type comes from the gravity–topography admittance (Ermakov et al., 2017a) under the assumption that Ceres' topography is isostatically compensated. Lastly, the third type of constraint comes from modeling the relaxation of Ceres' topography (Fu et al., 2017). While the first two constraints are primarily sensitive to the body's density structure, the third constraint is primarily sensitive to the viscosity structure and relatively insensitive to the density structure.

Park et al. (2016) derived a degree-6 gravity model from Dawn's High-Altitude Mapping Orbit (HAMO) data and inferred a partially differentiated interior of Ceres with a volatile-rich shell overlying a rocky core. This conclusion was primarily based on the high value of the normalized moment of inertia ($C/MR^2 \approx 0.37$). However, this value was not measured directly, but, rather, was inferred from the gravity and shape oblateness assuming hydrostatic equilibrium through the Radau-Darwin relationship (Park et al., 2016; also see page 257 in Zharkov & Trubitsyn, 1978). Alternatively, Ermakov et al. (2017a) computed hydrostatic equilibrium for two-layer models constrained by either the observed gravity or shape oblateness. The moment of inertia varies in the fourth digit for a two-layer model given a value of either $\bar{J}_2^{shape}$ or $\bar{J}_2^{gravity}$. The inferred normalized polar moment of inertia is equal to 0.37 from $\bar{J}_2^{gravity}$ and 0.39 from the $\bar{J}_2^{shape}$ for the present-day rotation period. Notably, the solutions for gravity and shape disagree. A lower moment of inertia ($C/MR^2 = 0.35$), implying a stronger differentiation, is inferred for the proposed paleo-rotation period (8.46 hours), at which the values of C/MR^2 inferred from gravity and shape coincide (Mao & McKinnon, 2018b).

The notion of partial differentiation is strengthened by the Gamma Ray and Neutron Detector (GRaND) data. GRaND observed abundant hydrogen on Ceres' surface consistent with a water ice-rich regolith. The non-icy portion of the regolith is compositionally similar to aqueously altered chondrites but is iron-depleted, supporting partial differentiation in the form of ice-rock fractionation (Prettyman et al., 2017).

The long-wavelength free-air, Bouguer, and isostatic anomalies are dominated by the degree-2 sectoral signal negatively correlated with topography. Ermakov et al. (2017a) hypothesized that a mantle plume could produce a bottom loading of a rigid and thick lithosphere, which would result in the observed negative correlation. The degree-2 isostatic gravity anomaly pattern has one of its two minima centered at Hanami Planum, a region of high-standing topography. Gravity inversion of the Hanami Planum region by Raymond et al. (2020) showed that the isostatic gravity anomaly could be accounted for if there is a large region of less dense material in the mid-to-upper mantle. Higher mantle temperatures enabled by

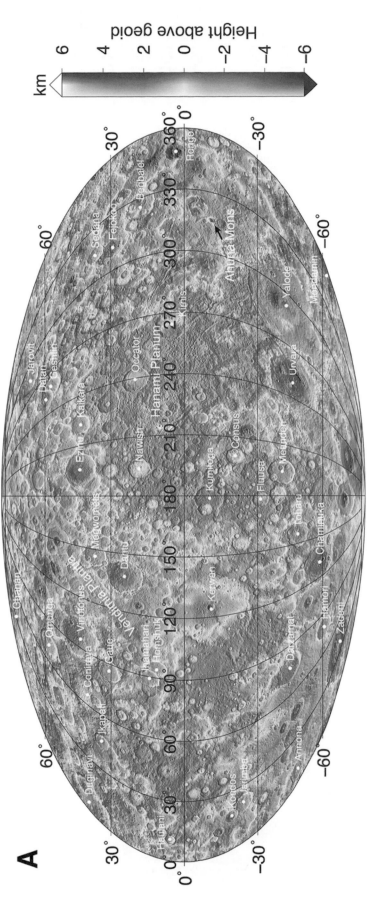

Figure 12.5 Maps of Ceres' (a) topography, (b) Bouguer anomaly, and (c) isostatic anomaly. The CERES70E gravity model (Park et al., 2020) and SPC shape model (Park et al., 2019) were used. The dynamic model of isostasy based on the viscous flow was used for computing the isostatic gravity anomaly for the parameters given in Table 4 of Ermakov et al. (2017a). The gravity anomalies were expanded from degree-3 up to the local degree strength (Park et al., 2020) to prevent overwhelming the maps by the sectoral degree-2 anomaly. A black and white version of this figure will appear in some formats. For the color version, refer to the plate section

185

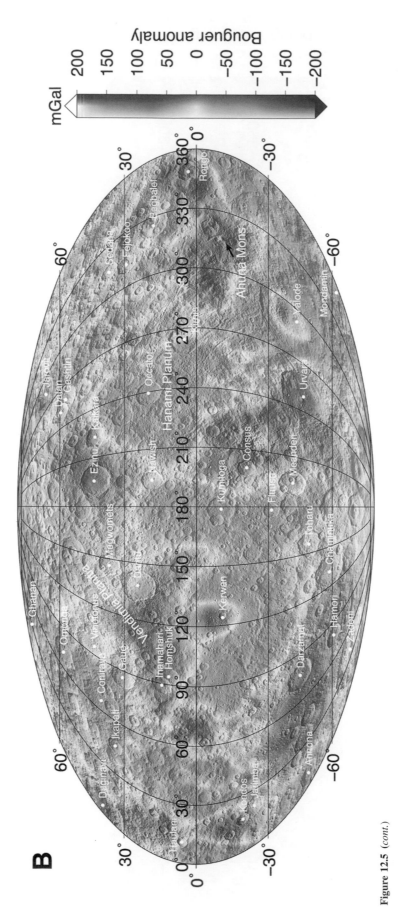

Bouguer anomaly

mGal

200 150 100 50 0 -50 -100 -150 -200

Figure 12.5 (*cont.*)

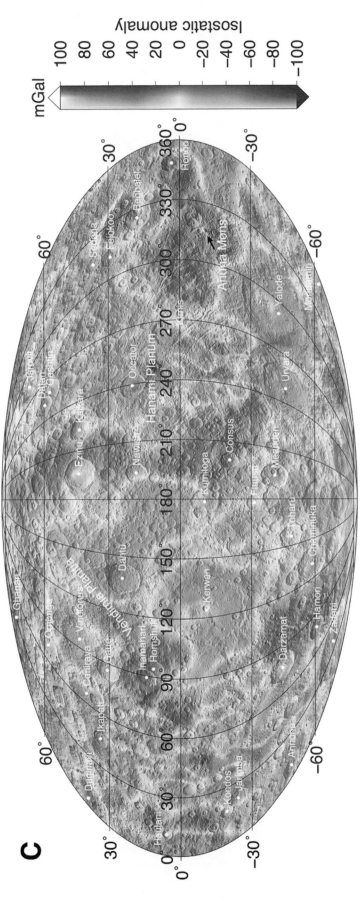

Figure 12.5 *(cont.)*

187

thicker Hanami crust could lead to a higher degree of brine–silicate differentiation. Thus, a higher regional concentration of low-density brines could be responsible for the observed negative gravity anomaly. Alternatively, the negative gravity–topography correlation and the Hanami negative gravity anomaly could be explained by a non-hydrostatic shape of the core. Fu et al. (2017) showed that the viscous relaxation models of Ceres' topography are not sensitive to depths over 100 km, thus leaving the possibility for a strong, rigid, and possibly non-hydrostatic core. King et al. (2018) showed that the presence of a high-density core, potentially due to rock dehydration, cannot be excluded based on the Dawn gravity data.

Ermakov et al. (2017a) used both the observed gravity oblateness and the gravity–topography admittance to constrain the parameters of the two-layer model of Ceres. Ermakov et al. (2017a) assumed that the gravity $\bar{J}_2$ is within 3% of its hydrostatic value. The 3% range for $\bar{J}_2$ comes from the ratio of the total amplitude of the degree-2 non-hydrostatic coefficients ($\sqrt{\bar{C}_{22}^2 + \bar{S}_{22}^2}$) to $\bar{J}_2$ (see Table 12.3). Thus, by varying $\bar{J}_2$ within 3% of its observed value, we allow the same level of non-hydrostaticity in $\bar{J}_2$ as in the sectoral terms. Sampling the family of solutions for a two-layer model defined by $\bar{J}_2$, Ermakov et al. (2017a) sought to minimize the admittance misfit by using an isostatic model (Figure 12.6(a)). The best fit two-layer isostatic model was found to have a crustal density of 1287^{+70}_{-87} kg/m^3 and a mean crustal thickness of $41.0^{+3.2}_{-4.7}$ km based on the dynamic model of isostasy (see the appendix of Ermakov et al. (2017a) and methods in Park et al. (2020)). The error bars correspond to the best-fit solutions found by varying the hydrostatic $\bar{J}_2$ value up and down by 3%. The mantle density was constrained to be $2{,}434^{+5}_{-8}$ kg/m^3. These values were later confirmed by the XM2 observations (Park et al., 2020). The low density of the mantle can be explained either by its being in a hydrated state (Fu et al., 2017) or by high porosity within a dehydrated mantle (Zolotov, 2020). A hydrated mantle would imply that the interior did not heat up to over 600 °C (Fu et al., 2017); such a thermal evolution is predicted based on thermochemical modeling (Neveu & Desch, 2015; Castillo-Rogez et al., 2018).

Bland (2013) argued that if Ceres was dominated by water ice, craters as small as 4 km in diameter should be significantly relaxed in the equatorial regions. In addition, Bland (2013) predicted that there should be a latitudinal gradient of the state of crater relaxation. Ceres' low obliquity (Bills & Scott, 2017; Ermakov et al., 2017b; Vaillant et al., 2019) leads to a systematic difference in temperature between the equator and the poles. Since viscous relaxation is strongly temperature-dependent, equatorial craters should be more relaxed than the polar craters. On the other hand, if Ceres' crust is dominated by rocky material, Bland (2013) predicted that crater relaxation should be negligible across the surface. Later, Bland et al. (2016) observed the lack of crater relaxation using the crater depths measured from the Dawn-derived shape models. This indicated that the rheology of Ceres' shell is not dominated by water ice.

Ermakov et al. (2017a) found that the RMS spectrum of Ceres topography deviates from a power law defined by the power at high degrees. This downward deviation of the topographic power spectrum was observed at wavelengths greater than 246 km. It was further shown that such deviation is not consistent with an elastic flexure signal. We note that there are only two impact basins, Kerwan and Yalode, of comparable scale, and one of

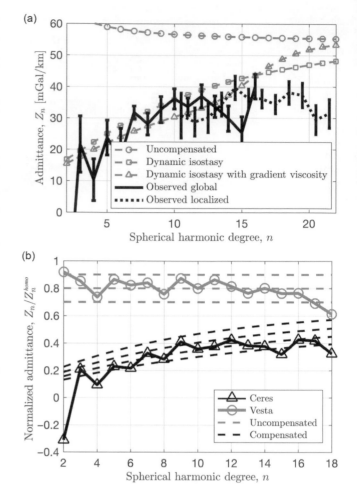

Figure 12.6 (a) Observed (black) and modeled (gray) gravity-topography admittances of Ceres, including an estimate of admittance localized within the XM2 high degree strength region (Park et al., 2020). (b) Gravity–topography admittance of Vesta and Ceres normalized to that of a homogeneous body. Normalization is needed in order to compare one body to another. Two sets of model admittances are shown for various crustal densities normalized to the mean body density. The first set of normalized modeled admittances (gray horizontal dashed lines at the top) is shown for uncompensated topography with crustal densities equal to 70%, 80%, and 90% of the body's mean density. The second set of normalized modeled admittances (four black dashed curves) is shown for isostatically compensated topography. The crustal densities were chosen to range from 1,100 (bottom line) to 1,400 kg/m^3 (top line). The crustal and mantle thicknesses were chosen to fall on the family of two-layer model solutions defined by the observed value of Ceres' $\bar{J}_2$. It can be seen that Vesta's observed admittance spectrum is more consistent with uncompensated model admittances, whereas Ceres' observed admittance is more consistent with the isostatically compensated admittances. $\bar{J}_2$ and $\bar{J}_4$ in gravity and shape were zeroed prior to computing admittances to capture only the non-hydrostatic signal.

them – Kerwan – appears to be morphologically relaxed (Hiesinger et al., 2016). Having only two basins large enough to appreciably relax makes it harder to draw conclusions about topographic relaxation by studying crater depths because of low number statistics.

Vesta and Ceres have experienced essentially the same impact flux (O'Brien & Sykes, 2011), which is assumed to be the primary topography building mechanism. Fu et al. (2017) compared the topography RMS spectra for Vesta and Ceres and suggested that the subdued long-wavelength topography of Ceres is due to viscous relaxation. This is corroborated by the observation of latitudinally varying topographic power (Ermakov et al., 2017a). The equatorial regions appear to have more subdued long-wavelength topography compared to the polar regions. Fu et al. (2017) modeled the relaxation of Ceres topography and, similarly to Bland et al. (2016), found that the observed topographic spectrum is not consistent with a crust dominated by water ice. Instead, a stronger, higher viscosity crust is required to explain the observed power spectrum of Ceres' topography. Fu et al. (2017) found that the viscosity has to be on the order of 10^{26} Pa s at the surface and should decrease by a factor of 10 every 10–15 km of depth to explain the observed topographic spectrum. The upper bound of water ice content was put at 35 vol%. In order to simultaneously satisfy the low crustal density required by admittance (Ermakov et al., 2017a) and high crustal viscosity (Fu et al., 2017), hydrated salts and gas (clathrate) hydrates were suggested to make up a significant portion of Ceres crust. Finally, Hiesinger et al. (2016) observed that the simple to complex crater transition follows the trend defined by bodies with icy crusts, indicating the presence of a weak, ice-like phase affecting the crater formation.

Presently Ceres rotates with a period of 9.07 hours (Chamberlain et al., 2007; Konopliv et al., 2018). The current rotation rate (Figure 12.2) is not consistent with the observed gravity and shape flattening. As shown by Mao and McKinnon (2018b), the two-layer model solutions inferred from gravity flattening and from the shape flattening become identical for the rotation period of 8.46 hours. The normalized moment of inertia for such internal structure is lower ($C/MR^2 \approx 0.35$) than that inferred from either the gravity or the shape implying stronger differentiation, which in the framework of a two-layer model leads to a higher density mantle and lower density crust. An obvious candidate to cause such despinning is a giant impact. However, it is hard to explain a 7.2% slowing of Ceres' rotation via a single giant impact since Ceres lacks evidence of large impact structures, except for a putative 750-km basin inferred mostly from a topographic depression (Marchi et al., 2016). Mao and McKinnon (2018a) argued that the cumulative effect of numerous impacts could decrease the rotation rate of Ceres through the angular momentum drain effect (Dobrovolskis & Burns, 1984).

Alternative models of Ceres interior exist. Zolotov (2020) proposed that Ceres' shape is consistent with an organic-rich interior and that abundant water or clathrates are not required to explain the low density of the body (also see Chapter 11). Zolotov (2020) pointed out that Ceres' surface chemistry is consistent with a rock–organic mixture, in which rock has the composition of CI chondrites and organic matter is similar to chondritic insoluble organic matter (IOM, Marchi et al., 2019). Assuming that a surface-like composition extends to the interior, it was suggested that high concentrations of gas hydrates, as was inferred by Fu et al. (2017) and Castillo-Rogez et al. (2018), are not needed to satisfy rheologic constraints from Fu et al. (2017). Instead, that study suggested that Ceres has a low degree of chemical differentiation and the increase of density with depth could be explained by a porosity gradient.

12.4.4 Ceres' Crustal Heterogeneity

Local analysis of the isostatic anomalies (Figures 12.5(c) and 12.7) yields heterogeneities within Ceres' interior, that is, deviations from isostatic equilibrium. In this subsection, we will go through several notable cerean features associated with gravity anomalies.

The high-resolution XM2 gravity data resolved regional structure in the vicinity of Occator (Figure 12.7(a)), which enables testing hypotheses of the origin of surface deposits. Neveu and Desch (2015) and Quick et al. (2019) hypothesized that ongoing freezing of subsurface liquid water or briny reservoirs could lead to their over-pressurization, driving cryovolcanic outflow to the surface through preexisting cracks, for example, in large craters. These outflows could form the observed bright deposits or faculae, which would be composed of aqueously altered minerals left after the water ice sublimated away. The gap between the age of the faculae and the Occator crater (Nathues et al., 2020) corroborated by the thermal modeling of impact heat dissipation (Hesse & Castillo-Rogez, 2019) indicates that the faculae cannot be sourced entirely from the impact melt chamber. In addition, the observation of hydrohalite (De Sanctis et al., 2020; Chapter 7), which is expected to dehydrate in less than 100 years, in Cerealia Facula indicates that the delivery of brines to the surface is recent and possibly ongoing. Raymond et al. (2020) found evidence in the XM2 gravity data (Figure 12.7(a)) for an extensive brine reservoir south-east of Occator, and heterogeneity in the deep mantle beneath Hanami Planum. Raymond et al. (2020) proposed that the endogenic brine reservoir was mobilized by the impact heating and fracturing, which led to a long-lived extrusion of brines and faculae formation. The Occator impact could have created a temporary hydrological system that represents an analogue for icy moon impact craters, with implications for the creation of transient habitable niches. Whether the brine enrichment below Occator is a local phenomenon or there is a global brine layer remains open to future investigation.

Ahuna Mons is a lone pyramid-shaped mountain hypothesized to have formed via cryovolcanic extrusion (Ruesch et al., 2016). It is associated with a strong positive Bouguer and isostatic anomaly, implying a dense, possibly salt-rich source. However, since the feature is only ≈ 20 km wide and ≈ 4 km tall, it is not possible to distinguish to what extent the mons itself is causing the anomaly or the observed anomaly is due to a more extended subsurface source. Ruesch et al. (2019) interpreted the anomaly source to be due to a regional mantle uplift caused by a plume.

Nar Sulcus is a system of extensional fractures in the western part of the Yalode basin (Figure 12.7(c); Hughson et al., 2019). The XM2 gravity data showed a negative gravity anomaly centered around the western part of the fractures. Park et al. (2020) hypothesized that the observed negative gravity anomaly could be explained by fracture-induced porosity, or by an increased abundance of light phases such as water ice.

Ermakov et al. (2017a) found evidence for lunar-like mass concentrations (or mascons) in the two biggest craters: Kerwan (280 km in diameter, Figure 12.7(b)) and Yalode (260 km in diameter, Figure 12.7(c)). Crustal inversion using gravity and shape data shows a superisostatic uplift (i.e., crust–mantle interface perturbation in excess of the prediction from an isostatic model) in the center of Kerwan. A similar structure is observed in Yalode. However, the Yalode anomaly pattern is less symmetric with respect to the crater center. The third crater by size, Urvara (160 km in diameter), does not have a superisostatic uplift. Instead, it appears to be only partially compensated. Thus, we observed the transition from supercompensated craters or mascons

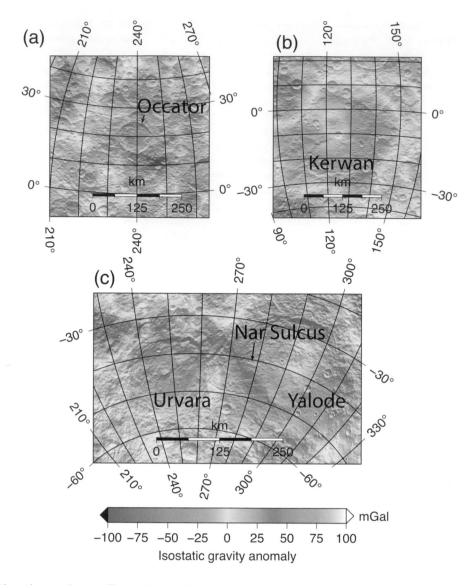

Figure 12.7 Regional isostatic anomaly maps. The gravity anomalies were expanded from degree-3 up to the local degree strength. (a) Hanami Planum and the Occator crater. A negative isostatic anomaly is seen southeast of Occator. (b) Kerwan basin. A bullseye is seen in the basin with a positive anomaly in the center and a negative annulus in the exterior. (c) Urvara and Yalode basins. Similarly to Kerwan, a bullseye pattern is seen in Yalode, whereas smaller Urvara is dominated by a negative isostatic anomaly.

A black and white version of this figure will appear in some formats. For the color version, refer to the plate section

(Kerwan and Yalode) to partially compensated craters (Urvara), and to uncompensated craters such as smaller Dantu and Zadeni.

Mascons were first observed on the Moon (Muller & Sjorgen, 1968). They have subsequently been observed on Mars (Neumann et al., 2004) and Mercury (Smith et al., 2012; Konopliv et al., 2020). Large lunar impact basins filled with volcanic deposits are associated with strong positive free-air gravity anomalies. Mantle uplift is superisostatic for a mascon basin, that is, the isostatic anomaly is positive at the basin's center (Neumann et al., 1996; Melosh et al., 2013; Freed et al., 2014). Unlike on the Moon, Ceres' mascons have a bullseye pattern (positive in the center, surrounded by a negative annulus) only in the isostatic anomaly map. The free-air anomaly is negative for Kerwan and Yalode – that is the mantle uplift, unlike on the Moon, creates a smaller signature compared to that of the crater's topographic depression. Therefore, Ceres' mascons appear to be light versions of lunar mascons. Melosh et al. (2013) proposed

a self-consistent mascon formation mechanism using the mascon patterns mapped by the Gravity Recovery And Interior Laboratory (GRAIL) mission (Zuber et al., 2013, 2016). The mascons were proposed to result from impact basin excavation and collapse of the impact cavity, followed by isostatic adjustment, cooling, and contraction of a voluminous melt pool. The central uplift required to produce the mascon pattern critically depends on the crustal thermal gradient. Hence, a similar investigation of cerean mascons could help constrain the range of thermal structures of the crust capable of yielding the observed mascon patterns.

Park et al. (2020) used the XM2 gravity data to derive localized estimates of the gravity–topography admittance. The localized admittance was observed to gradually shift to a lower crustal density solution for degrees between 15 and 21 (black dashed curve in Figure 12.6(b)). A greater level of isostatic compensation could produce lower admittance values under the same density and rheological structures. However, due to the longer isostatic adjustment timescale at higher degrees, the level of compensation is likely to decrease. Therefore, a change in the level of compensation is unlikely to explain the lower observed admittances. Park et al. (2020) suggested that the observed low admittances at high degrees could be explained by a vertical density gradient within the crust that is consistent with decreasing porosity with depth and/or increasing content of dense phases, such as rock and salts. Park et al. (2020) further suggested, as the cerean ocean was freezing, it incorporated impurities – rock particles and salts – and became progressively denser and muddier, which would explain the density gradient in Ceres' crust.

12.5 SYNTHESIS OF INTERNAL STRUCTURE EVOLUTION

Synthesizing the Dawn geophysical observations at Vesta and Ceres, we find that, in their current states, Vesta has a differentiated internal structure (Russell et al., 2012; Ermakov et al., 2014) and its topography is uncompensated, whereas Ceres is partially differentiated with compensated topography (Park et al., 2016, 2020; Ermakov et al., 2017a; King et al., 2018) as evidenced by the respective admittance spectra (Figure 12.6(b)). This leads to comparative evolutionary scenarios, illustrated schematically in Figure 12.8. Vesta and Ceres accreted with different volatile contents (see Chapter 14) in some primordial shapes. Vesta likely accreted within 2 My from the formation of CAIs (Neumann et al., 2014). Ceres accreted later than Vesta: its accretion time critically depends on the assumed location of accretion, ranging from 2 My to

7 My after CAIs, if it accreted in the asteroid belt or between Jupiter and Neptune (Neumann et al., 2020) and likely up to 10 My if it accreted in the Kuiper belt (e.g., Nesvorny et al., 2019; Morbidelli & Nesvorny, 2019). Ceres incorporated abundant volatile material during its accretion and subsequently experienced substantial aqueous alteration (see Chapter 7). While Vesta went through a magma ocean state in its early history (Righter & Drake, 1997; Formisano et al., 2013; Mandler & Elkins-Tanton, 2013; also see Chapter 4), efficient heat transfer due to hydrothermal circulation preserved a relatively cool interior of Ceres (Castillo-Rogez et al., 2018; Travis et al., 2018). Consequently, Ceres underwent only ice–rock fractionation, resulting in an aqueously altered surface composition and partially differentiated interior (see Chapter 11). Later, Vesta experienced giant impacts, responsible for creating the Rheasilvia and Veneneia basins (see Chapter 5). These basins formed when Vesta was effectively too rigid to relax and, thus, they presently dominate Vesta's non-hydrostatic topography. Giant impact events could have happened at Ceres as well, but a weaker crust resulted in no record of such impacts, with the potential exception of Vendimia Planitia – a 750-km diameter, roughly circular low-lying region (Marchi et al., 2016; also see Chapter 10). Ceres' topography is defined by a balance of viscous relaxation and ongoing impact cratering. Lower viscosities at depth allowed isostatic compensation of Ceres' long-wavelength topography. On the other hand, uncompensated topography is in agreement with the observed admittance of Vesta (Figure 12.6(b)). The high inferred abundance of water ice, hydrated salts, and/or clathrate phases in Ceres suggests the past presence of globally significant regions of solute-rich fluids, that froze from the surface inward (e.g., Castillo-Rogez et al., 2018; also see Chapter 11), leading to the vertical density gradient inferred from the Dawn XM2 gravity data (Park et al., 2020).

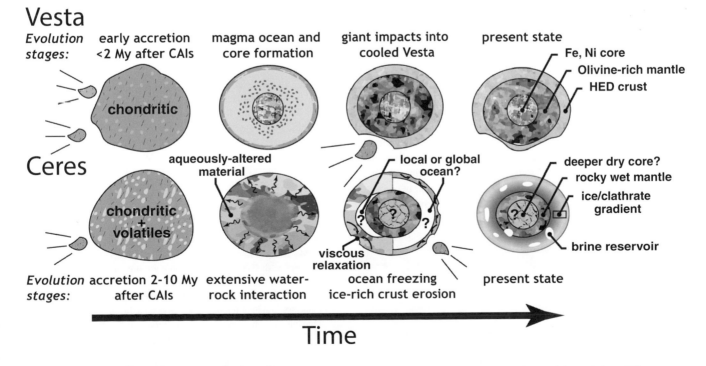

Figure 12.8 A comparison of Vesta and Ceres evolutionary scenarios. The drawing is not to scale, and individual phases of the evolution of Vesta are not meant to timewise correspond to the phases of evolution of Ceres.
A black and white version of this figure will appear in some formats. For the color version, refer to the plate section

12.6 OPEN QUESTIONS

12.6.1 Moment of Inertia of Vesta and Ceres

Further reanalysis of Vesta's landmark data could potentially improve the accuracy of the moment of inertia determination serving as an additional constraint on its differentiation state. The difference between the inferred normalized polar moments of inertia for the present and paleo-rotation for Ceres is on the order of 0.04. A direct measurement of C/MR^2 could be obtained from observing the precession of the Ceres' rotation axis and would provide an extra constraint on the differentiation state of Ceres without having to assume hydrostatic equilibrium.

12.6.2 Ceres' Despinning

Mao and McKinnon (2018b) proposed a way to reconcile the gravity and shape of Ceres with hydrostatic equilibrium by assuming that the rotation period changed instantaneously while the body was perfectly rigid and, therefore, did not relax by viscous flow to the hydrostatic shape dictated by the new rotation rate. However, as the body relaxes to a new hydrostatic shape (resulting in a vertical offset of the Ceres points plotted in Figure 12.2), it is expected that its polar moment of inertia changes, which leads to a change in the rotation rate, which feeds back in the hydrostatic response and the relaxation viscous flow. A study of Ceres' despinning coupled with ongoing relaxation is needed to provide an updated constraint on the paleo-rotation rate of Ceres, its moment of inertia, and differentiation state. Finally, the effect of solar tides Ceres' rotation needs to be explored.

12.6.3 Ceres' True Polar Wander

Tricarico (2018) found three lines of evidence for true polar wander (TPW) of Ceres. First, Tricarico (2018) found the location of the highest standing great circle on Ceres and argued that this region is akin to Iapetus' equatorial ridge (e.g., Castillo-Rogez et al., 2007) and, therefore, represents the cerean paleo-equator. Second, this study argued that the location and orientation of the observed extensional faults (Scully et al., 2017) match the pattern expected from the reorientation from the paleo-pole to the present pole position. Third, a general gravity inversion technique (Tricarico, 2013) was used to derive continuous density distributions consistent with the observed gravity field, from which a higher density crust in the equatorial regions was inferred, which presumably drove reorientation. However, Keane and Ermakov (2019) provided counterevidence to each of the three arguments by Tricarico (2018) and concluded that there is no evidence for TPW of Ceres from the Dawn data.

12.6.4 Ceres' Mascons

How do mascons form on Ceres? Bland et al. (2018) proposed that devolatilization of the central part of the Kerwan and Yalode craters leads to the observed positive isostatic anomaly in their centers. However, devolatilization does not explain the negative annulus in the outer part of the basins. Modeling mascon formation on Ceres with a two-step approach, in which the basin formation is modeled with a hydrocode simulation and subsequent isostatic adjustment, with cooling and thermal contraction of the basin modeled using finite element methods as was used by Melosh et al. (2013) for lunar basins, could constrain the thermal gradient at the time of basin formation. Such a constraint on the thermal gradient at the times and locations of the Kerwan and Yalode impacts would be important for understanding the thermal evolution of Ceres' crust with implications for potential long-term survivability of fluids in Ceres' interior.

12.6.5 Validation of Organic-Rich Models of Ceres

Ceres' organic content has several implications for geophysical modeling. Zolotov (2020) claimed that the viscosity of warm rock–organic mixtures at depth could account for the observed relaxation of long-wavelength topography. In addition, Zolotov (2020) required high crustal porosity to satisfy the low crustal density constraint. However, it is not clear that such porosity is consistent with the structure and composition of the crust derived from the entirety of the Dawn data. In addition, the low thermal conductivity of the upper porous layer of organic-rich materials could lead to elevated temperatures in the interior and, therefore, viscous closure of the required porous space (Castillo-Rogez & McCord, 2010; Lichtenberg et al., 2016), thus violating the crustal density constraint. Hence, a self-consistency check of the interior composition proposed by Zolotov (2020) via a coupled thermal and geodynamical evolution model is required. Finally, it remains to be assessed if the organic-rich composition (see Chapter 8) provides a viable alternative explanation for the surface observations from the Dawn mission, in particular geomorphological features that have been interpreted as water–ice related (see Sizemore et al., 2019 for review).

REFERENCES

Ammannito, E., De Sanctis, M. C., Palomba, E., et al. (2013) Olivine in an unexpected location on Vesta's surface. *Nature*, 504, 122–125.

Bangerth, W., Hartmann, R., & Kanschat, G. (2007) Deal. II – a general-purpose object-oriented finite element library. *ACM Transactions on Mathematical Software (TOMS)*, 33, 24-es.

Barrat, J. A., Yamaguchi, A., Zanda, B., Bollinger, C., & Bohn, M. (2010) Relative chronology of crust formation on asteroid Vesta: Insights from the geochemistry of diogenites. *Geochimica et Cosmochimica Acta*, 74, 6218–6231.

Beck, A. W., & McSween Jr, H. Y. (2010) Diogenites as polymict breccias composed of orthopyroxenite and harzburgite. *Meteoritics & Planetary Science*, 45, 850–872.

Beuthe, M., Rivoldini, A., & Trinh, A. (2016) Enceladus's and Dione's floating ice shells supported by minimum stress isostasy. *Geophysical Research Letters*, 43, 10088.

Bills, B. G., & Ermakov, A. I. (2019) Simple models of error spectra for planetary gravitational potentials as obtained from a variety of measurement configurations. *Planetary and Space Science*, 179, 104744.

Bills, B. G., & Scott, B. R. (2017) Secular obliquity variations of Ceres and Pallas. *Icarus*, 284, 59–69.

Bills, B. G., Asmar, S. W., Konopliv, A. S., Park, R. S., & Raymond, C. A. (2014) Harmonic and statistical analyses of the gravity and topography of Vesta. *Icarus*, 240, 161–173.

Bland, M. T. (2013) Predicted crater morphologies on Ceres: Probing internal structure and evolution. *Icarus*, 226, 510–521.

Bland, M. T., Ermakov, A. I., Raymond, C. A., et al. (2018) Morphological indicators of a mascon beneath Ceres's largest crater, Kerwan. *Geophysical Research Letters*, 45, 1297–1304.

Bland, M. T., Raymond, C. A., Schenk, P. M., et al. (2016) Composition and structure of the shallow subsurface of Ceres revealed by crater morphology. *Nature Geoscience*, 9, 538–542.

Buczkowski, D. L., Wyrick, D. Y., Toplis, M., et al. (2014) The unique geomorphology and physical properties of the Vestalia Terra plateau. *Icarus*, 244, 89–103.

Čadek, O., Souček, O., & Běhounková, M. (2019) Is Airy isostasy applicable to icy moons? *Geophysical Research Letters*, 46, 14299–14306.

Carry, B., Dumas, C., Fulchignoni, M., et al. (2008) Near-infrared mapping and physical properties of the dwarf-planet Ceres. *Astronomy & Astrophysics*, 478, 235–244.

Castillo-Rogez, J. C., Matson, D. L., Sotin, C., et al. (2007) Iapetus' geophysics: Rotation rate, shape, and equatorial ridge. *Icarus*, 190, 179–202.

Castillo-Rogez, J. C., & McCord, T. B. (2010) Ceres' evolution and present state constrained by shape data. *Icarus*, 205, 443–459.

Castillo-Rogez, J., Neveu, M., McSween, H. Y., et al. (2018) Insights into Ceres's evolution from surface composition. *Meteoritics & Planetary Science*, 53, 1820–1843.

Chamberlain, M. A., Sykes, M. V., & Esquerdo, G. A. (2007) Ceres light-curve analysis – Period determination. *Icarus*, 188, 451–456.

Clenet, H., Jutzi, M., Barrat, J. A., et al. (2014) A deep crust–mantle boundary in the asteroid 4 Vesta. *Nature*, 511, 303–306.

Consolmagno, G. J., Golabek, G. J., Turrini, D., et al. (2015) Is Vesta an intact and pristine protoplanet? *Icarus*, 254, 190–201.

De Sanctis, M. C., Ammannito, E., Raponi, A., et al. (2020) Fresh emplacement of hydrated sodium chloride on Ceres from ascending salty fluids. *Nature Astronomy*, 4, 786–793.

Dermott, S. F. (1979) Shapes and gravitational moments of satellites and asteroids. *Icarus*, 37, 575–586.

Dobrovolskis, A. R., & Burns, J. A. (1984) Angular momentum drain: A mechanism for despinning asteroids. *Icarus*, 57, 464–476.

Dobson, D. P., Crichton, W. A., Vocadlo, L., et al. (2000) In situ measurement of viscosity of liquids in the Fe–FeS system at high pressures and temperatures. *American Mineralogist*, 85, 1838–1842.

Durante, D., Hemingway, D. J., Racioppa, P., Iess, L., & Stevenson, D. J. (2019) Titan's gravity field and interior structure after Cassini. *Icarus*, 326, 123–132.

Ermakov, A. I. (2017) Geophysical Investigation of Vesta, Ceres and the Moon Using Gravity and Topography Data. Doctoral dissertation, Massachusetts Institute of Technology.

Ermakov, A. I., Fu, R. R., Castillo-Rogez, J. C., et al. (2017a) Constraints on Ceres' internal structure and evolution from its shape and gravity measured by the Dawn spacecraft. *Journal of Geophysical Research: Planets*, 122, 2267–2293.

Ermakov, A. I., Mazarico, E., Schröder, S. E., et al. (2017b) Ceres's obliquity history and its implications for the permanently shadowed regions. *Geophysical Research Letters*, 44, 2652–2661.

Ermakov, A. I., Zuber, M. T., Smith, D. E., et al. (2014) Constraints on Vesta's interior structure using gravity and shape models from the Dawn mission. *Icarus*, 240, 146–160.

Formisano, M., Federico, C., Turrini, D., et al. (2013) The heating history of Vesta and the onset of differentiation. *Meteoritics & Planetary Science*, 48, 2316–2332.

Freed, A. M., Johnson, B. C., Blair, D. M., et al. (2014). The formation of lunar mascon basins from impact to contemporary form. *Journal of Geophysical Research: Planets*, 119, 2378–2397.

Fu, R. R., Ermakov, A. I., Marchi, S., et al. (2017) The interior structure of Ceres as revealed by surface topography. *Earth and Planetary Science Letters*, 476, 153–164.

Fu, R. R., Hager, B. H., Ermakov, A. I., & Zuber, M. T. (2014) Efficient early global relaxation of asteroid Vesta. *Icarus*, 240, 133–145.

Fu, R. R., Weiss, B. P., Shuster, D. L., et al. (2012) An ancient core dynamo in asteroid Vesta. *Science*, 338, 238–241.

Gaskell, R. W. (2012) SPC shape and topography of Vesta from DAWN imaging data. *AAS/Division for Planetary Sciences Meeting Abstracts*, # 44 (Vol. 44), October, Reno, NV.

Ghosh, A., & McSween Jr, H. Y. (1998) A thermal model for the differentiation of asteroid 4 Vesta, based on radiogenic heating. *Icarus*, 134, 187–206.

Hemingway, D., Nimmo, F., Zebker, H., & Iess, L. (2013) A rigid and weathered ice shell on Titan. *Nature*, 500, 550–552.

Hesse, M. A., & Castillo-Rogez, J. C. (2019) Thermal evolution of the impact-induced cryomagma chamber beneath Occator crater on Ceres. *Geophysical Research Letters*, 46, 1213–1221.

Hiesinger, H., Marchi, S., Schmedemann, N., et al. (2016) Cratering on Ceres: Implications for its crust and evolution. *Science*, 353, aaf4759.

Hirth, G., & Kohlstedt, D. L. (1996) Water in the oceanic upper mantle: Implications for rheology, melt extraction and the evolution of the lithosphere. *Earth and Planetary Science Letters*, 144, 93–108.

Hughson, K. H., Russell, C. T., Schmidt, B. E., et al. (2019) Normal faults on Ceres: Insights into the mechanical properties and thermal history of Nar Sulcus. *Geophysical Research Letters*, 46, 80–88.

Ivanov, B. A., & Melosh, H. J. (2013) Two-dimensional numerical modeling of the Rheasilvia impact formation. *Journal of Geophysical Research: Planets*, 118, 1545–1557.

Jaumann, R., Williams, D. A., Buczkowski, D. L., et al. (2012) Vesta's shape and morphology. *Science*, 336, 687–690.

Johnson, T. V., & McGetchin, T. R. (1973) Topography on satellite surfaces and the shape of asteroids. *Icarus*, 18, 612–620.

Jutzi, M., & Asphaug, E. (2011) Mega-ejecta on asteroid Vesta. *Geophysical Research Letters*, 38, 1–5.

Jutzi, M., Asphaug, E., Gillet, P., Barrat, J. A., & Benz, W. (2013) The structure of the asteroid 4 Vesta as revealed by models of planet-scale collisions. *Nature*, 494, 207–210.

Keane, J. T., & Ermakov, A. I. (2019) No evidence for true polar wander of Ceres. *Nature Geoscience*, 12, 972–974.

King, S. D., Castillo-Rogez, J. C., Toplis, M. J., et al. (2018) Ceres internal structure from geophysical constraints. *Meteoritics & Planetary Science*, 53, 1999–2007.

Konopliv, A. S., Asmar, S. W., Bills, B. G., et al. (2011a) The Dawn gravity investigation at Vesta and Ceres. *Space Science Reviews*, 163, 461–486.

Konopliv, A. S., Asmar, S. W., Folkner, W. M., et al. (2011b) Mars high resolution gravity fields from MRO, Mars seasonal gravity, and other dynamical parameters. *Icarus*, 211, 401–428.

Konopliv, A. S., Asmar, S. W., Park, R. S., et al. (2014) The Vesta gravity field, spin pole and rotation period, landmark positions, and ephemeris from the Dawn tracking and optical data. *Icarus*, 240, 103–117.

Konopliv, A. S., Park, R. S., & Ermakov, A. I. (2020) The Mercury gravity field, orientation, love number, and ephemeris from the MESSENGER radiometric tracking data. *Icarus*, 335, 113386.

Konopliv, A. S., Park, R. S., Vaughan, A. T., et al. (2018) The Ceres gravity field, spin pole, rotation period and orbit from the Dawn radiometric tracking and optical data. *Icarus*, 299, 411–429.

Kovačević, A., & Kuzmanoski, M. (2007) A new determination of the mass of (1) Ceres. *Earth, Moon, and Planets*, 100, 117–123.

Lebofsky, L. A. (1978) Asteroid 1 Ceres: Evidence for water of hydration. *Monthly Notices of the Royal Astronomical Society*, 182, 17P–21P.

Lebofsky, L. A., Feierberg, M. A., Tokunaga, A. T., Larson, H. P., & Johnson, J. R. (1981) The 1.7-to 4.2-μm spectrum of asteroid 1 Ceres: Evidence for structural water in clay minerals. *Icarus*, 48, 453–459.

Li, S., & Milliken, R. E. (2015) Estimating the modal mineralogy of eucrite and diogenite meteorites using visible–near infrared reflectance spectroscopy. *Meteoritics & Planetary Science*, 50, 1821–1850.

Lichtenberg, T., Golabek, G. J., Gerya, T. V., & Meyer, M. R. (2016). The effects of short-lived radionuclides and porosity on the early thermo-mechanical evolution of planetesimals. *Icarus*, 274, 350–365.

Mandler, B. E., & Elkins-Tanton, L. T. (2013) The origin of eucrites, diogenites, and olivine diogenites: Magma ocean crystallization and shallow magma chamber processes on Vesta. *Meteoritics & Planetary Science*, 48, 2333–2349.

Mao, X., & McKinnon, W. B. (2018a) Effect of impacts on Ceres' spin evolution. *Lunar and Planetary Science Conference* (Vol. 49). The Woodlands, TX.

Mao, X., & McKinnon, W. B. (2018b) Faster paleospin and deep-seated uncompensated mass as possible explanations for Ceres' present-day shape and gravity. *Icarus*, 299, 430–442.

Marchi, S., Ermakov, A. I., Raymond, C. A., et al. (2016) The missing large impact craters on Ceres. *Nature Communications*, 7, 12257.

Marchi, S., Raponi, A., Prettyman, T. H., et al. (2019) An aqueously altered carbon-rich Ceres. *Nature Astronomy*, 3, 140–145.

McCord, T. B., Adams, J. B., & Johnson, T. V. (1970) Asteroid Vesta: Spectral reflectivity and compositional implications. *Science*, 168, 1445–1447.

McCord, T. B., & Sotin, C. (2005) Ceres: Evolution and current state. *Journal of Geophysical Research: Planets*, 110, 1–14.

McSween Jr, H. Y., Binzel, R. P., De Sanctis, M. C., et al. (2013) Dawn; the Vesta–HED connection; and the geologic context for eucrites, diogenites, and howardites. *Meteoritics & Planetary Science*, 48, 2090–2104.

Melosh, H. J. (2011) *Planetary Surface Processes*. Cambridge: Cambridge University Press.

Melosh, H. J., Freed, A. M., Johnson, B. C., et al. (2013) The origin of lunar mascon basins. *Science*, 340, 1552–1555.

Milliken, R. E., & Rivkin, A. S. (2009) Brucite and carbonate assemblages from altered olivine-rich materials on Ceres. *Nature Geoscience*, 2, 258–261.

Moore, W. B., & Webb, A. A. G. (2013) Heat-pipe earth. *Nature*, 501, 501–505.

Morbidelli, A., & Nesvorny, D. (2019) Kuiper belt: formation and evolution. In D. Prialnik, M. A. Barucci, & L. Young (eds.), *The Trans-Neptunian Solar System*. Amsterdam: Elsevier, p. 25.

Muller, P. M., & Sjogren, W. L. (1968) Mascons: Lunar mass concentrations. *Science*, 161, 680–684.

Nathues, A., Schmedemann, N., Thangjam, G., et al. (2020) Recent cryovolcanic activity at Occator crater on Ceres. *Nature Astronomy*, 4, 794–801.

Nesvorný, D., Li, R., Youdin, A. N., Simon, J. B., & Grundy, W. M. (2019) Trans-Neptunian binaries as evidence for planetesimal formation by the streaming instability. *Nature Astronomy*, 3, 808–812.

Neumann, G. A., Zuber, M. T., Smith, D. E., & Lemoine, F. G. (1996) The lunar crust: Global structure and signature of major basins. *Journal of Geophysical Research: Planets*, 101, 16841–16863.

Neumann, G. A., Zuber, M. T., Wieczorek, M. A., et al. (2004) Crustal structure of Mars from gravity and topography. *Journal of Geophysical Research: Planets*, 109, 1–18.

Neumann, W., Breuer, D., & Spohn, T. (2014) Differentiation of Vesta: Implications for a shallow magma ocean. *Earth and Planetary Science Letters*, 395, 267–280.

Neumann, W., Jaumann, R., Castillo-Rogez, J., Raymond, C. A., & Russell, C. T. (2020) Ceres' partial differentiation: Undifferentiated crust mixing with a water-rich mantle. *Astronomy & Astrophysics*, 633, A117.

Neveu, M., & Desch, S. J. (2015) Geochemistry, thermal evolution, and cryovolcanism on Ceres with a muddy ice mantle. *Geophysical Research Letters*, 42, 10–197.

O'Brien, D. P., & Sykes, M. V. (2011) The origin and evolution of the asteroid belt – Implications for Vesta and Ceres. *Space Science Reviews*, 163, 41–61.

Park, R. S., Konopliv, A. S., Asmar, S. W., et al. (2014) Gravity field expansion in ellipsoidal harmonic and polyhedral internal representations applied to Vesta. *Icarus*, 240, 118–132.

Park, R. S., Konopliv, A. S., Bills, B. G., et al. (2016) A partially differentiated interior for (1) Ceres deduced from its gravity field and shape. *Nature*, 537, 515–517.

Park, R. S., Konopliv, A. S., Ermakov, A. I., et al. (2020) Evidence of non-uniform crust of Ceres from Dawn's high-resolution gravity data. *Nature Astronomy*, 4, 748–755.

Park, R. S., Vaughan, A. T., Konopliv, A. S., et al. (2019) High-resolution shape model of Ceres from stereophotoclinometry using Dawn Imaging Data. *Icarus*, 319, 812–827.

Prettyman, T. H., Mittlefehldt, D. W., Yamashita, N., et al. (2013) Neutron absorption constraints on the composition of 4 Vesta. *Meteoritics & Planetary Science*, 48, 2211–2236.

Prettyman, T. H., Yamashita, N., Toplis, M. J., et al. (2017) Extensive water ice within Ceres' aqueously altered regolith: Evidence from nuclear spectroscopy. *Science*, 355, 55–59.

Preusker, F., Scholten, F., Matz, K. D., et al. (2016). Dawn at Ceres – Shape model and rotational state. *Lunar and Planetary Science Conference*, March, The Woodlands, TX, Vol. 47, Abstract 1954.

Quick, L. C., Buczkowski, D. L., Ruesch, O., et al. (2019) A possible brine reservoir beneath Occator crater: Thermal and compositional evolution and formation of the Cerealia Dome and Vinalia Faculae. *Icarus*, 320, 119–135.

Raymond, C. A., Ermakov, A. I., Castillo-Rogez, J. C., et al. (2020) Impact-driven mobilization of deep crustal brines on dwarf planet Ceres. *Nature Astronomy*, 4, 741–747.

Raymond, C. A., Jaumann, R., Nathues, A., et al. (2011) The Dawn topography investigation. In C. T. Russell, & C. A. Raymond (eds.), *The Dawn Mission to Minor Planets 4 Vesta and 1 Ceres*. New York: Springer, pp. 487–510.

Raymond, C. A., Park, R. S., Asmar, S. W., et al. (2013) Vestalia Terra: An ancient mascon in the southern hemisphere of Vesta. *Lunar and Planetary Science Conference*, March, The Woodlands, TX, Vol. 44, Abstract 2882.

Raymond, C. A., Russell, C. T., & McSween, H. Y. (2017) Dawn at Vesta: Paradigms and paradoxes. In L. T. Elkins-Tanton, & B. P. Weiss (eds.), *Planetesimals: Early Differentiation and Consequences for Planets*. Cambridge: Cambridge University Press, pp. 321–339.

Righter, K., & Drake, M. J. (1997) A magma ocean on Vesta: Core formation and petrogenesis of eucrites and diogenites. *Meteoritics & Planetary Science*, 32, 929–944.

Rivkin, A. S., Volquardsen, E. L., & Clark, B. E. (2006) The surface composition of Ceres: Discovery of carbonates and iron-rich clays. *Icarus*, 185, 563–567.

Ruesch, O., Genova, A., Neumann, W., et al. (2019) Slurry extrusion on Ceres from a convective mud-bearing mantle. *Nature Geoscience*, 12, 505–509.

Ruesch, O., Hiesinger, H., De Sanctis, M. C., et al. (2014) Detections and geologic context of local enrichments in olivine on Vesta with VIR/Dawn data. *Journal of Geophysical Research: Planets*, 119, 2078–2108.

Ruesch, O., Platz, T., Schenk, P., et al. (2016) Cryovolcanism on Ceres. *Science*, 353, aaf4286.

Russell, C. T., Raymond, C. A., Coradini, A., et al. (2012) Dawn at Vesta: Testing the protoplanetary paradigm. *Science*, 336, 684–686.

Ruzicka, A., Snyder, G. A., & Taylor, L. A. (1997) Vesta as the howardite, eucrite and diogenite parent body: Implications for the size of a core and for large-scale differentiation. *Meteoritics & Planetary Science*, 32, 825–840.

Scott, E. R., Greenwood, R. C., Franchi, I. A., & Sanders, I. S. (2009) Oxygen isotopic constraints on the origin and parent bodies of eucrites, diogenites, and howardites. *Geochimica et Cosmochimica Acta*, 73, 5835–5853.

Scully, J. E., Buczkowski, D. L., Schmedemann, N., et al. (2017) Evidence for the interior evolution of Ceres from geologic analysis of fractures. *Geophysical Research Letters*, 44, 9564–9572.

Scully, J. E., Yin, A., Russell, C. T., et al. (2014) Geomorphology and structural geology of Saturnalia Fossae and adjacent structures in the northern hemisphere of Vesta. *Icarus*, 244, 23–40.

Sizemore, H. G., Schmidt, B. E., Buczkowski, D. A., et al. (2019) A global inventory of ice-related morphological features on dwarf planet Ceres: Implications for the evolution and current state of the cryosphere. *Journal of Geophysical Research: Planets*, 124, 1650–1689.

Smith, D. E., Zuber, M. T., Phillips, R. J., et al. (2012) Gravity field and internal structure of Mercury from MESSENGER. *Science*, 336, 214–217.

Sterenborg, M. G., & Crowley, J. W. (2013) Thermal evolution of early Solar System planetesimals and the possibility of sustained dynamos. *Physics of the Earth and Planetary Interiors*, 214, 53–73.

Thomas, P. C., Binzel, R. P., Gaffey, M. J., et al. (1997) Impact excavation on asteroid 4 Vesta: Hubble space telescope results. *Science*, 277, 1492–1495.

Thomas, P. C., Parker, J. W., McFadden, L. A., et al. (2005) Differentiation of the asteroid Ceres as revealed by its shape. *Nature*, 437, 224–226.

Tkalcec, B. J., Golabek, G. J., & Brenker, F. E. (2013) Solid-state plastic deformation in the dynamic interior of a differentiated asteroid. *Nature Geoscience*, 6, 93–97.

Toplis, M. J., Mizzon, H., Monnereau, M., et al. (2013) Chondritic models of 4 Vesta: Implications for geochemical and geophysical properties. *Meteoritics & Planetary Science*, 48, 2300–2315.

Travis, B. J., Bland, P. A., Feldman, W. C., & Sykes, M. V. (2018) Hydrothermal dynamics in a CM-based model of Ceres. *Meteoritics & Planetary Science*, 53, 2008–2032.

Tricarico, P. (2013) Global gravity inversion of bodies with arbitrary shape. *Geophysical Journal International*, 195, 260–275.

Tricarico, P. (2014) Multi-layer hydrostatic equilibrium of planets and synchronous moons: Theory and application to Ceres and to Solar System moons. *The Astrophysical Journal*, 782, 99.

Tricarico, P. (2018) True polar wander of Ceres due to heterogeneous crustal density. *Nature Geoscience*, 11, 819–824.

Vaillant, T., Laskar, J., Rambaux, N., & Gastineau, M. (2019) Long-term orbital and rotational motions of Ceres and Vesta. *Astronomy & Astrophysics*, 622, A95.

Watts, A. B. (2001) *Isostasy and Flexure of the Lithosphere*. Cambridge: Cambridge University Press.

Wieczorek, M. A. (2015) Gravity and topography of the terrestrial planets. *Treatise on Geophysics*, 10, 165–206.

Wilson, L., & Keil, K. (2012) Volcanic activity on differentiated asteroids: A review and analysis. *Geochemistry*, 72, 289–321.

Zharkov, V. N., & Trubitsyn, V. P. (1978) *Physics of Planetary Interiors*. Astronomy and Astrophysics Series. Tucson, AZ: Pachart Pub House.

Zolotov, M. Y. (2009) On the composition and differentiation of Ceres. *Icarus*, 204, 183–193.

Zolotov, M. Y. (2020) The composition and structure of Ceres' interior. *Icarus*, 335, 113404.

Zuber, M. T., McSween, H. Y., Binzel, R. P., et al. (2011) Origin, internal structure and evolution of 4 Vesta. *Space Science Reviews*, 163, 77–93.

Zuber, M. T., Smith, D. E., Neumann, G. A., et al. (2016) Gravity field of the Orientale basin from the Gravity Recovery and Interior Laboratory Mission. *Science*, 354, 438–441.

Zuber, M. T., Smith, D. E., Watkins, M. M., et al. (2013) Gravity field of the Moon from the Gravity Recovery and Interior Laboratory (GRAIL) mission. *Science*, 339, 668–671.

Part III

IMPLICATIONS FOR THE FORMATION AND EVOLUTION OF THE SOLAR SYSTEM

Formation of Main Belt Asteroids

HUBERT KLAHR, MARCO DELBO, AND KONSTANTIN GERBIG

13.1 INTRODUCTION

Asteroids are remnants of the Solar System's formation process, and as such they are key to understanding the formation of planetesimals. Once asteroid collisional evolution is considered, the present-day size–frequency distribution of Main Belt asteroids (MBAs) is best explained if initial planetesimals formed with an initial size of ~100 km, and not through bottom-up growth (see Section 13.3.1). This is in line with the gravitational collapse scenario, where a cloud of centimeter to meter sized pebbles (or larger fluffy grains with the same aerodynamic properties) subject to tidal shear and turbulent diffusion contracts under its own self-gravity to form a set of planetesimals. The resulting theoretical size prediction depends on the strength of diffusion as well as on pebble size. At the asteroid Main Belt's location, using numerically obtained values for the streaming instability, this leads to the correct order of magnitude for the equivalent size (~100 km) of the gravitational unstable pebble cloud. The availability of a sufficient pebble supply suggests that initial planetesimals that would later make up the asteroid Main Belt formed rather rapidly (<1 Myr

The work of M.D. was partially supported by the ANR ORIGINS (ANR-18-CE31–0014). The research of H.K. and K.G. has been supported by the Deutsche Forschungsgemeinschaft Schwerpunktprogramm (DFG SPP) 1385 "The first ten million years of the Solar System" under contract KL 1469/4-(1-3) "Gravoturbulente Planetesimal Entstehung im frühen Sonnensystem" and by (DFG SPP) 1833 "Building a Habitable Earth" under contract KL 1469/13-1 "Der Ursprung des Baumaterials der Erde: Woher stammen die Planetesimale und die Pebbles? Numerische Modellierung der Akkretionsphase der Erde." This research was supported by the Munich Institute for Astro- and Particle Physics (MIAPP) of the DFG cluster of excellence "Origin and Structure of the Universe and in part at KITP Santa Barbara by the National Science Foundation under Grant No. NSF PHY11–25915. H.K. also acknowledges additional support from the DFG via the Heidelberg Cluster of Excellence STRUCTURES in the framework of Germany's Excellence Strategy (grant EXC-2181/1 – 390900948). The authors gratefully acknowledge the Gauss Centre for Supercomputing (GCS) for providing computing time for a GCS Large-Scale Project (additional time through the John von Neumann Institute for Computing (NIC)) on the GCS share of the supercomputer JUQUEEN at Jülich Supercomputing Centre (JSC). GCS is the alliance of the three national supercomputing centers HLRS (Universität Stuttgart), JSC (Forschungszentrum Jülich), and LRZ (Bayerische Akademie der Wissenschaften), funded by the German Federal Ministry of Education and Research (BMBF) and the German State Ministries for Research of Baden-Württemberg (MWK), Bayern (StMWFK), and Nordrhein-Westfalen (MIWF). Additional simulations were performed on the THEO and ISAAC clusters of the MPIA and the COBRA, HYDRA and DRACO clusters of the Max-Planck-Society, both hosted at the Max-Planck Computing and Data Facility in Garching (Germany). Figure 13.1 was produced using data from the Minor Planet Physical Property Catalog (mp3c.oca.eu).

after the start of the disk evolution). In order to match the ages of chondrules, which formed typically later, this initial formation phase likely was followed by a period of pebble accretion and planetesimal collisions, which likely also led to the formation of larger bodies such as Ceres and Vesta. Another possibility, which is not conflicting with the previous concept is that the parent bodies of meteorites and asteroids come from a later-generation of planetesimals (Morbidelli et al., 2020).

In order to shed light on how the aforementioned process shaped our Solar System one could think of studying planet formation disks around young stars. However, the processes within protoplanetary disks leading to the formation of planets, asteroids, and other bodies are in large part hidden from our telescopes. While dust strongly emits in the infrared, planetesimals – the precursors of planets defined as the smallest objects bound by their own self-gravity instead of their material binding forces – do neither provide enough collective surface density, nor are individually large enough to be detectable through observations. Nevertheless, the outcome of planetesimal formation can be informed by studying small size bodies in our Solar System, specifically Main Belt asteroids and trans-Neptunian objects, both of which are believed to constitute remnants of the Solar System's formation process.

The formation of MBAs such as Ceres and Vesta, and planetesimals in general, remains a central question in planet formation studies. A unique challenge arises once micrometer-sized dust particles have grown large enough to decouple from the gas. As a result, meter-sized particles drift inward so rapidly that they either fragment or fall victim to evaporation from the star before they can outgrow this fast drift regime (Birnstiel et al., 2012). While there are studies in which this so-called meter-barrier is overcome via efficient particle growth (e.g., Kokubo & Ida, 2012; Windmark et al., 2012), another convincing scenario is the direct formation of planetesimal through gravitational fragmentation (Safronov, 1969; Goldreich & Ward, 1973), which also happens to be the original mechanism to form planetesimals. Here, over-dense regions of dust and ice aggregates collapse under their own self-gravity and form a planetesimal in a spontaneous event instead of through continuous growth. Gravitational collapse can either be achieved through fragmentation of a particular massive particle at the mid-plane of the disk (Youdin & Shu, 2002) or through locally enhanced particle clumps (e.g. Johansen et al., 2007; Nesvorny et al., 2019; Gerbig et al., 2020). The latter case is also known as gravoturbulent planetesimal formation (Johansen et al., 2006b), as over-densities are typically either found in turbulent traps (Barge & Sommeria, 1995; Klahr & Henning, 1997; Klahr & Bodenheimer, 2006), such as for example vortices (Manger & Klahr, 2018) or zonal flows (Dittrich et al., 2013), or are directly created by local drag instabilities (e.g. Johansen & Youdin, 2007; Carrera et al., 2015; Yang et al., 2017).

What mechanism leads to the formation of planetesimals naturally strongly affects the size of initial planetesimals. In contrast to a bottom-up coagulation (e.g. Windmark et al., 2012), where dust grains outgrow the meter-barrier, the gravitational collapse scenario favors the direct formation of 10–1,000 km sized planetesimals, with details depending on the local dust-to-gas ratio. Numerical simulations have shown that the so-called streaming instability (Youdin & Goodman, 2005) can lead to sufficiently high particle concentrations for collapse to occur and planetesimals to form. The cumulative size–frequency distribution of initial planetesimals resulting from the fragmentation of streaming instability filaments was shown to follow an inverse power law (e.g., Johansen et al., 2015; Simon et al., 2016). For instance, as we will discuss, recently Abod et al. (2019) reported a mass distribution $dN/dM_p \propto M^{-p}$ with a value of $p \approx 1.6$ for a range of pressure gradients.

However, it is not clear whether this slope likewise applies for smaller sizes, which would imply the existence of many small objects, or if there is a minimal planetesimal size. A typical planetesimal size of D~100 km, is instead invoked by different works (Cuzzi et al., 2008; Klahr & Schreiber, 2016, 2020; Gerbig et al., 2020). There, turbulent diffusion leads to the dispersion of small clumps before they can collapse and form planetesimals.

As we discuss in this chapter, evidence has grown in the last years that planetesimals were formed as large bodies and not as small as collisional coagulation would suggest. However, the minimum size of original planetesimals remains a central question in planetology. In other words, for any asteroid (and other small bodies) how reliably can we say whether it has to be a fragment of a larger parent body or it could be an original planetesimal, depending on its size? In this article, we outline observational and theoretical constraints on the formation of MBAs. First, we discuss the current state of research on the size–frequency distribution, the composition and ages of asteroids, and the implications on the formation of asteroids. Then, we review planetesimal formation theory, specifically focusing on the initial sizes of primordial planetesimals. Lastly, we discuss the theoretical constraints on the timing of planetesimal formation as well as implications for MBAs, specifically Ceres and Vesta.

13.2 OBSERVATIONAL CONSTRAINTS FROM THE MAIN ASTEROID BELT

Asteroids, comets, trans-Neptunian objects, and irregular satellites constitute what is left over by the planet-formation processes of the original planetesimal disk. In this section we focus on what we can learn from asteroids of the original planetesimals and in particular of their size–frequency distribution; First of all, we note that not all asteroids are pristine survivors from the primordial times. Instead, a large amount are believed to be collisional fragments of bigger parent bodies. Although these families of fragments – the so-called asteroid collisional families – still carry the original composition of their progenitor(s), the sizes of these fragments and shapes do not necessarily provide information on the accretion process(es) that led to the formation of planetesimals.

13.2.1 Size–Frequency Distribution

Collisions between asteroids produce families of fragments which can reaccumulate to form newborn bodies (Michel et al., 2001; Michel & Richardson, 2013). This process, thus, affects the asteroid size–frequency distribution (SFD) by adding smaller objects. In addition, catastrophic collisions remove from the SFD the bodies that are destroyed. Thus, the study of the SFD of asteroids and asteroid collisional families is crucial for understanding collisional processes in the early Solar System, all along its history, and the sizes of asteroids involved therein. In some cases, the SFD can also give clues about the internal structure of asteroids (Durda et al., 2007). In an ideal scenario, a good understanding of the 4.5 Gyr long collisional evolution of the asteroid Main Belt (Figure 13.1), combined with the precise identification of the collisional families' size distribution, could be used to restore the original SFD. However, this exercise proved challenging as different asteroid collisional families were reported to have different SFD slopes as well as a widely different number of total objects. An additional problem is that the census of asteroid collisional families is incomplete.

Still, there are features of the current asteroid SFD that attracted the attention of several researchers since they cannot be easily explained as the by-product of the Main Belt's collisional evolution. While it was shown that the steep slope of the asteroid SFD for $D > 100$ km ($q = -2.87$) is likely original (i.e., the planetesimal SFD; Bottke et al., 2005), this slope could not extend all the way down to small sizes. Otherwise, the resulting large number of small asteroids would have created many collisions that subsequently left traces on larger asteroids. For instance, the resulting larger number of impacts would have removed the crust of Vesta (Davis et al., 1985; Bottke et al., 2005) or at least produced several large impact basins on this asteroid. This was not observed (Marchi et al., 2012). Morbidelli et al. (2009) pushed this argument to the extreme, proposing that there were essentially no $D < 100$ km planetesimals in the Main Belt region.

How can we distinguish asteroid fragments from primordial objects? In a break up process, fragments are launched into space at moderate velocities of ~1 kms^{-1}, and fragments keep orbital elements similar to that of their parent body in the asteroid Main Belt (but not elsewhere). Thus, fragments become themselves new asteroids, clustered in orbital space and with similar physical properties, such as albedo (p_V), colors, and spectra (Figure 13.1(a); except for the case of the disruption of a differentiated parent body, for which there exist some hints (e.g., Oszkiewicz et al., 2015). Asteroid families are typically identified by using the Hierarchical Clustering Method (HCM, Zappalà et al., 1990), which seeks asteroid groupings (some of which are visible in Figure 13.1(a)) in their three proper orbital elements, semi-major axis, a, eccentricity, e, and inclination, i. Thus, if one could identify all families and "clean" the Main Belt from their members, what would be left are the survivors of the original asteroid population: the planetesimals.

One of the first attempts at this "cleaning process" was carried out by Tsirvoulis et al. (2018). They studied the region of the asteroid Main Belt between 2.82 and 2.96 AU, which is also called the "pristine zone" (Figure 13.1(a)) as it has a low number density of asteroids and, as such, is deemed one of the most promising areas to reveal the original size–frequency distribution. Tsirvoulis et al. (2018) showed that current asteroid family catalogs are not

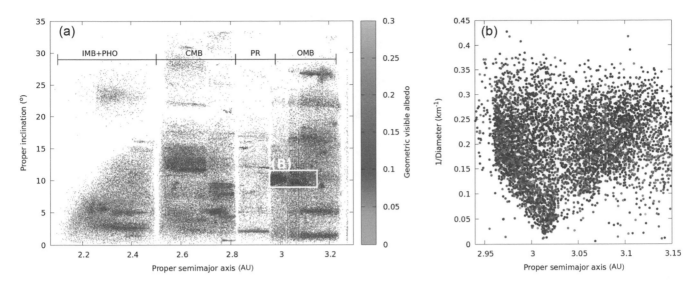

Figure 13.1 The asteroid Main Belt and its collisional families. (a) Proper orbital inclination versus proper orbital semi-major axis of all asteroids with a measured diameter larger or equal to 1 km. Each point represents an asteroid. The different regions of the Main Belt are presented with the following symbols: IMB = Inner Main Belt (below ≈18° of proper inclination); PHO = Phocaea group (above ≈18° of proper inclination); CMB = Central Main Belt; PR = pristine zone (as named by Brož et al., 2013); OMB = Outer Main Belt. If we extract asteroids with a proper semimajor axis between 2.94 and 3.15 AU, proper inclination between ≈9.8° and ≈11.2°, proper eccentricity between 0.03 and 0.1 (the white box in (a)), and we plot their inverse diameter as a function of their proper semimajor axis we obtain the plot in (b). This shows a typical V-shape (the Eos family), with its vertex on the x-axis at about 3.014 AU. The inclined inward border of the V-shape is very sharp and is cut at about 1/diameter 0.15 km^{-1} by the 7:3 major mean motion resonance with Jupiter (see text). The outward border is less clear and starts to overlap with background asteroids at 3.07 AU and 0.15 km^{-1}. Source of the data mp3c.oca.eu; download November 2020; For the reference of the proper elements, diameters and albedo see Delbo et al. (2019).
A black-and-white version of this figure will appear in some formats. For the color version, please refer to the plate section

suitable for the aforementioned cleaning. These catalogs are in general conservative, in order to have robust family membership and maintain a good level of separation between different families. Attempting to overcome said limitation, Tsirvoulis et al. (2018) proceeded by performing a new asteroid family identification in the pristine zone using a version of the HCM that was modified in order to allow the largest possible membership and then remove the majority of asteroid family members from the region.

The remaining, non-family asteroids were considered of primordial origin. This is a reasonable assumption also because the strong 5/2 and 7/3 mean-motion resonances with Jupiter, at 2.82 AU and 2.96 AU, respectively, border the pristine zone and thus inhibit diffusion of collisional fragment asteroids from the neighboring regions into the pristine zone. After removal of collisional family members, Tsirvoulis et al. (2018) found that the cumulative size–frequency distribution of asteroids in the size range $17 < D < 70$ km has a slope $q = -1$. After compensating for the collisional and dynamical loss of asteroids during 4.5 Gyr of evolution, an upper bound for the original size distribution slope was found to be $q = -1.43$ in said size range.

However, it has been speculated that the HCM has a bias against the detection of Gyr-old families, making it difficult to use for the removal of all collisional fragments from the Main Belt or subregions thereof in order to reveal the original asteroids, that is, those that accreted as planetesimals. Indeed, there is evidence of a severe deficit of known families older than 2 Gyr compared to what is expected assuming a constant asteroid break-up rate over the age of the Solar System (Brož et al., 2013; Spoto et al., 2015).

Families disperse over time due to the non-gravitational Yarkovsky effect (for reviews see Bottke et al., 2006; Vokrouhlický et al., 2015), which slowly changes asteroids' orbital semimajor axis, a, at a rate da/dt proportional to $1/D$. Prograde rotating asteroids have $da/dt > 0$, while for retrograde ones $da/dt < 0$. This creates correlations of points in the $1/D$ versus a plane called V-shapes, as they resemble the letter "V" (see Figure 13.1(b); Nesvorný et al., 2003, 2015; Spoto et al., 2015; Bolin et al., 2017, and references therein). During their Yarkovsky-driven orbital radial drift, family members encounter orbital resonances with the planets, which can randomly change the orbital eccentricity, e, and inclination, i. As a consequence, family members are subject to an increase in diffusion in e and i as a function of time, while their values of a also move further away from the family center, still keeping the correlation between their a and $1/D$ values. Therefore, older families become more difficult to identify since they are more and more dispersed in e and i and overlap each other. Recent works took advantage of the V-shape by searching for its correlation between the inverse diameter and proper semi-major axis of asteroids to identify old families with strongly dispersed (e, i), thus invisible to HCM (Walsh et al., 2013; Bolin et al., 2017; Delbo et al., 2017, 2019), resulting in the discovery of three Gyr-old families in the inner portion of the Main Belt (2.1 AU $< a < 2.5$ AU).

The identification of these vast families allowed these authors to assign an origin as collisional fragments to a population of asteroids of the inner Main Belt that were previously unlinked to asteroid families. When all family members were removed, Delbo et al. (2017, 2019) showed that the remaining asteroids are essentially larger than 35 km (only three asteroids are smaller than this value; Delbo et al., 2019) and have a very shallow slope of their cumulative SFD ($q < -0.86$) in the range $35 < D < 100$ km, even after the correction for collisional and dynamical depletion (see Figure 13.2). These studies find that the distribution of the leftover

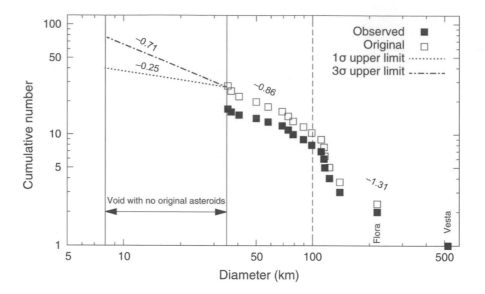

Figure 13.2 Observed cumulative size distribution of leftover planetesimals in the inner part of the Main Belt (2.1 AU < *a* < 2.5 AU). The distribution is then corrected for the maximum number of objects that were lost, accounting for 10 Gyr of collisional evolution in the present main-belt environment, following the prescription of Bottke et al. (2005) and dynamical loss as described by Delbo et al. (2017). This gives an upper limit for the distribution of the planetesimals (open squares). Functions of the form $N(D) = N_0 D^q$, where N is the cumulative number of asteroids, are fitted piecewise in the size ranges $D > 100$, $35 < D < 100$, and $D < 35$ km; The values of q are reported by the labels in the plot.
Source Delbo et al. (2019)

planetesimals in the inner portion of the Main Belt is steep for $D >$ 100 km, but becomes shallower for $D <$ 100 km, indicating that $D \sim 100$ km was a preferential size for planetesimals.

We recall that, according to the definition of Delbo et al. (2017, 2019), those asteroids that are inside V-shapes of families must have formed as collisional debris (or the re-accumulation thereof: Michel et al., 2001; Michel & Richardson, 2013) locally in the Main Belt, after the gas was left or after any event that could alter their orbital semimajor axis, such as close encounters with a planet (e.g., Brasil et al., 2016). The dispersion in e and i of the family members could be used to signal whether the family forming collision happened before or after the last event of planetary orbital instability (Brasil et al., 2016; Delbo et al., 2017). For a review about the latter, which is dated no later than 100 Myr after the start of the Solar System, see Nesvorný et al. (2018). On the other hand, leftover planetesimals are those asteroids that cannot be inscribed within V-shapes except for the parent bodies of each collisional family (e.g., the largest object within its own V-shape). These bodies cannot have been generated as collisional fragments within the Main Belt (otherwise that would be inside V-shapes). Hence, their origin is primordial, for example by accretion of dust in the protoplanetary disk.

13.2.2 Formation Timescale

According to the scenario described in Section 13.2.1, we can infer different formation times between asteroids included in the V-shapes of collisional families and those outside. The latter population of asteroids, which is composed by the original planetesimals, likely accreted from the dust while gas was still present in the protoplanetary disk. Timing of their formation can be constrained by the epoch of solidification of their constituent materials (e.g., CAIs and chondrules), the formation of the cores in the case

of differentiated planetesimals (e.g., Sugiura & Fujiya, 2014; Kruijer et al., 2020), the epoch of the last event of metamorphism, and interpretation of astronomical observations of disks around young stars (see also Chapter 14).

In this chronology it is essentially a "zero-point" that is established to be 4567.2 ± 0.2 Myr ago from the Pb–Pb radioisotopic chronometer age of the Calcium Aluminum-rich Inclusions (CAIs), the oldest solids in our Solar System (Amelin et al., 2002; Connelly et al., 2012; Kruijer et al., 2020, the latter for a review). This date is commonly taken as the start of our Solar System.

The Hf–W radioisotopic chronometry applied to iron meteorites indicates the time of core formation of differentiated planetesimals (e.g., Kruijer et al., 2020), which come in two distinct groups: those that have isotopic abundance ratios similar to carbonaceous chondrites (CC) and those similar to non-carbonaceous chondrites (NC; Kruijer et al., 2017, 2020). The ages of the CC irons are in the range $\sim$2.2 and $\sim$2.8 Myr after the CAIs, while those of the NC irons are $\sim$0.3 and $\sim$1.8 Myr (Kruijer et al., 2014, 2017).

On the other hand, chondritic parent bodies (which are probably the majority in the planetesimal list of Delbo et al., 2019) are thought to have never melted, but were substantially heated by the decay of radioactive elements (such the ^{26}Al; Gail & Trieloff, 2019). Chondrules, which are a constituent of chondrites, were molten at some point instead and must have formed and cooled before they could be incorporated into chondrites and the chondrite parent bodies. Chondrules have formation ages ranging between $\sim$2 and 4 Myr after the CAIs (see Kruijer et al., 2020, for a recent review).

Another method consists of using modeling the thermal evolution of the heating due to the ^{26}Al decay and cooling by radiation to space of planetesimals as a function of their sizes and accretion ages and match the model temperature–time curves with the metamorphic temperatures and closure times of meteorites (e.g., Henke et al., 2012; Sugiura & Fujiya, 2014; Gail & Trieloff, 2019). Results from these models indicate planetesimal accretion ages between 1 and 3.5 Myr after the CAIs, in rather good agreement with chondrules formation ages.

Moreover, astronomical observations of the presence of disks around stars in young star associations show an overall disk lifetime of $\sim 5-6$ Myr (Haisch et al., 2001), which could be considered the maximum time after the CAIs within which planetesimals can accrete.

13.3 PLANETESIMAL FORMATION THEORY

If the current asteroid SFD can indeed be best explained by an initial planetesimal size of ~ 100 km, then one must ask if this finding can be supported by theoretical models of planetesimal formation. Thus, we outline the framework of planetesimal formation theory today, starting with dust motion in protoplanetary disks and arising challenges for planetesimal formation theory, and concluding with recent developments and potential theoretical constraints for the initial mass-function of planetesimals.

13.3.1 Dust Settling, Coagulation, and Drift

Observations of protoplanetary disks suggest that micrometer-sized dust particles are abundantly present and constitute the initial condition for grain evolution and subsequent planetesimal formation. The vertical component of stellar gravity causes these dust particles to settle toward the mid-plane and form a dust layer that is embedded within the gas disk. It is in this particle mid-plane, where key processes such as dust coagulation and drift, and ultimately planetesimal formation occur.

Due to self-induced Kelvin–Helmholtz turbulence vertically exciting particles, the dust mid-plane is even in dead-zones, not razor-thin. Dead-zones refers to regions of the nebula in which neither magnetic fields nor hydro dynamical instabilities in the gas disk can drive turbulence and diffusion. The Kelvin–Helmholtz instability alone can produce a scale height for the dust of about $Hp \sim 10^{-2}H$ (e.g., Johansen et al., 2006a), where $H = c_s/\Omega$ is the scale height of the pressure-supported gas disk, given by the ratio of sound speed, c_s, and orbital angular velocity, Ω, following from gas in vertical hydrostatic balance.

Upon encounter, two dust particles can stick together to form a larger one. The outcome of an encounter, that is, sticking, bouncing, or fragmentation, strongly depends on the relative particle velocity (Windmark et al., 2012). Potential sources of this relative velocity are, for example, Brownian motion, turbulent gas velocities (Cuzzi et al., 2001), differential settling, or differential drift. This drift is caused by different preferred orbital velocities of gas and particles: while particles want to orbit with a Keplerian speed, gas is radially pressure-supported and thus feels an additional outwards force which leads to sub-Keplerian orbital velocities. As particles grow and start to decouple from the gas, this velocity difference increases and the lack of sufficient radial support causes grains to drift inwards (Nakagawa et al., 1986). For larger particles, this centrifugal deficiency transitions into a headwind, which likewise removes angular momentum and causes inward drift. Only very large particles that are fully decoupled from the gas will not

spiral inwards and remain on their Keplerian orbits. To assess the drift behavior of a particle, one typically invokes the dimensionless Stokes number $St = t_{stop}\Omega$, given by the product of stopping time, t_{stop}, and orbital angular velocity, Ω. Here, the stopping time is given by $t_{stop} = mw/F_D$ for a particle with mass, m, experiencing a drag force, F_D, when streaming through gas with a relative velocity of w. Particles are coupled to the gas for $St < 1$, while they are decoupled for $St > 1$. As such, for marginally coupled particles with $St \approx 1$, the two effects mentioned combine and the radial drift speed peaks. This not only imposes a challenging time scale constraint by rapidly drifting meter sized particles toward the central star, but also halts collisional growth due to increased relative velocities. These two limits, that is, drift and fragmentation, combine to the previously mentioned "meter-barrier," stating that collisional growth to planetesimal-sizes is severely impeded.

The meter-barrier may be circumvented by introducing long-lived particle trap-structures, such as large-scale pressure bumps. There, particle drift would slow down or even halt completely and bottom-up growth may be possible despite these challenges. Still, in addition to the requirement of long-lived trap structures, bottom-up growth cannot adequately explain the observational constraints from the asteroid Main Belt highlighted in Section 13.2. For these reasons, in recent years, the gravitational collapse scenario to directly form planetesimals in a spontaneous event has gained in popularity.

13.3.2 Requirements for Planetesimal Formation

In star formation, self-gravity of the gas has to overcome thermal gas pressure (for a review of star formation, see e.g., Mac Low & Klessen, 2004), while for gravitational instability of gas disks, fragmentation can additionally be inhibited by the disruptive effect of stellar tidal gravity (Toomre, 1964). In the latter, collapse can typically be assessed via the Toomre criterion. It states that if the Toomre-Q value, defined as

$$Q = \frac{c_s\Omega}{\pi G\Sigma_g} \tag{13.1}$$

falls below a critical value of order unity, i.e., if $Q < 1$, the gas disk is gravitationally unstable and can fragment. Here, Σ_g is the gas surface density and G the gravitational constant.

With these concepts in mind, one must naturally ask how the gravitational collapse of a pebble cloud embedded in (potentially) turbulent gas to form a planetesimal can be evaluated. While particles are typically assumed to be pressureless (e.g., Youdin & Goodman, 2005), they are coupled to the gas and therefore affected by turbulent gas motions. The resulting turbulent diffusion replaces the pressure as the process counteracting collapse on small scales. It is convenient to parameterize the diffusion strength D_s via the dimensionless diffusion coefficient, δ, that is, $D_s = \delta c_s H$. Note that the dimensionless diffusion coefficient δ is not to be confused with the Shakura-Sunyaev α parameter (Shakura & Sunyaev, 1973), quantifying the (gas) disk's (global) viscosity via $\nu = \alpha c_s H$, such that turbulent gas velocities are given by $v_{turb} = \sqrt{\alpha}c_s$ (Cuzzi et al., 2001). The two quantities are related via the dimensionless flow Schmidt number, $Sc = \alpha/\delta$. While local Schmidt numbers tend to be of order unity (Youdin & Lithwick, 2007; Schreiber & Klahr, 2018), it is important to highlight that,

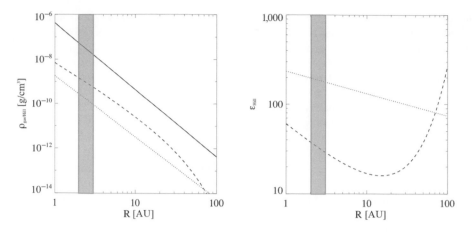

Figure 13.3 Gravitational collapse in the presence of the solar tides requires enhanced dust-to-gas ratios, that is, Hill density, when gravity is stronger than the solar tides. Left panel: Comparison of Hill density (solid line, Eq. (2)), that is, the required density for a cloud to be stable against tidal shear, with gas densities of the Minimum Mass Solar Nebula model (dotted line, Hayashi, 1981) and the most appealing Solar Nebula model from Lenz et al. (2020) (dashed line). Right panel: Required mid-plane dust-to-gas ratios to achieve Hill density. We highlight the region corresponding to the Main Belt by a gray shaded area. Figure reprinted from Klahr and Schreiber (2020)

unlike α, the diffusion coefficient δ is a local quantity that typically is self-generated in the particle layer by various dust–gas drag instabilities. Moreover, the strength of diffusion is highly scale dependent. While on large scales, gas instabilities very well may drive turbulent diffusion exceeding $\delta \sim 10^{-4}$, diffusivities are generally much lower on the scales of planetesimal formation, which are on the order of $\sim 10^{-3}H - 10^{-2}H$ (Gerbig et al., 2020). Streaming instability, which is the prime candidate to regulate turbulent diffusion on small scales, produces fiducial values of $\delta \sim 10^{-6}$ (see Schreiber & Klahr, 2018). Indeed, assessing whether the local diffusivity is dominated by effects related to the global viscosity α rather than local turbulent processes requires not only a reasonable estimate of α, but also knowledge of the appropriate turbulent cascade, and is therefore challenging. We refer to Klahr and Schreiber (2020, section 5.1), for a discussion of this.

For collapse to proceed, two criteria must be fulfilled: First, the diffusion time scale, which, following Youdin and Lithwick (2007), is inversely proportional to δ, of the dominating diffusive process must be longer than the collapse timescale of particles in-falling with their terminal velocity. Secondly, the pebble cloud must be stable against stellar tidal disruption, which is equivalent to requiring a cloud density that exceeds the local Hill-density) i.e., Klahr & Schreiber, 2016, 2020; Gerbig et al., 2020)

$$\rho_{Hill} = \frac{9}{4\pi} \frac{M_\odot}{R^3}. \tag{13.2}$$

The left panel of Figure 13.3 shows the Hill density (solid line) in comparison with mid-plane gas density in the Minimum Mass Solar Nebula (MMSN) (dotted line) (Hayashi, 1981) and a more recent model of the solar nebula from Lenz et al. (2020). The classical MMSN simply redistributes the mass of the planets to their local region and enhances it by volatiles and gas (H_2, He), thus it assumes "local" conversion of dust and ice into planetary cores. In contrast, Lenz et al. (2020) consider the radial drift of pebbles until they get "non-local" in trap converted into planetesimals as needed to build the Solar System. The resulting modern nebula is therefore shallower in density and considers an outer edge beyond

20 AU. Please note the conditions relevant for the Main Belt, as highlighted by a grey shaded area. The right panel of Figure 13.3 shows the required dust-to-gas ratio, ε_{Hill}, in order to reach Hill-density for the two models. At the Asteroid Belt's location, the required dust-to-gas ratio exceeds unity by a factor of about 200 for the MMSN disk, and a factor of ~ 12 for the model from Lenz et al. (2020), which is significantly easier to achieve.

Combining these two requirements, that is, tidal stability and diffusion, one arrives at a Toomre-like criterion for planetesimal formation, that is, collapse occurs if (Gerbig et al., 2020)

$$Q_p = \frac{Q}{Z} \sqrt{\frac{\delta}{St}}. \tag{13.3}$$

Here, we introduced the (local) height-integrated particle concentration as the ratio of particle surface density Σ_p and gas surface density, that is, $Z = \Sigma_p / \Sigma_g$, and assumed isotropic diffusion. Vertical sedimentation, which in addition to Z affects the mid-plane dust-to-gas ratio (compare to Figure 13.3), is implicitly accounted for in the diffusion coefficient as the particle scale height is given by $H_p = \sqrt{\frac{\delta}{St}}H$ (Youdin & Lithwick, 2007). In order for collapse to trigger, i.e., for Q_p to fall below unity, sufficiently high particle concentrations of Z are necessary. In the following we outline a variety of known processes of enhancing pebble concentration, thus favoring the conditions for gravitational collapse and planetesimal formation.

13.3.2.1 Global Enhancements in Pebble Concentration

There are multiple global mechanisms that can lead to significant enhancements in particle concentration. Pebble pile-up can, for example, be achieved in the local pressure maxima next to gaps in the radial gas surface density profile, where a pressure bump attracts particles. After all, ever since ALMA rendered resolved observations of protoplanetary disks possible, gaps and ring structures are believed to be a common occurrence therein (Andrews et al., 2018). However, the origin of these ring structures is often explained by planets carving out the material around their respective orbit, which does not explain how these planets were formed in the first place. Luckily, there are additional processes leading to the formation of pressure bumps, in particular large-scale instabilities in the gas. Prominent candidates are the vertical shear instability (Urpin & Brandenburg, 1998), and, if active, the magneto-rotational instability (Balbus & Hawley, 1991). As pointed out by, for example, Klahr et al. (2018), the turbulence generated by these instabilities leads to structure formation, in particular to the formation of zonal flows and vortices. These flow features display an equilibrium of gravity with Coriolis and Centrifugal forces and are therefore pressure maxima attracting particles. Dittrich et al. (2013) and Raettig et al. (2015) demonstrated numerically that this can lead to particle concentrations that favor gravitational collapse and planetesimal formation.

Even in the absence of turbulence-induced zonal flows and vortices, the pebble concentration can be increased by photoevaporation of gas (Throop & Bally, 2005) or by pile-ups due to radial drift (Youdin & Shu, 2002). The resulting high particle concentrations Z leads to a very massive dust mid-plane cusp, which cannot effectively be stirred up by self-sustained Kelvin–Helmholtz turbulence (see e.g., Sekiya & Ishitsu, 2000; Gómez & Ostriker, 2005; Gerbig et al., 2020). For $Z > 0.1$, this mid-plane cusp can fragment and planetesimals can form (Youdin & Shu, 2002).

13.3.2.2 Local Pebble Trapping via Streaming Instability

A key phenomenon in recent work on planetesimal formation theory is the streaming instability, which was first described analytically by Youdin and Goodman (2005) and successfully tested numerically in the following years (e.g., Johansen & Youdin, 2007; Bai & Stone, 2010; Carrera et al., 2015; Li et al., 2018; Sekiya & Onishi, 2018). Streaming instability arises when an epicyclic perturbation in the gas pressure, a small oscillation in pressure around the local semi-major axis R, is in resonance with the relative dust–gas streaming velocity (Squire & Hopkins, 2018). Key requirements for fast growth rates are marginally coupled particles (St > 0.1) and local dust-to-gas ratios of order unity. Then, streaming instability can produce significant local particle overdensities. In order for the mid-plane dust-to-gas ratio to reach unity, and assuming marginally coupled particles, height-integrated pebble concentrations of $Z > 0.02$ are sufficient (Carrera et al., 2015; Yang et al., 2017), which is significantly easier to achieve than the requirements for direct fragmentation in Youdin and Shu (2002). If the local over-densities exceed Hill density, that is, if the local dust-to-gas ratio ε exceeds $\varepsilon_{\text{Hill}}$, as depicted in Figure 13.3, and the diffusive properties of streaming instability itself (see Schreiber & Klahr, 2018) do not prevent collapse, overdense streaming instability filaments fragment into planetesimals, which was extensively investigated numerically (Johansen et al., 2012; Simon et al., 2016; Abod et al., 2019; Nesvorny et al., 2019; Gerbig et al., 2020). In general, simulations of planetesimal formation are simplified in terms of size distribution of pebbles and their initial distribution to a level that is numerically possible. The realistic size distribution of St ~ 0.01 relevant for the asteroid Main Belt regions could be shown to lead to planetesimals if sufficiently concentrated in local simulations (Klahr & Schreiber, 2020, 2021), yet no large box simulations have been possible so far to account for such small Stokes-Numbers and initial pebble-to-gas ratios as expected in the Main Belt. For instance, the simulation depicted in the snapshot in

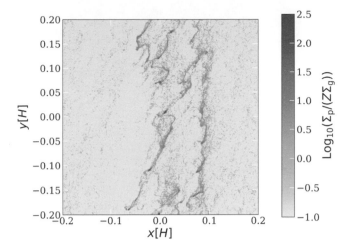

Figure 13.4 3D numerical simulation of planetesimal formation in the presence of streaming instability. Simulations are conducted in the shearing box approximation where x is the radial coordinate and y the azimuthal coordinate. Shown are vertically integrated, normalized particle densities projected onto the disk plane. The realistic conditions in the MB region (see Figure 13.3) are so far not accessible for numerical simulation, thus simulation results have to be extrapolated physical relevant conditions. This snapshot shows that self-gravity can lead to fragmentation of streaming instability filaments and the formation of distinct clumps which are commonly identified as planetesimals. See Gerbig et al. (2020) for a discussion of this particular setup.
A black and white version of this figure will appear in some formats. For the color version, refer to the plate section

Figure 13.4 uses pebbles of St $= 0.1$–0.2 and one then tries to extrapolate the findings to the conditions where the Main Belt asteroids were formed.

However, recently, the role of streaming instability has been put into question by findings suggesting that not only the diffusive properties of streaming instability itself, but also potentially existing external turbulence in the gas can prevent collapse (Gole et al., 2020) or damp streaming instability growth rates and thus particle concentration in the first place (Klahr & Schreiber 2016, 2020, 2021; Umurhan et al., 2020). Nevertheless, by concentrating particles on large scales and diffusing particles on small scales, streaming instability regulates planetesimal formation (Klahr & Schreiber, 2016; Gerbig et al., 2020) and therefore remains a key process, which, as we discuss in Section 13.3.3, sheds light on a variety of properties of initial planetesimals, in particular their initial sizes.

13.3.3 Initial Mass Function and Characteristic Sizes

In recent years, the properties of planetesimals formed in numerical simulations of streaming instability were studied with keen interest. As such, Nesvorny et al. (2019) were recently able to relate the spatial orientation of binary orbits of model planetesimals to that of trans-Neptunian objects. The tendency to form binary planetesimals in the MB is less pronounced, as the pebble clouds there contain a lower amount of spin angular momentum in comparison to trans-Neptunian pebble clouds, yet binaries are still possible. Nevertheless, most binaries will not survive the dynamical evolution of the MBA.

Of particular interest are the masses and sizes of initial planetesimals, as they allow for a direct comparison to the current

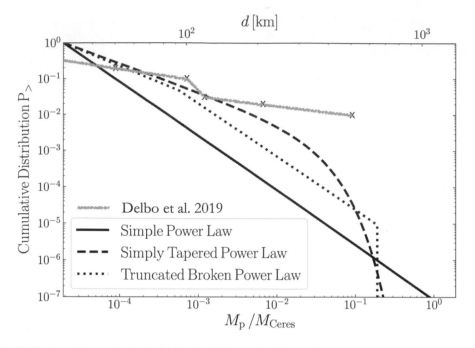

Figure 13.5 Cumulative distribution of planetesimals, that is, fraction of planetesimals larger than a given minimum mass (bottom x-axis) and size (upper x-axis). Shown are the qualitative shapes of the best fit power laws to three numerical studies of planetesimal formation in the presence of streaming instability: The solid line shows a simple power law $\propto M^{-1.5}$ found by Simon et al. (2016), the dashed line a simply tapered power law, as found in Abod et al. (2019), and the dotted line the truncated broken power law found in Li et al. (2019). The latter accounts for the reduced frequency of planetesimals for sizes above ∼100 km in their numerical simulations. We also added the initial size distribution from Delbo et al. (2019), see also Figure 13.2. As argued in this chapter, a narrow size distribution seems to better fit observations than any power law.

asteroid SFD and the inferred sizes of initial parent planetesimals at the start of the asteroid Main Belt's collisional evolution.

As shown in Figure 13.5, more recent studies modified the empiric initial mass function, for example, by adding exponential tapering (Abod et al., 2019) or a truncation (Li et al., 2019). Most recently Abod et al. (2019) report a mass distribution $dN/dM_p \propto M^{-p}$ with a value of $p \approx 1.6$ for a range of pressure gradients. Nevertheless, these rather shallow mass distributions would imply that most of the mass in the Main Belt is already initially in objects much larger than 100 km, which is in contradiction to the observational constraints from the asteroid Main Belt. In the quoted works, in order to obtain an initial mass function, masses of planetesimals formed in numerical simulations were fitted by a power law. The initial condition was set by using an enhanced dust-to-gas ratio globally in the entire simulation box, thus rendering planetesimal formation very efficient. In other words, the pebble concentration was gravitationally unstable from the beginning (Qp < 1). Hence, the shallow size distributions found numerically, leading to the formation of massive planetesimals, may be an artefact of the assumed initial conditions. Instead, one must consider how such an unstable situation, that is, local concentration or trapping of pebbles, can be achieved in the first place. Klahr and Schreiber (2020) argue here for the narrow size distribution around 100 km, that smaller pebble clouds cannot collapse in the presence of turbulent diffusion and larger pebble clouds take longer to accumulate than they would need to collapse. Especially the latter point is not considered in the large box simulations that find the extended power-laws.

Following, for example, Johansen et al. (2007) and Dittrich et al. (2013), dust enhancements are expected to occur on scales smaller than the typically simulated disk domains, and, an already much smaller pebble cloud may undergo gravitational collapse. As such, the key question is, what is the smallest pebble cloud that can gravitationally collapse and will therefore set the dominating initial size of planetesimals, and thus the primordial asteroid size.

This question was investigated in Klahr and Schreiber (2016, 2020), where a minimal planetesimal size was inferred analytically. By considering a marginally unstable system where the cloud density is exactly equal to the Hill-density ρ_H, which is equivalent to Qp = 1 (see Eq. (13.3) and Gerbig et al., 2020), Klahr and Schreiber (2016) derived a critical length scale for gravitational collapse and planetesimal formation (see Figure 13.6), given by (Klahr & Schreiber, 2020)

$$l_c = \frac{1}{3}\sqrt{\frac{\delta}{St}}H. \tag{13.4}$$

Clumps with radii smaller than l_c, (Qp > 1) will remain stable as contraction is sufficiently counteracted by internal diffusion (also see Gerbig et al., 2020). Only if the cloud is larger than l_c (Qp < 1) can self-gravity overcome turbulent diffusion. This concept of diffusion-limited planetesimal formation is visualized in Figure 13.6. Moreover, as pebble clumps are expected to collapse as soon as they grow beyond l_c, larger clumps are less likely to form, which subsequently sets an upper limit on planetesimal sizes, which is in contrast to the work depicted in Figure 13.5. If we denote the diameter of the resulting equivalent planetesimal size a_p in terms of scale height of the disk (H/R), a typical strength of streaming instability (δ) and pebble size (St) we find (Klahr & Schreiber 2016, 2020)

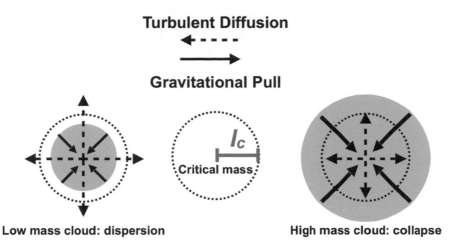

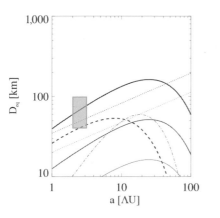

Figure 13.6 Visualization of diffusion limited planetesimal formation: Hill density is required to achieve tidal stability. The competition between self-gravity and turbulent diffusion then defines a critical length scale l_c. Then a large l_c will lead to a large planetesimal. Less massive clouds get dispersed (left), more massive clouds collapse (right).
For details see Klahr and Schreiber (2020)

Figure 13.7 Initial planetesimal diameter a_p according to Eq. (13.5) for a variety of models studied in Klahr and Schreiber (2020). The lower dotted line shows the initial size assuming a MMSN disk (Hayashi, 1981), while the thick solid line is for the most appealing disk in Lenz et al. (2020), which promisingly predicts an initial size of ∼80 km at the Asteroid Belt's location. The thinner solid lines (smaller sizes) result when removing 90% or 99% of the gas mass from the nebula, which occurs over the lifetime of the disk, that is a couple of million years.
This figure was reprinted from Klahr and Schreiber (2020)

$$D = \frac{H/R}{0.03} \sqrt{\frac{\delta}{10^{-6}}} \sqrt{\frac{0.1}{St}}\, 71 \text{ km.} \qquad (13.5)$$

Additionally, diffusivity scales with the Stokes number (Schreiber & Klahr 2018, Klahr & Schreiber, 2020) and with the pebble-to-gas ratio, thus eventually the pebble-to-gas ratio at reaching Hill density is the ultimate criterion determining the critical mass of a pebble cloud to undergo gravitational collapse. Using this scaling for the models of the Solar Nebula (see Figure 13.7) leads to reasonable equivalent sizes in the Main Belt and interestingly very similar sizes over wide ranges of the Solar System's heliocentric distances (Klahr & Schreiber, 2020). Figure 13.7 depicts the radial dependence of a_p according to Eq. (13.5) for a variety of different disk models (for details see Klahr & Schreiber, 2020). At the asteroid Main Belt's location, the most appealing disk model from Lenz

et al. (2020) promisingly predicts an equivalent diameter of ∼80 km.

It is worth noting that a_p, as given by Eq. (13.5), must be understood as an equivalent planetesimal size, that is, a_p would be the size of the initial planetesimal if the entire gravitationally unstable particle cloud were to contract to a single planetesimal with density ρ_p. However, numerical simulations suggest that gravitational collapse can also produce binary planetesimals rather than isolated planetesimals (e.g., Nesvorny et al., 2019). This consideration, of course, leads to on average slightly smaller planetesimals.

Schreiber and Klahr (2018) also found that when the dust-to-gas ratio increases, that is, when the disk eventually loses its gas mass, streaming instability weakens such that δ will become smaller. As the Hill density remains unchanged, this implies the formation of smaller planetesimals. As such, we may expect that over the course of the disk's evolution, the typical mass of initial planetesimals slowly declines, potentially explaining the formation of some 1–10 km sized comets at late times and a few small asteroids. Anyway, in both cases the number of these pristine small objects will be exceeded by many orders of magnitude by objects that formed in collisions from larger objects. Still, due to the (albeit relatively weak) time dependence of the initial planetesimal size, the question about sizes for initial asteroids is intertwined with the timing of the Asteroid Belt's formation (see also Chapters 14 and 15).

13.4 TIMING OF THE ASTEROID BELT'S FORMATION

The timing of the formation of the Asteroid Belt predominantly depends on the availability of a pebble supply: only if global dust evolution can provide a sufficiently massive stream of sufficiently large pebbles, planetesimals can form effectively and form the asteroid Main Belt. This concept is quantified by the paradigm of pebble flux regulated planetesimal formation (Lenz et al., 2019). Here, by studying growth, evolution, and drift of pebbles in an evolving solar nebula, Lenz et al. (2019) predict a time-dependent local pebble flux, which is then converted to planetesimals using an efficiency parameter ε_d – an ad-hoc description of the fraction of pebbles that are both trapped and converted into planetesimals over a drifting distance of d. Figure 13.8 shows the pebble flux for four different sets of parameters. In addition to the efficiency, ε_d, and disk viscosity, α, Lenz et al. (2019) varied the fragmentation velocity, v_{frag}, which is setting the size of the largest pebbles due to controlling the fragmentation barrier (also see e.g., Birnstiel et al., 2012).

To match constraints from the Solar System, Lenz et al. (2020) estimate that the appropriate efficiency for $d = 5H$ (Dittrich et al., 2013) is $\varepsilon_{d=5H} = 0.03$, which would have led to an initial asteroid Main Belt mass of about $\sim$5M_{Earth} consisting of on average 100 km

sized bodies, which took about a million years to gradually form (see Figure 13.8). Additionally, Lenz et al. (2020) found that this most appealing solar nebula is more massive ($M_{nebula} = 0.1M_{\odot}$) yet gravitationally stable, and shallower than the commonly used Minimum Mass Solar Nebula (MMSN) (Hayashi, 1981).

In the models of Lenz et al. (2019, 2020), planetesimal formation at the asteroid Main Belt's location occurs rather rapidly from $\sim$10^4 to $\sim$10^5 years after the start of the disk evolution, and has concluded after $\sim$10^6 years the latest. After this period the available pebble flux decreases, and planetesimals cannot form effectively (see Figure 13.8: local planetesimal formation is proportional to the local pebble flux). Although, if one entertains the possibility that planetesimal collisions produce small-size debris which then in turn

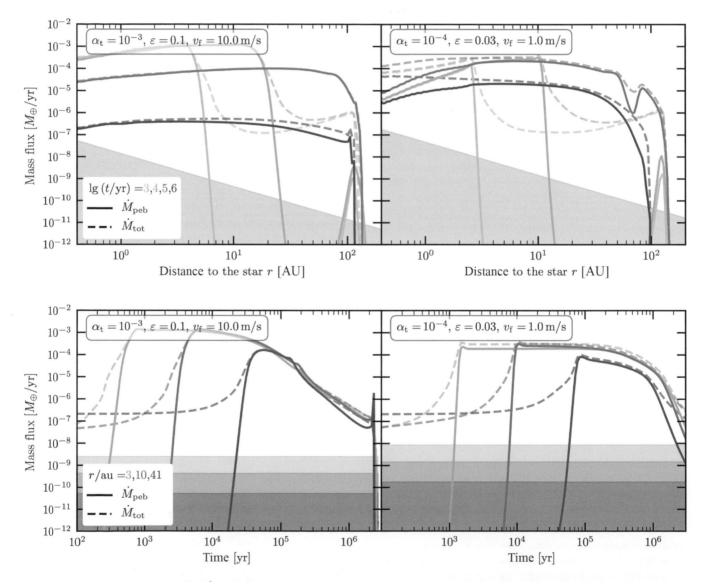

Figure 13.8 Evolution of the pebble flux $\dot{M}_{peb}$ (solid lines) in the solar nebula for two different sets of parameters. Since the availability of a sufficiently massive pebble flux is the key requirement for planetesimal formation (Lenz et al., 2019), the formation of the Asteroid Belt's initial planetesimals (see curves for 3 AU) is expected to last up to $\sim$10^5 to $\sim$10^6 years, which is when the pebble flux starts to decline. The right panels depict the scenario that is preferred by Lenz et al. (2020), with $\varepsilon_{d=5H} = 0.03$, $\alpha = 10^{-4}$, and a fragmentation speed of 1 m/s. The plot also shows the total mass flux $\dot{M}_{tot}$, also accounting for smaller dust grains that do not efficiently participate in planetesimal formation. The shaded areas indicate where the mass flux is too low to support the formation of planetesimals. We refer to figure 1 of Lenz et al. (2019) for a detailed discussion of a similar plot.

A black and white version of this figure will appear in some formats. For the color version, refer to the plate section leads to the formation of second generation planetesimals, the asteroid Main Belt's formation period may be significantly extended (Gerbig et al., 2019).

Interestingly, the Stokes numbers of the largest dust grains in the most appealing solar nebula in Lenz et al. (2020) at around 2 AU correspond to pebble sizes of several centimeters. If planetesimals predominantly form from these larger pebbles, then smaller material (St $\approx 10^{-3}$) is left behind. In the early, more gas-rich phase of the solar nebula's evolution, these small particles correspond to ~ 1 cm sized material, similar to CAIs (Connelly et al., 2012). During later stages, when the gas content is declining, it would correspond to approximately millimeter sized material, similar to typical chondrules, either before or after the flash heating (DeFelice et al., 2019). As such, this suggests that CAIs and chondrules were pebbles that were too small to be effectively involved in the formation of initial planetesimals. Instead, they may have been accreted onto already existing planetesimals via pebble accretion (Ormel & Klahr, 2010), once the available supply of larger pebbles was depleted. Alternatively, if chondrules were really the pebbles with St $\approx 10^{-2}$ then the gas density in the MB region would have to have been two orders of magnitude lower, which would require a very peculiar density profile for the solar nebula to still contain enough mass and also meet other constraints on the spatial distribution of planetesimals to form the Solar System.

13.5 IMPLICATIONS FOR CERES, VESTA, AND OTHER MAIN BELT ASTEROIDS

While the theoretically derived characteristic initial planetesimal diameter of D $\approx$ 100 km well describes the majority of MBAs, it does not fit the four largest asteroids, Ceres, Vesta, Pallas, and Hygiea, which make up about half of the total mass of the asteroid belt (see also Chapter 1). All of them are larger than the size prediction in Klahr and Schreiber (2020) by at least a factor of 4, and for Ceres, even an order of magnitude. Note that the size prediction is based on a minimal mass, thus of course a pebble cloud as massive as Ceres would also collapse. The question is then how it accumulated in the first place without collapsing when reaching the minimal mass, which was 1,000-times lower? So, it seems more likely that Ceres is the product of planetesimal collisions and pebble accretion.

If, instead of comparatively weak drag instabilities in the particle layer, stronger instabilities in the gas, like the magnetic or hydro turbulence, were to dominate turbulence at the asteroid Main Belt's location and the critical mass was on the order of that of Ceres, then the dominant size of asteroids would have initially also have been at 1,000 km, which is not likely considering the collisional evolution as represented in today's size distribution.

For these reasons, it seems much more likely that Ceres, Vesta, and other large asteroids are not primordial in the sense of stemming from one pebble cloud of similar large mass, but started out as smaller planetesimals and then grew to their present size via accretion of either other planetesimals (e.g., Liu et al., 2019) or smaller pebbles (Ormel & Klahr, 2010). These processes would produce compositional differentiated planetesimals, especially if the involved material originated from the outer parts of the Solar System.

When reproducing the present-day asteroid SFD, Morbidelli et al. (2009) did not exclude the possibility of a few 1,000 km size initial planetesimals in the Asteroid Belt. However, unlike the existence of $\sim$100 km sized objects, it also was not a requirement, and is, as such, in line with the herein presented theoretical framework.

To conclude, the concept of MBAs having a rather large initial size of $\sim$100 km with a narrow distribution in sizes is fully consistent with recent models of planetesimal formation in the gravitational collapse scenario, in particular, if it is streaming instability that is driving particle diffusion on small scales, limiting the possibility to form significantly smaller objects. In fact, it is the strength of this turbulent diffusion on the scales of planetesimal formation that is setting the minimum size of initial planetesimals. Whether or not direct formation of Ceres-like planetesimals is possible as a random event requires a better understanding of the accumulation phase of pebbles into a pebble cloud that is either faster than the contraction time or alternatively delayed by locally enhanced turbulence. Such a study of concentration of pebbles from global scales, while also resolving local turbulent diffusion, is subject to ongoing research.

The alternative to in-situ formation via a direct pebble cloud collapse is that Ceres, Vesta, and other large asteroids grew from $\sim$100-km sized planetesimals to their present-day size via collisional accretion processes, which is plausible and consistent with observational evidence (see Chapter 16). Finally, Ceres or other large asteroids may also have migrated from the outer regions during the dynamical evolution of the Solar System as discussed in Chapter 15, yet as displayed in Figure 13.7 this alone would not explain their larger size by itself and also has to invoke post formation evolution by collisions and pebble accretion. Conversely also a size of $\sim$100 km is not a proof for formation inside the MB region, as also at much larger distances to the sun, similar sizes are set by the streaming instability. Only composition may help to identify the location of initial aggregation from a pebble cloud.

REFERENCES

Abod, C. P., Simon, J. B., Li, R., et al. (2019) The mass and size distribution of planetesimals formed by the streaming instability. II. The effect of the radial gas pressure gradient. *The Astrophysical Journal*, 883, 192.

Amelin, Y., Krot, A. N., Hutcheon, I. D., & Ulyanov, A. A. (2002) Lead isotopic ages of chondrules and calcium-aluminum-rich inclusions. *Science*, 297, 1678–1683.

Andrews, S. M., Huang, J., Pérez, L. M., et al. (2018) The Disk Substructures at High Angular Resolution Project (DSHARP). I. Motivation, sample, calibration, and overview. *The Astrophysical Journal*, 869, L41.

Bai, X.-N., & Stone, J. M. (2010) Dynamics of solids in the midplane of protoplanetary disks: Implications for planetesimal formation. *The Astrophysical Journal*, 722, 1437–1459.

Balbus, S. A., Hawley, J. F. (1991) A powerful local shear instability in weakly magnetized disks. I. Linear analysis. *The Astrophysical Journal*, 376, 214.

Barge, P., & Sommeria, J. (1995) Did planet formation begin inside persistent gaseous vortices? *Astronomy and Astrophysics*, 295, L1–L4.

Birnstiel, T., Klahr, H., & Ercolano, B. (2012) A simple model for the evolution of the dust population in protoplanetary disks. *Astronomy and Astrophysics*, 539, A148.

Bolin, B. T., Delbo, M., Morbidelli, A., & Walsh, K. J. (2017). Yarkovsky V-shape identification of asteroid families. *Icarus*, 282, 290–312.

Bottke, W. F., Durda, D. D., Nesvorný, D., et al. (2005) The fossilized size distribution of the main asteroid belt. *Icarus*, 175, 111–140.

Bottke, W. F., Vokrouhlický, D., Rubincam, D. P., & Nesvorný, D. (2006) The Yarkovsky and Yorp effects: Implications for asteroid dynamics. *Annual Review of Earth and Planetary Sciences*, 34, 157–191.

Brasil, P. I. O., Roig, F., Nesvorný, D., et al. (2016) Dynamical dispersal of primordial asteroid families. *Icarus*, 266, 142–151.

Brož, M., Morbidelli, A., Bottke, W. F., et al. (2013) Constraining the cometary flux through the asteroid belt during the late heavy bombardment. *Astronomy and Astrophysics*, 551, A117.

Carrera, D., Johansen, A., & Davies, M. B. (2015) How to form planetesimals from mm-sized chondrules and chondrule aggregates. *Astronomy and Astrophysics*, 579, A43.

Connelly, J. N., Bizzarro, M., Krot, A. N., et al. (2012). The absolute chronology and thermal processing of solids in the solar protoplanetary disk. *Science*, 338, 651.

Cuzzi, J. N., Hogan, R. C., Paque, J. M., & Dobrovolskis, A. R. (2001) Size-selective concentration of chondrules and other small particles in protoplanetary nebula turbulence. *The Astrophysical Journal*, 546, 496–508.

Cuzzi, J. N., Hogan, R. C., & Shariff, K. (2008) Toward planetesimals: Dense chondrule clumps in the protoplanetary nebula. *The Astrophysical Journal*, 687, 1432–1447.

Davis, D. R., Chapman, C. R., Weidenschilling, S. J., & Greenberg, R. (1985) Collisional history of asteroids: Evidence from Vesta and the Hirayama families. *Icarus*, 62, 30–53.

DeFelice, J. D., Friedrich, J. M., Ebel, D. S., Flores, K. E., & Weisberg, M. K. (2019) Analysis of the shapes of CAIs in CV chondrites using 2D and 3D petrography. *Lunar and Planetary Science Conference*, 2919.

Delbo, M., Avdellidou, C., & Morbidelli, A. (2019) Ancient and primordial collisional families as the main sources of X-type asteroids of the inner Main Belt. *Astronomy and Astrophysics*, 624, A69.

Delbo, M., Walsh, K., Bolin, B., Avdellidou, C., & Morbidelli, A. (2017) Identification of a primordial asteroid family constrains the original planetesimal population. *Science*, 357, 1026–1029.

Dittrich, K., Klahr, H., & Johansen, A. (2013) Gravoturbulent planetesimal formation: The positive effect of long-lived zonal flows. *The Astrophysical Journal*, 763, 117.

Durda, D. D., Bottke, W. F., Nesvorný, D., et al. (2007) Size-frequency distributions of fragments from SPH/N-body simulations of asteroid impacts: Comparison with observed asteroid families. *Icarus*, 186, 498–516.

Gail, H.-P., & Trieloff, M. (2019) Thermal history modelling of the L chondrite parent body. *Astronomy and Astrophysics*, 628, A77.

Gerbig, K., Lenz, C. T., & Klahr, H. (2019) Linking planetesimal and dust content in protoplanetary disks via a local toy model. *Astronomy and Astrophysics*, 629, A116.

Gerbig, K., Murray-Clay, R. A., Klahr, H., & Baehr, H. (2020) Requirements for gravitational collapse in planetesimal formation – The impact of scales set by Kelvin-Helmholtz and nonlinear streaming instability. *The Astrophysical Journal*, 895, 91.

Goldreich, P., & Ward, W. R. (1973) The formation of planetesimals. *The Astrophysical Journal*, 183, 1051–1062.

Gole, D. A., Simon, J. B., Li, R., Youdin, A. N., & Armitage, P. J. (2020) Turbulence regulates the rate of planetesimal formation via gravitational collapse. *The Astrophysical Journal*, 904, 132.

Gómez, G. C., Ostriker, E. C. (2005) The effect of the coriolis force on Kelvin-Helmholtz-driven mixing in protoplanetary disks. *The Astrophysical Journal*, 630, 1093–1106.

Haisch, K. E., Lada, E. A., & Lada, C. J. (2001) Disk frequencies and lifetimes in young clusters. *The Astrophysical Journal*, 553, L153–L156.

Hayashi, C. (1981) Structure of the solar nebula, growth and decay of magnetic fields and effects of magnetic and turbulent viscosities on the nebula. *Progress of Theoretical Physics Supplement*, 70, 35–53.

Henke, S., Gail, H.-P., Trieloff, M., Schwarz, W. H., & Kleine, T. (2012) Thermal history modelling of the H chondrite parent body. *Astronomy and Astrophysics*, 545, A135.

Johansen, A., Henning, T., & Klahr, H. (2006a) Dust sedimentation and self-sustained Kelvin-Helmholtz turbulence in protoplanetary disk midplanes. *The Astrophysical Journal*, 643, 1219–1232.

Johansen, A., Klahr, H., & Henning, T. (2006b) Gravoturbulent formation of planetesimals. *The Astrophysical Journal*, 636, 1121–1134.

Johansen, A., Mac Low, M.-M., Lacerda, P., & Bizzarro, M. (2015) Growth of asteroids, planetary embryos, and Kuiper belt objects by chondrule accretion. *Science Advances*, 1, 1500109.

Johansen, A., Oishi, J. S., Mac Low, M.-M., et al. (2007) Rapid planetesimal formation in turbulent circumstellar disks. *Nature*, 448, 1022–1025.

Johansen, A., & Youdin, A. (2007) Protoplanetary disk turbulence driven by the streaming instability: Nonlinear saturation and particle concentration. *The Astrophysical Journal*, 662, 627–641.

Johansen, A., Youdin, A. N., & Lithwick, Y. (2012) Adding particle collisions to the formation of asteroids and Kuiper belt objects via streaming instabilities. *Astronomy and Astrophysics*, 537, A125.

Klahr, H., & Bodenheimer, P. (2006) Formation of giant planets by concurrent accretion of solids and gas inside an anticyclonic vortex. *The Astrophysical Journal*, 639, 432–440.

Klahr, H. H., & Henning, T. (1997) Particle-trapping eddies in protoplanetary accretion disks. *Icarus*, 128, 213–229.

Klahr, H., Pfeil, T., & Schreiber, A. (2018) Instabilities and flow structures in protoplanetary disks: Setting the stage for planetesimal formation. *Handbook of Exoplanets*, 138, 2251–2286.

Klahr, H., & Schreiber, A. (2016) Linking the origin of asteroids to planetesimal formation in the solar nebula. *Asteroids: New Observations, New Models*, 318, 1–8.

Klahr, H., & Schreiber, A. (2020) Turbulence sets the length scale for planetesimal formation: Local 2D simulations of streaming instability and planetesimal formation. *The Astrophysical Journal*, 901, 54.

Kokubo, E., & Ida, S. (2012) Dynamics and accretion of planetesimals. *Progress of Theoretical and Experimental Physics*, 2012, 01A308.

Kruijer, T. S., Burkhardt, C., Budde, G., & Kleine, T. (2017) Age of Jupiter inferred from the distinct genetics and formation times of meteorites. *Proceedings of the National Academy of Science (USA)*, 114, 6712–6716.

Kruijer, T. S., Kleine, T., & Borg, L. E. (2020) The great isotopic dichotomy of the early Solar System. *Nature Astronomy*, 4, 32–40.

Kruijer, T. S., Touboul, M., Fischer-Gödde, M., et al. (2014) Protracted core formation and rapid accretion of protoplanets. *Science*, 344, 1150–1154.

Lenz, C. T., Klahr, H., & Birnstiel, T. (2019) Planetesimal population synthesis: Pebble flux-regulated planetesimal formation. *The Astrophysical Journal*, 874, 36.

Lenz, C. T., Klahr, H., Birnstiel, T., Kretke, K., & Stammler, S. (2020) Constraining the parameter space for the solar nebula. The effect of disk properties on planetesimal formation. *Astronomy and Astrophysics*, 640, A61.

Li, R., Youdin, A. N., & Simon, J. B. (2018) On the numerical robustness of the streaming instability: Particle concentration and gas dynamics in protoplanetary disks. *The Astrophysical Journal*, 862, 14.

Li, R., Youdin, A. N., & Simon, J. B. (2019) Demographics of planetesimals formed by the streaming instability. *The Astrophysical Journal*, 885, 69.

Liu, B., Ormel, C. W., & Johansen, A. (2019) Growth after the streaming instability. From planetesimal accretion to pebble accretion. *Astronomy and Astrophysics*, 624, A114.

Mac Low, M.-M., & Klessen, R. S. (2004) Control of star formation by supersonic turbulence. *Reviews of Modern Physics*, 76, 125–194.

Manger, N., & Klahr, H. (2018) Vortex formation and survival in proto-planetary discs subject to vertical shear instability. *Monthly Notices of the Royal Astronomical Society*, 480, 2125–2136.

Marchi, S., McSween, H. Y., O'Brien, D. P., et al. (2012) The violent collisional history of asteroid 4 Vesta. *Science*, 336, 690.

Michel, P., Benz, W., Tanga, P., & Richardson, D. C. (2001) Collisions and gravitational reaccumulation: Forming asteroid families and satellites. *Science*, 294, 1696–1700.

Michel, P., & Richardson, D. C. (2013) Collision and gravitational reaccu-mulation: Possible formation mechanism of the asteroid Itokawa. *Astronomy and Astrophysics*, 554, L1.

Morbidelli, A., Bottke, W. F., Nesvorný, D., & Levison, H. F. (2009) Asteroids were born big. *Icarus*, 204, 558–573.

Morbidelli, A., Libourel, G., Palme, H., Jacobson, S. A., & Rubie, D. C. (2020) Subsolar Al/Si and Mg/Si ratios of non-carbonaceous chondrites reveal planetesimal formation during early condensation in the protoplanetary disk. *Earth and Planetary Science Letters*, 538, 116220.

Nakagawa, Y., Sekiya, M., & Hayashi, C. (1986) Settling and growth of dust particles in a laminar phase of a low-mass solar nebula. *Icarus*, 67, 375–390.

Nesvorný, D., Bottke, W. F., Levison, H. F., & Dones, L. (2003) Recent origin of the Solar System dust bands. *The Astrophysical Journal*, 591, 486–497.

Nesvorný, D., Brož, M., & Carruba, V. (2015) Identification and dynamical properties of asteroid families. In P. Michel, F. E. DeMeo, & W. F. Bottke (eds.), *Asteroids IV*. Tucson: University of Arizona Press, pp. 297–321.

Nesvorný, D., Li, R., Youdin, A. N., Simon, J. B., & Grundy, W. M. (2019) Trans-Neptunian binaries as evidence for planetesimal formation by the streaming instability. *Nature Astronomy*, 3, 808–812.

Nesvorný, D., Vokrouhlický, D., Bottke, W. F., & Levison, H. F. (2018) Evidence for very early migration of the Solar System planets from the Patroclus-Menoetius binary Jupiter Trojan. *Nature Astronomy*, 2, 878–882.

Ormel, C. W., & Klahr, H. H. (2010) The effect of gas drag on the growth of protoplanets. Analytical expressions for the accretion of small bodies in laminar disks. *Astronomy and Astrophysics*, 520, A43.

Oszkiewicz, D., Kankiewicz, P., Włodarczyk, I., & Kryszczyńska, A. (2015) Differentiation signatures in the Flora region. *Astronomy and Astrophysics*, 584, A18.

Raettig, N., Klahr, H., & Lyra, W. (2015) Particle trapping and streaming instability in vortices in protoplanetary disks. *The Astrophysical Journal*, 804, 35.

Safronov, V. S. (1969) *Evoliutsiia doplanetnogo oblaka. Evolution of the protoplanetary cloud and formation of the earth and planets*. Translated from Russian. Jerusalem: Israel Program for Scientific Translations, Keter Publishing House, 212 p.

Schreiber, A., & Klahr, H. (2018) Azimuthal and vertical streaming instability at high dust-to-gas ratios and on the scales of planetesimal formation. *The Astrophysical Journal*, 861, 47.

Sekiya, M., & Ishitsu, N. (2000) Shear instabilities in the dust layer of the solar nebula I. The linear analysis of a non-gravitating one-fluid model without the Coriolis and the solar tidal forces. *Earth, Planets, and Space*, 52, 517–526.

Sekiya, M., & Onishi, I. K. (2018) Two key parameters controlling particle clumping caused by streaming instability in the dead-zone dust layer of a protoplanetary disk. *The Astrophysical Journal*, 860, 140.

Shakura, N. I., & Sunyaev, R. A. (1973) Black holes in binary systems. Observational appearance. *Astronomy and Astrophysics*, 500, 33–51.

Simon, J. B., Armitage, P. J., Li, R., & Youdin, A. N. (2016) The mass and size distribution of planetesimals formed by the streaming instability. I. The role of self-gravity. *The Astrophysical Journal*, 822, 55.

Spoto, F., Milani, A., & Knežević, Z. (2015) Asteroid family ages. *Icarus*, 257, 275–289.

Squire, J., & Hopkins, P. F. (2018) Resonant drag instabilities in proto-planetary discs: The streaming instability and new, faster growing instabilities. *Monthly Notices of the Royal Astronomical Society*, 477, 5011–5040.

Sugiura, N., & Fujiya, W. (2014) Correlated accretion ages and ϵ^{54}Cr of meteorite parent bodies and the evolution of the solar nebula. *Meteoritics and Planetary Science*, 49, 772–787.

Throop, H. B., & Bally, J. (2005) Can photoevaporation trigger planet-esimal formation? *The Astrophysical Journal*, 623, L149–L152.

Toomre, A. (1964) On the gravitational stability of a disk of stars. *The Astrophysical Journal*, 139, 1217–1238.

Tsirvoulis, G., Morbidelli, A., Delbo, M., & Tsiganis, K. (2018) Reconstructing the size distribution of the primordial Main Belt. *Icarus*, 304, 14–23.

Umurhan, O. M., Estrada, P. R., & Cuzzi, J. N. (2020) Streaming instability in turbulent protoplanetary disks. *The Astrophysical Journal*, 895, 4.

Urpin, V., & Brandenburg, A. (1998) Magnetic and vertical shear instabil-ities in accretion discs. *Monthly Notices of the Royal Astronomical Society*, 294, 399–406.

Vokrouhlický, D., Bottke, W. F., Chesley, S. R., Scheeres, D. J., & Statler, T. S. (2015) The Yarkovsky and YORP effects. In P. Michel, F. E. DeMeo, & W. F. Bottke (eds.), *Asteroids IV*. Tucson: University of Arizona Press, pp. 509–531.

Walsh, K. J., Delbó, M., Bottke, W. F., Vokrouhlický, D., & Lauretta, D. S. (2013) Introducing the Eulalia and new Polana asteroid families: Re-assessing primitive asteroid families in the inner Main Belt. *Icarus*, 225, 283–297.

Windmark, F., Birnstiel, T., Güttler, C., et al. (2012) Planetesimal formation by sweep-up: how the bouncing barrier can be beneficial to growth. *Astronomy and Astrophysics*, 540, A73.

Yang, C.-C., Johansen, A., & Carrera, D. (2017) Concentrating small particles in protoplanetary disks through the streaming instability. *Astronomy and Astrophysics*, 606, A80.

Youdin, A. N., & Goodman, J. (2005) Streaming instabilities in protopla-netary disks. *The Astrophysical Journal*, 620, 459–469.

Youdin, A. N., & Lithwick, Y. (2007) Particle stirring in turbulent gas disks: Including orbital oscillations. *Icarus*, 192, 588–604.

Youdin, A. N., & Shu, F. H. (2002) Planetesimal formation by gravitational instability. *The Astrophysical Journal*, 580, 494–505.

Zappala, V., Cellino, A., Farinella, P., & Knezevic, Z. (1990) Asteroid families. I. Identification by hierarchical clustering and reliability assessment. *The Astronomical Journal*, 100, 2030.

Isotopic Constraints on the Formation of the Main Belt

KATHERINE R. BERMINGHAM AND THOMAS S. KRUIJER

14.1 INTRODUCTION

Nucleosynthetic and radiogenic isotope data from meteorites have enabled significant advancements to be made in understanding how the protoplanetary disk evolved into the Solar System. A potentially strict isotopic divide between the inner and outer protoplanetary disk has been inferred using data from some of the earliest formed asteroids. This isotope dichotomy has been interpreted to indicate that early Jupiter formation divided the protoplanetary disk, which significantly influenced the type of material available for terrestrial planet formation. These findings are discussed in this chapter, with an emphasis on Vesta and Ceres as these asteroids likely originated from the inner and outer disk, respectively.

As remnants of planet formation, meteorites provide the most direct samples of the protoplanetary disk. These cosmochemical materials are analyzed in detail to produce a wealth of chemical, isotopic, and petrographic data which provide some of the most stringent constraints on the composition of the disk. Synthesizing this information with constraints from astronomical observations, astrophysical models, disk dynamics, and planetary accretion models contextualizes the compositional data and generates a picture of Solar System and planet formation.

Particularly valuable sample-based data are mass-independent nucleosynthetic isotope compositions of bulk meteorites. Here, "bulk meteorite" refers to a whole rock sample of a meteorite, in contrast to meteorite components[1] or mineral separates which are isolated minerals from a bulk meteorite. The range of nucleosynthetic isotope compositions recorded by meteorites stems from parent bodies incorporating different proportions of presolar grains. Over the past 15 years, datasets of nucleosynthetic isotope variations[2] in meteorites have greatly expanded (for a review see Qin & Carlson, 2016). The magnitude and nature of nucleosynthetic

isotope variations are now commonly used to infer the "building blocks" or stellar precursors of the Solar System, the dynamical processes that controlled their distribution in the protoplanetary disk and, potentially, the compositional structure of the disk itself.

Based on the nucleosynthetic isotope signatures of different meteorites, an apparently strict grouping was identified (Warren, 2011). Meteorites, and thus their parent bodies, fall into one of two groups: the noncarbonaceous chondrite group (NC) and the carbonaceous chondrite group (CC). The terminology was based on the predominance of unmelted carbonaceous chondrites in the CC group, whereas the NC group was devoid of these materials. The use of these terms to describe the isotope dichotomy is somewhat of a misnomer, because both groups include iron and stony-iron meteorites (e.g., Warren, 2011; Kruijer et al., 2017; Bermingham et al., 2018b) which are chemically and petrographically distinct from carbonaceous chondrites. The NC and CC terminology, however, is used to describe an isotopic classification that supersedes the traditional classification of meteorites into chondrite or achondrite groups as based on the composition, mineralogy, and texture of a sample. The NC–CC terminology has become convention in the literature and will be followed here.

In addition to information provided by nucleosynthetic isotope variations in meteorites, radiometric isotope signatures of meteorites and their components date Solar System and parent body processes (for a review see Kleine & Wadhwa, 2017). For example, the first condensates to form in the protoplanetary disk are CAIs, which have an average absolute U-corrected Pb–Pb age of $4{,}567.30 \pm 0.16$ Ma (e.g., Connelly et al., 2012). Calcium aluminum inclusions are small (μm to cm-sized) inclusions composed of refractory[3] element oxides. The $4{,}567.30 \pm 0.16$ Ma CAI age is generally taken to be the cosmochemical start of the Solar System (t_0) and all relative ages defined by radiometric systems are compared to this start time. Less than 1 Ma after t_0, accretion of planetesimals (whose radii range from tens to hundreds of kilometers) and differentiation began. The timescales of early accretion and differentiation of planetesimals are determined using short-lived radioactive isotope systematics and thermal modeling of meteorite parent body accretion (e.g., Kruijer et al., 2014, 2017; Sugiura & Fujiya, 2014; Desch et al., 2018; Hilton et al., 2019).

By comparing the nucleosynthetic isotope compositions of NC and CC meteorites with their accretion ages, parent body accretion locations in the disk and the timing of Jupiter's formation were

Our sincere thanks to the editors, S. Marchi, C. A. Raymond, and C. T. Russell, for their invitation to contribute to this edition and thorough reviews. We also thank E. R. D. Scott for his insightful review. K. R. B. was supported by NASA Emerging Worlds grant 80NSSC18K0496, NASA SSERVI grant NNA14AB07A, NASA Emerging Worlds grant NNX16AN07G, and the Department of Earth and Planetary Sciences, Rutgers University. T. S. K. was supported by the Laboratory Directed Research and Development Program at Lawrence Livermore National Laboratory (grant 20-ERD-001). This study was in part performed under the auspices of the US DOE by Lawrence Livermore National Laboratory under contract DE-AC52–07NA2734.

[1] Calcium aluminum inclusions (CAIs), ameboid olivine aggregates (AOAs), and chondrules (which are millimeter-sized igneous spherules).

[2] Isotope variations are defined as isotope ratios that are in excess or depletion relative to a terrestrial reference value, where terrestrial materials are generally defined as "normal" (or equal to zero) in isotopic composition.

[3] Refractory elements have 50% condensation temperatures above 1,335 K, moderately volatiles between 1,335 and 665 K, volatile elements below 665 K, and highly volatile elements below 371 K (for a gas of solar composition at a total pressure of 10^{-4} bar; Lodders, 2003).

inferred (Kruijer et al., 2017). Noncarbonaceous chondrite parent bodies were suggested to originate from the inner disk and CC parent bodies from the outer disk (Warren, 2011; Kruijer et al., 2017), where inner disk refers to the region inbound of the heliocentric distance where Jupiter formed and outer disk refers to the outbound region from Jupiter. Warren (2011) and Kruijer et al. (2017) postulated that the inner and outer disk regions were separated by an early formed Jupiter which inhibited mixing material between the two regions until giant planets migration toward the end of the disk's lifetime (following Walsh et al., 2011).

This chapter discusses NC–CC isotope dichotomy and its role in advancing the reconstruction of early Solar System evolution. The chapter begins by reviewing Main Belt asteroids and meteorites. Following this is an introduction to presolar grains and how their heterogeneous distribution in the protoplanetary disk produced nucleosynthetic isotope variability, and the significance of radiometric dating. Then there is a discussion about how the NC–CC isotope dichotomy is defined, its proposed causes, and implications for protoplanetary disk evolution. The chapter concludes with a discussion about the history of Ceres and Vesta in the context of the NC–CC isotope dichotomy.

14.2 SOLAR SYSTEM ORIGINS AND EARLY EVOLUTION

The protoplanetary disk developed from a segment of the parental molecular cloud, composed of presolar gas and dust grains, which had spun into a disk around the nascent Sun. The cause of the separation of the segment from the parental molecular cloud remains debated, but a shockwave from a nearby supernova may have triggered the collapse, or it may have occurred because of a weakening of magnetic field support following ambipolar diffusion (Cameron & Truran, 1977; Shu et al., 1987). If a supernova triggered the collapse of the disk, it is a potential source of short-lived radioactive isotopes for the Solar System (e.g., Dauphas & Chaussidon, 2011). Alternatively, the Solar System may have been formed by a triggered star formation at the edge of a Wolf-Rayet bubble (Dwarkadas et al., 2017). A supernova trigger event, however, is not required as the observed abundances of short-lived radioactive isotopes may be what is expected in average star-forming clouds (Young, 2014). The original protoplanetary disk was a common by-product of the star forming process, where such disks typically range in mass from 0.001 to 0.3 Solar masses (10^{27}–10^{29} kg) and in size from 10^{12} to 10^{14} m (Hogerheijde, 2011). The protoplanetary disk comprised gas (~99% by mass) and solids (~1%). The gas was predominately H and He, with volatile compounds such as CO and H_2O also present. The dust grains included material from different stellar sources added to the interstellar medium at least 3 ± 2 Ga before the start of the Solar System (e.g., Heck et al., 2020).

Shortly after CAIs formed, small bodies began to rapidly accrete to form planetesimals, possibly aided by electrostatic and magnetic forces and the formation of pebbles (e.g., Blum & Wurm, 2000; Lambrechts & Johansen, 2012). Collisional accretion of planetesimals and subsequent oligarchic growth produced the observed planetary system. Planetary accretion and its relationship to the dynamical evolution of the protoplanetary disk are described in detail in Chapters 13 and 15.

14.3 THE MAIN ASTEROID BELT AND METEORITES

14.3.1 The Main Asteroid Belt

All known meteorites (~40,000 specimens),[4] except those from Mars and the Moon, originate from the asteroid belt. Meteorites are estimated to come from ~100 parent bodies (e.g., Delbo' et al., 2017; Dermott et al., 2018; Greenwood et al., 2020) and many meteorites have been linked to specific asteroid types (for a review see Vernazza & Beck, 2017; Chapter 1). The extent to which the main asteroid belt samples the protoplanetary disk, however, remains uncertain, and it is likely that much of the asteroid belt has not been sampled by meteorites that fall to Earth. Nevertheless, the mineralogical, textural, elemental, and isotopic studies of meteorites reveal significant compositional diversity in the protoplanetary disk. Moreover, as meteorites are among the only available hand samples of the protoplanetary disk, they provide the most robust constraints on the composition of planetary building blocks and planetary accretion.

The main asteroid belt houses >1,000,000 asteroids, but the total mass of the main asteroid belt is ~0.05% of Earth's mass. Two of its largest bodies, Vesta and Ceres, account for ~9% and ~25% of the total mass of the asteroid belt, respectively. Presently, the asteroid belt is radially zoned, with S-type asteroids ("dry" asteroids) dominating the inner belt and C-type asteroids ("wet" asteroids) dominating the outer belt (Gradie & Tedesco, 1982; DeMeo & Carry, 2014; Chapter 1). Asteroid scattering caused by giant planet migration during the final stages of the disk's lifetime, however, likely disrupted the asteroid belt significantly (e.g., Walsh et al., 2011; Raymond & Izidoro, 2017b). Hence, the present distribution of asteroids in the Main Belt is unlikely to reflect their original accretion locations in the Solar System.

An example of the compositional diversity in the main asteroid belt is evident when comparing Vesta (V-type asteroid, 525 km diameter) and Ceres (C-type asteroid, 946 km in diameter). Vesta has been linked to howardite, eucrite, and diogenite (HED) meteorites (Binzel & Xu, 1993). Chapter 3 discusses this relationship using a synthesis of data collected during the Dawn mission and from HED meteorites. Carbonaceous chondrites have been linked to C-type asteroids, of which Ceres is an example (Burbine, 1998), but, unlike Vesta, there are no confirmed meteorite samples of Ceres.

Vesta and Ceres could not be more different in composition. Vesta is a differentiated rocky asteroid with a metallic core and silicate mantle and crust. This asteroid has a density of 3,456 kg/m^3 (Russell et al., 2012), is water- and organic-poor compared to Ceres, and may have accumulated hydrogen on the surface via the addition of carbonaceous chondrite material after accretion (Prettyman et al., 2012; Sarafian et al., 2014). By contrast, Ceres is the largest asteroid in the main asteroid belt but it has a lower density of 2,162 kg/m^3 (Russell et al., 2016; Konopliv et al., 2018) and a C-rich surface. Ceres is modeled to have 12–29 vol% of macromolecular organic matter and Ceres-forming materials may have been more water-rich than some of the most hydrated meteorites documented (e.g., CI chondrites; Prettyman et al., 2017; Marchi et al., 2019). Although gravity data of Ceres suggest a

---

[4] The meteorites discussed in this contribution do not include micrometeorites.

two-layer structure with an abrupt density change, this body did not undergo igneous differentiation like Vesta (see Chapter 12 for a comparative discussion about the structure of Ceres and Vesta).

14.3.2 Meteorites and Their Components

Although meteorites fall into one of two petrographic types, melted or unmelted, they vary significantly in composition which reflects the diversity in asteroid types and the different parts of an asteroid sampled by meteorites. Melted meteorites (also known as achondrites or differentiated meteorites) originate from rocky asteroids that underwent wholescale melting and differentiation to form a parent body with a metallic core, and silicate mantle and crust (e.g., Vesta). The melting process of what would have begun as an undifferentiated asteroid was likely caused by the presence of short-lived radioactive isotopes which produced heat via the decay of an unstable parent to a stable daughter species (short-lived species have half-lives ($t_{1/2}$) < 100 Ma; Dauphas & Chaussidon, 2011). Efficient differentiation of asteroids only occurred for those larger than a minimum threshold and at the same time contained sufficient heat-producing radioactive isotopes (capable of producing high temperatures >1,773 K in some cases; Taylor et al., 1993). The short-lived radioactive isotope ^{26}Al ($^{26}Al \rightarrow {}^{26}Mg$, $t_{1/2} = 0.73$ Ma; Lee et al., 1977; Hevey & Sanders, 2006) is considered to be the primary heat-producing isotope that caused some parent bodies to undergo wholescale melting, where after $\sim$3 Ma this heat source for melting rocky material was extinct (e.g., Elkins-Tanton, 2012). Tang and Dauphas (2012) concluded the current best estimate for the Solar System initial $^{60}Fe/^{56}Fe$ ratio ($1.0 \pm 0.3 \times 10^{-8}$) indicates heating from ^{60}Fe decay ($t_{1/2} \sim 2.6$ Ma) would have been negligible and thus this ^{60}Fe did not significantly contribute to the heat required for planetesimal differentiation. Short-lived radioactive isotope systematics indicate that differentiation of a planetesimal to form a core ($^{182}Hf-^{182}W$; Kleine et al., 2009), mantle, and crust ($^{53}Mn-^{53}Cr$; Trinquier et al., 2008) occurred relatively quickly (<3 Ma). As outlined in Chapter 3, however, some meteorites from Vesta have had longer (>10 Ma) magmatic or cooling histories.

While achondrites sample a range of different parent bodies, they also sample distinct domains of individual asteroids. For instance, some achondrites are samples of the crust, mantle, or a mixture of these silicate layers. Of these, Vesta is the most well-sampled parent body, Vesta's mantle, crust, or a mixture of the two are identified in diogenites, eucrites, or howardites, respectively. Other achondrites, iron meteorites, are samples of the core or metal-dominated portion of an asteroid. Based on their compositions, iron meteorites were originally classified into 14 different chemical groups based on their trace element compositions, where $\sim$15% are ungrouped (for review see Krot et al., 2014). Iron meteorites, however, have not yet been chemically or isotopically linked to known achondrites, which suggests that the silicate mantles of iron meteorite parent bodies are "missing," possibly destroyed by impacts in the early Solar System. Magmatic iron meteorites are FeNi alloys with ≤1 wt.% Co, S, P, and C (Buchwald, 1975; Scott & Wasson, 1976; Goldstein et al., 2009; Krot et al., 2014), and likely sample different stages of fractional crystallization of a core (Lovering, 1957; Scott, 1972). Non-magmatic iron meteorites have macroscopic silicates, graphite, and carbides, indicating a different history to magmatic iron meteorites. Unlike magmatic iron meteorites, the chemical composition of non-magmatic iron meteorites cannot be produced by simple fractionation crystallization of a core, but likely sample more complex, multi-stage metal–silicate segregation events on differentiated parent bodies (e.g., Scott, 1972; Benedix et al., 2000; Wasson & Kallemeyn, 2002; Worsham et al., 2016, 2017). Stony iron meteorites (mesosiderites or pallasites) contain mixtures of silicates, Fe-Ni metal, and troilite (for a review see Krot et al., 2014). Main group pallasites may represent the transition zone between the core and mantle of differentiated asteroids (Scott, 1977), however, updated metallographic cooling rates suggest an impact origin for these meteorites (Yang et al., 2007, 2010), possibly including an influx of core–metal from the impactor (Walte et al., 2020).

Unmelted meteorites, also known as chondrites or undifferentiated meteorites, come from asteroids that have not undergone wholescale melting. Chondrites are the most common undifferentiated meteorite type (for a review see Krot et al., 2014). Unlike melted meteorites, unmelted meteorites preserve a variety of disk-derived components that are held together by a matrix of fine-grained dust. Disk-derived components include CAI, chondrules, AOAs, and presolar grains (Krot et al., 2009; MacPherson, 2014). It is likely that both melted and unmelted meteorites accreted meteorite components, but the melting process which differentiated parent bodies altered the components beyond visual recognition. Volatile species can be more abundant in chondrites as compared to achondrites, although the volatile content varies significantly between different chondrite groups (for a recent review see Bermingham et al., 2020). Although there are no confirmed samples from Ceres, it is the largest object in the main asteroid belt and has been classified as a "wet" unmelted asteroid. The exact reason why Ceres remained undifferentiated may well be related to the absence of sufficient ^{26}Al to cause widespread melting of the parent body. Why Ceres might have lacked ^{26}Al is not yet clear, but it could be related to a relatively late accretion time (i.e., after extinction of ^{26}Al) and/or processes that internally redistributed ^{26}Al on Ceres (see Chapter 11 for more details).

14.3.3 Presolar Grains

As chondrites have not undergone wholescale melting, they preserve remnants of material that was present in the protoplanetary disk. In addition to meteorite components such as CAIs, chondrules, and AOAs, chondrites preserve small (≤µm-scale) presolar grains. Presolar grains are defined as "stardust that formed in stellar outflows or ejecta and remained intact throughout its journey into the solar system where it was preserved in meteorites" (Lodders & Amari, 2005, p. 94). Through laboratory studies of residues that remained after dissolving undifferentiated meteorites in concentrated oxidizing acid, part of the $\sim$1% (by mass) of the protoplanetary disk's solid material inherited from the parental molecular cloud was identified. Through isotopic and structural analysis of the residual slurries and individual grains, these chemically robust solids were recognized as presolar in origin (Lewis et al., 1987; Zinner et al., 1987). Presolar materials are characterized by highly variable ($\sim$1 to 0.1%) isotopic compositions which develop by virtue of their formation in stellar environments (for a review see Zinner, 2014).

Lodders and Amari (2005) and Huss and McSween (2010) provide thorough reviews of presolar grains, their origins, composition, and significance in cosmochemistry. To summarize these reviews, there are two types of presolar grains: circumstellar condensates and interstellar grains. Circumstellar condensates are *bone fide* presolar grains. They condensed from hot gas ejected from dying stars in the immediate vicinity of their parent stars and remained intact until they were preserved in meteorites (Huss & McSween, 2010). The elemental and isotopic composition of these grains can be directly linked to their stellar source, and presolar grains have markedly different compositions to materials formed within the Solar System. Interstellar grains are distinct from circumstellar grains because they formed in interstellar space. These grains did not survive the long journey from their stars, rather they are vaporized circumstellar condensates which eventually recondensed into grains within molecular clouds. Interstellar grains often lack a mineral structure and have variable elemental and isotopic compositions which reflects their mixture of material from different stellar sources (Huss & McSween, 2010).

lithophile isotope composition is best represented by whole rock analyses of meteorites originating from the silicate portion of the body. Bulk parent body siderophile isotope compositions of melted bodies can most reliably be obtained by analyzing material that sample well-mixed siderophile element compositions, such as from the core (magmatic iron meteorites) or metal portions of large impact melt sheets (non-magmatic iron meteorites) (Worsham et al., 2016; Bermingham et al., 2018b).

Collectively, the small-scale nucleosynthetic isotope variations among bulk meteorites reflect that (i) the protoplanetary disk contained presolar grains with strongly variable isotopic compositions, and (ii) processes occurring in the protoplanetary disk reasonably but not perfectly, mixed these grains into the disk. Moreover, the nucleosynthetic isotope variations are not only useful for identifying the stellar building blocks of the Solar System, but may also provide valuable information about how and to what extent material was transported and mixed in the protoplanetary disk.

14.4 NUCLEOSYNTHETIC ISOTOPE ANOMALIES

A comparison between meteorites and terrestrial samples reveals isotopic variability in the following elements: Ca, Ti, Cr, Ni, Sr, Zr, Mo, Ru, Pd, Ba, W, Nd, Sm, HREE, and noble gases (Ne, Ar, Kr, Xe) (e.g., Warren, 2011; Ott, 2014; Mayer et al., 2015; Qin & Carlson, 2016; Bermingham et al., 2016, 2018a, 2018b; Dauphas & Schauble, 2016; Kruijer et al., 2017; Nanne et al., 2019). These isotope variations are independent of radiogenic decay, mass-dependent anomalies, cosmic ray exposure effects, and nuclear field shift effects, although some of these effects can overprint nucleosynthetic isotope compositions. Nucleosynthetic isotope compositions vary from terrestrial compositions by 10s to 1,000s of parts per million (ppm). The variations are reported in either epsilon (ε), which reflects the compositional deviation from a terrestrial standard in parts per 10,000, or mu (μ), the parts per 1000,000 deviation.

Nucleosynthetic isotope anomalies in bulk meteorites and planetary bodies arise as a consequence of the heterogeneous distribution of highly isotopically variable presolar grains in the protoplanetary disk. Nucleosynthetic isotope variations are similar to those seen in presolar grains, but they are of smaller magnitude ($\sim$0.001–0.01%). The heterogeneous distribution of presolar grains is thought to be the result of poor mixing of these grains into the protoplanetary disk (Clayton, 1982, and references therein) and/or their selective destruction via thermal chemical processing in the protoplanetary disk (Trinquier et al., 2009).

Bulk meteorite isotopic variations are generally considered to reflect the average composition of the parent body and thus the nebular region from which it accreted (e.g., Walker et al., 2015). Nucleosynthetic isotope variations are used to assess the "genetic" relationships between meteorites and their parent bodies, where the term "genetic" refers to the average isotopic compositions of the primary nebular materials from which an object was built (Walker et al., 2015).

The nucleosynthetic isotope composition of an unmelted parent body is determined by bulk rock lithophile and siderophile isotope measurements of a meteorite. In melted parent bodies, the bulk

14.5 LONG-LIVED AND SHORT-LIVED ISOTOPE CHRONOMETERS

Radiometric isotope dating is a method that determines the chronology of events. An age can be determined by comparing the initial abundance of a radioactive isotope to the abundance of its decay product in material present during the event, where the rate of decay occurs at a known and constant rate (International Union of Pure and Applied Chemistry, 2006). Using this technique, formation age limits are placed on components found in unmelted meteorites (e.g., CAIs and chondrules), as well as on the timescales of differentiation on the parent bodies of melted meteorites. Information on the timescales of parent-body accretion is obtained indirectly, either by dating the formation of a specific component (e.g., chondrules) that is closely linked in time to the accretion of their parent body or, alternatively, by dating a specific chemical differentiation process (e.g., core formation), which can be linked to the time of parent-body accretion through thermal modeling (for reviews see Kleine & Wadhwa, 2017; Kleine & Walker, 2017).

There are two types of chronometers (long-lived and short-lived) that are used to date Solar System and parent body processes. These chronometers are distinguished based on the half-life of the parent isotope. As their parent isotopes have been decaying until the present-day and beyond, long-lived decay systems can provide *absolute* ages. An important example of a long-lived chronometer is the U-Pb system, which provides precise absolute ages of meteorites and their components, provided that they are corrected for variable $^{235}U/^{238}U$ in early Solar System materials (Brennecka et al., 2010). For instance, the Pb–Pb age of CAIs defines the start of the Solar System.

The half-lives of short-lived systems used in cosmochemistry range from ca. 0.1 Ma ($^{41}Ca \rightarrow {}^{41}K$) to ca. 103 Ma ($^{146}Sm \rightarrow {}^{142}Nd$). Short-lived chronometers produce *relative* ages and are usually expressed relative to an age anchor for which precise absolute ages are available. Commonly, this absolute age is taken to be the age of CAIs, as the formation of these materials is taken to reflect the start of the Solar System (Amelin et al., 2002, 2010; Connelly et al., 2012). Short-lived systems are capable of recording events that occurred during the first few 100 Ma of Solar System history because ingrowth of the daughter nuclide cannot occur after the extinction of the parent nuclide. Long-lived systems may also record these early events; however, these records may be over-printed by younger events, thereby diminishing the utility of long-lived systems for recording events during the first few 100 Ma.

Examples of short-lived chronometers relevant to reconstructing the chronology of events in the early Solar System include the ^{26}Al–^{26}Mg and ^{182}Hf–^{182}W decay systems. Because of the very short half-life of ^{26}Al ($t_0 \sim 0.73$ Ma), the Al–Mg system provides very precise relative ages for meteorites and their components. The ^{182}Hf–^{182}W ($t_0 \sim 8.9$ Ma; Vockenhuber et al., 2004) system can also be used to obtain relative ages for meteorites and their components, but it is most well-known for its ability to date metal–silicate segregation (e.g., core formation). This application is based on the different geochemical nature of the refractory parent (Hf) and daughter (W) species, where W is moderately siderophile and Hf strongly lithophile. Thus, during core–mantle differentiation a high Hf/W ratio develops in the silicate mantle, whereas a Hf/W of essentially zero develops in the Fe–Ni core. If this segregation of Hf and W occurs during the effective lifetime of ^{182}Hf (~ 50 Ma, six half-lives), the ^{182}Hf–^{182}W system can provide model ages for the timing of core formation in planetary bodies (for a recent review see Kleine & Walker, 2017).

The application of short-lived chronometers requires robust initial abundances of each parent isotope and their homogeneous distribution in the early Solar System. While most radionuclides (e.g.,^{182}Hf) were homogeneously distributed in the early Solar System, at least at the level relevant for chronological studies, this is debated for ^{26}Al (e.g., Krot et al., 2012; Schiller et al., 2015; Van Kooten et al., 2016). The agreement of Al–Mg and Hf–W ages of meteorites, however, is inconsistent with the presence of significant ^{26}Al heterogeneity at the bulk meteorite scale (Kruijer et al., 2014; Budde et al., 2018).

14.6 THE CARBONACEOUS (CC)–NONCARBONACEOUS (NC) DICHOTOMY

The chemistry and chronology of planet formation can be constrained by documenting the composition and age of a planet's meteoritic building blocks. If these characteristics can be linked to accretion locations in the protoplanetary disk, a time-dependent compositional and spatial "map" of the early Solar System could be constructed. Such information would provide constraints on the composition of the disk, its influence on the composition of the planets, the extent of inner and outer disk mass transfer during planet building, and potentially the origin of volatiles on the terrestrial planets.

The most promising means of obtaining such information is through combined elemental and isotopic analysis of meteorites. A major hurdle, however, is being able to robustly link these (volatile) compositions to physical accretion locations of their parent bodies in the evolving dynamic, three-dimensional protoplanetary disk (Bermingham et al., 2020). As outlined in Chapter 15, although there is a general compositional trend observed in the asteroid belt (dry bodies in the inner belt, more wet bodies in the outer belt), it is unlikely that the asteroids originally accreted in their current location in the belt. On the contrary, the asteroid belt was likely composed of planetesimals that accreted in vastly different regions of the disk but were subsequently implanted in the asteroid belt during gas giant migration and subsequent scattering of planetary bodies (e.g., Walsh et al., 2011; Raymond & Izidoro, 2017; see Chapter 15). As such, the cosmochemical map that the asteroid belt and meteorites seemingly provides may not be as informative as initially envisioned. Recent interpretation of the NC–CC classifications of meteorites and their accretion age, however, has seen the inference of accretion locations of parent bodies, which has led to an integration of meteoritic constraints into large-scale models of disk evolution and planet formation (for recent reviews see Desch et al., 2018; Scott et al., 2018; Kleine et al., 2020; Kruijer et al., 2020).

14.6.1 The NC–CC Dichotomy

To understand the compositional or genetic relationships between meteorites, earlier studies primarily used O isotopes (e.g., Clayton, 1993), where O isotopes variations are likely not nucleosynthetic in origin and instead reflect photochemical processes in the protoplanetary disk or the molecular cloud (for a review of O isotopes, see Ireland et al., 2020). Over the past few decades, several studies demonstrated that carbonaceous chondrites are isotopically distinct in O isotopes and in heavier isotopes (e.g., ^{50}Ti, ^{54}Cr, ^{58}Ni) from ordinary and enstatite chondrites, several achondrite groups (including the HED meteorites), and the Earth, Moon, and Mars (Niemeyer, 1988; Rotaru et al., 1992; Clayton, 1993; Trinquier et al., 2007; Leya et al., 2008; Regelous et al., 2008). The fundamental nature of this finding with regards to the evolution of the protoplanetary disk was first recognized by Warren (2011). The author coined the terms NC and CC and suggested that, given the Earth, Moon, and Mars have compositions which fall within the NC field, NC meteorites probably come from an inner Solar System reservoir. In contrast, because some of the more volatile-rich CC meteorites were considered to come from the outer Solar System, CC meteorites probably come from an outer Solar System reservoir (for a recent review of some of the complexities involved, see Bermingham et al., 2020). Warren (2011) postulated that the NC and CC regions were separated by proto-Jupiter. The genetic dichotomy has since been interpreted as a pervasive division of meteorites into NC and CC meteorites.

The NC–CC dichotomy is most clearly observed when different nucleosynthetic tracers (e.g., ^{50}Ti versus ^{54}Cr) are plotted against each other (Figure 14.1) and two distinct compositional fields become apparent. Recently, it has become apparent that the NC–CC dichotomy extends to several other isotope systems (e.g., siderophile elements Mo, Ru, W) and meteorite groups (achondrites, grouped and ungrouped iron meteorites, pallasites) (e.g., Budde et al., 2016; Kruijer et al., 2017; Poole et al., 2017; Worsham et al., 2017, 2019; Bermingham et al., 2018b; Hilton et al., 2019; Tornabene et al., 2020). Despite isotopic variations within the individual NC and CC reservoirs, there is a gap in between the two groups which complicates the possibility of grouping representing a mixing continuum in compositions. Rather it implies minimal to absent mixing between the NC and CC reservoirs during the formation times of the meteorite parent bodies.

Several publications have investigated the NC–CC isotope dichotomy using nucleosynthetic isotope variations defined by

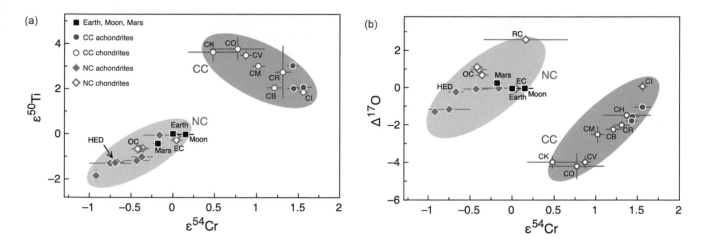

Figure 14.1 NC–CC meteorite dichotomy inferred from isotopic signatures of bulk meteorites. (a) ε^{50}Ti versus ε^{54}Cr, (b) Δ^{17}O versus ε^{54}Cr. Note that 1 ε-unit represents the 0.01% deviation (and 1 δ-unit the 0.1% deviation) in the isotopic ratio of a sample relative to terrestrial rock standards. Mass-independent O isotope variations are expressed in Δ^{17}O (Δ^{17}O $\equiv \delta^{17}$O $- 0.52\ \delta^{18}$O, where 0.52 is the slope of mass-dependent mass fractionation). Errors bars denote external uncertainties (2σ) reported in respective studies. The isotopic data shown are based on that initially summarized in Warren (2011) with several updates (see Kruijer et al., 2020, and references therein).

lithophile and siderophile elements (e.g., Desch et al., 2018; Scott et al., 2018; Alexander, 2019a, 2019b). The rest of this section primarily focuses on recent advances made through the isotopic analysis of siderophile elements Ni, Mo, Ru, and W, which provide information on the origin and dynamical implications of the NC–CC dichotomy.

Nucleosynthetic isotope variations observed for Mo are particularly illustrative because, unlike for elements such as Ti and Cr, the isotopic composition of Mo has been precisely analyzed in almost all classes of meteorites. Moreover, Mo isotopes permit isotopic anomalies of distinct nucleosynthetic origins to be distinguished. Specifically, the heterogeneous distribution of carriers enriched in nuclides produced in the *s*-process of stellar nucleosynthesis (often associated with asymptotic giant branch, AGB, stars) and the rapid neutron capture process *r*-process (associated with neutron-rich stellar environments, such as supernovae or neutron star merger events) yields different patterns of Mo isotope anomalies within individual samples (Dauphas et al., 2002; Burkhardt et al., 2011).

Although nucleosynthetic Mo isotope variations have received interest for almost two decades, several recent high-precision Mo isotope studies have greatly expanded the available data for bulk meteorites (e.g., Kruijer et al., 2017; Poole et al., 2017; Worsham et al., 2017, 2019; Bermingham et al., 2018a, 2018b; Budde et al., 2019; Hilton et al., 2019; Tornabene et al., 2020). For Mo, the NC–CC dichotomy is most clearly seen in a plot of ε^{94}Mo versus ε^{95}Mo, where stony, stony iron, and iron NC and CC meteorites define two approximately parallel *s*-process mixing lines which are offset from each other (Figure 14.2) (Budde et al., 2016; Kruijer et al., 2017; Poole et al., 2017; Worsham et al., 2017; Hilton et al., 2019; Yokoyama et al., 2019; Tornabene et al., 2020). Why this difference in NC and CC Mo isotope compositions persists remains debated. This offset has been considered to reflect an approximately homogeneous enrichment in *r*-process (and possibly *p*-process) nuclides in the CC compared to the NC reservoir (Budde et al., 2016; Kruijer et al., 2017; Poole et al., 2017; Worsham et al., 2017). Regardless of the cause of the isotopic

difference between the NC and CC groups, the bimodality is found in many bulk meteorite samples and thus appears to record a fundamental property of the protoplanetary disk.

Whereas the *s*-process variability in each reservoir can potentially be controlled by individual presolar carrier phases (or derivatives thereof), the homogeneous enrichment in *r*-process Mo in the CC over the NC reservoir cannot. It has been interpreted to be a characteristic composition of the entire CC reservoir (Budde et al., 2016; Worsham et al., 2017, 2019; Burkhardt et al., 2019; Nanne et al., 2019). A key finding based on the Mo isotope data is that the NC and CC reservoirs do not only contain chondrites, but also iron meteorites and pallasites (Kruijer et al., 2017; Poole et al., 2017; Worsham et al., 2017; Bermingham et al., 2018b; Hilton et al., 2019; Tornabene et al., 2020) and other achondrites (Worsham et al., 2017, 2019; Budde et al., 2019). For both the NC and CC reservoirs, iron meteorites and chondrites plot on single s-process mixing lines (i.e., the NC- and CC-lines), implying that CC irons, achondrites, and carbonaceous chondrites have the same characteristic *r*-process excess over the NC irons and enstatite and ordinary chondrites.

The dichotomy exists for elements covering a wide range of condensation temperatures (non-refractory elements like Cr and Ni, as well as refractory elements such as Ni, Mo, Ru, and W) and both for lithophile (Ti, Cr) and siderophile (Ni, Mo, Ru, W) elements. As the NC–CC dichotomy exists for elements covering a broad range of geochemical and cosmochemical behavior, the dichotomy may indicate a ubiquitous characteristic of the protoplanetary disk. For some element pairs, nucleosynthetic isotope signatures are correlated in one reservoir but not the other (Worsham et al., 2019). For example, W and Mo isotope signatures show correlations for CC, but not for CC meteorites, whereas Mo and Ru isotope signatures are correlated in the NC but not the CC reservoir. These more complex inter-element isotope relationships may reflect distinct redox and thermal conditions at which presolar carrier phases were processed in the individual reservoirs (Burkhardt et al., 2012; Fischer-Gödde et al., 2015). This observation suggests that the CC reservoir may have been characterized by more oxidizing conditions

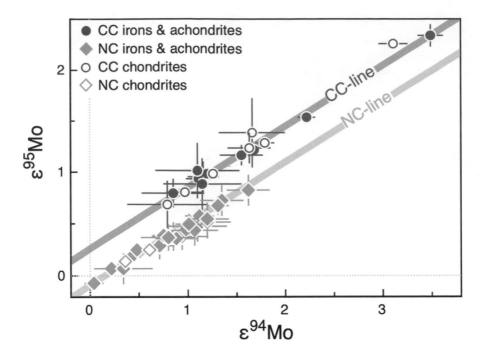

Figure 14.2 ε^{95}Mo versus ε^{94}Mo data for bulk meteorites. NC (grey) and CC (black) meteorites define two approximately parallel *s*-process mixing lines with identical slopes, but distinct intercept values (Budde et al., 2016, 2019; Kruijer et al., 2017). Error bars denote external uncertainties reported in respective studies (2σ).

than the NC reservoir (Worsham et al., 2019). If this differential volatility model is supported by elemental compositions remains an open question exceptions include iron meteorite Wiley and members of the IIC iron meteorite group (Tornabene et al., 2020). These samples have chemical and isotopic compositions that reflect either selective thermal processing of nucleosynthetic carriers, or are genetically distinct from the CC and NC precursor materials (Tornabene et al., 2020).

Collectively, isotopic data reveal that the NC–CC dichotomy exists widely, demonstrating that the NC–CC dichotomy is a fundamental characteristic of the meteorite record. The dichotomy indicates that the NC and CC meteorites stem from different reservoirs that did not mix, or only very minimally mixed during the formation period of meteorite parent bodies. As will be illustrated, the discovery of the NC–CC dichotomy, combined with a precise chronology for the accretion of meteorite parent bodies, has led to new large-scale models of disk evolution and planet formation.

14.7 NC–CC DICHOTOMY: IMPLICATIONS FOR THE STRUCTURE AND DYNAMICAL EVOLUTION OF THE EARLY SOLAR SYSTEM

Given the predominance of the NC–CC grouping in meteorites, the dichotomy may provide a fundamental constraint on the structure and dynamical evolution of the early Solar System. The dichotomy has been interpreted to reflect the formation of meteorite parent bodies in two spatially separated regions in the disk. Understanding the significance of the NC-CC grouping requires knowledge of the accretion times of NC and CC meteorite parent bodies. Such information can be obtained from the chronology of meteorites and their

components. This section focuses on those ages that provide the most precise constraints on the accretion timescales of NC and CC meteorite parent bodies, rather than providing a comprehensive summary of the chronology of meteorites.

14.7.1 Chronology of NC and CC Meteorites

14.7.1.1 Differentiated (or Melted) Meteorites

Collectively, meteorite ages demonstrate that planetesimal differentiation occurred within the first few million years after CAI formation, consistent with heating driven mainly by ^{26}Al decay. Direct evidence for early planetesimal differentiation comes from the ^{182}Hf–^{182}W chronometry of magmatic iron meteorites.

Model ages obtained using the ^{182}Hf–^{182}W system indicate that magmatic iron meteorite parent bodies segregated their cores very early in Solar System history, between ca. 0.5 and 3 Ma after Solar System formation (Kruijer et al., 2014, 2017). Although the ^{182}Hf–^{182}W ages of NC and CC iron meteorite groups are not clearly resolved outside their uncertainties (Hellmann et al., 2019; Hilton et al., 2019; Kruijer & Kleine, 2019), the Hf–W chronology of iron meteorites demonstrate that core formation on their parent bodies occurred around ~0.5–2 Ma (NC irons) and ~2–3 Ma (CC irons) after CAI formation (Kruijer et al., 2014, 2017, 2020). Combining the ^{182}Hf–^{182}W ages of iron meteorites with thermal modeling for small bodies heated by ^{26}Al decay yields accretion ages for iron meteorite parent bodies of ~0.5 Ma (NC irons) and ~1 Ma (CC irons) after CAI formation, with some overlap in accretion age between NC and CC parent bodies (Hilton et al., 2019). Iron meteorite parent bodies, therefore, accreted within <1 Ma after Solar System formation and are among the first planetesimals formed in the Solar System preserved in the meteorite collection.

In principle, accretion and core formation timescales can also be inferred for the parent bodies of differentiated achondrites (e.g., eucrites, angrites, ureilites). These accretion ages, however, are less well constrained than those of iron meteorites, because additional parent–daughter (e.g., $^{182}Hf-^{182}W$ or $^{26}Al-^{26}Mg$) chemical fractionation events in the silicate mantles after core formation can complicate the chronology recorded by the radiometric systems. The isotopic compositions of these samples may, therefore, reflect more than one differentiation event, thereby adding substantial uncertainty to the model ages for core formation. Nevertheless, the available data are most consistent with early core formation on the angrite and eucrite parent bodies, well within the first ~1–2 Ma of the Solar System (Bizzarro et al., 2005; Kleine et al., 2012; Touboul et al., 2015), and thus approximately as early as the iron meteorite parent bodies. This is consistent with thermal models for the early evolution of Vesta that also predict a very early formation (Formisano et al., 2013; Neumann et al., 2014). Collectively the meteorite data indicate that Vesta and the angrite and ureilite parent bodies likely accreted and segregated its core very early in the history of the Solar System. Note that the timing of subsequent minor additions of CC materials to Vesta remains unconstrained by the meteorite data.

14.7.1.2 Undifferentiated (or Unmelted) Meteorites: Chondrules

The accretion of undifferentiated parent bodies of chondrites cannot be dated directly but must instead be inferred by dating individual meteorite components that formed just prior to or during accretion of their host chondrite (Alexander et al., 2008). Of these components, chondrules are the most abundant as well as most extensively dated component (see e.g., Krot et al., 2018, and references therein). Radiometric ages of chondrules are obtained either by pooling multiple chondrules (Pb–Pb, Hf–W) or by dating individual chondrules (Pb–Pb, Al–Mg,). Perhaps the strongest constraint comes from $^{26}Al-^{26}Mg$ chronometry of individual chondrules from the least altered chondrites, which yields clear age peaks at ~2–3 Ma (for chondrules from ordinary, CV, and CO chondrites) and at ~3.7 Ma (CR chondrites), after CAI formation (Villeneuve et al., 2009; Kita & Ushikubo, 2012; Nagashima et al., 2017; Schrader et al., 2017; Pape et al., 2019). These ages are in excellent agreement with $^{182}Hf-^{182}W$ (Budde et al., 2016, 2018) and Pb–Pb ages of pooled chondrule separates from CV and CR chondrites (Amelin et al., 2002; Amelin & Krot, 2007; Connelly et al., 2008; Connelly & Bizzarro, 2009). Thus, the vast majority of chondrules formed between ~2 and ~4 Ma after CAI formation. Moreover, chondrules from a given chondrite group formed in a narrow time span of <1 Ma, suggesting they rapidly accreted into their parent bodies. This inferred narrow time interval of chondrule formation is also supported by independent evidence from the chronology of secondary alteration products (for example, carbonates and secondary fayalites; e.g., Doyle et al., 2015; Jogo et al., 2017) as well as with studies combining meteorite chronology with thermal modeling of bodies by ^{26}Al decay (Henke et al., 2012; Blackburn et al., 2017; Hellmann et al., 2019). This suggests that the $^{26}Al-^{26}Mg$ chondrule ages are chronologically meaningful, which implies that the chondrule ages closely approximate the time of chondrite parent body accretion.

The Pb–Pb ages for individual chondrules from ordinary, CV, and CR chondrites, however, are more variable and some even appear to be as old as CAIs (Connelly et al., 2012; Bollard et al., 2017). These authors suggested that the discrepancy between the Al–Mg and Pb–Pb ages reflects local variations of $^{26}Al/^{27}Al$ in the disk, and specifically, a reduced initial $^{26}Al/^{27}Al$ in the inner Solar System (Bollard et al., 2019). This interpretation, however, is difficult to reconcile with the good agreement between Hf–W and Al–Mg ages for several meteorites and meteorite components. This includes samples with formation ages varying by almost 5 Ma and materials stemming from both the inner (NC) and outer (CC) disk (Budde et al., 2018; Kleine et al., 2020; Kruijer et al., 2020). Thus, the discrepancy between Al–Mg and Pb–Pb ages for single chondrules cannot be caused by spatial heterogeneities of $^{26}Al/^{27}Al$. Another possibility to explain this discrepancy is that the Pb–Pb ages were shifted toward older ages due to the loss of intermediate decay products within the U–Pb decay chains (Pape et al., 2019). Nevertheless, regardless of the variable Pb–Pb ages reported for some individual chondrules and the exact origin for these variations, there is strong evidence that the vast majority of chondrules formed between ~2 and ~4 Ma after CAI formation. Thus, collectively, the available Al–Mg, U–Pb, and Hf–W chronology of chondrules indicates that chondrite parent body accretion in both the NC and CC reservoirs occurred later than that of iron meteorites, around ~2 Ma in the NC reservoir (ordinary chondrites) and continuing until at least ~4 Ma after CAI formation in the CC reservoir (CR chondrites).

14.7.2 Time of Formation and Lifetime of the NC–CC Dichotomy

Combined, the chronology of meteorite parent body accretion indicates that meteorite parent body accretion in both NC and CC reservoirs (i) commenced within <1 Ma after CAI formation and (ii) lasted several Ma, up to at least 2 Ma in the NC reservoir and up to ~4 Ma in the CC reservoir. A key observation is that the NC- or CC-characteristic isotopic signatures of their respective nebular reservoirs were not significantly modified during this period. This is evident in the observation that early formed iron meteorites and later-formed chondrites plot on two individual s-process mixing lines in Mo isotope space (Figure 14.2). Even though the Mo isotopic data reveal small deviations from each line (Yokoyama et al., 2019), these differences are small compared to the overall offset between the NC- and CC-lines (Spitzer et al., 2020). Combined, the nucleosynthetic signatures and ages of NC and CC meteorites thus suggest that there was minimal influx of CC matter into the NC reservoir between ~0.5 and ~2 Ma after CAI formation, and negligible influx of NC matter into the CC reservoir between ~1 and ~4 Ma after CAIs. Collectively, the isotopic data indicate the presence of two co-existing and spatially separated NC and CC reservoirs for several Ma (Kruijer et al., 2017).

14.7.3 Possible Mechanisms for the NC–CC Isotopic Divide and the Jupiter Barrier

The spatial separation of NC and CC reservoirs inferred from the isotopic data requires an efficient mechanism preventing the radial exchange of NC and CC materials (Figure 14.3). One way to retain

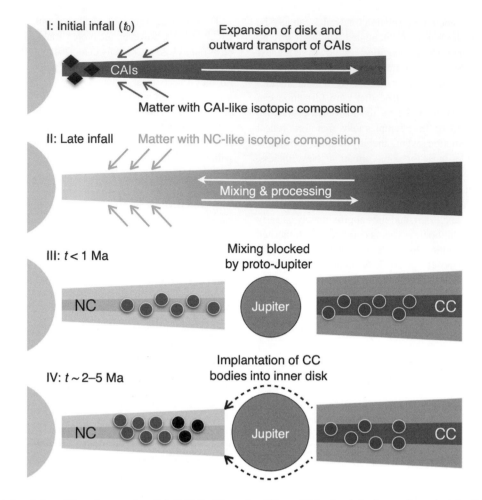

Figure 14.3 Postulated evolution of the solar accretion disk. Initially (Stages I and II, $t \sim 0$ Ma after Solar System formation) the disk grew and facilitated relatively rapid transport and mixing of nebular dust. The growth of proto-Jupiter (Stage III, $t < 1$ Ma), however, may have acted as a barrier against material transport, preserving any pre-existing isotope difference between NC and CC materials. This would have been followed (Stage IV, $t \sim 2$–5 Ma) by the opening of a gap within the disk and/or Jupiter's migration. In either case, Jupiter's growth likely facilitated the inward scattering of planetesimals and their implantation into the main asteroid belt region.

Figure modified after Nanne et al. (2019) and Kruijer et al. (2020)

the NC–CC isotope dichotomy over several Ma is the rapid accretion of dust into planetesimals with more stable orbits, which might have prevented exchange of dust between the NC and CC reservoirs. This mechanism, however, does not explain the observation that meteorite parent bodies with the same characteristic NC–CC isotopic difference continued to accrete for several Ma. Rapid radial transport and mixing of dust in the disk would have homogenized the NC–CC isotopic difference in a much shorter period (Weidenschilling, 1977; Birnstiel et al., 2013). Alternatively, if it is assumed that the NC group formed in the inner Solar System and the CC formed in the outer Solar System, a spatial separation of NC and CC reservoir may have developed due to the early formation of proto-Jupiter. The formation of Jupiter could have prevented material transport between the inner and outer disk in two ways, either by the growth of Jupiter itself (Warren, 2011; Morbidelli et al., 2016; Kruijer et al., 2017) or alternatively through a pressure maximum or dust trap at the orbital distance where Jupiter subsequently formed (Brasser & Mojzsis, 2020). The early presence of proto-Jupiter could have prevented the inward and outward transport of dust grains after it had reached ca. 10–20 Earth masses (Lambrechts & Johansen, 2012; Alibert et al., 2018), and thus eventually divided

the early disk into two separate NC and CC regions with limited mixing between them (Kruijer et al., 2017). If correct, the ancient accretion ages obtained for NC and CC iron meteorites indicate that the Jupiter barrier was present very early, within <1 Ma after CAI formation (Figure 14.3, Stage 3). This would imply that Jupiter is the oldest planet of the Solar System, and the solid core of Jupiter must have formed well within ~ 1 Ma after CAI formation.

In detail, the efficacy of the Jupiter barrier depends on the grain size of the dust drifting inwards and the size and growth history of Jupiter. The Jupiter barrier may have resulted in a filtering effect, that allowed small (ca. 100 μm sized) dust grains to pass through, but efficiently hampered the drift of larger (ca. mm-sized) grains (Weber et al., 2018). This filtering process may have imparted small isotopic changes within the NC reservoir, as seen in ε^{50}Ti versus ε^{54}Cr space (Figure 14.1) as well as in nucleosynthetic Mo (Spitzer et al., 2020) and Ca isotope signatures (Schiller et al., 2018). Hence, one possibility is that the isotopic composition of the inner Solar System might have changed continuously through the addition of CC dust. Alternatively, the small isotopic variations among NC meteorites record a rapidly changing composition of the disk during infall from the Sun's parental molecular cloud, where each planetesimal locks the instant composition of the disk when it forms (Nanne et al., 2019; Spitzer et al., 2020). An important implication of the infall model is that later accreted planetesimals in the inner disk primarily formed from secondary dust produced by collisions among pre-existing NC planetesimals.

The early growth of Jupiter is consistent with a previously proposed mechanism for the implantation of carbonaceous chondrites parent bodies from the outer disk into the inner disk (e.g., Walsh et al., 2011). This is because the subsequent growth of Jupiter (to ca. 50 Earth masses) would eventually have led to the formation of a gap within the disk (Crida et al., 2006). This likely ultimately led to inward migration of Jupiter and gravitational scattering of outer Solar System parent bodies from beyond its orbit into the inner Solar System (Walsh et al., 2011; Raymond & Izidoro, 2017a) (Figure 14.3, Stage 4). Although the exact timing remains somewhat uncertain, the available isotopic data of NC and CC meteorites suggest that migration and inward scattering of CC bodies occurred between ~2 and ~5 Ma after CAI formation (Kruijer et al., 2017, 2020). Consistent with this, a recent study identified one meteorite with a mixed NC–CC isotopic composition that was established relatively late, implying that it may have formed by collisional mixing of NC and CC planetesimals (Spitzer et al., in press). The implantation of CC bodies into the inner Solar System probably led to mixing of CC and NC materials from this time onward. Jupiter subsequently grew to its final size (~318 Earth masses) because of gas accretion onto its core (Pollack et al., 1996), which must have happened prior to dissipation of the nebular gas, probably ~4–5 Ma. The inferred timescale for Jupiter's growth based on cosmochemical data is consistent with predictions from the core accretion model for giant planet formation (Pollack et al., 1996; Alibert et al., 2018).

14.7.4 Origin of the NC–CC Dichotomy

Although the cause of the isotopic difference between the NC and CC reservoirs is still debated, three key characteristics of the NC–CC meteorite dichotomy provide important clues about its origin. First, the dichotomy exists for elements with a wide range of condensation temperatures, including refractory (e.g., Ti, Mo, W) and non-refractory and/or volatile elements (e.g., Cr, Ni). Second, the CC reservoir is relatively enriched in nuclides produced in neutron-rich stellar environments, as demonstrated by excesses in ^{50}Ti, ^{54}Cr, ^{58}Ni, and *r*-process Mo. Third, similar but more pronounced enrichments in neutron rich isotopes are found in CAIs (Papanastassiou, 1986; Burkhardt et al., 2011; Brennecka et al., 2013; Davis et al., 2018), the oldest objects of the Solar System. From these observations, it seems unlikely the NC–CC dichotomy reflects preferential destruction and volatilization of isotopically anomalous material through locally elevated temperatures within the disk (e.g., Trinquier et al., 2009; Burkhardt et al., 2011), because this would have resulted in disparate effects on carriers of elements with different volatilities (Nanne et al., 2019). In addition, such thermal processing would not selectively affect carriers from specific neutron-rich stellar environments. Instead, it has been suggested that the NC–CC dichotomy reflects an isotopic heterogeneity in the disk that was imparted during heterogeneous infall from the molecular cloud (Burkhardt et al., 2019; Jacquet et al., 2019; Nanne et al., 2019; Figure 14.3, Stages I and II). In these models, the earliest solar accretion disk, which formed by viscous spreading of early infalling matter (Yang & Ciesla, 2012; Desch et al., 2018;

Jacquet et al., 2019), would have had a CAI-like isotopic composition, and be enriched in nuclides produced in neutron rich stellar environments. Note that this earliest disk would not only have contained refractory materials (e.g., CAIs) but also other, non-refractory, dust particles. In contrast, later infalling matter primarily was added to the inner disk and the inner disk thus retained its obtained NC-like isotopic composition. Mixing of these two end member sources within the disk subsequently generated the CC reservoir with an isotopic composition that is intermediate between early (CAI-like) and late-infalling (NC-like) material (Burkhardt et al., 2019; Nanne et al., 2019; Spitzer et al., 2020). After infall had ended, the proto-Sun continued to accrete material from the outer disk (Yang & Ciesla, 2012). Thus, to preserve the isotopic difference between the inner (NC) and outer (CC) reservoirs a barrier is required that efficiently inhibits radial transport of CC material into the inner disk. As mentioned in Section 14.7.3, the most plausible mechanism for such an efficient separation of NC and CC disk reservoirs is the growth of Jupiter or, alternatively, the presence of a pressure bump where Jupiter subsequently formed.

14.8 GENETIC HERITAGE OF VESTA AND CERES

Clues about the genetic heritage of Vesta come from the nucleosynthetic signatures of HED meteorites. The ε^{50}Ti and ε^{54}Cr isotope signatures reveal that HED meteorites fall within the field of NC meteorites (Figures 14.1 and 14.2), making the HED meteorites and Vesta a very early accreted and differentiated NC body. Previous studies have identified exogenic carbonaceous chondrite-like signatures in the regolith of Vesta (e.g., De Sanctis et al., 2012; Jaumann et al., 2012; McCord et al., 2012; Prettyman et al., 2012; Reddy et al., 2012; Lunning et al., 2016; Chapter 6). These CC materials are not endogenous to Vesta and may have come from an impactor that formed Vesta's Veneneia (diameter 400 ± 20 km) impact basin which is superimposed on Rheasilvia (diameter 500 ± 20 km) (Jaumann et al., 2012; Reddy et al., 2012; Schenk et al., 2012; Lunning et al., 2016).

Searching for endogenous water on Vesta and the angrite parent body, Sarafian et al., (2014, 2017a, 2017b) concluded that the similarity between the H isotope compositions of HED (and angrite) meteorites (Barrett et al., 2016) and bulk carbonaceous chondrites indicated that water was acquired by the accretion of carbonaceous chondrite material onto these parent bodies while they were still partially molten (see review by Alexander et al., 2018). Sarafian et al. (2017a, 2017b) concluded that Vesta and the angrite parent body acquired the majority of their water content earlier than 5 Ma. Sarafian et al. (2019) recently reported a bulk H_2O (10–70 ppm) and fluorine (0.3–2 ppm) composition of Vesta determined from unequilibrated eucrites, where the D/H of these samples also match CC material. These authors built on work by Humayun and Clayton (1995), Righter (2007), and Day and Moynier (2014) to investigate the idea that inner Solar System material accumulated water during accretion and retained water and possibly other volatile elements proportional to their size. If so, this would indicate that there was some indigenous water in the inner (NC) Solar System that imparted low volatile concentrations on some asteroids including Vesta. Authors, however, conclude that the isotopic composition of unequilibrated eucrites is consistent with a chondritic source of volatile elements to early forming Vesta.

As Ceres has not been linked to any known meteorite group, it is currently not possible to confirm if Ceres belongs to the NC or CC meteorite reservoir. Nevertheless, given that Ceres has a C-rich and ammonia-rich surface, is modeled to have relatively large amounts of macromolecular organic matter, and is more water-rich than known "wet" meteorites (e.g., CI chondrites) (see e.g., Prettyman et al., 2012, 2017; De Sanctis et al., 2016; Carrozzo et al., 2018; as well as Chapters 7, 8–9), it is likely that Ceres is a CC asteroid which initially formed in the outer Solar System and was then implanted into the inner Solar System during giant planet migration (see Chapter 15). Alternatively, Ceres may be an asteroid which originally accreted in the NC reservoir and either accreted significantly more water and volatiles that Vesta, or Ceres' chemical composition was subsequently modified through late accretion of more organic and volatile rich material derived from the CC reservoir. These later accreted materials may then either have been added through inward drift of small CC dust during the lifetime of the NC–CC bodies or, alternatively, may reflect the later implantation of CC-like bodies into the inner Solar System that mixed with NC bodies at a later stage.

14.9 CONCLUDING REMARKS

Nucleosynthetic and radiogenic isotope signatures of melted and unmelted meteorites are currently used to infer a compositional and temporal map of protoplanetary disk's evolution into the planets. Although questions remain about the confidence with which these meteorite compositions can be linked to accretion locations in the disk, the combination of nucleosynthetic and radiogenic isotope data suggest that the inner and outer protoplanetary disk underwent early segregation to form the NC–CC regions, possibly by the formation of proto-Jupiter. This segregation event likely limited the movement of material from the inner and outer disk, thereby strongly influencing the diversity of material available for planetary formation in the different regions. Although the communication between the NC and CC reservoirs appears to be limited, the volatile content of some asteroids suggests that some outer Solar System material may have breached this early divide. Although the details remain to be investigated, such additions of CC bodies to the inner Solar System could either have been accommodated by inward drift of small CC dust during the lifetime of NC–CC dichotomy or, alternatively, by the later implantation of CC bodies into inner Solar System. How this movement of outer Solar System material may have happened without significant modification of the NC–CC isotopic indicia and if this played a role in the origin of Earth's volatile content remain open questions that can be addressed by future analysis of main asteroid belt materials.

REFERENCES

Alexander, C. M. O. (2019a) Quantitative models for the elemental and isotopic fractionations in chondrites: The carbonaceous chondrites. *Geochimica et Cosmochimica Acta*, 254, 277–309.

Alexander, C. M. O. (2019b) Quantitative models for the elemental and isotopic fractionations in the chondrites: The non-carbonaceous chondrites. *Geochimica et Cosmochimica Acta*, 254, 246–276.

Alexander, C. M. O., Grossman, J. N., Ebel, D. S., & Ciesla, F. J. (2008) The formation conditions of chondrules and chondrites. *Science*, 320, 1617–1619.

Alexander, C. M. O., McKeegan, K. D., & Altwegg, K. (2018) Water reservoirs in small planetary bodies: Meteorites, asteroids, and comets. *Space Science Reviews*, 214, 36.

Alibert, Y., Venturini, J., Helled, R., et al. (2018) The formation of Jupiter by hybrid pebble–planetesimal accretion. *Nature Astronomy*, 2, 873–877.

Amelin, Y., Kaltenbach, A., Iizuka, T., et al. (2010) U–Pb chronology of the Solar System's oldest solids with variable $^{238}U/^{235}U$. *Earth and Planetary Science Letters*, 300, 343–350.

Amelin, Y., & Krot, A. (2007) Pb isotopic age of the Allende chondrules. *Meteoritics & Planetary Science*, 42, 1321–1335.

Amelin, Y., Krot, A. N., Hutcheon, I. D., & Ulyanov, A. A. (2002) Lead isotopic ages of chondrules and calcium-aluminum-rich inclusions. *Science*, 297, 1678.

Barrett, T. J., Barnes, J. J., Tartèse, R., et al. (2016) The abundance and isotopic composition of water in eucrites. *Meteoritics & Planetary Science*, 51, 1110–1124.

Benedix, G. K., McCoy, T. J., Keil, K., & Love, S. G. (2000) A petrologic study of the IAB iron meteorites: Constraints on the formation of the IAB-winonaite parent body. *Meteoritics & Planetary Science*, 35, 1127–1141.

Bermingham, K. R., Füri, E., Lodders, K., & Marty, B. (2020) The NC–CC isotope dichotomy: Implications for the chemical and isotopic evolution of the early Solar System. *Space Science Reviews*, 216, 133.

Bermingham, K. R., Gussone, N., Mezger, K., & Krause, J. (2018a) Origins of mass-dependent and mass-independent Ca isotope variations in meteoritic components and meteorites. *Geochimica et Cosmochimica Acta*, 226, 206–223.

Bermingham, K. R., Mezger, K., Scherer, E. E., et al. (2016) Barium isotope abundances in meteorites and their implications for early Solar System evolution. *Geochimica et Cosmochimica Acta*, 175, 282–298.

Bermingham, K. R., Worsham, E. A., & Walker, R. J. (2018b) New insights into Mo and Ru isotope variation in the nebula and terrestrial planet accretionary genetics. *Earth and Planetary Science Letters*, 487, 221–229.

Binzel, R. P., & Xu, S. (1993) Chips off of asteroid 4 Vesta: Evidence for the parent body of basaltic achondrite meteorites. *Science*, 260, 186.

Birnstiel, T., Dullemond, C. P., & Pinilla, P. (2013) Lopsided dust rings in transition disks. *Astronomy & Astrophysics*, 550, L8.

Bizzarro, M., Baker, J. A., Haack, H., & Lundgaard, K. L. (2005) Rapid timescales for accretion and melting of differentiated planetesimals inferred from ^{26}Al–^{26}Mg chronometry. *Astrophysics Journal*, 632, L41–L44.

Blackburn, T., Alexander, C. M. O., Carlson, R., & Elkins-Tanton, L. T. (2017) The accretion and impact history of the ordinary chondrite parent bodies. *Geochimica et Cosmochimica Acta*, 200, 201–217.

Blum, J., & Wurm, G. (2000) Experiments on sticking, restructuring, and fragmentation of preplanetary dust aggregates. *Icarus*, 143, 138–146.

Bollard, J., Connelly, J. N., Whitehouse, M. J., et al. (2017) Early formation of planetary building blocks inferred from Pb isotopic ages of chondrules. *Science Advances*, 3, e1700407.

Bollard, J., Kawasaki, N., Sakamoto, N., et al. (2019) Combined U-corrected Pb–Pb dating and ^{26}Al–^{26}Mg systematics of individual chondrules – Evidence for a reduced initial abundance of ^{26}Al amongst inner Solar System chondrules. *Geochimica et Cosmochimica Acta*, 260, 62–83.

Brasser, R., & Mojzsis, S. J. (2020) The partitioning of the inner and outer Solar System by a structured protoplanetary disk. *Nature Astronomy*, 4, 492–499.

Brennecka, G. A., Borg, L. E., & Wadhwa, M. (2013) Evidence for supernova injection into the solar nebula and the decoupling of r-process nucleosynthesis. *Proceedings of the National Academy of Sciences (USA)*, 110, 17241.

Brennecka, G. A., Weyer, S., Wadhwa, M., et al. (2010) 238U/235U Variations in meteorites: Extant 247Cm and implications for Pb–Pb dating. *Science*, 327, 449–451.

Buchwald, V. F. (1975) *Handbook of Iron Meteorites: Their History, Distribution, Composition, and Structure, in 3 volumes.* Berkeley: University of California Press.

Budde, G., Burkhardt, C., Brennecka, G. A., et al. (2016) Molybdenum isotopic evidence for the origin of chondrules and a distinct genetic heritage of carbonaceous and non-carbonaceous meteorites. *Earth and Planetary Science Letters*, 454, 293–303.

Budde, G., Burkhardt, C., & Kleine, T. (2019) Molybdenum isotopic evidence for the late accretion of outer Solar System material to Earth. *Nature Astronomy*, 3, 736–741.

Budde, G., Kruijer, T. S., & Kleine, T. (2018) Hf–W chronology of CR chondrites: Implications for the timescales of chondrule formation and the distribution of ^{26}Al in the solar nebula. *Geochimica et Cosmochimica Acta*, 222, 284–304.

Burbine, T. H. (1998) Could G-class asteroids be the parent bodies of the CM chondrites? *Meteoritics & Planetary Science*, 33, 253–258.

Burkhardt, C., Dauphas, N., Hans, U., Bourdon, B., & Kleine, T. (2019) Elemental and isotopic variability in Solar System materials by mixing and processing of primordial disk reservoirs. *Geochimica et Cosmochimica Acta*, 261, 145–170.

Burkhardt, C., Kleine, T., Dauphas, N., & Wieler, R. (2012) Origin of isotopic heterogeneity in the solar nebula by thermal processing and mixing of nebular dust. *Earth and Planetary Science Letters*, 357–358, 298–307.

Burkhardt, C., Kleine, T., Oberli, F., et al. (2011) Molybdenum isotope anomalies in meteorites: Constraints on solar nebula evolution and origin of the Earth. *Earth and Planetary Science Letters*, 312, 390–400.

Cameron, A. G. W., & Truran, J. W. (1977) The supernova trigger for formation of the Solar System. *Icarus*, 30, 447–461.

Carrozzo, F. G., De Sanctis, M. C., Raponi, A., et al. (2018) Nature, formation, and distribution of carbonates on Ceres. *Science Advances*, 4, e1701645.

Clayton, D. D. (1982) Cosmic chemical memory: a new astronomy. *Quarterly Journal of the Royal Astronomical Society*, 23, 174–212.

Clayton, R. N. (1993) Oxygen isotopes in meteorites. *Annual Review of Earth and Planetary Sciences*, 21, 115–149.

Connelly, J. N., Amelin, Y., Krot, A. N., & Bizzarro, M. (2008) Chronology of the Solar System's oldest solids. *Astrophysics Journal*, 675, L121–L124.

Connelly, J. N., & Bizzarro, M. (2009) Pb–Pb dating of chondrules from CV chondrites by progressive dissolution. *Chemical Geology*, 259, 143–151.

Connelly, J. N., Bizzarro, M., Krot, A. N., et al. (2012) The absolute chronology and thermal processing of solids in the solar protoplanetary disk. *Science*, 338, 651–655.

Crida, A., Morbidelli, A., & Masset, F. (2006) On the width and shape of gaps in protoplanetary disks. *Icarus*, 181, 587–604.

Dauphas, N., & Chaussidon, M. (2011) A perspective from extinct radionuclides on a young stellar object: The sun and its accretion disk. *Annual Review of Earth and Planetary Sciences*, 39, 351–386.

Dauphas, N., Marty, B., & Reisberg, L. (2002) Molybdenum nucleosynthetic dichotomy revealed in primitive meteorites. *Astrophysics Journal*, 569, L139–L142.

Dauphas, N., & Schauble, E. A. (2016) Mass fractionation laws, mass-independent effects, and isotopic anomalies. *Annual Review of Earth and Planetary Sciences*, 44, 709–783.

Davis, A. M., Zhang, J., Greber, N. D., et al. (2018) Titanium isotopes and rare earth patterns in CAIs: Evidence for thermal processing and gas-dust decoupling in the protoplanetary disk. *Geochimica et Cosmochimica Acta*, 221, 275–295.

Day, J. M. D., & Moynier, F. (2014) Evaporative fractionation of volatile stable isotopes and their bearing on the origin of the Moon. *Philosophical Transactions of the Royal Society A: Mathematical, Physical and Engineering Sciences*, 372, 20130259.

De Sanctis, M. C., Ammannito, E., Capria, M. T., et al. (2012) Spectroscopic characterization of mineralogy and its diversity across Vesta. *Science*, 336, 697.

De Sanctis, M. C., Raponi, A., Ammannito, E., et al. (2016) Bright carbonate deposits as evidence of aqueous alteration on (1) Ceres. *Nature*, 536, 54–57.

Delbo, M., Walsh, K., Bolin, B., Avdellidou, C., & Morbidelli, A. (2017) Identification of a primordial asteroid family constrains the original planetesimal population. *Science*, 357, 1026.

DeMeo, F. E., & Carry, B. (2014) Solar System evolution from compositional mapping of the asteroid belt. *Nature*, 505, 629–634.

Dermott, S. F., Christou, A. A., Li, D., Kehoe Thomas, J. J., & Robinson, J. M. (2018) The common origin of family and non-family asteroids. *Nature Astronomy*, 2, 549–554.

Desch, S. J., Kalyaan, A., & Alexander, C. M. O. (2018) The effect of Jupiter's formation on the distribution of refractory elements and inclusions in meteorites. *Astrophysics Journal Supplementary Series*, 238, 11.

Doyle, P. M., Jogo, K., Nagashima, K., et al. (2015) Early aqueous activity on the ordinary and carbonaceous chondrite parent bodies recorded by fayalite. *Nature Communications*, 6, 7444.

Dwarkadas, V. V., Dauphas, N., Meyer, B., Boyajian, P., & Bojazi, M. (2017) Triggered star formation inside the shell of a Wolf–Rayet bubble as the origin of the Solar System. *Astrophysics Journal*, 851, 147.

Elkins-Tanton, L. T. (2012) Magma oceans in the inner Solar System. *Annual Review of Earth and Planetary Sciences*, 40, 113–139.

Fischer-Gödde, M., Burkhardt, C., Kruijer, T. S., & Kleine, T. (2015) Ru isotope heterogeneity in the solar protoplanetary disk. *Geochimica et Cosmochimica Acta*, 168, 151–171.

Formisano, M., Federico, C., Turrini, D., et al. (2013) The heating history of Vesta and the onset of differentiation. *Meteoritics & Planetary Science*, 48, 2316–2332.

Goldstein, J. I., Scott, E. R. D., & Chabot, N. L. (2009) Iron meteorites: Crystallization, thermal history, parent bodies, and origin. *Geochemistry*, 69, 293–325.

Gradie, J., & Tedesco, E. (1982) Compositional structure of the asteroid belt. *Science*, 216, 1405–1407.

Greenwood, R. C., Burbine, T. H., & Franchi, I. A. (2020) Linking asteroids and meteorites to the primordial planetesimal population. *Geochimica et Cosmochimica Acta*, 277, 377–406.

Heck, P. R., Greer, J., Kööp, L., et al. (2020) Lifetimes of interstellar dust from cosmic ray exposure ages of presolar silicon carbide. *Proceedings of the National Academy of Sciences (USA)*, 117, 1884.

Hellmann, J. L., Kruijer, T. S., Orman, J. A. V., Metzler, K., & Kleine, T. (2019) Hf–W chronology of ordinary chondrites. *Geochimica et Cosmochimica Acta*, 258, 290–309.

Henke, S., Gail, H.-P., Trieloff, M., Schwarz, W. H., & Kleine, T. (2012) Thermal history modelling of the H chondrite parent body. *Astronomy & Astrophysics*, 545, A135.

Hevey, P. J., & Sanders, I. S. (2006) A model for planetesimal meltdown by ^{26}Al and its implications for meteorite parent bodies. *Meteoritics & Planetary Science*, 41, 95–106.

Hilton, C. D., Bermingham, K. R., Walker, R. J., & McCoy, T. J. (2019) Genetics, crystallization sequence, and age of the South Byron Trio iron meteorites: New insights to carbonaceous chondrite (CC) type parent bodies. *Geochimica et Cosmochimica Acta*, 251, 217–228.

Hogerheijde, M. R. (2011) Protoplanetary disk. In M. Gargaud, R. Amils, J. C. Quintanilla, et al. (eds.), *Encyclopedia of Astrobiology*. Berlin: Springer, pp. 1357–1366.

Humayun, M., & Clayton, R. N. (1995) Potassium isotope cosmochemistry: Genetic implications of volatile element depletion. *Geochimica et Cosmochimica Acta*, 59, 2131–2148.

Huss, G. R., & McSween, J. H Y. (eds.) (2010) Presolar grains: A record of stellar nucleosynthesis and processes in interstellar space. In *Cosmochemistry* Cambridge: Cambridge University Press, pp. 120–156.

International Union of Pure and Applied Chemistry (2006) *IUPAC Compendium of Chemical Terminology: The Gold Book*. Research Triangle Park, NC: International Union of Pure and Applied Chemistry.

Ireland, T. R., Avila, J., Greenwood, R. C., Hicks, L. J., & Bridges, J. C. (2020) Oxygen isotopes and sampling of the Solar System. *Space Science Reviews*, 216, 25.

Jacquet, E., Pignatale, F. C., Chaussidon, M., & Charnoz, S. (2019) Fingerprints of the protosolar cloud collapse in the Solar System. II. Nucleosynthetic anomalies in meteorites. *Astrophysics Journal*, 884, 32.

Jaumann, R., Williams, D. A., Buczkowski, D. L., et al. (2012) Vesta's shape and morphology. *Science*, 336, 687.

Jogo, K., Nakamura, T., Ito, M., et al. (2017) Mn–Cr ages and formation conditions of fayalite in CV3 carbonaceous chondrites: Constraints on the accretion ages of chondritic asteroids. *Geochimica et Cosmochimica Acta*, 199, 58–74.

Kita, N. T., & Ushikubo, T. (2012) Evolution of protoplanetary disk inferred from 26Al chronology of individual chondrules: Disk evolution and ^{26}Al chronology of chondrules. *Meteoritics & Planetary Science*, 47, 1108–1119.

Kleine, T., Budde, G., Burkhardt, C., et al. (2020) The non-carbonaceous–carbonaceous meteorite dichotomy. *Space Science Reviews*, 216, 55.

Kleine, T., Hans, U., Irving, A. J., & Bourdon, B. (2012) Chronology of the angrite parent body and implications for core formation in protoplanets. *Geochimica et Cosmochimica Acta*, 84, 186–203.

Kleine, T., Touboul, M., Bourdon, B., et al. (2009) Hf–W chronology of the accretion and early evolution of asteroids and terrestrial planets. *Geochimica et Cosmochimica Acta*, 73, 5150–5188.

Kleine, T., & Wadhwa, M. (2017) Chronology of planetesimal differentiation. In L. T. Elkins-Tanton, & B. P. Weiss (eds.), *Planetesimals: Early Differentiation and Consequences for Planets*. Cambridge: Cambridge University Press, pp. 224–245.

Kleine, T., & Walker, R. J. (2017) Tungsten isotopes in planets. *Annual Review of Earth and Planetary Sciences*, 45, 389–417.

Konopliv, A. S., Park, R. S., Vaughan, A. T., et al. (2018) The Ceres gravity field, spin pole, rotation period and orbit from the Dawn radiometric tracking and optical data. *Icarus*, 299, 411–429.

Krot, A. N., Amelin, Y., Bland, P., et al. (2009) Origin and chronology of chondritic components: A review. *Geochimica et Cosmochimica Acta*, 73, 4963–4997.

Krot, A. N., Keil, K., Scott, E. R. D., Goodrich, C. A., & Weisberg, M. K. (2014) Classification of meteorites and their genetic relationships. In H. D. Holland, & K. K. Turekian (eds.), *Treatise on Geochemistry*. Amsterdam: Elsevier, pp. 1–63.

Krot, A. N., Makide, K., Nagashima, K., et al. (2012) Heterogeneous distribution of ^{26}Al at the birth of the Solar System: Evidence from refractory grains and inclusions: ^{26}Al heterogeneity in the early Solar System. *Meteoritics & Planetary Science*, 47, 1948–1979.

Krot, A. N., Nagashima, K., Libourel, G., & Miller, K. E. (2018) Multiple mechanisms of transient heating events in the protoplanetary disk: Evidence from precursors of chondrules and igneous Ca, Al-rich inclusions. In A. N. Krot, H. C. Connolly Jr., & S. S. Russell (eds.), *Chondrules: Records of Protoplanetary Disk Processes*. Cambridge: Cambridge University Press, pp. 11–56.

Kruijer, T. S., Burkhardt, C., Budde, G., & Kleine, T. (2017) Age of Jupiter inferred from the distinct genetics and formation times of meteorites. *Proceedings of the National Academy of Sciences (USA)*, 114, 6712–6716.

Kruijer, T. S., & Kleine, T. (2019) Age and origin of IIE iron meteorites inferred from Hf-W chronology. *Geochimica et Cosmochima Acta*, 262, 92–103.

Kruijer, T. S., Kleine, T., & Borg, L. E. (2020) The great isotopic dichotomy of the early Solar System. *Nature Astronomy*, 4, 32–40.

Kruijer, T. S., Touboul, M., Fischer-Gödde, M., et al. (2014) Protracted core formation and rapid accretion of protoplanets. *Science*, 344, 1150.

Lambrechts, M., & Johansen, A. (2012) Rapid growth of gas-giant cores by pebble accretion. *Astronomy & Astrophysics*, 544, A32.

Lee, T., Papanastassiou, D. A., & Wasserburg, G. J. (1977) Aluminum-26 in the early Solar System: Fossil or fuel? *Astrophysics Journal*, 211, L107–L110.

Lewis, R. S., Ming, T., Wacker, J. F., Anders, E., & Steel, E. (1987) Interstellar diamonds in meteorites. *Nature*, 326, 160–162.

Leya, I., Schönbächler, M., Wiechert, U., Krähenbühl, U., & Halliday, A. N. (2008) Titanium isotopes and the radial heterogeneity of the Solar System. *Earth and Planetary Science Letters*, 266, 233–244.

Lodders, K. (2003) Solar System abundances and condensation temperatures of the elements. *Astrophysics Journal*, 591, 1220–1247.

Lodders, K., & Amari, S. (2005) Presolar grains from meteorites: Remnants from the early times of the Solar System. *Chemie der Erde – Geochemistry*, 65, 93–166.

Lovering, J. F. (1957) Differentiation in the iron-nickel core of a parent meteorite body. *Geochimica et Cosmochimica Acta*, 12, 238–252.

Lunning, N. G., Corrigan, C. M., McSween, H. Y., et al. (2016) CV and CM chondrite impact melts. *Geochimica et Cosmochimica Acta*, 189, 338–358.

Marchi, S., Raponi, A., Prettyman, T. H., et al. (2019) An aqueously altered carbon-rich Ceres. *Nature Astronomy*, 3, 140–145.

Mayer, B., Wittig, N., Humayun, M., & Leya, I. (2015) Palladium isotopic evidence for nucleosynthetic and cosmogenic isotope anomalies in IVB iron meteorites. *Astrophysics Journal*, 809, 180.

McCord, T. B., Li, J.-Y., Combe, J.-P., et al. (2012) Dark material on Vesta from the infall of carbonaceous volatile-rich material. *Nature*, 491, 83–86.

Morbidelli, A., Bitsch, B., Crida, A., et al. (2016) Fossilized condensation lines in the Solar System protoplanetary disk. *Icarus*, 267, 368–376.

Nagashima, K., Krot, A. N., & Komatsu, M. (2017) ^{26}Al–^{26}Mg systematics in chondrules from Kaba and Yamato 980145 CV3 carbonaceous chondrites. *Geochimica et Cosmochimica Acta*, 201, 303–319.

Nanne, J. A. M., Nimmo, F., Cuzzi, J. N., & Kleine, T. (2019) Origin of the non-carbonaceous–carbonaceous meteorite dichotomy. *Earth and Planetary Science Letters*, 511, 44–54.

Neumann, W., Breuer, D., & Spohn, T. (2014) Differentiation of Vesta: Implications for a shallow magma ocean. *Earth and Planetary Science Letters*, 395, 267–280.

Niemeyer, S. (1988) Titanium isotopic anomalies in chondrules from carbonaceous chondrites. *Geochimica et Cosmochimica Acta*, 52, 309–318.

Ott, U. (2014) Planetary and pre-solar noble gases in meteorites. *Geochemistry*, 74, 519–544.

Papanastassiou, D. A. (1986) Chromium isotopic anomalies in the Allende meteorite *Astrophysics Journal*, 308, L27–L30.

Pape, J., Mezger, K., Bouvier, A.-S., & Baumgartner, L. P. (2019) Time and duration of chondrule formation: Constraints from 26Al–26Mg ages of individual chondrules. *Geochimica et Cosmochimica Acta*, 244, 416–436.

Pollack, J. B., Hubickyj, O., Bodenheimer, P., et al. (1996) Formation of the giant planets by concurrent accretion of solids and gas. *Icarus*, 124, 62–85.

Poole, G. M., Rehkämper, M., Coles, B. J., Goldberg, T., & Smith, C. L. (2017) Nucleosynthetic molybdenum isotope anomalies in iron meteorites – new evidence for thermal processing of solar nebula material. *Earth and Planetary Science Letters*, 473, 215–226.

Prettyman, T. H., Mittlefehldt, D. W., Yamashita, N., et al. (2012) Elemental mapping by Dawn reveals exogenic H in Vesta's regolith. *Science*, 338, 242.

Prettyman, T. H., Yamashita, N., Toplis, M. J., et al. (2017) Extensive water ice within Ceres' aqueously altered regolith: Evidence from nuclear spectroscopy. *Science*, 355, 55.

Qin, L., & Carlson, R. W. (2016) Nucleosynthetic isotope anomalies and their cosmochemical significance. *Geochemical Journal*, 50, 43–65.

Raymond, S. N., & Izidoro, A. (2017a) The empty primordial asteroid belt. *Science Advances*, 3, e1701138.

Raymond, S. N., & Izidoro, A. (2017b) Origin of water in the inner Solar System: Planetesimals scattered inward during Jupiter and Saturn's rapid gas accretion. *Icarus* 297, 134–148.

Reddy, V., Corre, L. L., O'Brien, D. P., et al. (2012) Delivery of dark material to Vesta via carbonaceous chondritic impacts. *Icarus*, 221, 544–559.

Regelous, M., Elliott, T., & Coath, C. D. (2008) Nickel isotope heterogeneity in the early Solar System. *Earth and Planetary Science Letters*, 272, 330–338.

Righter, K. (2007) Not so rare Earth? New developments in understanding the origin of the Earth and Moon. *Geochemistry*, 67, 179–200.

Rotaru, M., Birck, J. L., & Allègre, C. J. (1992) Clues to early Solar System history from chromium isotopes in carbonaceous chondrites. *Nature*, 358, 465–470.

Russell, C. T., Raymond, C. A., Ammannito, E., et al. (2016) Dawn arrives at Ceres: Exploration of a small, volatile-rich world. *Science*, 353, 1008.

Russell, C. T., Raymond, C. A., Coradini, A., et al. (2012) Dawn at Vesta: Testing the protoplanetary paradigm. *Science*, 336, 684.

Sarafian, A. R., Hauri, E. H., McCubbin, F. M., et al. (2017a) Early accretion of water and volatile elements to the inner Solar System: Evidence from angrites. *Philosophical Transactions of the Royal Society A: Mathematical, Physical and Engineering Sciences*, 375, 20160209.

Sarafian, A. R., Nielsen, S. G., Marschall, H. R., et al. (2017b) Angrite meteorites record the onset and flux of water to the inner Solar System. *Geochimica et Cosmochimica Acta*, 212, 156–166.

Sarafian, A. R., Nielsen, S. G., Marschall, H. R., et al. (2019) The water and fluorine content of 4 Vesta. *Geochimica et Cosmochimica Acta*, 266, 568–581.

Sarafian, A. R., Nielsen, S. G., Marschall, H. R., McCubbin, F. M., & Monteleone, B. D. (2014) Early accretion of water in the inner Solar System from a carbonaceous chondrite-like source. *Science*, 346, 623–626.

Schenk, P., O'Brien, D. P., Marchi, S., et al. (2012) The geologically recent giant impact basins at Vesta's south pole. *Science*, 336, 694.

Schiller, M., Bizzarro, M., & Fernandes, V. A. (2018) Isotopic evolution of the protoplanetary disk and the building blocks of Earth and the Moon. *Nature*, 555, 507–510.

Schiller, M., Paton, C., & Bizzarro, M. (2015) Evidence for nucleosynthetic enrichment of the protosolar molecular cloud core by multiple supernova events. *Geochimica et Cosmochimica Acta*, 149, 88–102.

Schrader, D. L., Nagashima, K., Krot, A. N., et al. (2017) Distribution of ^{26}Al in the CR chondrite chondrule-forming region of the protoplanetary disk. *Geochimica et Cosmochimica Acta*, 201, 275–302.

Scott, E. R. D. (1972) Chemical fractionation in iron meteorites and its interpretation. *Geochimica et Cosmochimica Acta*, 36, 1205–1236.

Scott, E. R. D. (1977) Formation of olivine-metal textures in pallasite meteorites. *Geochimica et Cosmochimica Acta*, 41, 693–710.

Scott, E. R. D., Krot, A. N., & Sanders, I. S. (2018) Isotopic dichotomy among meteorites and its bearing on the protoplanetary disk. *Astrophysics Journal*, 854, 164.

Scott, E. R. D., & Wasson, J. T. (1976) Chemical classification of iron meteorites – VIII. Groups IC. IIE, IIIF and 97 other irons. *Geochimica et Cosmochimica Acta*, 40, 103–115.

Shu, F. H., Adams, F. C., & Lizano, S. (1987) Star formation in molecular clouds: Observation and theory. *Annual Review of Astronomy & Astrophysics*, 25, 23–81.

Spitzer, F., Burkhardt, C., Budde, G., et al. (2020) Isotopic evolution of the inner Solar System inferred from molybdenum isotopes in meteorites. *Astrophysics Journal*, 898, L2.

Spitzer, F., Burkhardt, C., Pape, J., & Kleine, T. (in press) Collisional mixing between inner and outer solar system planetesimals inferred from the Nedagolla iron meteorite. *Meteoritics & Planetary Science*.

Sugiura, N., & Fujiya, W. (2014) Correlated accretion ages and ε ^{54} Cr of meteorite parent bodies and the evolution of the solar nebula. *Meteoritics & Planetary Science*, 49, 772–787.

Tang, H., & Dauphas, N. (2012) Abundance, distribution, and origin of 60Fe in the solar protoplanetary disk. *Earth and Planetary Science Letters*, 359–360, 248–263.

Taylor, G. J., Keil, K., McCoy, T., Haack, H., & Scott, E. R. D. (1993) Asteroid differentiation: Pyroclastic volcanism to magma oceans. *Meteoritics*, 28, 34–52.

Tornabene, H. A., Hilton, C. D., Bermingham, K. R., Ash, R. D., & Walker, R. J. (2020) Genetics, age and crystallization history of group IIC iron meteorites. *Geochimica et Cosmochimica Acta*, 288, 36–50.

Touboul, M., Sprung, P., Aciego, S. M., Bourdon, B., & Kleine, T. (2015) Hf–W chronology of the eucrite parent body. *Geochimica et Cosmochimica Acta*, 156, 106–121.

Trinquier, A., Birck, J., & Allegre, C. J. (2007) Widespread ^{54}Cr heterogeneity in the inner Solar System. *Astrophysics Journal*, 655, 1179–1185.

Trinquier, A., Birck, J.-L., Allègre, C. J., Göpel, C., & Ulfbeck, D. (2008) ^{53}Mn–^{53}Cr systematics of the early Solar System revisited. *Geochimica et Cosmochimica Acta*, 72, 5146–5163.

Trinquier, A., Elliott, T., Ulfbeck, D., et al. (2009) Origin of nucleosynthetic isotope heterogeneity in the solar protoplanetary disk. *Science*, 324, 374–376.

Van Kooten, E. M. M. E., Wielandt, D., Schiller, M., et al. (2016) Isotopic evidence for primordial molecular cloud material in metal-rich carbonaceous chondrites. *Proceedings of the National Academy of Sciences (USA)*, 113, 2011–2016.

Vernazza, P., & Beck, P. (2017) Composition of Solar System small bodies. In B. P. Weiss, & L. T. Elkins-Tanton (eds.), *Planetesimals: Early Differentiation and Consequences for Planets*. Cambridge: Cambridge University Press, pp. 269–297.

Villeneuve, J., Chaussidon, M., & Libourel, G. (2009) Homogeneous distribution of 26Al in the Solar System from the Mg isotopic composition of chondrules. *Science*, 325, 985–988.

Vockenhuber, C., Oberli, F., Bichler, M., et al. (2004) New half-life measurement of ^{182}Hf: Improved chronometer for the early Solar System. *Physical Review Letters*, 93, 172501.

Walker, R. J., Bermingham, K., Liu, J., et al. (2015) In search of late-stage planetary building blocks. *Chemical Geology*, 411, 125–142.

Walsh, K. J., Morbidelli, A., Raymond, S. N., O'Brien, D. P., & Mandell, A. M. (2011) A low mass for Mars from Jupiter's early gas-driven migration. *Nature*, 475, 206–209.

Walte, N. P., Solferino, G. F. D., Golabek, G. J., Souza, D. S., & Bouvier, A. (2020) Two-stage formation of pallasites and the evolution of their parent bodies revealed by deformation experiments. *Earth and Planetary Science Letters*, 546, 116419.

Warren, P. H. (2011) Stable-isotopic anomalies and the accretionary assemblage of the Earth and Mars: A subordinate role for carbonaceous chondrites. *Earth and Planetary Science Letters*, 311, 93–100.

Wasson, J. T., & Kallemeyn, G. W. (2002) The IAB iron-meteorite complex: A group, five subgroups, numerous grouplets, closely related, mainly formed by crystal segregation in rapidly cooling melts. *Geochimica et Cosmochimica Acta*, 66, 2445–2473.

Weber, P., Benítez-Llambay, P., Gressel, O., Krapp, L., & Pessah, M. E. (2018) Characterizing the variable dust permeability of planet-induced gaps. *Astrophysics Journal*, 854, 153.

Weidenschilling, S. J. (1977) Aerodynamics of solid bodies in the solar nebula. *Monthly Notices of the Royal Astronomical Society*, 180, 57–70.

Worsham, E. A., Bermingham, K. R., & Walker, R. J. (2017) Characterizing cosmochemical materials with genetic affinities to the Earth: Genetic and chronological diversity within the IAB iron meteorite complex. *Earth and Planetary Science Letters*, 467, 157–166.

Worsham, E. A., Bermingham, K. R., & Walker, R. J. (2016) Siderophile element systematics of IAB complex iron meteorites: New insights into the formation of an enigmatic group. *Geochimica et Cosmochimica Acta*, 188, 261–283.

Worsham, E. A., Burkhardt, C., Budde, G., et al. (2019) Distinct evolution of the carbonaceous and non-carbonaceous reservoirs: Insights from Ru, Mo, and W isotopes. *Earth and Planetary Science Letters*, 521, 103–112.

Yang, J., Goldstein, J. I., & Scott, E. R. D. (2007) Iron meteorite evidence for early formation and catastrophic disruption of protoplanets. *Nature*, 446, 888–891.

Yang, J., Goldstein, J. I., & Scott, E. R. D. (2010) Main-group pallasites: Thermal history, relationship to IIIAB irons, and origin. *Geochimica et Cosmochimica Acta*, 74, 4471–4492.

Yang, L., & Ciesla, F. J. (2012) The effects of disk building on the distributions of refractory materials in the solar nebula. *Meteoritics & Planetary Science*, 47, 99–119.

Yokoyama, T., Nagai, Y., Fukai, R., & Hirata, T. (2019) Origin and evolution of distinct molybdenum isotopic variabilities within carbonaceous and noncarbonaceous reservoirs. *Astrophysics Journal*, 883, 62.

Young, E. D. (2014) Inheritance of solar short- and long-lived radionuclides from molecular clouds and the unexceptional nature of the Solar System. *Earth and Planetary Science Letters*, 392, 16–27.

Zinner, E. (2014) Presolar grains. In A. M. Davis (ed.), *Treatise on Geochemistry*. Amsterdam: Elsevier, pp. 181–213.

Zinner, E., Ming, T., & Anders, E. (1987) Large isotopic anomalies of Si, C, N and noble gases in interstellar silicon carbide from the Murray meteorite. *Nature*, 330, 730–732.

Origin and Dynamical Evolution of the Asteroid Belt

SEAN N. RAYMOND AND DAVID NESVORNÝ

15.1 THE ASTEROID BELT IN THE CONTEXT OF SOLAR SYSTEM FORMATION

The asteroid belt marks the boundary between the rocky and gaseous planets. It is the widest piece of Solar System real estate between Mercury and Neptune that does not contain a planet (as measured in dynamical terms: the orbital period ratio between inner and outer stable orbits within the belt is >2, wider than the dynamical spacing between many planets). The belt's total mass is less than one one-thousandth of an Earth mass (Krasinsky et al., 2002; DeMeo & Carry, 2013; Kuchynka & Folkner, 2013). Asteroids' orbits are dynamically excited, with eccentricities ranging from zero to above 0.3 and inclinations from zero to >20°. The belt's composition varies radially, with S-class objects most common in the inner Main Belt, C-class objects dominating the outer main and substantial mixing between the two populations (Gradie & Tedesco, 1982; DeMeo and Carry, 2013, 2014). S-types are spectroscopically linked (Burbine et al., 2002) with ordinary chondrite meteorites, which have little water (Robert et al., 1977; Alexander et al., 2018) and C-types with carbonaceous chondrites, which typically have ~10% water by mass (Kerridge, 1985; Alexander et al., 2018).

Understanding the asteroid belt's dynamical history is challenging. In general, the more mass in a dynamical system, the more excited it becomes by self-stirring. For instance, a higher-mass disk of planetesimals excites itself due to mutual gravitational interactions faster and to a higher degree than a lower-mass disk (see, for example, Wetherill & Stewart, 1989; Ida & Makino, 1992; Kokubo & Ida, 2000). The excitation is driven mainly by the most massive bodies, which grow faster and more massive in higher-mass disks.

Models for the asteroid belt's origins make assumptions about the belt's birth conditions. The classical view is that the belt was born with several Earth masses in planetesimals and was subsequently depleted by a factor of ~1,000. That view is a legacy of a terrestrial planet formation that assumes that disks follow smooth radial surface density profiles (Wetherill, 1980; Chambers & Wetherill, 1998; Raymond et al., 2014). An alternate view suggests

We thank Rogerio Deienno for a careful review that improved this chapter, and the Editors for several important comments. We are grateful to a long list of colleagues and collaborators including Andre Izidoro, Matt Clement, Alessandro Morbidelli, Seth Jacobson, Nate Kaib, Rogerio Deienno, Bill Bottke, Fernando Roig, Arnaud Pierens, Kevin Walsh, Marco Delbo, Julie Castillo-Rogez, Franck Selsis, Karen Meech, Simone Marchi, Hal Levison, Luke Dones, and David Vokrouhlicky. S. N. R. thanks the CNRS's PNP program and the Agence Nationale pour la Recherche (grant ANR-13-BS05–0003-002 – MOJO) for funding and support. D. N.'s work was supported by NASA's SSW program. Figure 15.1 was prepared with data downloaded from mp3c.oca.eu.

that the belt was born with very little mass – perhaps none at all – and was later populated by planetesimals that originated in other parts of the Sun's planet-forming disk (e.g., Hansen, 2009; Raymond & Izidoro, 2017b). Such an assumption is motivated by observations of ringed substructure seen in the dust component of protoplanetary disks (e.g., ALMA Partnership et al., 2015; Andrews et al., 2016), as well as models of planetesimal formation in localized concentrations of dust and pebbles (e.g., Johansen et al., 2014; Birnstiel et al., 2016).

Here we review the dynamical processes that may have sculpted the asteroid belt over its history. In Sections 15.1.1 and 15.1.2 we describe the observed structure of the asteroid belt and present a rough Solar System timeline as inferred from cosmochemical constraints (see also Chapter 13). In Section 15.2 we examine dynamical processes that may have affected the asteroid belt during the gaseous disk phase. In Section 15.3 we describe the key processes affecting the belt after gas disk dispersal. In Section 15.4 we show how the asteroid belt's story fits within global models of Solar System formation, discuss the origins of Ceres and Vesta in the context of different models, and highlight outstanding problems.

15.1.1 The Asteroid Belt's Observed Structure

Figure 15.1 shows the orbital distribution of asteroids with diameters $D > 10$ km. We can distill the asteroid belt's characteristics down to four broad constraints that must be matched by any model of Solar System evolution. These are:

1. The belt's very low total mass ($\sim 4.5 \times 10^{-4} M\oplus$; Krasinsky et al., 2002; DeMeo & Carry, 2013; Kuchynka & Folkner, 2013). Most of the mass is contained within just four asteroids: Ceres (31% of the total mass of the belt), Vesta (9%), Pallas (7%), and Hygiea (3%).

2. The orbital structure of the belt. Asteroid orbits are excited, with eccentricities $0 < e < 0.3$ and inclinations $0 < i < 20°$ (Figure 15.1) and modest variations across the Main Belt (see figure 1 from Morbidelli et al., 2015). There is additional substructure within the belt such as the Kirkwood gaps at 3:1, 5:3, and 2:1 resonances and clumps associated with asteroid families (see Nesvorný et al., 2015, for a review).

3. The compositional structure of the belt. Taxonomic classification show that the inner Main Belt is dominated by S-types and the outer Main Belt by C-types (Gradie & Tedesco, 1982; Bus & Binzel, 2002; DeMeo & Carry, 2013, 2014). D-types are prevalent in the outer Main Belt and Jupiter's 1:1 resonant Trojan swarms (Emery et al., 2015).

4. The asteroid size-frequency distribution (SFD) is top-heavy with the bulk of the belt's mass contained in the most massive objects. Given that the asteroid SFD is likely a signature of

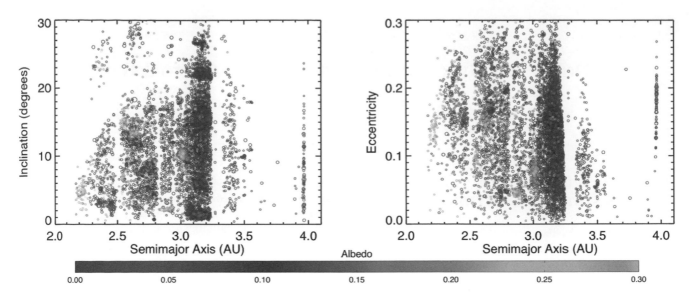

Figure 15.1 Orbital and compositional distribution of 8,237 asteroids larger than 10 km in diameter. The symbol size is proportional to actual size and the colors represent asteroid albedos. Typical albedos of C-types are ∼ 5% and of S-types ∼ 20% (Pravec et al., 2012). Asteroids with albedos above 0.3 were given colors corresponding to albedos of 0.3. Asteroid data from mp3c.oca.eu[1]

A black and white version of this figure will appear in some formats. For the color version, refer to the plate section.

accretion via the streaming instability (e.g. Youdin & Goodman, 2005; Morbidelli et al., 2009; Johansen et al., 2015; Nesvorný et al., 2019) and subsequent sculpting by collisional processes rather than dynamical ones (e.g., Bottke et al., 2005), we will not discuss it further in this chapter.

15.1.2 A Rough Solar System Timeline

The timeline of events that took place during Solar System formation can be roughly inferred from a combination of astrophysical and cosmochemical constraints (see also figure 9 of Nittler & Ciesla, 2016, and their associated discussion). Here is a simplified summary of the current thinking:

- The inner Solar System's first solids – Calcium–Aluminum rich Inclusions, or CAIs – have a narrow range of formation ages centered on 4.568 Gyr (e.g., Bouvier & Wadhwa, 2010; Connelly et al., 2012). This sets the Solar System's time zero, the start of planet formation.
- Planetesimals likely formed over roughly the subsequent five million years, during the gaseous disk phase. Analysis of iron meteorites indicates that their parent bodies formed within ∼ 1 Myr of CAIs (Kruijer et al., 2014, 2020; Schiller et al., 2015), meaning that planetesimal formation was underway very quickly. Current thinking (see reviews by Johansen et al., 2014, 2015; Birnstiel et al., 2016) suggests that planetesimals formed as drifting dust was locally concentrated to a sufficient degree to trigger gas-particle instabilities such as the streaming instability (Youdin & Goodman, 2005; Johansen et al., 2007; Squire & Hopkins, 2018). The resulting planetesimals had sizes extending to hundreds of kilometers in diameter, with a characteristic size of ∼100 km (Johansen et al., 2015; Schäfer et al., 2017; Simon et al., 2017; Abod et al., 2019). Indeed, the bump

in the size distribution of asteroids at $D \simeq 100$ km can be interpreted as a sign of efficient growth of 100-km-class asteroids (Morbidelli et al., 2009; but see Weidenschilling, 2011). Planetesimals that formed within 1–2 Myr of CAIs contained a high enough concentration of the active short-lived radionuclide ^{26}Al (half life of 717,000 years) to be strongly heated and dried out (Grimm & McSween, 1993; Lichtenberg et al., 2016; Monteux et al., 2018). These are the parent bodies of iron meteorites.

- Mars' growth was nearly complete during the gaseous disk phase. Isotopic (Hf/W/Th) analyses of Martian meteorites indicate that it was mostly formed within a few Myr (Nimmo & Kleine, 2007; Dauphas & Pourmand, 2011), although a new result indicates that this timeline may be less well constrained (Marchi et al., 2020). The giant planet ∼ $10M\oplus$ cores may also have formed within 1–2 Myr. Circumstantial evidence supporting this idea comes from the overlap in ages of noncarbonaceous and carbonaceous chondrite meteorites (Budde et al., 2016; Kruijer et al., 2017, 2020,; see also Chapter 13). As the components of these meteorites – chondrules – are at the size scale to drift rapidly within the disk, something must have kept them spatially separated. A candidate mechanism is the pressure trap generated exterior to the orbit of a large core (Lambrechts & Johansen, 2014; Bitsch et al., 2018). Another possibility is that these two chondrule reservoirs were kept apart by a structure within the disk itself, unrelated to the formation of giant planet cores (Brasser & Mojzsis, 2020).
- The gaseous disk dissipated within roughly 5 Myr. Studies of disk frequencies around star clusters of different ages indicate that disks usually last a few Myr (Haisch et al., 2001; Mamajek, 2009; Pfalzner et al., 2014). The latest-forming chondrites (CB chondrites) formed ∼5 Myr after CAIs (Krot et al., 2005; Johnson et al., 2016), possibly indicating that is when the Sun's gas disk dissipated. If true, this would set an upper limit on the timescale of the formation of gas giants.

---

[1] Website last accessed 2020 July 1.

- Hf/W analyses suggest that the Earth's final giant collision – the Moon-forming impact – happened 50–100 Myr after CAIs (Kleine et al., 2009). This timeframe is consistent with estimates using different methods, for example, by calibrating Earth's late veneer to N-body simulations (Jacobson et al., 2014) and from a spike in the distribution of shock degassing ages of meteorites, presumed to originate from collisions between asteroids and debris from the impact (Bottke et al., 2015). The constraint on the late veneer's total mass inferred from highly siderophile elements (Day et al., 2007; Walker, 2009) disappears if the magma ocean phase lasted for a sufficient time (see Rubie et al., 2016; Morbidelli et al., 2018).

- The giant planets underwent a dynamical instability no later than 100 Myr after CAIs. The so-called Late Heavy Bombardment (LHB) hypothesis was based on an inferred spike in the age distribution of craters on the Moon that implied a large flux of impactors in the inner Solar System roughly 500 Myr after CAIs (Tera et al., 1974; Bottke & Norman, 2017). The Nice model invokes an instability in the orbits of the giant planets to explain the LHB (Gomes et al., 2005; Levison et al., 2010; Deienno et al., 2017). However, new analyses of crater ages – coupled with cosmochemical constraints and modeling – suggest that the bombardment flux more likely simply followed a smooth decay (Boehnke & Harrison, 2016; Zellner, 2017; Morbidelli et al., 2018; Hartmann, 2019). The giant planet instability – which is needed to explain the giant planets' orbits as well as the orbital distribution of irregular satellites, the giant planets' Trojan asteroids, and the Kuiper belt (see Nesvorný, 2018, for a review) – could have happened anytime within the first 100 Myr (Nesvorný et al., 2018; Mojzsis et al., 2019), perhaps during terrestrial planet formation (Clement et al., 2018).

15.2 ASTEROID BELT EVOLUTION DURING THE GASEOUS DISK PHASE

The asteroid belt's story starts with the first planetesimals. Simulations of planetesimal formation by the streaming instability find that planetesimals form with most of the mass in the largest objects, which are generally hundreds of kilometers in size (Johansen et al., 2015; Simon et al., 2017; Abod et al., 2019). The largest planetesimals grow by accreting other planetesimals (e.g., Greenberg et al., 1978; Wetherill & Stewart, 1989; Kokubo & Ida, 1998) as well as millimeter- to centimeter-sized "pebbles" that continually drift inward through the disk due to aerodynamic gas drag (e.g., Weidenschilling, 1977a; Ormel & Klahr, 2010; Johansen & Lambrechts, 2017). The varying strength of gas drag can also cause size-sorting of planetesimals. The smallest planetesimals undergo radial drift on timescales shorter than the disk's lifetime, but planetesimals larger than roughly 10 km do not lose enough energy from gas drag to significantly alter their orbital radii (although this depends on the disk properties; Adachi et al., 1976).

A massive population of planetesimals undergoes gravitational self-stirring, which acts to excite their orbital eccentricities and inclinations (e.g., Ida & Makino, 1992). While the level of excitation is low, the collisional cross-sections of the largest planetesimals are augmented by gravitational focusing, and these objects can undergo runaway growth (Greenberg et al., 1978; Wetherill & Stewart, 1989, 1993; Kokubo & Ida, 1998; Rafikov, 2003). However, when the largest objects become massive enough they excite the random velocities of planetesimals to such a degree that gravitational focusing is shut down and their growth slows and transitions to the so-called oligarchic regime (Ida & Makino, 1993; Kokubo & Ida, 1998, 2000; Leinhardt & Richardson, 2005). The most massive objects are typically Moon- to Mars-mass (in the inner Solar System) and are called planetary embryos.

A well-studied model proposes that the asteroids' eccentricities and inclinations were excited by planetary embryos that formed within the belt (Wetherill, 1992; Chambers & Wetherill, 2001; Petit et al., 2001; O'Brien et al., 2007). For historical reasons, and because of the connection between Mars' accretion and the asteroid belt's excitation (Izidoro et al., 2015b), we will discuss this theory in the context of gas-free dynamics in Section 15.3.1.

15.2.1 Implantation of Planetesimals Driven by the Giant Planets' Growth

Two gas giants formed next door to the asteroid belt. Let us consider how this process affected the belt.

The prevailing model for gas giant formation proposes that giant planets first grow cores of $\sim 5 - 10 M_\oplus$, which gravitationally accrete gas from the disk (Mizuno, 1980; Pollack et al., 1996; Ida & Lin, 2004; Piso & Youdin, 2014). The early phase of gas accretion is slow and limited by cooling and contraction of the gaseous envelope (Ikoma et al., 2000; Hubickyj et al., 2005) as well as the gas supply from the disk (Fung et al., 2014; Lambrechts & Lega, 2017; Lambrechts et al., 2019). Most growing giant planets may not get past this phase, which would explain why Neptune-mass planets are much more common than Jupiter-mass planets among exoplanets (e.g., Suzuki et al., 2016). When the mass in the planet's gaseous envelope becomes comparable to the core mass, gas accretion accelerates (Pollack et al., 1996; Ikoma et al., 2001; Lissauer et al., 2009). The growing planet also carves an annular gap in the gas disk (Lin & Papaloizou, 1986; Bryden et al., 1999; Crida et al., 2006).

The phase of rapid gas accretion onto a giant planet core destabilizes nearby orbits. A planetesimal orbiting the Sun in the vicinity of a giant planet core is stable as long as its orbit remains well-separated from that of the core. The critical distance for stability of a small object orbiting the Sun near a massive core is $2\sqrt{3} \approx 3.5$-times the core's Hill radius, $R_H = a(m/3M_\odot)^{1/3}$, where a is the core's orbital radius, m is its mass, and $M_\odot$ is the Sun's mass (Marchal & Bozis, 1982; Gladman, 1993). As the core rapidly increases in mass, its Hill sphere increases in size. This means that nearby objects that were once on stable orbits are destabilized.

Figure 15.2 shows how Jupiter and Saturn's growth invariably implants C-types into the asteroid belt. This is a simple consequence of the combined effects of orbital destabilization, gravitational scattering, and aerodynamic gas drag. There are two populations of implanted objects: those trapped on stable orbits within the asteroid belt and those scattered past the belt toward the terrestrial planets (perhaps to deliver water to the growing Earth; see Meech & Raymond, 2020). The balance between these two outcomes depends on the strength of gas drag (Raymond & Izidoro, 2017a). For stronger gas drag, implantation is favored, and for weaker gas drag, scattering toward the terrestrial planets dominates.

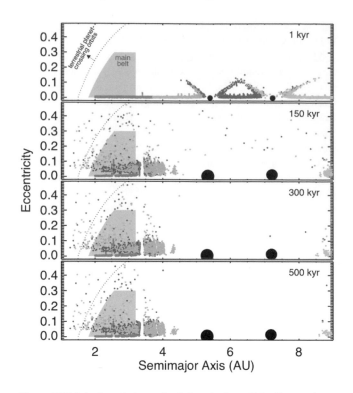

Figure 15.2 Injection of planetesimals into the inner Solar System due to Jupiter and Saturn's rapid gas accretion (adapted from Raymond & Izidoro, 2017a). In this simulation the giant planets' cores (initially $3M\oplus$) underwent rapid gas accretion from 100–200 kyr (for Jupiter) and 300–400 kyr (for Saturn). The gaseous disk structure responded to the giant planets' growth (from hydrodynamical models of Morbidelli & Crida, 2007). As the giant planets increased in mass, the orbits of nearby planetesimals (whose colors correspond to their starting orbital radii) were destabilized and gravitationally scattered by the giant planets onto eccentric orbits. Under the action of gas drag (assuming $D = 100$ km), some planetesimals were trapped on stable orbits within the Main Belt, originating in a region extending roughly from 4–9 AU. These may be the present-day C-type asteroids. In this simulation giant planet migration was not accounted for, so the 5 AU-wide source region for C-types is the narrowest it could possibly be; when migration is included planetesimals can be implanted into the belt from as far out as 20–30 AU (Raymond & Izidoro, 2017a; Pirani et al., 2019a).

A black and white version of this figure will appear in some formats. For the color version, refer to the plate section.

The strength of gas drag is a function of both (1) the density of the gaseous disk and therefore the timing of scattering, as the gas disk's density decays in time and (2) the planetesimal size, with smaller planetesimals feeling stronger drag.

Planetesimals are implanted throughout the asteroid belt but preferentially in the outer Main Belt (Kretke et al., 2017; Raymond & Izidoro, 2017a; Ronnet et al., 2018). Implanted planetesimals roughly match the distribution of C-type asteroids. The source region for C-types implanted in the belt in the simulation from Figure 15.2 was ~5 AU wide, from 4 to 9 AU. That represents an absolute minimum width for the source region because migration was not included. When migration of Jupiter and Saturn is accounted for, the source region widens to ~10–20 AU (Walsh et al., 2011; Raymond & Izidoro, 2017a; Pirani et al., 2019a; this will be discussed in more detail in Section 15.2.2). This broad source region, as well as the multiple different epochs of

implantation (corresponding to, at a minimum, the rapid gas accretion of Jupiter and Saturn and the inward migration of the ice giants), may help to explain the compositional diversity seen in carbonaceous chondrites and among C-types (e.g., Vernazza et al., 2017; Alexander et al., 2018). And if the C-types and Earth's water share a common dynamical source then this naturally explains the isotopic (e.g., D/H and [15/14]N) match between carbonaceous chondrites and Earth's water (Marty & Yokochi, 2006; Marty, 2012; Meech & Raymond, 2020).

The results of Figure 15.2 imply that a large fraction of the asteroid belt – perhaps all of the C-types and even other classes in the outer belt – was implanted from the Jupiter–Saturn region and beyond. This includes Ceres, the largest asteroid ($D = 946$ km), which is located at the (admittedly broad) peak in the distribution of $D = 1,000$ km planetesimals implanted in the simulations of Raymond and Izidoro (2017a). The idea that Ceres was implanted from the Jupiter–Saturn region appears consistent with its inferred composition, which is broadly similar to carbonaceous chondrites (e.g., Bland et al., 2016; Prettyman et al., 2017; McSween et al., 2018; Marchi et al., 2019).

This process cannot completely explain the C-types. The implanted objects nicely match the radial distribution of C-types, but the model does not match their observed eccentricity and inclination distributions, and so another source of excitation is required. In addition, the population of planetesimals that existed in the Jupiter–Saturn region when they underwent rapid gas accretion is uncertain. Given the high efficiency of implantation (Raymond & Izidoro, 2017a), it is possible that they did not represent a large mass. For example, in the simulation from Figure 15.2 about 10% of 100 km planetesimals from the Jupiter–Saturn region were implanted onto stable orbits within the Main Belt. Allowing for an order of magnitude depletion, that would imply $<0.01M\oplus$ in planetesimals in the source region. Such a low mass in planetesimals would be consistent with giant planet cores growing by pebble accretion (Lambrechts & Johansen, 2012; Levison et al., 2015a) but not by planetesimal accretion. Finally, Jupiter's growth may also have affected the collisional environment within the belt, by increasing the typical collision velocity between asteroids (Turrini et al., 2012; Turrini & Svetsov, 2014).

15.2.2 Effect of Giant Planet Migration

Migration is a central mechanism of planet formation (Kley & Nelson, 2012; Baruteau et al., 2014). Planets with masses as low as Earth's and Mars' exchange angular momentum with the gas disk, which acts both to damp their orbital eccentricities and inclinations (Papaloizou & Larwood, 2000; Tanaka & Ward, 2004) and to induce radial migration (Goldreich & Tremaine, 1980; Ward, 1986). Low-mass planets – in the so-called type I regime – migrate through the disk because of torques from nonaxisymmetric density perturbations, whereas more massive planets open radial gaps in the disk and their migration is linked with the disk's viscous evolution in the type II regime (Ward, 1997).

15.2.2.1 Effect of Migrating Giant Planet Cores

The giant planet cores formed while the gas disk was still dense and likely migrated. The effect of the cores' migration on the inner Solar System has not been extensively studied because of large uncertainties. Some simulations show that the inward migration of large cores would have devastated the growing terrestrial planets (Izidoro et al., 2014). Other simulations find that the absence of planets interior to Mercury could be explained if the seed of Jupiter's core formed very close to the Sun and migrated outward (Raymond et al., 2016). Both scenarios fall within the realm of possibility given the uncertainty in the underlying structure and evolution of the Sun's protoplanetary disk (see Morbidelli & Raymond, 2016). Yet the fact that we have a system of outer giant planets rather than close-in super-Earths suggests that the migration of the giant planets' cores was limited, although the reason behind their staying in the outer Solar System remains a central problem in planet formation (see discussion in Raymond et al., 2020). A model based on migration and pebble accretion found that Jupiter's core may have originated at 20–25 AU and undergone large-scale inward migration to end up at ∼ 5 AU (Bitsch et al., 2015). This begs the question of the fate of the solids that formed interior to 20–25 AU. Another possibility is that Jupiter quickly transitioned to the slower, type II migration regime (perhaps because the disk's viscosity was very low) and acted as a barrier to the inward migration of the ice giants and Saturn's core (Izidoro et al., 2015a).

15.2.2.2 The Grand Tack

The Grand Tack model (Walsh et al., 2011) relies on a particular type of migration pathway for Jupiter and Saturn that comes directly from hydrodynamical simulations by Masset and Snellgrove (2001). The scenario is as follows. After it opened a gap in the disk Jupiter migrated inward. Saturn accreted gas, started to open a gap, and migrated inward rapidly. Saturn caught up to Jupiter and became trapped in the outer 2:3 or 1:2 mean motion resonance with Jupiter. The two planets shared a common gap within the disk, which changed the flow of gas and the associated torque balance. The two planets subsequently migrated outward (this mechanism has been demonstrated by a number of authors – Masset & Snellgrove, 2001; Morbidelli et al., 2007; Pierens & Nelson, 2008; Crida et al., 2009; Zhang & Zhou, 2010; Pierens & Raymond, 2011; Pierens et al., 2014). Outward migration stopped as the disk dissipated or perhaps at an asymptotic radius if the disk was flared.

The Grand Tack model invokes this inward-then-outward migration of Jupiter to sculpt the inner Solar System (Walsh et al., 2011, 2012; Jacobson & Morbidelli, 2014; Raymond & Morbidelli, 2014; Brasser et al., 2016; Walsh & Levison, 2016; Deienno et al., 2019). The model requires that Jupiter's turnaround point was at 1.5–2 AU, in which case Jupiter would deplete Mars' feeding zone but not Earth's, thus explaining the large Earth/Mars mass ratio (sometimes called the "small Mars" problem; Wetherill, 1991; Raymond et al., 2009).

In the Grand Tack scenario, the asteroid belt was emptied then re-populated from multiple reservoirs (Figure 15.3). During Jupiter's inward migration, planetesimals interior to Jupiter's starting orbit were either shepherded inward or scattered outward. Planetesimals shepherded inward became building blocks of the terrestrial planets while those that were scattered outward were stranded on eccentric orbits exterior to Jupiter. When Jupiter and Saturn migrated back outward, they encountered the scattered planetesimals. Most planetesimals were ejected but a fraction was scattered inward and stranded on stable orbits as the giant planets migrated away. The gas giants then encountered planetesimals that

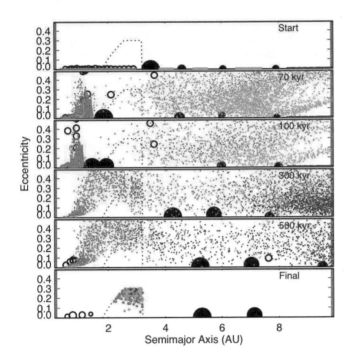

Figure 15.3 Snapshots in the evolution of the Grand Tack model. The simulation starts a few million years into the gaseous disk phase: Jupiter is fully grown, Saturn and the ice giants are large cores, and planetary embryos have formed in the inner Solar System. Jupiter migrates inward through rocky planetesimals and embryos, some of which are shepherded by inner resonances and some of which are scattered outward (see Raymond et al., 2006a; Mandell et al., 2007). Saturn grows and migrates inward, is trapped in 2:3 resonance with Jupiter, and both planets migrate back outward. Despite Jupiter's traversing, the main asteroid belt is populated by S-types (red) originating interior to Jupiter's initial orbit, and C-types from belts originally located between the giant planets' cores (light blue) and from a disk of planetesimals initially exterior to the giant planets (dark blue). The inner belt is dominated by S-types and the outer belt by C-types (Walsh et al., 2011, 2012), consistent with the compositional (Gradie & Tedesco, 1982; DeMeo & Carry, 2014) and orbital (Deienno et al., 2016) structure of the present-day belt.
A black and white version of this figure will appear in some formats. For the color version, refer to the plate section.

had originated farther out in the disk, between and beyond the giant planets. Most of these were again ejected but a fraction were stranded on orbits interior to the outward-migrating giant planets.

S-type planetesimals that survived Jupiter's migration followed a dynamic path. Planetesimals were scattered outward then back inward, finishing on orbits similar to their initial ones. However, only ∼0.1% of planetesimals followed this evolutionary path; most were accreted by the terrestrial planets or ejected by Jupiter. The middle panel shows how, after the giant planets' migration, the inner Main Belt is dominated by inner disk planetesimals and the outer Main Belt by planetesimals that originated between and beyond the giant planets' orbits. This is a simple consequence of timing. The orbital radii of scattered planetesimals correlates with Jupiter's orbital radius at the time of scattering. Because the inner-disk planetesimals were encountered first during the giant planets' outward migration when Jupiter was closer-in, they were trapped on closer-in orbits than the outer-disk planetesimals that were encountered later when Jupiter was farther out (Walsh et al., 2011, 2012).

The bottom panel of Figure 15.3 shows the orbital distribution of planetesimals trapped within the belt after the giant planets' migration was complete. While the asteroids' orbits are quite excited, they become consistent with the present-day belt, as high-eccentricity ($e \gtrsim 0.2$) planetesimals are destabilized over Solar System history (Deienno et al., 2016). The Grand Tack has the strength of both exciting and depleting the belt. It can also match the belt's compositional dichotomy if one makes simple assumptions about the nature of planetesimals that formed in different regions – that inner-disk planetesimals are linked with S-types and, as in the implantation simulation from Figure 15.2, outer disk planetesimals are linked with C-types.

Of course, the Grand Tack is built on a single migration pathway for the giant planets, among many (e.g., Pierens et al., 2014). While long-range outward migration of Jupiter and Saturn has been shown to be robust under certain assumptions related to the disk properties, it only holds for a range of mass ratios for Jupiter and Saturn (Masset & Snellgrove, 2001). It remains to be seen whether the appropriate mass ratio can be maintained during long-range outward migration when gas accretion onto the planets is taken into account in a realistic fashion.

15.2.2.3 Inward-Migrating Jupiter and Saturn

Giant planets typically migrate inward (e.g., Ward, 1997). As it migrates, the planetesimals a giant planet encounters may be shepherded inward by mean motion resonances, accreted, or scattered (Tanaka & Ida, 1999; Fogg and Nelson, 2005, 2007; Raymond et al., 2006a, 2016; Mandell et al., 2007; this is like Jupiter's inward migration in Figure 15.3). The balance between outcomes depends on the planet's mass and migration rate as well as the strength of the dissipation (aerodynamic or other drag) felt by the planetesimals. For inward-migrating gas giants in current disk models, shepherding dominates.

Bitsch et al. (2015) showed that, including the effects of pebble accretion and migration in an evolving disk, Jupiter's core may have originated at ∼20 AU. Pirani et al. (2019a) imposed this growth and migration history and studied its effect on a population of planetesimals. They found that planetesimals were shepherded by strong resonances, as expected, and that certain populations of near-resonant asteroids (e.g., the 3:2 resonant Hildas) were broadly matched. They also matched the observed asymmetry among Jupiter's 1:1 resonant populations (the so-called Trojans), for which the leading population outnumbers the trailing one. In addition to specific populations of asteroids, inward migration also implants planetesimals into the outer asteroid belt, expanding the source region for C-types to be 5–20 AU in width (Raymond & Izidoro, 2017a; Pirani et al., 2019a), although a large fraction of implanted outer disk planetesimals originate in the present-day Jupiter–Saturn region. It remains to be seen whether this model can fully explain the Trojan population (see Pirani et al., 2019b). Specific challenges include matching the Trojans' broad inclination distribution, reconciling the fact that Jupiter's Trojans are mainly D-types rather than C-types (Emery et al., 2015) with their source location, and that most Trojans should not survive the giant planet instability (Morbidelli et al., 2005; Nesvorný et al., 2013).

A purely inward migration of Jupiter would not significantly deplete the belt, so another mechanism is required to explain the belt's very low total mass. For example, the belt may simply never have contained much mass in planetesimals; this is the foundation of the Low-mass Asteroid belt model (Hansen, 2009; Drążkowska et al., 2016; Raymond & Izidoro, 2017b). Global models that match all of the asteroid belt constraints will be discussed in Section 15.4 (see Raymond et al., 2020, for building global models for Solar System- and exoplanet formation).

15.2.3 Effect of Secular Resonance Sweeping during Disk Dissipation

The gaseous disk's dissipation – and its changing gravitational potential – caused the locations of secular resonances with the giant planets to shift (Heppenheimer, 1980; Ward, 1981). Certain strong resonances may have swept across the asteroid belt (Lemaitre & Dubru, 1991; Lecar & Franklin, 1997). Secular resonances such as the ν_6 and ν_{16} strongly excite eccentricities and inclinations of small bodies and, in the present-day Solar System, excite asteroids to the point that they become unstable and are lost from the belt (see, e.g., Morbidelli & Henrard, 1991; Gladman et al., 1997).

Sweeping secular resonances have been invoked to explain the primordial excitation of the asteroid belt (Lecar & Franklin, 1997; Nagasawa et al., 2000, 2001, 2002). When gas drag (either aerodynamic or tidal) is accounted for, inward-sweeping secular resonances may even act to clear large planetary embryos from the belt, and this process has even been invoked to explain the terrestrial planets' orbital structure (Nagasawa et al., 2005; Thommes et al., 2008; Bromley & Kenyon, 2017). However, the secular resonance sweeping model suffers from a self-consistency problem. Secular resonances require noncircular, noncoplanar giant planet orbits. The studies presented up to this point in this section assumed that the giant planets were on their present-day orbits during the late parts of the gaseous disk phase. But we know this was not the case because planet–disk interactions would very rapidly have decreased their orbital eccentricities and inclinations and driven the planets into a resonant configuration. This is the generic outcome of hydrodynamical simulations (e.g., Morbidelli et al., 2007; Kley & Nelson, 2012; Baruteau et al., 2014; Pierens et al., 2014). When plausible orbits for the giant planets are used, sweeping secular resonances are far too weak to excite the asteroids (O'Brien et al., 2007). Future work may revive the sweeping secular resonance model if it is shown that giant planets can indeed maintain significant orbital eccentricities and/or inclinations during the gaseous disk phase.

15.3 EVOLUTION OF THE ASTEROID BELT AFTER DISPERSAL OF THE GASEOUS DISK

We now turn our attention to the asteroid belt's dynamical evolution after the dispersal of the gaseous disk. This covers the vast majority of our system's history. We first discuss gravitational self-stirring of the belt (Section 15.3.1). Next, we show how asteroids are implanted from the terrestrial planet region (Section 15.3.2) and how the belt may have been chaotically excited by the giant planets (Section 15.3.3). The giant planet instability is likely to have depleted and excited the belt and implanted outer disk planetesimals (Section 15.3.4).

15.3.1 Gravitational Self-Stirring of the Belt

The so-called classical model of terrestrial planet formation assumes that the terrestrial planets formed locally from a massive disk of planetesimals (see Wetherill, 1980; Chambers, 2001; Raymond et al., 2014). There is a rich literature on the growth of rocky planetary embryos by accretion in swarms of planetesimals (e.g., Safronov, 1969; Greenberg et al., 1978; Wetherill & Stewart, 1989; Kokubo & Ida, 2000), with a possible contribution from pebbles (Levison et al., 2015b; Schiller et al., 2018, 2020; Budde et al., 2019). When planetary embryos enter the oligarchic regime the distribution of planetesimals' random velocities (and thus their eccentricities and inclinations) becomes dominated by gravitational scattering by embryos (Ida & Makino, 1993; Kokubo & Ida, 2000). The level of excitation is related to the escape velocity from the largest embryos relative to the escape velocity from the Sun at the same orbital distance; this ratio is sometimes called the Safronov number (Safronov, 1969).

Early studies showed that the primordial asteroid belt could have been excited to its current level in eccentricity and inclination by a population of Moon- to Mars-mass planetary embryos (Wetherill, 1992; Chambers & Wetherill, 2001; Petit et al., 2001; O'Brien et al., 2007). There are constraints on this model. First, embryos could not have survived in the belt for too long because they would open gaps in the distribution of asteroids that are not seen (see figure 1 in Raymond et al., 2009). These gaps could have been smeared out to some degree during the giant planet instability but only for certain evolutionary scenarios (Brasil et al., 2016). Simulations of terrestrial planet formation in the context of the classical model often strand embryos in the belt (e.g. Raymond et al., 2006b, 2009; Fischer & Ciesla, 2014). Second, the total amount of mass contained in the asteroid belt must have remained consistent with the mass distribution of the terrestrial planets (see Izidoro et al., 2015b).

Figure 15.4 shows how, in the self-stirring model, the level of excitation of the asteroids is linked with the mass of Mars (Izidoro et al., 2015b). This is a consequence of the mass distribution of the belt. In simulations in a relatively shallow disk (i.e., with large asteroidal mass), massive embryos in the belt stirred up the eccentricities and inclinations of the asteroids, although embryos were often stranded within the belt. While this is inconsistent with the present-day Solar System, the more severe problem for shallow-disk simulations comes from the terrestrial planets. These simulations suffer from the "small Mars problem" (Wetherill, 1991; Raymond et al., 2009): they systematically form Mars analogues that are far more massive than the real one. There is simply too much mass in Mars' feeding zone.

In simulations in a very steep disk (right-hand panels of Figure 15.4) much less mass starts off in the asteroid belt and

no embryos more massive than $10^{-3} M \oplus$ form past roughly 2 AU (see Izidoro et al., 2015b, for details). Given the mass deficit in Mars' feeding zone these simulations form terrestrial planets that match the real ones, with Earth/Mars mass ratios comparable to the real one. However, the mass deficit in the asteroid belt suppresses self-stirring such that the asteroids' eccentricities and inclinations are too low, especially in the outer Main Belt.

This argument illustrates the no-win nature of the self-stirring model (at least in the context of the classical model of terrestrial planet formation; we will discuss alternatives in Section 15.4). Either the belt is self-excited but Mars is far too massive or the correct Earth/Mars mass ratio is matched but the asteroid belt is under-excited (Izidoro et al., 2015b).

15.3.2 Implantation of Asteroids from the Terrestrial Planet Region

There is ample circumstantial evidence that the terrestrial planets grew from a population of $\sim$ Mars-mass planetary embryos (see discussion in Morbidelli et al., 2012a). During this process – which lasted at least until Earth's last giant impact at 50–100 Myr (Kleine et al., 2009; Jacobson et al., 2014; Bottke et al., 2015) – planetesimals were frequently scattered in all directions by planetary embryos. Most planetesimals ended up being accreted by a growing planet or ejected after a close encounter with Jupiter (Raymond et al., 2006b), but a small fraction were implanted into the asteroid belt.

Bottke et al. (2006) proposed that the parent bodies of iron meteorites were implanted from the terrestrial planet-forming region. They found that roughly one in a thousand terrestrial planetesimals was scattered outward by successive encounters with embryos and ended up on a stable orbit within the belt (see also Haghighipour & Scott, 2012; Mastrobuono-Battisti & Perets, 2017). While the mechanism of implantation was conceptually sound, these studies piggybacked on the classical model of terrestrial planet formation and therefore suffered from the same fatal flaw. Implantation relied on a chain of embryos that scatter planetesimals outward from roughly 1 AU to the Main Belt. Those same chains of embryos formed Mars analogues that were far too massive (once again, the dreaded "small Mars" problem; Wetherill, 1991; Raymond et al., 2009).

Hansen (2009) proposed that the initial conditions being used in classical model simulations were not the real ones. He showed that the large Earth/Mars mass ratio was naturally matched if the terrestrial planets did not form from a broad disk but from a narrow annulus from 0.7–1 AU. Mars and Mercury were scattered out of the annulus and starved, whereas Earth and Venus grew massive within the annulus (see also Kaib & Cowan, 2015; Raymond & Izidoro, 2017b). More detailed simulations find that this evolution may require embryos to form very quickly within the annulus (or the annulus to have shaped by external forces as in the Grand Tack model; see Jacobson & Morbidelli, 2014); otherwise, a narrow annulus of planetesimals rapidly spreads out and systems that emerge often fail to match the real terrestrial planets using metrics such as the *radial mass concentration* (Walsh & Levison, 2016; Deienno et al., 2019).

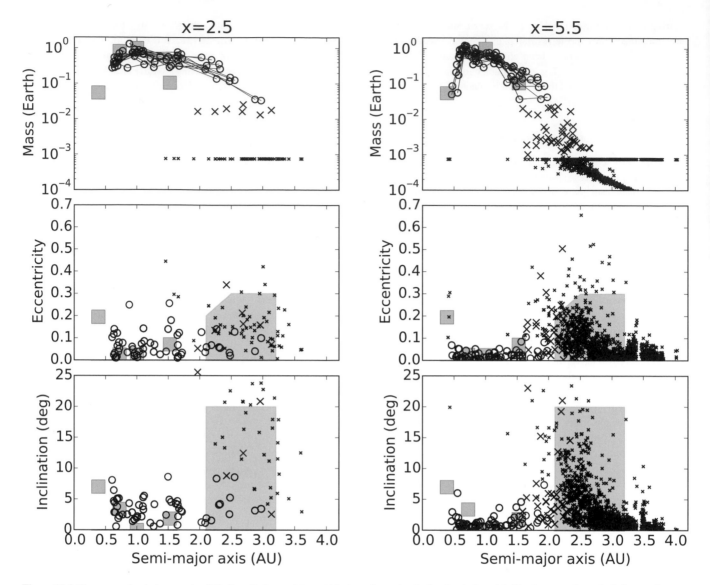

Figure 15.4 The connection between asteroid belt excitation and terrestrial planet formation in the classical model. The three panels on the left come from a suite of simulations containing rocky planetary embryos and planetesimals within a disk following a surface density profile $\Sigma \propto r^{-x}$, where $x = 2.5$, and the ones on the right are from analogous simulations in a disk with a much steeper radial surface density gradient ($x = 5.5$). Note that the minimum-mass solar nebula model has $x = 1.5$ (Weidenschilling, 1977b; Hayashi, 1981) and observations of the outer parts of protoplanetary disks find $x = 0.5 - 1$ (Andrews et al., 2009; Williams & Cieza, 2011). Top: Mass versus orbital separation for the surviving planets after 700 Myr of evolution. Middle: Eccentricity versus orbital radius of all surviving bodies. Bottom: Inclination versus orbital radius for surviving bodies. Circles represent surviving planetary embryos or planets and dots are planetesimals. The main asteroid belt is shaded. The actual terrestrial planets are shown with solid triangles.
Adapted from Izidoro et al. (2015b)

Within the framework of a narrow annulus, Raymond and Izidoro (2017b) showed that planetesimals are implanted into the asteroid belt from the terrestrial planet-forming region. While the efficiency of implantation was only $\sim 10^{-3}$ this was sufficient to implant all of the S-types, even accounting for later dynamical losses (e.g., during the giant planet instability). It is interesting to note that the rate of implantation of terrestrial planetesimals from a narrow annulus (as in Raymond & Izidoro, 2017b) is only modestly smaller than from a broad disk (as in Bottke et al., 2006).

Implantation of asteroids from a terrestrial annulus is slightly different than from a broad disk. Figure 15.5 shows the distribution

of implanted terrestrial planetesimals from Raymond and Izidoro (2017b). Planetesimals are continually scattered by growing embryos onto eccentric orbits that may cross the asteroid belt. These orbits have semimajor axes within the asteroid region, but their eccentricities are too high. Two mechanisms can drop their eccentricities (under gas-free conditions). First, a planetesimal can be scattered into a strong mean motion resonance with Jupiter, which causes an angular momentum exchange that can drop the planetesimal's eccentricity. Planetesimals in or near the 3:1 resonance with Jupiter can be seen in Figure 15.5 as the peak at ~ 2.55 AU. Resonant planetesimals with high eccentricities are not stable for long timescales unless another event can drop their

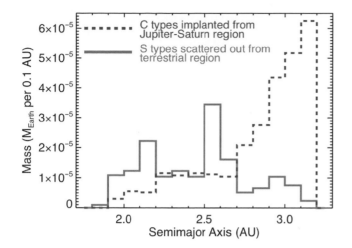

Figure 15.5 Radial distribution of a 100% implanted asteroid belt (adapted from Raymond & Izidoro, 2017b). The light grey curve indicates asteroids scattered into the Main Belt from the terrestrial planet-forming region, and the dark grey curve are asteroids scattered inward during Jupiter and Saturn's growth (by the mechanism illustrated in Figure 15.2).

eccentricities further, for example the Kozai resonance within mean motion resonances, which offers a robust path to low eccentricity. A planetesimal trapped on a resonant orbit with Jupiter when the giant planet instability took place may have ended up on a long-term stable orbit if it was removed from resonance as Jupiter's orbit shifted (see Section 15.3.4).

The second mechanism for trapping scattered planetesimals relies on embryo scattering. During accretion embryos scatter each other onto asteroid belt-crossing orbits, sometimes crossing Jupiter's orbit (often leading to ejection from the Solar System). When a scattered embryo encounters a scattered planetesimal with a random phase, the planetesimal's eccentricity can either increase or decrease. If the eccentricity increases the planetesimal crosses Jupiter's orbit and is quickly ejected. But if the eccentricity decreases the planetesimal may be trapped on a stable, non-resonant orbit within the Main Belt. This mechanism was the most effective in implanting terrestrial planetesimals into the belt in the simulations from Figure 15.5. Scattered embryos do not survive in the belt but end up either colliding with a growing terrestrial planet or being ejected.

A primordial empty asteroid belt would have been populated with scattered planetesimals that roughly match the present-day belt's radial distribution. Figure 15.5 shows the radial distribution of planetesimals implanted as a byproduct of the giant planets' growth from a suite of simulations, like the one from Figure 15.2. This matches the compositional distribution of the belt if we associate terrestrial planet-forming planetesimals with S-types and giant planet–region planetesimals with C-types.

It is not clear that planetesimals from the terrestrial planet-forming region should be associated with S-types. Ordinary chondrite meteorites (sourced from S-type asteroids; Bus & Binzel, 2002) are compositionally different than Earth (e.g., Warren, 2011). While chemical models for Earth's growth allow for a contribution from ordinary chondrites they generally find that Earth is mostly made from Enstatite chondrite-like material

(e.g., Dauphas, 2017). Given that the dynamical mechanism of planetesimal implantation is robust, it is unclear whether S-types are indeed representative of the terrestrial planet-forming region but the terrestrial planetesimals' compositions varied in time and/or with radial distance (e.g., Schiller et al., 2018) or whether implanted planetesimals simply only represent a fraction of inner belt asteroids (perhaps Enstatites).

15.3.3 Chaotic Excitation by Jupiter and Saturn

Izidoro et al. (2016) showed that chaos in the orbits of Jupiter and Saturn could have excited the asteroids' orbits. Emerging from the gaseous disk, Jupiter and Saturn's orbits sometimes exhibit chaotic evolution. This is more likely if Jupiter and Saturn were in 2:1 resonance rather than in 3:2, although chaotic configurations exist throughout the relevant parameter space.

Figure 15.6 shows how the asteroids are excited when the giant planets exhibit regular or chaotic motion. In the left-hand simulation, the peaked frequency spectrum indicates that the system is dominated by a small number of fundamental modes and that Jupiter and Saturn's orbits are evolving regularly. Secular resonances within the belt – corresponding to specific locations at which asteroids' precession rates match these frequencies – remain fixed. Asteroids' eccentricities and inclinations are only excited in those locations. This does not reproduce the observed distribution (shown as the grey points in the bottom two panels), with asteroids excited across the width of the belt (see also Figure 15.1).

In contrast, when Jupiter and Saturn evolve chaotically they excite the entire width of the asteroid belt (Izidoro et al., 2016). In the right-hand simulation of Figure 15.6, Jupiter and Saturn's initial orbits evolved chaotically, as shown by the broad (not peaked) precession frequency spectrum. The planets' precession frequencies underwent chaotic jumps, causing the locations of secular resonances to jump to different orbital radii. This acts to excite the full width of the belt. In the example from Figure 15.6 the belt may even have been over-excited, as there is a large population of asteroids with inclinations above 20° that cannot easily be removed (Roig & Nesvorný, 2015; Deienno et al., 2016, 2018).

For this chaos-driven mechanism to be responsible for exciting the belt, two conditions must apply. First, the belt's mass must initially have been low. If the primordial belt was massive then perturbations between objects would overwhelm the secular resonant perturbations. Of course, if the primordial belt contained a lot of mass then it would simply have self-excited (Figure 15.4). Second, the giant planets must have spent enough time in a chaotic configuration to excite the whole belt. It is plausible to imagine that the giant planets emerged from the gaseous disk in a chaotic configuration (with a low-mass asteroid belt) and then excited the asteroids' orbits during an interval of at least a few Myr before the onset of the giant planet instability (Izidoro et al., 2016).

15.3.4 Effect of the Giant Planet Instability: Dynamical Excitation, Radial Mixing, and Implantation

It is now well-accepted that the giant planets underwent a dynamical instability after their formation (see review by Nesvorný, 2018). The giant planet instability offers an elegant explanation

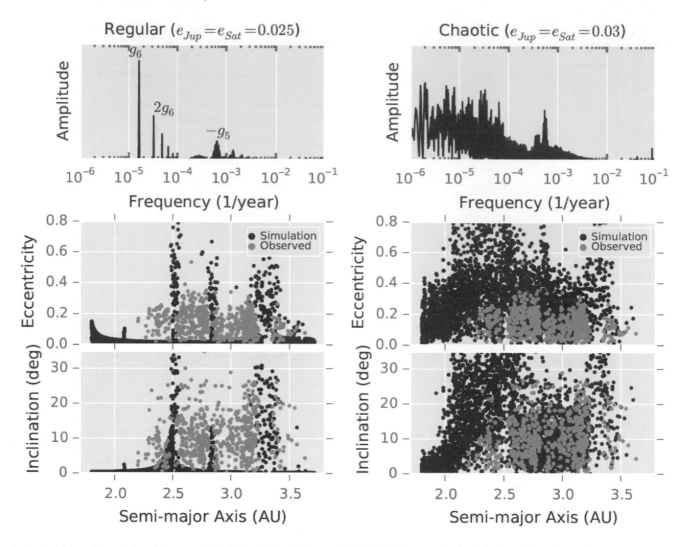

Figure 15.6 Chaotic excitation of the asteroid belt (adapted from Izidoro et al., 2016). Left: An example simulation in which Jupiter and Saturn's evolution is regular, with the planets locked in 2:1 resonance as an outcome of their migration within the gaseous disk (Pierens et al., 2014). Right: An example simulation in which Jupiter and Saturn's eccentricity is only slightly higher (0.03 rather than 0.025) than in the panels on the left but their evolution is chaotic. The top panels show the frequency spectrum of each system and the bottom panels show the asteroid belt's inclinations and eccentricities, as compared with the present-day Main Belt (in gray). Chaos in the giant planets' orbits causes secular resonances to have a broad frequency spectrum and act to excite asteroids across the entire belt rather than only at specific orbital radii.

for many small body Solar System populations and has important consequences for the asteroid belt. In the upcoming subsections we first discuss the instability itself and then the implications of the instability for the asteroid belt. We discuss the implantation of outer Solar System planetesimals, the depletion and excitation of the belt, and the smearing out of asteroid families. Finally, we discuss how the final phase of spreading out of Jupiter and Saturn's orbits appears to have left its imprint in the inclination distribution of the inner belt asteroids.

15.3.4.1 The Giant Planet Instability

The idea that the giant planets' orbits have changed since their formation was inferred by Fernandez and Ip (1984) by considering the back-reaction of planetesimals scattered into the inner Solar System. This is called planetesimal-driven migration. Neptune, Uranus, or Saturn are not massive enough to scatter planetesimals

to hyperbolic orbits (e.g. Duncan et al., 1987). Planetesimals instead are scattered inward by Neptune, Uranus, and Saturn, and then Jupiter scatters them out to interstellar space. Orbital energy conservation dictates that Neptune, Uranus, and Saturn migrate outward, whereas Jupiter migrates in. Malhotra (1993, 1995) showed that Pluto's resonant orbit with Neptune could be taken as evidence for Neptune's outward migration.

Planetesimal-driven migration likely took place in the early Solar System. It was originally suggested (Levison & Stewart, 2001; Gomes et al., 2005) that migration was delayed in order to explain the LHB (Late Heavy Bombardment), an apparent spike in the bombardment in the inner Solar System (Tera et al., 1974; Gomes et al., 2005; Bottke et al., 2012; Marchi et al., 2012; Morbidelli et al., 2012b). Constraints on the collisional grinding of the outer disk indicate, however, that there probably was no delay and migration started no later than 100 Myr after the gas disk dispersal (Nesvorný et al., 2018). Lunar craters were probably produced by impactors from the terrestrial planet region (Morbidelli et al., 2018).

The timescale of planetesimal-driven migration can be constrained from the structure of the asteroid and Kuiper belts. If Jupiter's planetesimal-driven migration were slow, secular resonances would produce a large population of asteroids on highly inclined orbits ($i > 15°$; Morbidelli et al., 2010; Walsh & Morbidelli, 2011), and would destabilize the orbits of the terrestrial planets (Brasser et al., 2009; Agnor & Lin, 2012). The absence of any large high-i population in the present asteroid belt indicates that Jupiter's (and Saturn's) migration was fast (e-folding time $\tau < 1$ Myr). In contrast, the distribution of orbital inclinations in the Kuiper belt is broad, reaching above 30°. This implies that Neptune's migration was slow (with e-folding time $\tau > 5$ Myr, Nesvorný, 2015). This is a problem because the migration timescales of Jupiter and Neptune cannot be so different, as they are both linked to the outer disk mass (more massive disks produce faster migration; Gomes et al., 2004).

The tension between asteroid and Kuiper belt constraints can be resolved if a dynamical instability took place. This idea is sometimes called the Nice model because it was first developed in Nice, France (Tsiganis et al., 2005). Its most recent incarnation is sometimes called the Jumping Jupiter model.

The current paradigm for the giant planet instability is as follows (see Nesvorný, 2018, for a review). The giant planets likely emerged from the gaseous disk in a multi-resonant chain, a generic outcome of planet-disk interactions (e.g., Morbidelli et al., 2007). Jupiter and Saturn were either in 3:2 or 2:1 resonance (Pierens et al., 2014). An outer disk of planetesimals extended out past 30 AU and contained a total of $\sim 15 - 20 M \oplus$. The instability may have been triggered by a resonance between a pair of migrating planets (see Levison et al., 2011; Deienno et al., 2017; Nesvorný, 2018; Quarles & Kaib, 2019), although it is possible that the giant planet system was unstable on its own (meaning that interactions with the planetesimal disk may not have been the trigger of instability (Ribeiro de Sousa et al., 2020), although the initial conditions, and thus the formation models, are critical in determining self-instability timescales). While the exact timing of the instability is unknown, it almost certainly took place within 100 Myr of CAIs (Nesvorný et al., 2018; Mojzsis et al., 2019). During the instability at least one ice giant was scattered by Jupiter and likely ejected (meaning the Solar System likely formed one or two extra ice giants; Nesvorný, 2011; Batygin, 2012; Nesvorný & Morbidelli, 2012). The semimajor axis of Jupiter nearly instantaneously changed during the scattering event and the secular resonances jumped from >3.5 AU to ~ 2 AU (this is the "jumping Jupiter" variant of the Nice model). This resolves the problem with the high-i population because the ν_{16} resonance never spent too much time between 2 and 3.5 AU (Morbidelli et al., 2010). The instability can also explain why Jupiter has a significant orbital eccentricity (proper $e = 0.044$). The outer planetesimal disk was destabilized, and the Oort cloud and Kuiper belt were populated (Levison et al., 2008; Brasser & Morbidelli, 2013; Nesvorný, 2015). As the last dregs of the planetesimal disk were cleared, giant planets reached their present-day orbital configuration. In summary, the planetesimal-driven migration could have been slow as required from the Kuiper belt constraints, but the jumping Jupiter instability is needed for the asteroid belt (see Nesvorný, 2015; Deienno et al., 2017).

15.3.4.2 Implantation of D-types from the Trans-Neptunian Disk

During the gaseous disk phase, the giant planets' growth and migration acted to implant bodies from ~ 5–20 AU into the asteroid belt (see discussion in Section 15.2.1). A second stage of implantation followed after the gas disk dispersed, as a consequence of the giant planet instability.

By examining simulations of the original giant planet instability model, Levison et al. (2009) noticed that a fraction of outer disk planetesimals ended on stable orbits in the asteroid belt. They hypothesized that planetesimals were captured and subsequently released from migrating resonances with Jupiter, leaving planetesimals on low-e orbits in the Main Belt. The results suggested that the implanted population could be huge, some ~ 1–2 orders in magnitude larger than the current population of main-belt asteroids. Vokrouhlický et al. (2016) used a large statistical sample (effectively $\sim 10^9$ bodies) to revisit this issue. Using the best migration/instability models from Nesvorný and Morbidelli (2012), they confirmed the general process of implantation proposed by Levison et al. (2009) but implanted a smaller fraction of the outer disk planetesimals. Vokrouhlický et al. (2016) assumed the outer disk to start with $\sim 5 \times 10^7$ bodies with $D > 100$ km and a Trojan-like size distribution for $D < 100$ km (differential power law index $q \simeq 2$). Figure 15.7 shows the orbital distribution of asteroids captured in the simulations from Vokrouhlický et al. (2016).

The outer disk planetesimals are thought to be represented by P/D types in asteroid taxonomy. This is because simulations of the giant planet instability also explain Jupiter's Trojan asteroids – which are predominantly D/P types – as captured outer disk planetesimals (Morbidelli et al., 2005; Nesvorný et al., 2013). By comparing the implanted population with the Main Belt P/D-types, Vokrouhlický et al. (2016) found that: (1) the number of *large* P/D bodies implanted in the model matches observations, but (2) the population of implanted P/D bodies with $D \sim 10$ km is roughly 10-times larger than observed. Vokrouhlický et al. (2016) argued that collisional grinding, thermal destruction of bodies before their implantation, and/or removal of small bodies by the Yarkovsky effect would resolve this issue. This remains to be demonstrated. The implantation hypothesis described here raises an interesting possibility that some D-type near-Earth asteroids, potential targets of sample-return space missions, are compositionally related to Kuiper belt objects.

15.3.4.3 Dynamical Excitation and Depletion

The giant planet instability's effect on the asteroid belt depends on the belt's initial distribution and the exact evolution of the planets. A common assumption is that the primordial asteroid belt was massive and dynamically cold, in contrast with the present-day belt's low mass and excited orbits (see Section 15.1.1). However, it is entirely possible that the primordial belt was already depleted and that a weaker instability could still explain the present-day belt.

Recent studies have simulated the dynamics of the inner Solar System during and after the giant planet instability. Roig and Nesvorný (2015) modeled the asteroid belt's dynamical evolution during the instability including just the giant planets, and Nesvorný

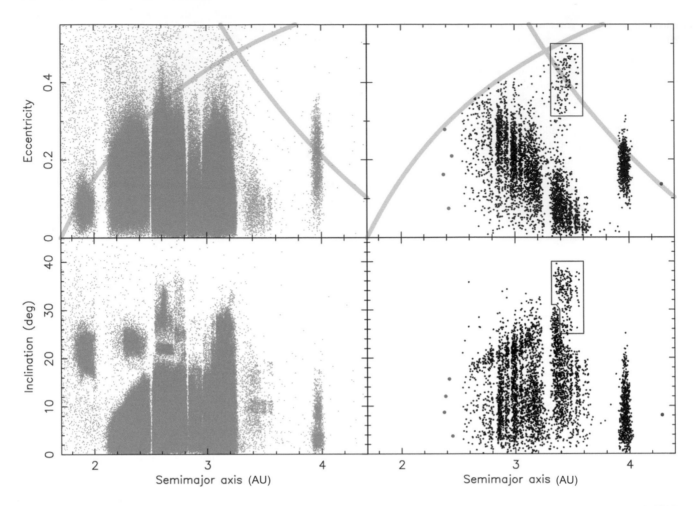

Figure 15.7 Population of asteroids captured from the outer planetesimal disk (right) in the giant planet instability simulations of Vokrouhlický et al. (2016), compared with the actual distribution of all numbered asteroids (left). The four innermost (larger) particles were implanted interior to the 3:1 resonance with Jupiter. The rectangle/polygon in the right panels indicates the population of particles captured in Kozai states (for details, see Vokrouhlický et al., 2016).

et al. (2017) extended this analysis to also include the terrestrial planets, assuming them to be fully-formed at the time of instability. Both treated the asteroids as massless test particles initially spread across a broad range of parameters space (a, e, and i) without specifying the initial mass. Instead, the asteroid orbits were propagated over 4.5 Gyr and matched against the present asteroid belt. The results were used to estimate the overall depletion and the initial mass. A small number of orbital histories for the giant planets were carefully selected from a vast number of simulations (Nesvorný & Morbidelli, 2012) to match various Solar System constraints (e.g., the outer planet orbits, Kuiper belt, planetary satellites, Jupiter Trojans). The planets' migration histories were strictly controlled to only consider orbital evolutions compatible with constraints. Both studies accounted for the effects of an additional ice giant whose orbit may have briefly ($\sim$10,000 yr) overlapped with the asteroid belt during the instability. The selected migration/instability models were compatible with the dynamical structure of the terrestrial planets, including the excited orbit of Mercury (Roig et al., 2016). Thus, whereas the terrestrial planet system is in general fragile and susceptible to excessive dynamical excitation during planetary migration and instability (Agnor & Lin, 2012; Kaib & Chambers, 2016), the cases

considered in Roig and Nesvorný (2015) and Nesvorný et al. (2017) did not have this problem.

Figure 15.8 shows the final distribution of simulated asteroids, which provide a good match to the present-day belt. Roig and Nesvorný (2015) and Nesvorný et al. (2017) did not find a difference in the final orbital distribution of asteroids as a function of the timing of the instability. In their setup there was not much difference between the effects of dynamical depletion over $\sim$ 4.5 Gyr (for the early instability) or over $\sim$ 4 Gyr (for the late instability). The early instability led to a slightly smaller number of asteroids surviving at the present time, because the dynamical erosion with Jupiter on its current orbit has more time available. The original (i.e., at the time of the solar nebula dispersal) inclination distribution of Main Belt asteroids was inferred to have been relatively narrow with a great majority of bodies starting with $i < 20°$ (Roig & Nesvorný, 2015). This is because objects starting with $i > 20°$ would have survived, but a large population of Main Belt asteroids with $i > 20°$ do not exist today (Morbidelli et al., 2010). The nominal Grand Tack model produces an inclination distribution that is slightly too wide to satisfy this constraint (Deienno et al., 2016) but it is possible that a minor modification of the original Grand Tack model would yield a more satisfactory result.

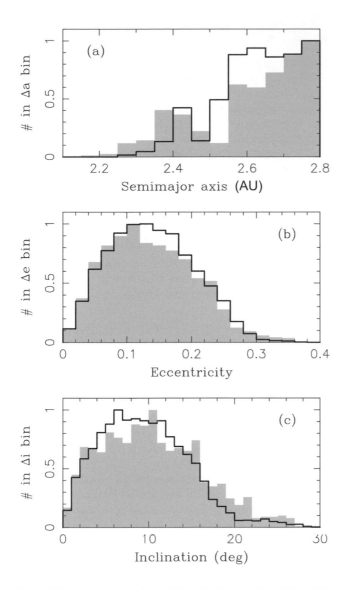

Figure 15.8 A comparison of the orbital distribution of $D > 30$ km Main Belt asteroids (histograms) with the distribution from simulations by Nesvorný et al. (2017) of asteroid belt sculpting during the giant planet instability and for an additional 4.5 Gyr (dark lines).

The original eccentricity distribution is less well constrained because planets act, through overlapping orbital resonances, to erode the population of high-e objects. For example, Deienno et al. (2016) showed that the broad eccentricity distribution produced by the Grand Tack model would erode toward a final state similar to the present asteroid belt. Indeed, Roig and Nesvorný (2015) favored cases where the belt was at least slightly excited in eccentricities before the planetary migration/instability happened. Minton and Malhotra (2011) considered the eccentricity evolution of asteroids during planetary migration (without instability) and found that either the belt was initially dynamically cold and Jupiter/Saturn migrated fast, or the belt was dynamically hot and Jupiter/Saturn migrated slow. It is not clear what the meaning of this result is in the jumping Jupiter paradigm. Finally, Deienno et al. (2018) showed that a subset of jumping Jupiter models with more chaotic evolution of Jupiter and Saturn can strongly excite the Main Belt. The mechanism demonstrated in Deienno et al. (2018) is similar to the

chaotic excitation mechanism of Izidoro et al. (2016) discussed in Section 15.3.3; it relies on secular excitation of asteroids by Jupiter and Saturn while their orbits are themselves in flux during the instability.

The degree of depletion of the belt remains an area of contention. Nesvorný et al. (2017) found that ~20% of the original population of asteroids (starting at ~2−3.5 AU) would survive in the Main Belt today. This estimate applies to the model of planetary migration/instability that best satisfies Solar System constraints, includes the dynamical erosion of the asteroid belt over 4.5 Gyr, and treats the asteroids as test particles. Models without planetary migration/instability indicate larger surviving fractions. For example, Minton and Malhotra (2011) estimated that ~50% of Main Belt asteroids would have survived since ~1 Myr after the establishment of the current Solar System architecture. This is consistent with the results obtained in Nesvorný et al. (2017) in the models *without* planetary migration/instability. Most removed asteroids are therefore expected to be removed early.

A higher fraction (~90%) of asteroids is removed during the instability starting from the orbital distributions obtained from the Grand Tack model (see Section 15.2.2.2) (Deienno et al., 2016). This is due to a large unstable population of asteroids on high-e orbits. Still, this depletion factor is not large enough to explain the Main Belt depletion, if the Main Belt started as massive. The combined depletion from the Grand Tack (99%) and subsequent migration/instability would work, within an order of magnitude, to reduce an initially massive Main Belt to the present population (Deienno et al., 2016). The asteroid belt depletion is not a problem if the asteroid belt started with a low mass (see discussion in Section 15.4.1). It is interesting that the inner belt ($a < 2.5$ AU) contains ~10-times less mass that the rest of the Main Belt, despite containing roughly one third of the orbital real estate. This can be explained if the secular resonances such as ν_6 and ν_{16} spent some time at 2–2.5 AU and strongly depleted the inner belt, which indeed takes place in the jumping-Jupiter model (Nesvorný et al., 2017).

Simulations in which the asteroids carry mass sometimes find larger depletion factors than those that treat asteroids as massless test particles. Clement et al. (2019a) showed that depletion of up to three-to-four orders of magnitude is possible when taking into account the pre-excitation of the belt by self-stirring. A caveat is that the pre-excitation in Clement et al. (2019a) was probably an upper limit on the plausible range as it was driven by gravitational self-stirring of very massive planetary embryos. In addition, the strongest depletion factors came from simulations that did not provide good matches to the Solar System, for example with the giant planets on orbits that are much more excited or spread out than the actual ones. Yet depletion factors of two orders of magnitude (~99% of asteroids removed) were a common outcome among systems that matched constraints. This also fits into models of terrestrial planet formation, as Clement et al. (2018, 2019a) showed that if the giant planet instability took place early enough it could also explain the terrestrial planets' mass distribution (see Section 15.4).

What are we to make of the differences between models of the inferred asteroid depletion during the giant planet instability? There are two fundamental differences between the studies. First, Roig and Nesvorný (2015), Nesvorný et al. (2017), and Deienno et al. (2018) imposed a specific orbital history for the giant planets that was drawn from simulations of the instability itself that best matched constraints (Nesvorný & Morbidelli, 2012). They then studied the effects of the giant planets' dynamics on the asteroids. In contrast, Clement et al. (2019b) did not prescribe the giant planets' evolution but only triggered their instability, then simulated its effects. Second, while Roig and Nesvorný (2015) and Nesvorný et al. (2017) treated the asteroids as massless test particles, Clement et al. (2019b) treated them as massive objects that interacted gravitationally with each other (assuming them to be equal-mass and quite massive) and with the giant planets.

Dynamical instabilities are inherently chaotic. Infinitesimal changes in the configuration of individual close encounters between planets may completely change the outcome. There is no single evolution that can produce the Solar System but rather a wide range of possible pathways. The studies of Roig and Nesvorný (2015), Nesvorný et al. (2017), and Deienno et al. (2018) only included giant planet evolutionary pathways that are consistent with constraints such as avoiding the over-excitation of the asteroids and insuring the survival of the (assumed to be already-formed) terrestrial planets. However, they only included a tiny subset of possible giant planet evolutionary pathways. In contrast, Clement et al. (2019b) sampled the effects of a much broader range of instabilities, but that included many that were not consistent with the present-day Solar System. The differences in outcome between the studies show how sensitive the asteroid belt's dynamics is to the exact evolution of the giant planets during the instability. Each of these approaches is valid: Roig and Nesvorný (2015), Nesvorný et al. (2017), and Deienno et al. (2018)'s approach goes deep to study the detailed consequences of a small number of plausible instabilities, whereas Clement et al. (2019b) goes wide and captures the outcome of more general instabilities, essentially putting the Solar System's instability in a larger context.

The two sets of studies' outcomes may in principle be reconciled by simply taking into account the asteroid belt's mass distribution. Imagine a scenario in which the early asteroid belt included ten large planetary embryos that contained 90% of the belt's mass. In this idealized case, and assuming that all 10 embryos were lost during the instability, removal of 90% of objects within the belt would produce a mass depletion of two orders of magnitude.

The mixing of asteroids in semimajor axis during the planetary migration/instability was insufficient (characteristic change $\Delta a <$ 0.1 AU) to explain the compositional mixing of the asteroid belt (Roig & Nesvorný, 2015; Clement et al., 2019b). Compositional mixing therefore requires a stronger effect such as the ones discussed in Section 15.2. Collisional families that should have formed during an intense period of collisional activity after the gas nebula removal would have been dispersed (Brasil et al., 2016, 2017). The dispersal of asteroid families in e and i is especially significant, suggesting that most ancient families cannot be identified by the usual clustering techniques. The V-shape method seems to be more promising (Delbo' et al., 2017). Minton and Malhotra (2009) identified gaps just outside the 5:2, 7:3, and 2:1 resonances and explained them by invoking a dynamical model with fast migration of Jupiter (one e-fold $\simeq$0.5 Myr; no instability). The gaps are created in their model when orbital resonances with Jupiter move inward and remove objects in their path. It may be difficult to obtain the same result with the jumping-Jupiter model where resonances jump to their final locations.

15.3.4.4 Effect of Jupiter and Saturn's Final Migration Phase

A common shortcoming of models for inner Solar System formation is that they produce too many high-inclination asteroids in the inner Main Belt (O'Brien et al., 2007; Morbidelli et al., 2010; Walsh & Morbidelli, 2011; Roig & Nesvorný, 2015; Deienno et al., 2016, 2018; Clement et al., 2018, 2019a, 2019b). Clement et al. (2020) showed that this deficit may be a signature of the final phase of migration of Jupiter and Saturn. This phase took place after the giant planets' instability, as the last remnants of the outer planetesimal disk were scattered away, causing the giant planets' orbits to slowly spread apart. This phase left an imprint on the asteroid belt via the ν_6 secular resonance, where asteroids' orbits precess at the same rate as Saturn. The ν_6 is a function of asteroid semimajor axis, eccentricity, and inclination (Morbidelli & Henrard, 1991). In the inner Main Belt, the ν_6 is located at $\sim$10–15° in inclination. Asteroids that fall in the ν_6 are removed by being driven to high enough eccentricity to collide with the Sun (Gladman et al., 1997).

Saturn's current precession rate is 28.22 arc seconds per year. Figure 15.9 (left panel) shows how Saturn's precession rate would have evolved during its last phase of migration, as it distanced itself from Jupiter. As Jupiter and Saturn's orbits spread apart, Saturn's precession rate naturally slowed. However, after reaching a minimum of 26 arcsec yr^{-1} its precession rate sped back up as it approached the 5:2 mean motion resonance with Jupiter (Milani & Knezevic, 1990; Morbidelli & Henrard, 1991). The right-hand panel of Figure 15.9 shows the distribution of asteroid precession rates. No asteroids can survive long at the same precession rate as Saturn because that would put them in the unstable ν_6 secular resonance. If Saturn had always maintained its current orbit, then there should be a narrow gap in the distribution with no asteroids having precession rates of 28.2 arcsec yr^{-1}. Clement et al. (2020) interpreted the large empty swath at precession rates between 26 and 28 arcsec yr^{-1} as evidence that Saturn's precession rate had swept between those values during its final phase of migration.

The asteroids that were destabilized during the final phase of migration – with precession rates within that empty range – include those on high-inclination orbits in the inner parts of the Main Belt. Taking the final phase of migration into account reconciles models of inner Solar System dynamics (that did not force the giant planets to end up on their exact current orbits) with the orbital structure of the present-day belt (Clement et al., 2020).

15.4 DISCUSSION

We now discuss how the dynamical mechanisms described in Sections 15.2 and 15.3 fit in a larger context. Different combinations of processes form the basis of three global models that each broadly match the Solar System: the Low-mass Asteroid belt, Grand Tack, and Early Instability models (see Figure 15.10 and extensive discussion in Raymond et al., 2020). These models are not mutually exclusive. For example, a Grand Tack-like migration of Jupiter and Saturn could have preceded an early giant planet instability. Likewise, a pre-depleted asteroid belt would be consistent with an early instability.

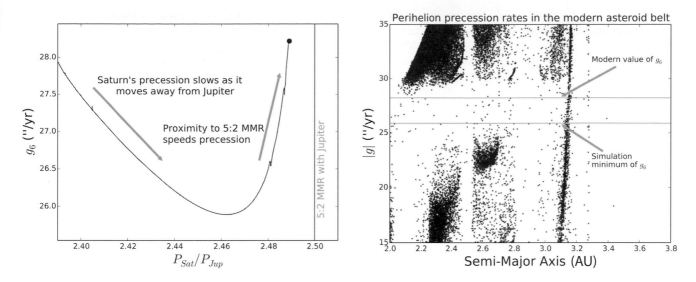

Figure 15.9 How Jupiter and Saturn's final phase of migration shaped the asteroid belt (adapted from Clement et al., 2020). Left: Saturn's orbital precession rate as a function of its separation from Jupiter. During the final phase of planetesimal scattering, Saturn moved away from Jupiter (from left to right in the figure). The black dot represents the present-day configuration. Right: Precession rates of known asteroids as a function of semimajor axis (from Knežević & Milani, 2003). The horizontal lines denote the minimum and maximum precession frequencies from the left panel. The fact that almost no asteroids have precession rates within that range is evidence that they were dynamically removed by the ν_6 secular resonance with Saturn during the giant planets' migration. The grouping of asteroids near 3.1 AU that span the gap represents the Euphrosyne family (Novaković et al., 2011) that is dynamically spreading due to the Yarkovsky effect (e.g., Bottke et al., 2001).

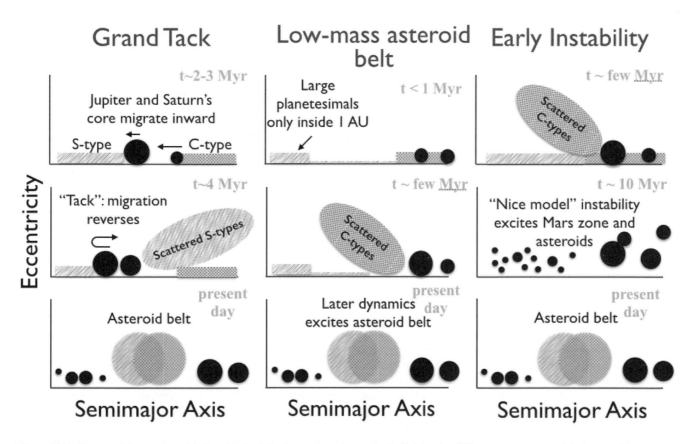

Figure 15.10 Cartoon of three current global models for Solar System formation, each of which invokes different processes to explain the asteroid belt. In the Grand Tack model the asteroid belt is depleted by Jupiter's migration, then re-populated by objects from different regions (Walsh et al., 2011, 2012; O'Brien et al., 2014). In the Low-mass asteroid belt model (e.g., Hansen, 2009; Drążkowska et al., 2016), the asteroids are implanted from both the terrestrial planet-forming and giant planet regions (Bottke et al., 2006; Raymond & Izidoro, 2017a, 2017b). In the Early Instability model (Clement et al., 2018, 2019a), the belt is depleted and excited by the giant planets' instability.
Adapted from Raymond et al. (2020).

In Section 15.4.1 we discuss how different combinations of processes can match the asteroid belt constraints laid out in Section 15.1.1, how they compare with global models of Solar System formation, and open questions. To conclude, we discuss the likely evolutionary paths of Vesta and Ceres (Section 15.4.2).

15.4.1 Mechanisms That Match the Asteroid Belt Constraints

Let us consider combinations of dynamical processes that can explain the observed asteroid belt. Based on current thinking, there are three plausible assumptions for the asteroid belt's initial mass distribution. The first is that the belt started with an Earth-mass or more in planetesimals. The second is that the belt was born low in mass, with perhaps a few percent of an Earth-mass in planetesimals. The third is that no planetesimals ever formed within the belt. In this subsection we discuss how the present-day belt could have been sculpted, starting from each of these starting assumptions. We will focus on three constraints: the asteroid belt's very low mass, its orbital excitation, and its compositional diversity (see Section 15.1.1).

A high-mass primordial asteroid belt requires a depletion of at least three orders of magnitude to match the current one. The asteroids likely started out on dynamically cold orbits. Self-stirring by planetary embryos within the belt (in the framework of the classical model of terrestrial planet formation) has been shown to deplete the belt by roughly two orders of magnitude (see Section 15.3.1). However, for self-stirring to match the asteroid eccentricity and inclination distribution requires such a large mass in asteroidal embryos that simulations cannot avoid producing Mars analogues that are far more massive than the real planet (the so-called small Mars problem), and often stranding Mars-mass embryos in the belt (see Section 15.3.3). Jupiter's inward-then-outward migration in the Grand Tack model can deplete the belt sufficiently and also explain the belt's compositional and orbital structure (see Section 15.2.2.2). Its main weakness is that it is unclear whether long-range outward migration can be sustained when gas accretion is taken into account. A third mechanism for strongly depleting the asteroid belt is a violent giant planet instability, which may also explain the belt's excitation (Section 15.3.4). In this case, implantation during giant planet growth would be needed to match the belt's compositional structure (see Section 15.2.1). However, it remains to be seen whether a strong enough instability to deplete the belt by three orders of magnitude can remain consistent with the Solar System as a whole.

A low-mass primordial asteroid belt would only need to be depleted by an order of magnitude to match the current one. Such depletion is a natural consequence of a relatively gentle giant planet instability (Section 15.3.4). The terrestrial planets' masses and orbits are reproduced if they accreted from a narrow annulus initially located between the orbits of Venus and Earth (Hansen, 2009; Kaib & Cowan, 2015; Walsh & Levison, 2016; Raymond & Izidoro, 2017b; Clement et al., 2019a; Deienno et al., 2019). The C-types would have been implanted during the giant planets' growth (Section 15.2.1) or migration (Section 15.2.2). A Grand Tack-like migration of Jupiter and Saturn (Section 15.2.2.2) or purely inward migration of the giant planets is consistent with a low-mass primordial belt (Section 15.2.2.3). Some planetesimals from the terrestrial region would still be implanted (Section 15.3.2), which would especially be needed in a Grand Tack scenario given

Jupiter's strong depletion of asteroidal planetesimals during migration. The asteroids' orbits could have been excited by chaotic phase in the giant planets' orbits (Section 15.3.3) or possibly during the giant planet instability (Section 15.3.4.2).

An empty primordial asteroid belt must be dynamically populated rather than depleted. The C-types would naturally be implanted from the Jupiter–Saturn region and beyond during the giant planets' growth (Section 15.2.1) and migration (Section 15.2.2). The D-types would have been implanted during the giant planet instability (Section 15.3.4.1), regardless of the belt's initial mass. As for a low-mass asteroid belt, the terrestrial planets are reproduced if they grew from a narrow annulus. Although implantation of planetesimals from the terrestrial region into the belt is efficient enough to explain the S-types (Section 15.3.2), it remains to be seen whether this process is consistent with cosmochemical constraints. The asteroid orbits – in particular the C-types' – could have been excited by the giant planets during a chaotic phase (Section 15.3.3) or the giant planet instability (Section 15.3.4.2).

Moving forward, there remain three central uncertainties for models of the asteroid belt's origin and dynamical evolution. First, the initial mass distribution of the belt in large planetesimals remains. Studies of planetesimal formation in evolving disks (e.g. Drążkowska et al., 2016; Carrera et al., 2017; Drążkowska & Dullemond, 2018) and of the interpretation of dust rings in planet-forming disks (e.g. ALMA Partnership et al., 2015; Andrews et al., 2016, 2018) will address this issue in the coming years. Two other uncertainties are related to the giant planets' orbital evolution. The degree to which – and in what direction – the giant planets and their cores migrated has a strong impact on the early depletion or implantation and excitation of the belt (see Section 15.2.2). This issue will continue to be addressed with hydrodynamical simulations that include realistic prescriptions for gas accretion. Finally, simulations of the giant planet instability show a spectrum of outcomes (Section 15.3.4.2); limiting which pathways are plausible will constrain the belt's excitation and depletion, and feed back on the belt's starting mass.

Future constraints will come from continually improving numerical simulations and observations. Upcoming large surveys such as LSST will improve the size and quality of asteroid data. Some of the most helpful constraints are those that are not anticipated. For example, Dermott et al. (2018) showed that the entire inner asteroid belt may come from just a handful of parent bodies. What this represents in terms of the asteroids' origin and dynamical evolution is not immediately clear.

15.4.2 The Dynamical Histories of Ceres and Vesta

We now turn our attention to Ceres and Vesta, the targets of the Dawn spacecraft (Russell et al., 2016). They are the two most massive asteroids and between them comprise more than 40% of the belt's total mass. While Ceres and Vesta currently occupy relatively close orbits, it is likely that they formed in vastly different environments in different parts of the Sun's protoplanetary disk.

Ceres has long been thought to be rich in water ice (e.g., Lebofsky et al., 1981; Fanale & Salvail, 1989; McCord & Sotin, 2005). Dawn confirmed this view and detected a number of volatile species including water ice (e.g., Küppers et al., 2014; Nathues et al., 2015; Bland et al., 2016; De Sanctis et al., 2017; Prettyman et al., 2017). Carbonaceous chondrites appear to be the closest meteorite analogues to Ceres (despite some differences – Milliken & Rivkin, 2009; Bland et al., 2016; McSween et al., 2018; Marchi et al., 2019; see also Chapters 7 and 8). The Dawn spacecraft also detected ammoniated phyllosilicates, thought to have an outer Solar System origin (De Sanctis et al., 2015). This evidence suggests that Ceres may been implanted by the same dynamical processes that are thought to have implanted the C-type asteroids (to our knowledge, the idea that Ceres was implanted from the outer Solar System was first proposed by McKinnon, 2008). During the giant planets' gas accretion, planetesimals from the Jupiter–Saturn region and beyond were scattered and implanted into the outer parts of the belt, on orbits that match the C-types' (see Section 15.2.1). The distribution of implanted Ceres-sized planetesimals has a broad peak at $\sim$2.7 AU, corresponding to roughly half of Jupiter's orbital radius (Raymond & Izidoro, 2017a). This peak is a result of the dynamics of scattering by Jupiter. As a result of weak gas drag felt by large planetesimals their orbits are slow to decouple from Jupiter's (by a drop in eccentricity and therefore aphelion) and so they are scattered repeatedly. This creates a much broader distribution in the orbital eccentricities and semimajor axes of large planetesimals, which is eroded from the outside-in as objects with Jupiter-crossing orbits are eventually ejected. This creates a broad peak at half of Jupiter's orbital distance (at the time of scattering), close to Ceres' orbital position. Migration may also have played a role; implantation into the outer Main Belt is a common outcome in both the Grand Tack model (Walsh et al., 2011, 2012) and in scenarios that invoke inward migration of the giant planets (Raymond & Izidoro, 2017a; Pirani et al., 2019a).

Vesta is a large ($D = 525$ km) differentiated asteroids in the inner Main Belt whose surface is extensively cratered (Marchi et al., 2012). Vesta's Oxygen isotopic composition does not fit within a clear Solar System gradient (Δ^{17}O $\approx$ $-$240 ppm, compared with zero and 320 ppm for Earth and Mars, respectively; e.g., Clayton & Mayeda, 1996; Franchi et al., 1999; Scott, 2002; Zhang et al., 2019). This argues that Vesta may have been scattered onto its current orbit from closer to the Sun. Bottke et al. (2006) proposed that Vesta was scattered out from the terrestrial planet-forming region. Multiple studies have shown that this is dynamically plausible (Bottke et al., 2006; Mastrobuono-Battisti & Perets, 2017; Raymond & Izidoro, 2017b, see also Section 15.3.2). Future studies of Vesta's detailed composition may further constrain where in the planet-forming disk it could have accreted (e.g. Toplis et al., 2013).

While their present-day orbits cross, Ceres and Vesta likely formed many astronomical units apart. The Jupiter–Saturn region represents the source for most C-types and therefore Ceres' most likely formation region (McKinnon, 2008; Walsh et al., 2011; Raymond & Izidoro, 2017a). If Vesta was implanted from closer-in then its most likely formation distance is exterior to Earth's orbit, perhaps at $\sim$1.5 AU, because the implantation rate is higher for objects originating near Mars' orbit than Earth's (Bottke et al., 2006; Raymond & Izidoro, 2017b). In addition, the collisional environment at that distance was less violent than at Earth's orbit (Bottke et al., 2006).

On a dynamical note, Ceres and Vesta have a non-negligible probability of collision in the future (a probability of roughly 0.2% per Gyr; Laskar et al., 2011). In fact, orbital chaos induced by close encounters between Ceres and Vesta is the limiting factor in predicting the long-term future evolution of the terrestrial planets' orbits. Thus, in addition to providing insights on our Solar System's early history, Ceres and Vesta prevent us from seeing our distant future.

REFERENCES

Abod, C. P., Simon, J. B., Li, R., et al. (2019) The mass and size distribution of planetesimals formed by the streaming instability. II. The effect of the radial gas pressure gradient. *The Astrophysical Journal*, 883, 192.

Adachi, I., Hayashi, C., & Nakazawa, K. (1976) The gas drag effect on the elliptical motion of a solid body in the primordial solar nebula. *Progress in Theoretical Physics*, 56, 1756–1771.

Agnor, C. B., & Lin, D. N. C. (2012) On the migration of Jupiter and Saturn: Constraints from linear models of secular resonant coupling with the terrestrial planets. *The Astrophysical Journal*, 745, 143.

Alexander, C. M. O., McKeegan, K. D., & Altwegg, K. (2018) Water reservoirs in small planetary bodies: Meteorites, asteroids, and comets. *Space Science Reviews*, 214, 36.

ALMA Partnership, Brogan, C. L., Pérez, L. M., et al. (2015) The 2014 ALMA long baseline campaign: First results from high angular resolution observations toward the HL tau region. *The Astrophysical Journal*, 808, L3.

Andrews, S. M., Huang, J., Pérez, L. M., et al. (2018) The Disk Substructures at High Angular Resolution Project (DSHARP). I. Motivation, sample, calibration, and overview. *The Astrophysical Journal*, 869, L41.

Andrews, S. M., Wilner, D. J., Hughes, A. M., Qi, C., & Dullemond, C. P. (2009) Protoplanetary disk structures in ophiuchus. *The Astrophysical Journal*, 700, 1502–1523.

Andrews, S. M., Wilner, D. J., Zhu, Z., et al. (2016) Ringed substructure and a gap at 1 au in the nearest protoplanetary disk. *The Astrophysical Journal*, 820, L40.

Baruteau, C., Crida, A., Paardekooper, S.-J., et al. (2014) Planet–disk interactions and early evolution of planetary systems. In H. Beuther, R. S. Klessen, C. P. Dullemond, & T. K. Henning (eds.), *Protostars and Planets VI*. Tucson: University of Arizona Press, pp. 667–689.

Batygin, K. (2012) A primordial origin for misalignments between stellar spin axes and planetary orbits. *Nature*, 491, 418–420.

Birnstiel, T., Fang, M., & Johansen, A. (2016) Dust evolution and the formation of planetesimals. Space Science Reviews, 205, 41–75.

Bitsch, B., Johansen, A., Lambrechts, M., & Morbidelli, A. (2015) The structure of protoplanetary discs around evolving young stars. *Astronomy & Astrophysics*, 575, A28.

Bitsch, B., Morbidelli, A., Johansen, A., et al. (2018) Pebble-isolation mass: Scaling law and implications for the formation of super-Earths and gas giants. *Astronomy & Astrophysics*, 612, A30.

Bland, M. T., Raymond, C. A., Schenk, P. M., et al. (2016) Composition and structure of the shallow subsurface of Ceres revealed by crater morphology. *Nature Geoscience*, 9, 538–542.

Boehnke, P., & Harrison, T. M. (2016) Illusory late heavy bombardments. *Proceedings of the National Academy of Sciences (USA)*, 113, 10802–10806.

Bottke, W. F., Durda, D. D., Nesvorný, D., et al. (2005) The fossilized size distribution of the main asteroid belt. *Icarus*, 175, 111–140.

Bottke, W. F., Nesvorný, D., Grimm, R. E., Morbidelli, A., & O'Brien, D. P. (2006) Iron meteorites as remnants of planetesimals formed in the terrestrial planet region. *Nature*, 439, 821–824.

Bottke, W. F., & Norman, M. D. (2017) The late heavy bombardment. *Annual Review of Earth and Planetary Sciences*, 45, 619–647.

Bottke, W. F., Vokrouhlický, D., Broz, M., Nesvorný, D., & Morbidelli, A. (2001) Dynamical spreading of asteroid families by the Yarkovsky effect. *Science*, 294, 1693–1696.

Bottke, W. F., Vokrouhlický, D., Marchi, S., et al. (2015) Dating the Moon-forming impact event with asteroidal meteorites. *Science*, 348, 321–323.

Bottke, W. F., Vokrouhlický, D., Minton, D., et al. (2012) An Archaean heavy bombardment from a destabilized extension of the asteroid belt. *Nature*, 485, 78–81.

Bouvier, A., & Wadhwa, M. (2010) The age of the Solar System redefined by the oldest Pb–Pb age of a meteoritic inclusion. *Nature Geoscience*, 3, 637–641.

Brasil, P. I. O., Roig, F., Nesvorný, D., & Carruba, V. (2017) Scattering V-type asteroids during the giant planet instability: a step for Jupiter, a leap for basalt. *Monthly Notices of the Royal Astronomical Society*, 468, 1236–1244.

Brasil, P. I. O., Roig, F., Nesvorný, D., et al. (2016) Dynamical dispersal of primordial asteroid families. *Icarus*, 266, 142–151.

Brasser, R., Matsumura, S., Ida, S., Mojzsis, S. J., & Werner, S. C. (2016) Analysis of terrestrial planet formation by the Grand Tack model: System architecture and tack location. *The Astrophysical Journal*, 821, 75.

Brasser, R., & Mojzsis, S. J. (2020) The partitioning of the inner and outer Solar System by a structured protoplanetary disk. *Nature Astronomy*, 4, 492–499.

Brasser, R., & Morbidelli, A. (2013) Oort cloud and scattered disc formation during a late dynamical instability in the Solar System. *Icarus*, 225, 40–49.

Brasser, R., Morbidelli, A., Gomes, R., Tsiganis, K., & Levison, H. F. (2009) Constructing the secular architecture of the Solar System II: the terrestrial planets. *Astronomy & Astrophysics*, 507, 1053–1065.

Bromley, B. C., & Kenyon, S. J. (2017) Terrestrial planet formation: Dynamical shake-up and the low mass of Mars. *The Astronomical Journal*, 153, 216.

Bryden, G., Chen, X., Lin, D. N. C., Nelson, R. P., & Papaloizou, J. C. B. (1999) Tidally induced gap formation in protostellar disks: Gap clearing and suppression of protoplanetary growth. *The Astrophysical Journal*, 514, 344–367.

Budde, G., Burkhardt, C., & Kleine, T. (2019) Molybdenum isotopic evidence for the late accretion of outer Solar System material to Earth. *Nature Astronomy*, 3, 736–741.

Budde, G., Kleine, T., Kruijer, T. S., Burkhardt, C., & Metzler, K. (2016) Tungsten isotopic constraints on the age and origin of chondrules. *Proceedings of the National Academy of Sciences (USA)*, 113, 2886–2891.

Burbine, T. H., McCoy, T. J., Meibom, A., Gladman, B., & Keil, K. (2002) Meteoritic parent bodies: Their number and identification. In W. F. Bottke Jr., A. Cellino, P. Paolicchi, & R. P. Binzel (eds.), *Asteroids III*. Tucson: University of Arizona Press, pp. 653–667.

Bus, S. J., & Binzel, R. P. (2002) Phase II of the small main-belt asteroid spectroscopic survey. A feature-based taxonomy. *Icarus*, 158, 146–177.

Carrera, D., Gorti, U., Johansen, A., & Davies, M. B. (2017) Planetesimal formation by the streaming instability in a photoevaporating disk. *The Astrophysical Journal*, 839, 16.

Chambers, J. E. (2001) Making more terrestrial planets. *Icarus*, 152, 205–224.

Chambers, J. E., & Wetherill, G. W. (1998) Making the terrestrial planets: N-body integrations of planetary embryos in three dimensions. *Icarus*, 136, 304–327.

Chambers, J. E., & Wetherill, G. W. (2001) Planets in the asteroid belt. *Meteoritics & Planetary Science*, 36, 381–399.

Clayton, R. N., & Mayeda, T. K. (1996) Oxygen isotope studies of achondrites. *Geochimica et Cosmochimica Acta*, 60, 1999–2017.

Clement, M. S., Kaib, N. A., Raymond, S. N., Chambers, J. E., & Walsh, K. J. (2019a) The early instability scenario: Terrestrial planet formation during the giant planet instability, and the effect of collisional fragmentation. *Icarus*, 321, 778–790.

Clement, M. S., Kaib, N. A., Raymond, S. N., & Walsh, K. J. (2018) Mars' growth stunted by an early giant planet instability. *Icarus*, 311, 340–356.

Clement, M. S., Morbidelli, A., Raymond, S. N., & Kaib, N. A. (2020) A record of the final phase of giant planet migration fossilized in the asteroid belt's orbital structure. *Monthly Notices of the Royal Astronomical Society*, 492, L56–L60.

Clement, M. S., Raymond, S. N., & Kaib, N. A. (2019b) Excitation and depletion of the asteroid belt in the early instability scenario. *The Astronomical Journal*, 157, 38.

Connelly, J. N., Bizzarro, M., Krot, A. N., et al. (2012) The absolute chronology and thermal processing of solids in the solar protoplanetary disk. *Science*, 338, 651.

Crida, A., Masset, F., & Morbidelli, A. (2009) Long range outward migration of giant planets, with application to Fomalhaut b. *The Astrophysical Journal*, 705, L148–L152.

Crida, A., Morbidelli, A., & Masset, F. (2006) On the width and shape of gaps in protoplanetary disks. *Icarus*, 181, 587–604.

Dauphas, N. (2017) The isotopic nature of the Earth's accreting material through time. *Nature*, 541, 521–524.

Dauphas, N., & Pourmand, A. (2011) Hf-W-Th evidence for rapid growth of Mars and its status as a planetary embryo. *Nature*, 473, 489–492.

Day, J. M. D., Pearson, D. G., & Taylor, L. A. (2007) Highly siderophile element constraints on accretion and differentiation of the Earth–Moon system. *Science*, 315, 217.

De Sanctis, M. C., Ammannito, E., McSween, H. Y., et al. (2017) Localized aliphatic organic material on the surface of Ceres. *Science*, 355, 719–722.

De Sanctis, M. C., Ammannito, E., Raponi, A., et al. (2015) Ammoniated phyllosilicates with a likely outer Solar System origin on (1) Ceres. *Nature*, 528, 241–244.

Deienno, R., Gomes, R. S., Walsh, K. J., Morbidelli, A., & Nesvorný, D. (2016) Is the Grand Tack model compatible with the orbital distribution of Main Belt asteroids? *Icarus*, 272, 114–124.

Deienno, R., Izidoro, A., Morbidelli, A., et al. (2018) Excitation of a primordial cold asteroid belt as an outcome of planetary instability. *The Astrophysical Journal*, 864, 50.

Deienno, R., Morbidelli, A., Gomes, R. S., & Nesvorný, D. (2017) Constraining the giant planets' initial configuration from their evolution: Implications for the timing of the planetary instability. *The Astrophysical Journal*, 153, 153.

Deienno, R., Walsh, K. J., Kretke, K. A., & Levison, H. F. (2019) Energy dissipation in large collisions – No change in planet formation outcomes. *The Astrophysical Journal*, 876, 103.

Delbo', M., Walsh, K., Bolin, B., Avdellidou, C., & Morbidelli, A. (2017) Identification of a primordial asteroid family constrains the original planetesimal population. *Science*, 357, 1026–1029.

DeMeo, F. E., & Carry, B. (2013) The taxonomic distribution of asteroids from multi-filter all-sky photometric surveys. *Icarus*, 226, 723–741.

DeMeo, F. E., & Carry, B. (2014) Solar System evolution from compositional mapping of the asteroid belt. *Nature*, 505, 629–634.

Dermott, S. F., Christou, A. A., Li, D., Kehoe, T. J. J., & Robinson, J. M. (2018) The common origin of family and non-family asteroids. *Nature Astronomy*, 2, 549–554.

Drążkowska, J., Alibert, Y., & Moore, B. (2016) Close-in planetesimal formation by pile-up of drifting pebbles. *Astronomy & Astrophysics*, 594, A105.

Drążkowska, J., & Dullemond, C. P. (2018) Planetesimal formation during protoplanetary disk buildup. *Astronomy & Astrophysics*, 614, A62.

Duncan, M., Quinn, T., & Tremaine, S. (1987) The formation and extent of the Solar System comet cloud. *The Astronomical Journal*, 94, 1330–1338.

Emery, J. P., Marzari, F., Morbidelli, A., French, L. M., & Grav, T. (2015) The complex history of trojan asteroids. In P. Michel, F. E. DeMeo, & W. F. Bottke (eds.), *Asteroids IV*. Tucson: University of Arizona Press, pp. 203–220.

Fanale, F. P., & Salvail, J. R. (1989) The water regime of asteroid (1) Ceres. *Icarus*, 82, 97–110.

Fernandez, J. A., & Ip, W. (1984) Some dynamical aspects of the accretion of Uranus and Neptune – The exchange of orbital angular momentum with planetesimals. *Icarus*, 58, 109–120.

Fischer, R. A., & Ciesla, F. J. (2014) Dynamics of the terrestrial planets from a large number of N-body simulations. *Earth and Planetary Science Letters*, 392, 28–38.

Fogg, M. J., & Nelson, R. P. (2005) Oligarchic and giant impact growth of terrestrial planets in the presence of gas giant planet migration. *Astronomy & Astrophysics*, 441, 791–806.

Fogg, M. J., & Nelson, R. P. (2007) On the formation of terrestrial planets in hot-Jupiter systems. *Astronomy & Astrophysics*, 461, 1195–1208.

Franchi, I. A., Wright, I. P., Sexton, A. S., & Pillinger, C. T. (1999) The oxygen-isotopic composition of Earth and Mars. *Meteoritics & Planetary Science*, 34, 657–661.

Fung, J., Shi, J.-M., & Chiang, E. (2014) How empty are disk gaps opened by giant planets? *The Astrophysical Journal*, 782, 88.

Gladman, B. (1993) Dynamics of systems of two close planets. *Icarus*, 106, 247.

Gladman, B. J., Migliorini, F., Morbidelli, A., et al. (1997) Dynamical lifetimes of objects injected into asteroid belt resonances. *Science*, 277, 197–201.

Goldreich, P., & Tremaine, S. (1980) Disk-satellite interactions. *The Astrophysical Journal*, 241, 425–441.

Gomes, R., Levison, H. F., Tsiganis, K., & Morbidelli, A. (2005) Origin of the cataclysmic Late Heavy Bombardment period of the terrestrial planets. *Nature*, 435, 466–469.

Gomes, R. S., Morbidelli, A., & Levison, H. F. (2004) Planetary migration in a planetesimal disk: why did Neptune stop at 30 AU? *Icarus*, 170, 492–507.

Gradie, J., & Tedesco, E. (1982) Compositional structure of the asteroid belt. *Science*, 216, 1405–1407.

Greenberg, R., Hartmann, W. K., Chapman, C. R., & Wacker, J. F. (1978) Planetesimals to planets – Numerical simulation of collisional evolution. *Icarus*, 35, 1–26.

Grimm, R. E., & McSween, H. Y. (1993) Heliocentric zoning of the asteroid belt by aluminum-26 heating. *Science*, 259, 653–655.

Haghighipour, N., & Scott, E. R. D. (2012) On the effect of giant planets on the scattering of parent bodies of iron meteorite from the terrestrial planet region into the asteroid belt: A concept study. *The Astrophysical Journal*, 749, 113.

Haisch, K. E., Jr., Lada, E. A., & Lada, C. J. (2001) Disk frequencies and lifetimes in young clusters. *The Astrophysical Journal*, 553, L153–L156.

Hansen, B. M. S. (2009) Formation of the terrestrial planets from a narrow annulus. *The Astrophysical Journal*, 703, 1131–1140.

Hartmann, W. K. (2019) The collapse of the terminal cataclysm paradigm … and where we go from here. *50th Lunar and Planetary Science Conference*, March 18–22, The Woodlands, TX, p. 1064.

Hayashi, C. (1981) Structure of the solar Nebula, growth and decay of magnetic fields and effects of magnetic and turbulent viscosities on the Nebula. *Progress of Theoretical Physics Supplement*, 70, 35–53.

Heppenheimer, T. A. (1980) Secular resonances and the origin of eccentricities of Mars and the asteroids. *Icarus*, 41, 76–88.

Hubickyj, O., Bodenheimer, P., & Lissauer, J. J. (2005) Accretion of the gaseous envelope of Jupiter around a 5 10 Earth-mass core. *Icarus*, 179, 415–431.

Ida, S., & Lin, D. N. C. (2004) Toward a deterministic model of planetary formation. I. A desert in the mass and semimajor axis distributions of extrasolar planets. *The Astrophysical Journal*, 604, 388–413.

Ida, S., & Makino, J. (1992) N-body simulation of gravitational interaction between planetesimals and a protoplanet. I – Velocity distribution of planetesimals. *Icarus*, 96, 107–120.

Ida, S., & Makino, J. (1993) Scattering of planetesimals by a protoplanet – Slowing down of runaway growth. *Icarus*, 106, 210.

Ikoma, M., Emori, H., & Nakazawa, K. (2001) Formation of giant planets in dense nebulae: Critical core mass revisited. *The Astrophysical Journal*, 553, 999–1005.

Ikoma, M., Nakazawa, K., & Emori, H. (2000) Formation of giant planets: Dependences on core accretion rate and grain opacity. *The Astrophysical Journal*, 537, 1013–1025.

Izidoro, A., Morbidelli, A., & Raymond, S. N. (2014) Terrestrial planet formation in the presence of migrating super-Earths. *The Astrophysical Journal*, 794, 11.

Izidoro, A., Morbidelli, A., Raymond, S. N., Hersant, F., & Pierens, A. (2015a) Accretion of Uranus and Neptune from inward-migrating planetary embryos blocked by Jupiter and Saturn. *Astronomy & Astrophysics*, 582, A99.

Izidoro, A., Raymond, S. N., Morbidelli, A., & Winter, O. C. (2015b) Terrestrial planet formation constrained by Mars and the structure of the asteroid belt. *Monthly Notices of the Royal Astronomical Society*, 453, 3619–3634.

Izidoro, A., Raymond, S. N., Pierens, A., et al. (2016) The asteroid belt as a relic from a chaotic early Solar System. *The Astrophysical Journal*, 833, 40.

Jacobson, S. A., & Morbidelli, A. (2014) Lunar and terrestrial planet formation in the Grand Tack scenario. *Philosophical Transactions of the Royal Society of London Series A*, 372, 0174.

Jacobson, S. A., Morbidelli, A., Raymond, S. N., et al. (2014) Highly siderophile elements in Earth's mantle as a clock for the Moon-forming impact. *Nature*, 508, 84–87.

Johansen, A., Blum, J., Tanaka, H., et al. (2014) The multifaceted planetesimal formation process. In H. Beuther, R. S. Klessen, C. P. Dullemond, & T. K. Henning (eds.), *Protostars and Planets VI*. Tucson: University of Arizona Press, pp. 547–570.

Johansen, A., & Lambrechts, M. (2017) Forming planets via pebble accretion. *Annual Review of Earth and Planetary Sciences*, 45, 359–387.

Johansen, A., Mac Low, M.-M., Lacerda, P., & Bizzarro, M. (2015) Growth of asteroids, planetary embryos, and Kuiper belt objects by chondrule accretion. *Science Advances*, 1, 1500109.

Johansen, A., Oishi, J. S., Mac Low, M.-M., et al. (2007) Rapid planetesimal formation in turbulent circumstellar disks. *Nature*, 448, 1022–1025.

Johnson, B. C., Walsh, K. J., Minton, D. A., Krot, A. N., & Levison, H. F. (2016) Timing of the formation and migration of giant planets as constrained by cb chondrites. *Science Advances*, 2.

Kaib, N. A., & Chambers, J. E. (2016) The fragility of the terrestrial planets during a giant-planet instability. *Monthly Notices of the Royal Astronomical Society*, 455, 3561–3569.

Kaib, N. A., & Cowan, N. B. (2015) The feeding zones of terrestrial planets and insights into Moon formation. *Icarus*, 252, 161–174.

Kerridge, J. F. (1985) Carbon, hydrogen and nitrogen in carbonaceous chondrites abundances and isotopic compositions in bulk samples. *Geochimica et Cosmochimica Acta*, 49, 1707–1714.

Kleine, T., Touboul, M., Bourdon, B., et al. (2009) Hf-W chronology of the accretion and early evolution of asteroids and terrestrial planets. *Geochimica et Cosmochimica Acta*, 73, 5150–5188.

Kley, W., & Nelson, R. P. (2012) Planet–disk interaction and orbital evolution. *Annual Review of Astronomy & Astrophysics*, 50, 211–249.

Knežević, Z., & Milani, A. (2003) Proper element catalogs and asteroid families. *Astronomy & Astrophysics*, 403, 1165–1173.

Kokubo, E., & Ida, S. (1998) Oligarchic growth of protoplanets. *Icarus*, 131, 171–178.

Kokubo, E., & Ida, S. (2000) Formation of protoplanets from planetesimals in the Solar Nebula. *Icarus*, 143, 15–27.

Krasinsky, G. A., Pitjeva, E. V., Vasilyev, M. V., & Yagudina, E. I. (2002) Hidden mass in the asteroid belt. *Icarus*, 158, 98–105.

Kretke, K. A., Bottke, W., Kring, D. A., & Levison, H. F. (2017) Effect of giant planet formation on the compositional mixture of the asteroid belt. *AAS/Division of Dynamical Astronomy Meeting #48*, June 11–15, Queen Mary University of London, London, p. 103.02.

Krot, A. N., Amelin, Y., Cassen, P., & Meibom, A., 2005. Young chondrules in CB chondrites from a giant impact in the early Solar System. *Nature*, 436, 989–992.

Kruijer, T. S., Burkhardt, C., Budde, C., & Kleine, T. (2017) Age of Jupiter inferred from the distinct genetics and formation times of meteorites. *Proceedings of the National Academy of Sciences (USA)*, 114, 6712–6716.

Kruijer, T. S., Kleine, T., & Borg, L. E. (2020) The great isotopic dichotomy of the early Solar System. *Nature Astronomy*, 4, 32–40.

Kruijer, T. S., Touboul, M., Fischer-Gödde, M., et al. (2014) Protracted core formation and rapid accretion of protoplanets. *Science*, 344, 1150–1154.

Kuchynka, P., & Folkner, W. M. (2013) A new approach to determining asteroid masses from planetary range measurements. *Icarus*, 222, 243–253.

Küppers, M., O'Rourke, L., Bockelée-Morvan, D., et al. (2014) Localized sources of water vapour on the dwarf planet (1) Ceres. *Nature*, 505, 525–527.

Lambrechts, M., & Johansen, A. (2012) Rapid growth of gas-giant cores by pebble accretion. *Astronomy & Astrophysics*, 544, A32.

Lambrechts, M., & Johansen, A. (2014) Forming the cores of giant planets from the radial pebble flux in protoplanetary discs. *Astronomy & Astrophysics*, 572, A107.

Lambrechts, M., & Lega, E. (2017) Reduced gas accretion on super-Earths and ice giants. *Astronomy & Astrophysics*, 606, A146.

Lambrechts, M., Lega, E., Nelson, R. P., Crida, A., & Morbidelli, A. (2019) Quasi-static contraction during runaway gas accretion onto giant planets. *Astronomy & Astrophysics*, 630, A82.

Laskar, J., Gastineau, M., Delisle, J. B., Farrés, A., & Fienga, A. (2011) Strong chaos induced by close encounters with Ceres and Vesta. *Astronomy & Astrophysics*, 532, L4.

Lebofsky, L. A., Feierberg, M. A., Tokunaga, A. T., Larson, H. P., & Johnson, J. R. (1981) The 1.7- to 4.2-μ m spectrum of asteroid 1 Ceres: Evidence for structural water in clay minerals. *Icarus*, 48, 453–459.

Lecar, M., & Franklin, F. (1997) The Solar Nebula, secular resonances, gas drag, and the asteroid belt. *Icarus*, 129, 134–146.

Leinhardt, Z. M., & Richardson, D. C. (2005) Planetesimals to protoplanets. I. Effect of fragmentation on terrestrial planet formation. *The Astrophysical Journal*, 625, 427–440.

Lemaitre, A., & Dubru, P. (1991) Secular resonances in the primitive solar nebula. *Celestial Mechanics and Dynamical Astronomy*, 52, 57–78.

Levison, H. F., Bottke, W. F., Gounelle, M., et al. (2009) Contamination of the asteroid belt by primordial trans-Neptunian objects. *Nature*, 460, 364–366.

Levison, H. F., Kretke, K. A., & Duncan, M. J. (2015a) Growing the gas-giant planets by the gradual accumulation of pebbles. *Nature*, 524, 322–324.

Levison, H. F., Kretke, K. A., Walsh, K. J., & Bottke, W. F. (2015b) Growing the terrestrial planets from the gradual accumulation of sub-meter sized objects. *Proceedings of the National Academy of Sciences (USA)*, 112, 14180–14185.

Levison, H. F., Morbidelli, A., Tsiganis, K., Nesvorný, D., & Gomes, R. (2011) Late orbital instabilities in the outer planets induced by interaction with a self-gravitating planetesimal disk. *The Astronomical Journal*, 142, 152.

Levison, H. F., Morbidelli, A., Vanlaerhoven, C., Gomes, R., & Tsiganis, K. (2008) Origin of the structure of the Kuiper belt during a dynamical instability in the orbits of Uranus and Neptune. *Icarus*, 196, 258–273.

Levison, H. F., & Stewart, G. R. (2001) Remarks on modeling the formation of Uranus and Neptune. *Icarus*, 153, 224–228.

Levison, H. F., Thommes, E., & Duncan, M. J. (2010) Modeling the formation of giant planet cores. I. Evaluating key processes. *The Astronomical Journal*, 139, 1297–1314.

Lichtenberg, T., Golabek, G. J., Gerya, T. V., & Meyer, M. R. (2016) The effects of short-lived radionuclides and porosity on the early thermo-mechanical evolution of planetesimals. *Icarus*, 274, 350–365.

Lin, D. N. C., & Papaloizou, J. (1986) On the tidal interaction between protoplanets and the protoplanetary disk. III – Orbital migration of protoplanets. *The Astrophysical Journal*, 309, 846–857.

Lissauer, J. J., Hubickyj, O., D'Angelo, G., & Bodenheimer, P. (2009) Models of Jupiter's growth incorporating thermal and hydrodynamic constraints. *Icarus*, 199, 338–350.

Malhotra, R. (1993) The origin of Pluto's peculiar orbit. *Nature*, 365, 819–821.

Malhotra, R. (1995) The origin of Pluto's orbit: Implications for the Solar System beyond Neptune. *The Astronomical Journal*, 110, 420.

Mamajek, E. E. (2009) Initial conditions of planet formation: Lifetimes of primordial disks. In T. Usuda, M. Tamura, & M. Ishii (eds.), *American Institute of Physics Conference Series*, Volume 1158 of American Institute of Physics Conference Series. AIP Publishing, pp. 3–10.

Mandell, A. M., Raymond, S. N., & Sigurdsson, S. (2007) Formation of Earth-like planets during and after giant planet migration. *The Astrophysical Journal*, 660, 823–844.

Marchal, C., & Bozis, G. (1982) Hill stability and distance curves for the general three-body problem. *Celestial Mechanics*, 26, 311–333.

Marchi, S., Bottke, W. F., Kring, D. A., & Morbidelli, A. (2012) The onset of the lunar cataclysm as recorded in its ancient crater populations. *Earth and Planetary Science Letters*, 325, 27–38.

Marchi, S., Raponi, A., Prettyman, T. H., et al. (2019) An aqueously altered carbon-rich Ceres. *Nature Astronomy*, 3, 140–145.

Marchi, S., Walker, R. J., & Canup, R. M. (2020) A compositionally heterogeneous martian mantle due to late accretion. *Science Advances*, 6, eaay2338.

Marty, B. (2012) The origins and concentrations of water, carbon, nitrogen and noble gases on Earth. *Earth and Planetary Science Letters*, 313, 56–66.

Marty, B., & Yokochi, R. (2006) Water in the early Earth. *Reviews in Mineralogy and Geophysics*, 62, 421–450.

Masset, F., & Snellgrove, M. (2001) Reversing type II migration: resonance trapping of a lighter giant protoplanet. *Monthly Notices of the Royal Astronomical Society*, 320, L55–L59.

Mastrobuono-Battisti, A., & Perets, H. B. (2017) The composition of Solar System asteroids and Earth/Mars moons, and the Earth–Moon composition similarity. *Monthly Notices of the Royal Astronomical Society*, 469, 3597–3609.

McCord, T. B., & Sotin, C. (2005) Ceres: Evolution and current state. *Journal of Geophysical Research (Planets)*, 110, E05009.

McKinnon, W. B. (2008) Could Ceres be a refugee from the Kuiper Belt? *Asteroids, Comets, Meteors 2008*, July 14–18, Baltimore, MD, Vol. 1405, 8389.

McSween, H. Y., Emery, J. P., Rivkin, A. S., et al. (2018) Carbonaceous chondrites as analogs for the composition and alteration of Ceres. *Meteoritics & Planetary Science*, 53, 1793–1804.

Meech, K., & Raymond, S. N. (2020) Origin of Earth's water: Sources and constraints. In V. Meadows, G. Arney, D. D. Marais, & B. Schmidt (eds.), *Planetary Astrobiology*. Tucson: University of Arizona Press.

Milani, A., & Knezevic, Z. (1990) Secular perturbation theory and computation of asteroid proper elements. *Celestial Mechanics and Dynamical Astronomy*, 49, 347–411.

Milliken, R. E., & Rivkin, A. S. (2009) Brucite and carbonate assemblages from altered olivine-rich materials on Ceres. *Nature Geoscience*, 2, 258–261.

Minton, D. A., & Malhotra, R. (2009) A record of planet migration in the main asteroid belt. *Nature*, 457, 1109–1111.

Minton, D. A., & Malhotra, R. (2011) Secular resonance sweeping of the main asteroid belt during planet migration. *The Astrophysical Journal*, 732, 53.

Mizuno, H. (1980) Formation of the giant planets. *Progress of Theoretical Physics*, 64, 544–557.

Mojzsis, S. J., Brasser, R., Kelly, N. M., Abramov, O., & Werner, S. C. (2019) Onset of giant planet migration before 4480 million years ago. *The Astrophysical Journal*, 881, 44.

Monteux, J., Golabek, G. J., Rubie, D. C., Tobie, G., & Young, E. D. (2018) Water and the interior structure of terrestrial planets and icy bodies. *Space Science Reviews*, 214, 39.

Morbidelli, A., Bottke, W. F., Nesvorný, D., & Levison, H. F. (2009) Asteroids were born big. *Icarus*, 204, 558–573.

Morbidelli, A., Brasser, R., Gomes, R., Levison, H. F., & Tsiganis, K. (2010) Evidence from the asteroid belt for a violent past evolution of Jupiter's orbit. *The Astronomical Journal*, 140, 1391–1401.

Morbidelli, A., & Crida, A. (2007) The dynamics of Jupiter and Saturn in the gaseous protoplanetary disk. *Icarus*, 191, 158–171.

Morbidelli, A., & Henrard, J. (1991) The main secular resonances ν_6, vs and ν_{16} in the asteroid belt. *Celestial Mechanics and Dynamical Astronomy*, 51, 169–197.

Morbidelli, A., Levison, H. F., Tsiganis, K., & Gomes, R. (2005) Chaotic capture of Jupiter's Trojan asteroids in the early Solar System. *Nature*, 435, 462–465.

Morbidelli, A., Lunine, J. I., O'Brien, D. P., Raymond, S. N., & Walsh, K. J. (2012a) Building terrestrial planets. *Annual Review of Earth and Planetary Sciences*, 40, 251–275.

Morbidelli, A., Marchi, S., Bottke, W. F., & Kring, D. A. (2012b) A sawtooth-like timeline for the first billion years of lunar bombardment. *Earth and Planetary Science Letters*, 355, 144–151.

Morbidelli, A., Nesvorny, D., Laurenz, V., et al. (2018) The timeline of the lunar bombardment: Revisited. *Icarus*, 305, 262–276.

Morbidelli, A., & Raymond, S. N. (2016) Challenges in planet formation. *Journal of Geophysical Research (Planets)*, 121, 1962–1980.

Morbidelli, A., Tsiganis, K., Crida, A., Levison, H. F., & Gomes, R. (2007) Dynamics of the giant planets of the Solar System in the gaseous protoplanetary disk and their relationship to the current orbital architecture. *The Astronomical Journal*, 134, 1790–1798.

Morbidelli, A., Walsh, K. J., O'Brien, D. P., Minton, D. A., & Bottke, W. F. (2015) The dynamical evolution of the asteroid belt. In P. Michel, F. E. DeMeo, & W. F. Bottke (eds.), *Asteroids IV*. Tucson: University of Arizona Press, pp. 493–507.

Nagasawa, M., Ida, S., & Tanaka, H. (2001) Origin of high orbital eccentricity and inclination of asteroids. *Earth, Planets, and Space*, 53, 1085–1091.

Nagasawa, M., Ida, S., & Tanaka, H. (2002) Excitation of orbital inclinations of asteroids during depletion of a protoplanetary disk: Dependence on the disk configuration. *Icarus*, 159, 322–327.

Nagasawa, M., Lin, D. N. C., & Thommes, E. (2005) Dynamical shake-up of planetary systems. I. Embryo trapping and induced collisions by the sweeping secular resonance and embryo-disk tidal interaction. *The Astrophysical Journal*, 635, 578–598.

Nagasawa, M., Tanaka, H., & Ida, S. (2000) Orbital evolution of asteroids during depletion of the Solar Nebula. *The Astronomical Journal*, 119, 1480–1497.

Nathues, A., Hoffmann, M., Schaefer, M., et al. (2015) Sublimation in bright spots on (1) Ceres. *Nature*, 528, 237–240.

Nesvorný, D. (2011) Young Solar System's fifth giant planet? *The Astrophysical Journal*, 742, L22.

Nesvorný, D. (2015) Evidence for slow migration of Neptune from the inclination distribution of Kuiper Belt objects. *The Astronomical Journal*, 150, 73.

Nesvorný, D. (2018) Dynamical evolution of the early Solar System. *Annual Review of Astronomy & Astrophysics*, 56, 137–174.

Nesvorný, D., Brož, M., & Carruba, V. (2015) Identification and dynamical properties of asteroid families. In P. Michel, F. E. DeMeo, & W. F. Bottke (eds.), *Asteroids IV*. Tucson: University of Arizona Press, pp. 297–321.

Nesvorný, D., Li, R., Youdin, A. N., Simon, J. B., & Grundy, W. M. (2019) Trans-Neptunian binaries as evidence for planetesimal formation by the streaming instability. *Nature Astronomy*, 3, 808–812.

Nesvorný, D., & Morbidelli, A. (2012) Statistical study of the early Solar System's instability with four, five, and six giant planets. *The Astronomical Journal*, 144, 117.

Nesvorný, D., Roig, F., & Bottke, W. F. (2017) Modeling the historical flux of planetary impactors. *The Astronomical Journal*, 153, 103.

Nesvorný, D., Vokrouhlický, D., Bottke, W. F., & Levison, H. F. (2018) Evidence for very early migration of the Solar System planets from the Patroclus-Menoetius binary Jupiter Trojan. *Nature Astronomy*, 2, 878–882.

Nesvorný, D., Vokrouhlický, D., & Morbidelli, A. (2013) Capture of Trojans by jumping Jupiter. *The Astrophysical Journal*, 768, 45.

Nimmo, F., & Kleine, T. (2007) How rapidly did Mars accrete? Uncertainties in the Hf W timing of core formation. *Icarus*, 191, 497–504.

Nittler, L. R., & Ciesla, F. (2016) Astrophysics with extraterrestrial materials. *Annual Review of Astronomy & Astrophysics*, 54, 53–93.

Novaković, B., Cellino, A., & Knežević, Z. (2011) Families among high-inclination asteroids. *Icarus*, 216, 69–81.

O'Brien, D. P., Morbidelli, A., & Bottke, W. F. (2007) The primordial excitation and clearing of the asteroid belt – Revisited. *Icarus*, 191, 434–452.

O'Brien, D. P., Walsh, K. J., Morbidelli, A., Raymond, S. N., & Mandell, A. M. (2014) Water delivery and giant impacts in the Grand Tack scenario. *Icarus*, 239, 74–84.

Ormel, C. W., & Klahr, H. H. (2010) The effect of gas drag on the growth of protoplanets. Analytical expressions for the accretion of small bodies in laminar disks. *Astronomy & Astrophysics*, 520, A43.

Papaloizou, J. C. B., & Larwood, J. D. (2000) On the orbital evolution and growth of protoplanets embedded in a gaseous disc. *Monthly Notices of the Royal Astronomical Society*, 315, 823–833.

Petit, J., Morbidelli, A., & Chambers, J. (2001) The primordial excitation and clearing of the asteroid belt. *Icarus*, 153, 338–347.

Pfalzner, S., Steinhausen, M., & Menten, K. (2014) Short dissipation times of proto-planetary disks: An artifact of selection effects? *The Astrophysical Journal*, 793, L34.

Pierens, A., & Nelson, R. P. (2008) Constraints on resonant-trapping for two planets embedded in a protoplanetary disc. *Astronomy & Astrophysics*, 482, 333–340.

Pierens, A., & Raymond, S. N. (2011) Two phase, inward-then-outward migration of Jupiter and Saturn in the gaseous solar nebula. *Astronomy & Astrophysics*, 533, A131.

Pierens, A., Raymond, S. N., Nesvorny, D., & Morbidelli, A. (2014) Outward migration of Jupiter and Saturn in 3:2 or 2:1 resonance in radiative disks: Implications for the Grand Tack and Nice models. *The Astrophysical Journal*, 795, L11.

Pirani, S., Johansen, A., Bitsch, B., Mustill, A. J., & Turrini, D. (2019a) Consequences of planetary migration on the minor bodies of the early Solar System. *Astronomy & Astrophysics*, 623, A169.

Pirani, S., Johansen, A., & Mustill, A. J. (2019b) On the inclinations of the Jupiter Trojans. *Astronomy & Astrophysics*, 631, A89.

Piso, A.-M. A., & Youdin, A. N. (2014) On the minimum core mass for giant planet formation at wide separations. *The Astrophysical Journal*, 786, 21.

Pollack, J. B., Hubickyj, O., Bodenheimer, P., et al. (1996) Formation of the giant planets by concurrent accretion of solids and gas. *Icarus*, 124, 62–85.

Pravec, P., Harris, A. W., Kušnirák, P., Galád, A., & Hornoch, K. (2012) Absolute magnitudes of asteroids and a revision of asteroid albedo estimates from WISE thermal observations. *Icarus*, 221, 365–387.

Prettyman, T. H., Yamashita, N., Toplis, M. J., et al. (2017) Extensive water ice within Ceres' aqueously altered regolith: Evidence from nuclear spectroscopy. *Science*, 355, 55–59.

Quarles, B., & Kaib, N. (2019) Instabilities in the early Solar System due to a self-gravitating disk. *The Astronomical Journal*, 157, 67.

Rafikov, R. R. (2003) The growth of planetary embryos: Orderly, runaway, or oligarchic? *The Astronomical Journal*, 125, 942–961.

Raymond, S. N., & Izidoro, A. (2017a) Origin of water in the inner Solar System: Planetesimals scattered inward during Jupiter and Saturn's rapid gas accretion. *Icarus*, 297, 134–148.

Raymond, S. N., & Izidoro, A. (2017b) The empty primordial asteroid belt. *Science Advances*, 3, e1701138.

Raymond, S. N., Izidoro, A., Bitsch, B., & Jacobson, S. A. (2016) Did Jupiter's core form in the innermost parts of the Sun's protoplanetary disc? *Monthly Notices of the Royal Astronomical Society*, 458, 2962–2972.

Raymond, S. N., Izidoro, A., & Morbidelli, A. (2020) Solar System formation in the context of extra-solar planets. In V. Meadows, G. Arney, D. D. Marais, & B. Schmidt (eds.), *Planetary Astrobiology*. Tucson: University of Arizona Press.

Raymond, S. N., Kokubo, E., Morbidelli, A., Morishima, R., & Walsh, K. J. (2014) Terrestrial planet formation at home and abroad. In H. Beuther, R. S. Klessen, C. P. Dullemond, & T. K. Henning (eds.), *Protostars and Planets VI*. Tucson: University of Arizona Press, pp. 595–618.

Raymond, S. N., Mandell, A. M., & Sigurdsson, S. (2006a) Exotic Earths: Forming habitable worlds with giant planet migration. *Science*, 313, 1413–1416.

Raymond, S. N., & Morbidelli, A. (2014) The Grand Tack model: A critical review. In *Complex Planetary Systems, Proceedings of the International Astronomical Union*, Volume 310 of IAU Symposium. Cambridge: Cambridge University Press, pp. 194–203.

Raymond, S. N., O'Brien, D. P., Morbidelli, A., & Kaib, N. A. (2009) Building the terrestrial planets: Constrained accretion in the inner Solar System. *Icarus*, 203, 644–662.

Raymond, S. N., Quinn, T., & Lunine, J. I. (2006b) High-resolution simulations of the final assembly of Earth-like planets I. Terrestrial accretion and dynamics. *Icarus*, 183, 265–282.

Ribeiro de Sousa, R., Morbidelli, A., Raymond, S. N., et al. (2020) Dynamical evidence for an early giant planet instability. *Icarus*, 339, 113605.

Robert, F., Merlivat, L., & Javoy, M. (1977) Water and deuterium content in eight condrites. *Meteoritics*, 12, 349.

Roig, F., & Nesvorný, D. (2015) The evolution of asteroids in the jumping-Jupiter migration model. *The Astronomical Journal*, 150, 186.

Roig, F., Nesvorný, D., & DeSouza, S. R. (2016) Jumping Jupiter can explain Mercury's orbit. *The Astrophysical Journal*, 820, L30.

Ronnet, T., Mousis, O., Vernazza, P., Lunine, J. I., & Crida, A. (2018) Saturn's formation and early evolution at the origin of Jupiter's massive moons. *The Astronomical Journal*, 155, 224.

Rubie, D. C., Laurenz, V., Jacobson, S. A., et al. (2016) Highly siderophile elements were stripped from Earth's mantle by iron sulfide segregation. *Science*, 353, 1141–1144.

Russell, C. T., Raymond, C. A., Ammannito, E., et al. (2016) Dawn arrives at Ceres: Exploration of a small, volatile-rich world. *Science*, 353, 1008–1010.

Safronov, V. S. (1969) *Evoliutsiia doplanetnogo oblaka*.

Schäfer, U., Yang, C.-C., & Johansen, A. (2017) Initial mass function of planetesimals formed by the streaming instability. *Astronomy & Astrophysics*, 597, A69.

Schiller, M., Bizzaro, M., & Fernandes, V. A. (2018) Isotopic evolution of the protoplanetary disk and the building blocks of Earth and the Moon. *Nature*, 555, 507–510.

Schiller, M., Bizzaro, M., & Siebert, J. (2020) Iron isotope evidence for very rapid accretion and differentiation of the proto-earth. *Science Advances*, 6, eaay7604.

Schiller, M., Connelly, J. N., Glad, A. C., Mikouchi, T., & Bizzarro, M. (2015) Early accretion of protoplanets inferred from a reduced inner Solar System ^{26}Al inventory. *Earth and Planetary Science Letters*, 420, 45–54.

Scott, E. R. D. (2002) Meteorite evidence for the accretion and collisional evolution of asteroids. In W. F. Bottke Jr., A. Cellino, P. Paolicchi, & R. P. Binzel (eds.), *Asteroids III*. Tucson: University of Arizone Press, pp. 697–709.

Simon, J. B., Armitage, P. J., Youdin, A. N., & Li, R. (2017) Evidence for universality in the initial planetesimal mass function. *The Astrophysical Journal*, 847, L12.

Squire, J., & Hopkins, P. F. (2018) Resonant drag instabilities in protoplanetary discs: The streaming instability and new, faster growing instabilities. *Monthly Notices of the Royal Astronomical Society*, 477, 5011–5040.

Suzuki, D., Bennett, D. P., Sumi, T., et al. (2016) The exoplanet mass-ratio function from the MOA-II survey: Discovery of a break and likely peak at a Neptune mass. *The Astrophysical Journal*, 833, 145.

Tanaka, H., & Ida, S. (1999) Growth of a migrating protoplanet. *Icarus*, 139, 350–366.

Tanaka, H., & Ward, W. R. (2004) Three-dimensional interaction between a planet and an isothermal gaseous disk. II. Eccentricity waves and bending waves. *The Astrophysical Journal*, 602, 388–395.

Tera, F., Papanastassiou, D. A., & Wasserburg, G. J. (1974) Isotopic evidence for a terminal lunar cataclysm. *Earth and Planetary Science Letters*, 22, 1.

Thommes, E., Nagasawa, M., & Lin, D. N. C. (2008) Dynamical shake-up of planetary systems. II. N-body simulations of Solar System terrestrial planet formation induced by secular resonance sweeping. *The Astrophysical Journal*, 676, 728–739.

Toplis, M. J., Mizzon, H., Monnereau, M., et al. (2013) Chondritic models of 4 Vesta: Implications for geochemical and geophysical properties. *Meteoritics & Planetary Science*, 48, 2300–2315.

Tsiganis, K., Gomes, R., Morbidelli, A., & Levison, H. F. (2005) Origin of the orbital architecture of the giant planets of the Solar System. *Nature*, 435, 459–461.

Turrini, D., Coradini, A., & Magni, G. (2012) Jovian early bombardment: Planetesimal erosion in the inner asteroid belt. *The Astrophysical Journal*, 750, 8.

Turrini, D., & Svetsov, V. (2014) The formation of Jupiter, the Jovian early bombardment and the delivery of water to the asteroid belt: The case of (4) Vesta. *Life*, 4, 4–34.

Vernazza, P., Castillo-Rogez, J., Beck, P., et al. (2017) Different origins or different evolutions? Decoding the spectral diversity among C-type asteroids. *The Astronomical Journal*, 153, 72.

Vokrouhlický, D., Bottke, W. F., & Nesvorný, D. (2016) Capture of trans-Neptunian planetesimals in the main asteroid belt. *The Astronomical Journal*, 152, 39.

Walker, R. J. (2009) Highly siderophile elements in the Earth, Moon and Mars: Update and implications for planetary accretion and differentiation. *Chemie der Erde / Geochemistry*, 69, 101–125.

Walsh, K. J., & Levison, H. F. (2016) Terrestrial planet formation from an annulus. *The Astronomical Journal*, 152, 68.

Walsh, K. J., & Morbidelli, A. (2011) The effect of an early planetesimal-driven migration of the giant planets on terrestrial planet formation. *Astronomy & Astrophysics*, 526, A126.

Walsh, K. J., Morbidelli, A., Raymond, S. N., O'Brien, D. P., & Mandell, A. M. (2011) A low mass for Mars from Jupiter's early gas-driven migration. *Nature*, 475, 206–209.

Walsh, K. J., Morbidelli, A., Raymond, S. N., O'Brien, D. P., & Mandell, A. M. (2012) Populating the asteroid belt from two parent source regions due to the migration of giant planets – "The Grand Tack." *Meteoritics & Planetary Science*, 47, 1941–1947.

Ward, W. R. (1981) Solar nebula dispersal and the stability of the planetary system I. Scanning secular resonance theory. *Icarus*, 47, 234–264.

Ward, W. R. (1986) Density waves in the solar nebula – Differential Lindblad torque. *Icarus*, 67, 164–180.

Ward, W. R. (1997) Protoplanet migration by nebula tides. *Icarus*, 126, 261–281.

Warren, P. H. (2011) Stable-isotopic anomalies and the accretionary assemblage of the Earth and Mars: A subordinate role for carbonaceous chondrites. *Earth and Planetary Science Letters*, 311, 93–100.

Weidenschilling, S. J. (1977a) Aerodynamics of solid bodies in the solar nebula. *Monthly Notices of the Royal Astronomical Society*, 180, 57–70.

Weidenschilling, S. J. (1977b) The distribution of mass in the planetary system and solar nebula. *Astrophysics & Space Science*, 51, 153–158.

Weidenschilling, S. J. (2011) Initial sizes of planetesimals and accretion of the asteroids. *Icarus*, 214, 671–684.

Wetherill, G. W. (1980) Formation of the terrestrial planets. *Annual Review of Astronomy & Astrophysics*, 18, 77–113.

Wetherill, G. W. (1991) Why isn't Mars as big as Earth? In *Lunar and Planetary Institute Science Conference Abstracts*, Volume 22 of Lunar and Planetary Inst. Technical Report, March 18–22, Houston, TX, pp. 1495.

Wetherill, G. W. (1992) An alternative model for the formation of the asteroids. *Icarus*, 100, 307–325.

Wetherill, G. W., & Stewart, G. R. (1989) Accumulation of a swarm of small planetesimals. *Icarus*, 77, 330–357.

Wetherill, G. W. & Stewart, G. R. (1993) Formation of planetary embryos – Effects of fragmentation, low relative velocity, and independent variation of eccentricity and inclination. *Icarus*, 106, 190.

Williams, J. P., & Cieza, L. A. (2011) Protoplanetary disks and their evolution. *Annual Review of Astronomy & Astrophysics*, 49, 67–117.

Youdin, A. N., & Goodman, J. (2005) Streaming instabilities in protoplanetary disks. *The Astrophysical Journal*, 620, 459–469.

Zellner, N. E. B. (2017) Cataclysm no more: New views on the timing and delivery of lunar impactors. *Origins of Life and Evolution of the Biosphere*, 47, 261–280.

Zhang, C., Miao, B., & He, H. (2019) Oxygen isotopes in HED meteorites and their constraints on parent asteroids. *Planetary and Space Science*, 168, 83–94.

Zhang, H., & Zhou, J.-L. (2010) On the orbital evolution of a giant planet pair embedded in a gaseous disk. I. Jupiter–Saturn configuration. *The Astrophysical Journal*, 714, 532–548.

Collisional Evolution of the Main Belt as Recorded by Vesta

WILLIAM F. BOTTKE AND MARTIN JUTZI

16.1 INTRODUCTION

Vesta is one of the most fascinating and best studied asteroids in the main asteroid belt. Our insights into its nature come not only from the Dawn mission but also from decades of remote observations, in-depth studies of the howardite, eucrite, and diogenite (HED) meteorites that are likely from Vesta, numerical studies of how large impacts affected Vesta, numerical studies of Main Belt evolution, and numerical studies of Vesta's family, a population of small asteroids ejected from Vesta (see Chapter 3). Collectively, they provide powerful constraints describing how Vesta and the main asteroid belt reached their current states. The issue facing collisional and dynamical modelers is to assemble a self-consistent story from these clues that can plausibly match what we know of the origin and evolution of the asteroid belt, meteorites, and the surfaces of inner Solar System bodies.

Vesta is the second most massive Main Belt asteroid and is considered a primordial planetesimal. It is located in the middle of the inner Main Belt with proper semimajor axis, eccentricity, and inclination (a, e, i) of 2.36 AU, 0.099, and 6.4°, respectively (Knežević & Milani, 2003). Some key physical parameters for Vesta include dimensions of $572.6 \times 557.2 \times 446.4$ km, a bulk density of 3.456 ± 0.035 g/cm^3, and a surface gravity of 0.25 m/s^2 (Russell et al., 2012).

Vesta has also differentiated into a metallic core, silicate mantle, and basaltic crust (e.g., Russell et al., 2012; McSween et al., 2019; Chapter 3). Studies of the HED meteorites indicate differentiation took place within a few Myr of CAI formation (Chapter 4), while studies of the eucrite meteorites indicate the emplacement of Vesta's basaltic crust lasted for a few million years to a few tens of millions of years after CAIs (e.g., Hopkins et al., 2015; Jourdan et al., 2020). Vesta is the only primordial asteroid left in the Main Belt that shows clear evidence of a basaltic crust.

The primary source of impacts on Vesta over most of its history have been Main Belt asteroids (e.g., Bottke et al., 2015a). Vesta is old enough, however, that during the planet formation era, it may have also been struck by planetesimals from the terrestrial planet region and cometesimals from the outer Solar System (e.g., Brož et al., 2013; Morbidelli et al., 2015). This means

Vesta's surface history – and the presence or absence of impacts – can potentially be used to place limits on the nature of these bombarding populations, provided one can filter out Main Belt impact signatures.

Vesta's impact history is dominated by two enormous impact structures, Rheasilvia, a 505 km diameter basin, and Veneneia, a 395 km basin (Marchi et al., 2012; Schenk et al., 2012; Chapter 5). Together, they define Vesta's southern hemisphere. Rheasilvia, being younger, overlaps with and has largely obscured Veneneia (Jaumann et al., 2012; Schenk et al., 2012). The basins are offset by ~40° and have sizes comparable to the diameter of Vesta itself. The formation of each basin also produced a set of fracture-like troughs, or graben. Studies of each trough group show they form planes that are orthogonal to the basin centers. Those for Rheasilvia are located near Vesta's equator, while those for Veneneia are at an angle of ~40° to the equator in the northern hemisphere (Buczkowski et al., 2012; Yingst et al., 2014; see also Chapter 5 and Section 16.2.2).

Vesta exhibits no indications that basins similar in size to Rheasilvia or Veneneia were erased or buried after its basaltic crust was emplaced. Vesta's topography and shape show no obvious signs for such impacts, there are no unaccounted sets of troughs that could be linked with a missing or erased basin, and there is no evidence that interior material was dredged up in some unusual place on Vesta. In addition, the northern hemisphere of Vesta is compatible with an equilibrium ellipsoid (Ermakov et al., 2014), a shape that does not suggest additional large craters formed in the past.

With that said, we cannot rule out the possibility that some very early basins were erased by viscous relaxation when Vesta's interior was still hot (e.g., Marchi et al., 2016). Indeed, some large C-complex families have been shown to have parent bodies without evidence for any large associated impact structure (e.g., (10) Hygiea; Vernazza et al., 2020). Until better evidence is available, though, such erased early basins on Vesta will remain speculative.

For the purposes of our work in interpreting Vesta's impact history, we will take a minimalist approach and assume that Vesta is complete in Rheasilvia- or Veneneia-sized basins. If so, these basins can be considered a constraint on the earliest evolution of the Main Belt. These large craters also help inform us on the collisional evolution of large planetesimals which could have contributed to the formation of larger rocky planets.

In our review of the inferred history of Vesta using these basins, we will start by probing their likely ages.

We thank Miroslav Brož for his helpful and constructive comments. M. J. acknowledges support from the Swiss National Center of Competence in Research Planets.

16.2 THE AGES OF RHEASILVIA AND VENENEIA BASINS

16.2.1 Rheasilvia Basin Age

The youngest largest impact structure on Vesta is Rheasilvia (Marchi et al., 2012, 2013b, 2015). It is covered by craters, yet crater spatial densities in the $0.15 < D_{crater} < 35$ km diameter range from Marchi et al. (2015) do not appear to have reached saturation. Given what we know of the crater spatial densities of other asteroids observed by spacecraft, we infer that Rheasilvia was not formed in the primordial era of Vesta, but instead formed within the last one or two billion years.

To estimate Rheasilvia's approximate formation age, one needs to model and reproduce the size–frequency distribution of superposed craters found within its rim or on its ejecta blanket (e.g., Schenk et al., 2012). Our calculation requires that we know the following components: the size frequency distribution (SFD) of impactors in the Main Belt that can strike Vesta, the collision probabilities and impact velocities that Main Belt asteroids have with Vesta, and a scaling law that can turn projectiles into crater sizes.

All of these components were recently recalculated by Bottke et al. (2020). In brief, they determined the following components:

1. A Main Belt size frequency distribution (SFD) that maintains a cumulative power law slope of $q = -1.3$ between $\sim 0.2 < D_{ast} < 2$ km before steepening to $q = -2.7$ for $D_{ast} < 0.2$ km. This differs from the SFD presented in Bottke et al. (2005), which moves to $q = -2.7$ for $D_{ast} < 0.5$ km.

2. A mean intrinsic collision probability and mean impact velocity between Main Belt asteroids and Vesta of $P_i = 2.88 \times 10^{-18}/km^2/yr$ and $V_{imp} = 4.7$ km/s, respectively.

3. Two different crater scaling laws; a Holsapple and Housen (2007) scaling law for cohesive soils, assuming $Y = 2 \times 10^7$ dynes/cm^2, and a second crater scaling law where the ratio of crater to projectile sizes was a ratio of ~ 10.

When assembled together, it allowed them to create a model crater production function and fit it to small craters superposed on Rheasilvia. Testing the best fits between model and data, they found that Rheasilvia's mean age was between ~ 0.6 and ~ 1.5 Ga (Figure 16.1). These values are consistent with the ~ 1 Ga crater retention age of Rheasilvia derived by Marchi et al. (2012).

We consider our inferred formation age to be reasonable because crater spatial densities on or near Rheasilvia basin are modestly lower than those on (951) Gaspra (Marchi et al., 2013b, 2015), an asteroid observed by the Galileo spacecraft whose formation age is arguably well constrained. Three separate age indicators suggest Gaspra is ~ 1.3 Ga; its crater-derived surface age, the dynamical age of Gaspra's source asteroid family Flora, and the ^{40}Ar/^{39}Ar ages of LL chondrite grains returned from (25143) Itokawa, a probable Flora family member (Park et al., 2015; Vokrouhlický et al., 2017; Bottke et al., 2020).

The Rheasilvia basin–forming event also ejected numerous fragments onto trajectories that allowed them to escape Vesta. These bodies likely comprise Vesta's color-, spectral-, and albedo-distinctive asteroid family (e.g., Parker et al., 2008; Masiero et al., 2015; Nesvorný et al., 2015). Asteroids with spectral characteristics similar to the basaltic surface of Vesta are called V-type asteroids (Binzel & Xu, 1993). They have been investigated

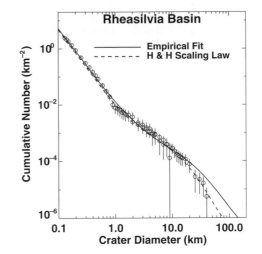

Figure 16.1 A comparison between the observed crater size distribution found on the floor and ejecta blanket of Vesta's Rheasilvia basin and various crater models. The observed crater counts are from Marchi et al. (2015). Both lines are model crater production functions that were created using a Main Belt asteroid size frequency distribution from Bottke et al. (2020) and a model crater scaling law. The solid line assumes the crater over impactor diameter ratio is ~ 10, and it was found by empirical methods. The dashed line uses a crater scaling law from Holsapple and Housen (2007). Using results from both methods, we find the crater retention age of Rheasilvia basin is between 0.6 and 1.5 Ga, with a preferred model age near ~ 1 Gyr.

in a number of numerical impact modeling studies (Asphaug, 1997; Jutzi & Asphaug, 2011; Bowling et al., 2013; Ivanov & Melosh, 2013; Jutzi et al., 2013; Stickle et al., 2015).

Using a collisional evolution model, Bottke (2014) examined the steep size distribution of Vesta's family from Rheasilvia. It is composed of $D_{ast} < 10$ km fragments, with a steep slope down to $D_{ast} < 3$ km, a shallower cumulative power law slope of –2.5 between $1.5 < D_{ast} < 3$ km, and a still shallower slope for $D_{ast} < 1.5$ km, at least in part because of observational incompleteness. See DellaGiustina et al. (2021) for a discussion of the best estimate of the present-day Vesta family size distribution. Their work showed no strong indication of any change away from these slopes produced by collisional grinding. On this basis, they estimated that the Vesta family has an 80% probability of being <1 Ga old. The orbital distribution of the family members, and how they have likely been influenced by the Yarkovsky thermal forces (Bottke et al., 2006a; Vokrouhlický et al., 2015) also suggests an age of ~ 1 Ga (Spoto et al., 2015), though we caution that the high ejection velocity of the family members makes it difficult to precisely determine the family's dynamical age (e.g., Nesvorný et al., 2015).

16.2.2 Veneneia Basin Age

The age of the Veneneia basin is more uncertain. Rheasilvia landed on top of Veneneia and obliterated more than half of its surface. The surviving portion of Veneneia stretches into the northern hemisphere and has sufficient superposed craters to suggest an age of ~ 2.1 Ga (Schenk et al., 2012). It seems likely, though, that many superposed craters were partially erased by the formation of Rheasilvia, which would make this crater retention age a lower limit.

An alternative way to estimate Veneneia's formation age is to examine the crater spatial densities of troughs and ridges produced by this basin-forming event. They are located between 0° and 40° latitude on Vesta, with the units collectively called Saturnalia Fossae (Yingst et al., 2014; see also Chapter 5). For reference, the most prominent trough, called Saturnalia Fossa, is longer than ~460 km and has widths of up to ~40 km. It seems probable that pre-existing craters at those locations were erased when these troughs and ridges were formed. This means the approximate crater retention age of Veneneia can be estimated from superposed craters on Saturnalia Fossae.

Using models similar to those discussed in Schenk et al. (2012) and Bottke et al. (2020), Yingst et al. (2014) estimated that the crater retention age of Saturnalia Fossae (and Veneneia) was ~3.2–3.5 Ga. This age calculation may have been influenced to an unknown degree by the formation of Rheasilvia basin, but the removal of any putative superposed secondary craters from the observed crater population would only make the age of Saturnalia Fossae (and Veneneia basin) younger than ~3.2–3.5 Ga.

An age of ~3.2–3.5 Ga for Veneneia is consistent with the orbits of V-type asteroids located outside the Vesta family but within the inner Main Belt ($2.2 < a < 2.5$ AU) (Nesvorný et al., 2008). Several V-types with the same colors as Vesta-family members have been identified at inclinations that are lower than those of standard family members (i.e., 3 to 5° versus 5 to 9°). The most plausible source of these objects is Vesta. For Vesta family members to reach these orbits, they would need to undergo billions of years of dynamical evolution via Yarkovsky thermal forces and interactions with resonances. These ages imply that these particular V-type asteroids did not come from the Rheasilvia formation event, but instead were created by the older Veneneia formation event, where they took ~3 Gyr to reach their current orbits.

16.3 VARIOUS REGIMES OF COLLISIONS: THE CASE OF VESTA

Asteroid collisions lead to a large spectrum of outcomes, depending on the impact regime they take place in. The average impact velocity V_{imp} in today's asteroid belt is around 5 km/s (Bottke et al., 1994). This value is much higher than Vesta's escape velocity of $V_{esc} \sim 360$ m/s, so that means the ratio $V_{imp}/V_{esc} \gg 1$. Impacts with $V_{imp}/V_{esc} \gg 1$ are erosive and always lead to a net reduction of the mass of the target body. This regime includes cratering impacts as well as disruptive collisions. Cratering events range from small, local impact craters to planet-scale craters that can affect the target body on a global scale (Asphaug et al., 2015). Rheasilvia and Veneneia can be considered global-scale collisions on Vesta.

Disruptive impacts are more energetic than cratering events. They lead to the destruction of the target and to a significant loss of mass, with half of the initial mass in the case of catastrophic disruptions. An important aspect is the gravitational re-accumulation of a part of the ejected material on what is left of the remaining body as well as among the smaller fragments (Michel et al., 2001, 2003).

While $V_{imp}/V_{esc} \gg 1$ are strongly erosive, low-velocity collisions with V_{imp}/V_{esc} of the order of one can lead to accretion and a net increase of the target's mass.

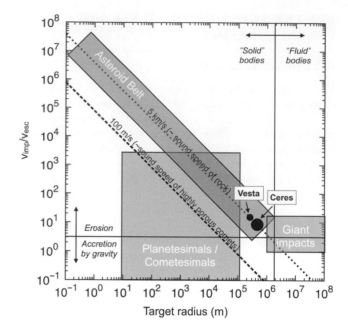

Figure 16.2 Collision regimes, distinguished by the ratio V_{imp}/V_{esc} as a function of target size. In this simplified picture, where only "gravitational sticking" is considered, collisions transition from accretionary to erosive when the impact velocities become higher than a few times V_{esc}. At very large scales, bodies behave like a "fluid" because of the large overburden pressure due to self-gravitation. The dashed lines of constant velocity correspond to the typical sound speed of rocks and highly porous materials, respectively, marking the transition between subsonic and supersonic impacts. Some examples are indicated by the gray areas. Adapted from Asphaug et al. (2015)

An overview of various collision regimes is presented in Figure 16.2. According to this simplified picture, which ignores the effect of oblique impact angles, collisions transition from accretionary to erosive when the impact velocities become higher than a few times V_{esc}. Together with the roughly linear dependence of V_{esc} on the target size this leads to interesting consequences.

At small scales, there is a large range of V_{imp}/V_{esc} with impact velocities below the sound speed, which means that accretionary collisions do not involve shocks (at very small scales, this is even true for erosive events). At large scales, on the other hand, accretionary collisions take place with velocities above the sound speed and involve strong shocks. Furthermore, at very large scales, planetary bodies behave like a "fluid" because the overburden pressure due to self-gravitation is large compared to the material strength. According to recent studies (Emsenhuber et al., 2018), however, material strength is still important at target body sizes of up to 1,000–2,000 km.

In Figure 16.2, we indicate the regimes for typical collisions taking place in the main asteroid belt. For comparison, we also show the corresponding regions for early planetesimal collisions and giant impacts. At the size scale of Vesta, both material strength effects and gravitational forces determine the outcome of collision events concurrently. Impacts on Vesta take place in the erosion regime, ejecting significantly more material than is accreted. As discussed in Section 16.4, the most energetic impacts lead to the excavation of Vesta's huge basins in the southern hemisphere and the formation of its asteroid family. We note, however, that the specific impact energies involved in the formation of these global scale craters are a factor $\sim$100 below the specific impact energy that would be needed for Vesta's catastrophic disruption (based on a scaling obtained from the disruption simulations of Jutzi et al., 2019). Although they cause global scale effects, these events are far from catastrophic.

16.4 VESTA'S GLOBAL EVOLUTION UNDER PLANET-SCALE COLLISIONS

Global-scale impacts modify the physical and thermal state of a substantial fraction of a target asteroid and alter its global shape (Asphaug et al., 2015). Vesta represents a prime example of a body that has experienced multiple such global-scale collisions, as suggested by the presence of its huge impact basins.

As discussed in Section 16.3, even for this large asteroid, the ratio V_{imp}/V_{esc} is still high for typical impact velocities of a few km/s, leading to net erosion (Figure 16.3). Still, re-accretion of ejecta for large collisions on Vesta is significant and takes place on a global scale. Importantly, the dynamical timescale of re-accretion is comparable to Vesta's rotation period of 5.3 hours. This can lead to complex features in deposition of ejecta and it also affects the dynamics of the crater collapse (see Chapter 5).

The overlap of two planet-scale basins, Rheasilvia and Veneneia, produces additional complications for modeling efforts. Vesta's non-spherical shape and gravity potential, rapid rotation, and differentiated structure further complicate the situation. Because the problem is strongly non-axisymmetric, a 3D approach is required.

16.4.1 Modeling the Subsequent Formation of Two Overlapping Basins

The sequential formation of Veneneia followed by Rheasilvia was studied by Jutzi et al. (2013) using a 3D Smooth Particle Hydrodynamics (SPH) impact code. The simulations started with a head-on impact into a non-rotating differentiated $D_{ast} = 550$ km sphere to form Veneneia. Then, the resulting basin was placed on axis and the body was spun up to Vesta's period of 5.3 hours. Finally, a second impact on the rotating cratered body 40° off-axis led to the observed topography of Vesta.

It is clear that the creation of Vesta's topography involved a very complex process of ejection and deposition. The modeling of these events therefore requires a realistic treatment of the ejecta gravitational dynamics and the post-impact rheological response. In the Jutzi et al. (2013) simulations, a pressure-dependent strength model, a tensile fracture model, a friction model, and self-gravity were included. In addition, a block-model approximation of acoustic fluidization was used to reproduce the central peak formation.

The best match to the observed topography was obtained using projectiles with impact velocities of $V_{imp} = 5.4$ km/s and diameters $D_{ast} = 64$ km for the first impact (to form Veneneia) and $D_{ast} = 66$ km for the second one (to form Rheasilvia on top of pre-existing Veneneia). A comparison of model outcome, observed topography, and reaccumulated ejecta is shown in Figure 16.3.

For comparison, using a 2D modeling approach, Ivanov and Melosh (2013) obtained a best fit model for Rheasilvia corresponding to a 52 km diameter projectile in the case of a 45° impact.

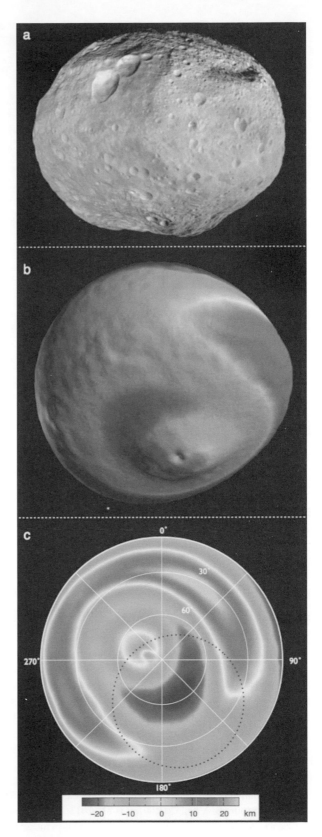

Figure 16.3 (a) Asteroid Vesta as seen by Dawn; (b) Final result of the simulation of two large impacts in the southern hemisphere (Jutzi et al., 2013); and (c) Lambert azimuthal projection (equal area) of the southern hemisphere in the model. Shading indicates the elevation (in km) with respect to a reference ellipsoid.

Photo credit for (a), NASA/JPL/UCLA/MPS/DLR/IDA

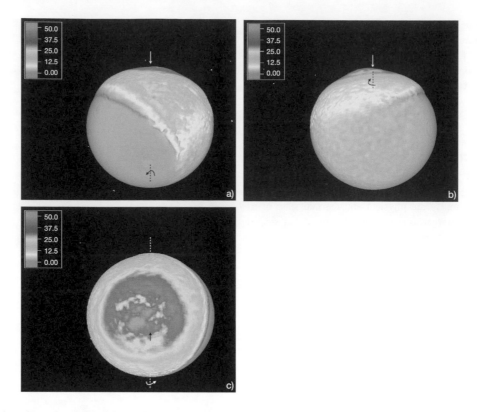

Figure 16.4 Distribution of re-accumulated ejecta resulting from an impact into a rotating target (adapted from Jutzi & Asphaug, 2011, figure 2). Shading corresponds to the initial depth of the ejecta (0–50 km). Vesta is shown (a, b) from two opposite sides and (c) from the top. The dashed line and the arrow indicate the rotation axis and the impact direction, respectively.

16.4.2 Rotational Effects

The rotation of Vesta $(P = 5.3\,\text{h})$ is rapid and leads to curious effects during the process of basin formation and ejecta re-accumulation. In a pre-Dawn study of the formation of the Rheasilvia basin on an already-spinning Vesta, Jutzi and Asphaug (2011) found that the ejecta is draped back on the asteroid in complex overlapping lobes, falling back in the rotating frame (Figure 16.4).

Coriolis forces are also acting during the formation and collapse of the large basins. On the rapidly rotating Vesta, the crater rebound timescale is comparable to the rotation timescale and the Rossby number of the flow $R_o \sim 1$. This can lead to crater rebound morphologies that form a spiral pattern relative to the impact locus (Figure 16.5), possibly explaining the features observed in the Rheasilvia crater on Vesta (as confirmed by Otto et al., 2016).

16.4.3 Constraints of Vesta's Structure and Composition

The formation of Vesta's giant craters and the provenance and specific distribution of their ejecta provide important constraints for the internal structure of Vesta (McSween et al., 2013; Clenet et al., 2014; see also Chapters 5 and 12). The key issue here is to model the formation of Veneneia and Rheasilvia basins in tandem, thereby allowing the Rheasilvia event to sample layers exposed by Veneneia.

By studying the sequential formation of both basins, it was found that a significant fraction of the rocks exposed in the South Pole region should come from >50 km depths, while the central mound of Rheasilvia and various basin overlap regions may sample depths

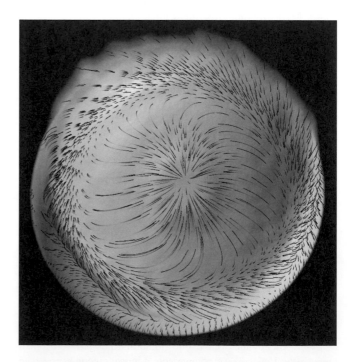

Figure 16.5 Simulating the opening and collapse of the Rheasilvia complex crater on asteroid Vesta; from Jutzi et al. (2013). The rapid rotation of the asteroid leads to strong Coriolis forces in the rebounding ejecta, as indicated by the vectors and in agreement with the patterns on the crater floor.

of up to ~60–100 km. For impacts into a pre-existing basin, the 2D modeling by Ivanov and Melosh (2013) also suggests that a significant amount of material from initial depths >50 km was exposed on the surface of Vesta.

These findings are puzzling at face value because olivine-rich mantle rock has not yet been detected in the Veneneia/Rheasilvia region (e.g., De Sanctis et al., 2012; Ammannito et al., 2013; McSween et al., 2013; see also Chapter 6). According to certain numerical models, olivine is by far the dominant phase at depths >50 km (Neumann et al., 2014; see also Chapter 4). The pyroxene dominated layer extends down to depths between 40 and 50 km, and the "eucrite" layer has a thickness of order of ~13 km.

The dilution of olivine in the regolith to below the detection limit of ~25 vol.% (Ammannito et al., 2013) may reconcile models and observations. Alternatively, the olivine-rich layers within Vesta are even deeper than predicted by standard models. For instance, a non-chondritic bulk composition has been suggested to explain the paucity of olivine-rich and other putative mantle rocks on Vesta's surface (Consolmagno et al., 2015).

Further constraints are given by the properties of the HED meteorites. Clenet et al. (2014) found that a significant fraction of the material that escapes Vesta during the more recent Rheasilvia impact, which would be the dominant source for HEDs, comes from greater than 40 km in depth (Figure 16.6). As previously suggested, the deep excavation depth is a consequence of the subsequent formation of two overlapping basins. The lack of mantle samples among the HED meteorites is difficult to explain and may again suggest that either the olivine-rich layers within Vesta are even deeper than predicted or that a different impact geometry may have excavated to shallower depths.

Alternatively, olivine may have been limited to Vesta's deep mantle (Mandler & Elkins-Tanton, 2013), a result that finds support in Dawn's compositional and gravity data (Raymond et al., 2017; see also Chapter 12). Models fitting the thicknesses and densities of the crust and mantle using gravity/topography constraints prefer a crustal thickness of ~20 km, and a fairly modest density contrast between crust and mantle. This supports the idea that Vesta's magmatic history was more complex than a body simply freezing after global melting and differentiation (e.g., Mizzon et al., 2015).

16.5 USING RHEASILVIA AND VENENEIA TO CONSTRAIN THE EARLY EVOLUTION OF THE MAIN BELT

The ages of Rheasilvia and Veneneia basins potentially provide us with insights into the early evolution of the asteroid belt. Given the lack of evidence for very early basins, we will postulate here that these two basins were made by the largest impactors to ever strike Vesta. As suggested in previous sections, there is no evidence for comparable impact events that would have affected Vesta's shape, nor are there any unaccounted-for ridges and troughs.

Jutzi et al. (2013) found projectile diameters of 64 and 66 km were needed to form Rheasilvia and Veneneia basins, respectively. Rheasilvia was modeled as forming on top of the pre-existing Veneneia. The ratio of the basin size to projectile size for these two cases are 7.6 and 6.3, respectively.

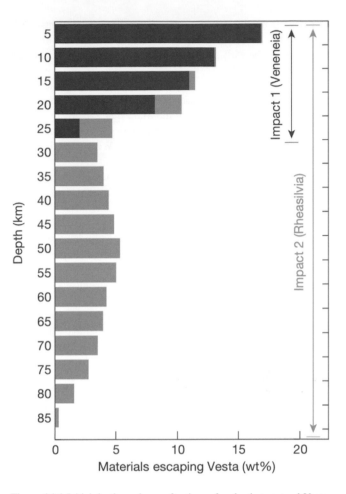

Figure 16.6 Initial depths and mass fractions of rocks that escaped Vesta. Shown is the relative proportion of material that escaped Vesta, compared to the total mass loss, as a function of its original depth before the two basin forming impacts (modeled in Jutzi et al., 2013). The first impact (Veneneia, in black) ejected mostly material from shallow depths (25 km), while the second one (Rheasilvia, in gray) ejected material from greater depths (mainly between 20 and 80 km).

From Clenet et al. (2014)

The next issue is to determine the probability of getting either a Rheasilvia or a Veneneia formation event over the course of Main Belt history. To keep things simple, we will assume that the Main Belt has always been close to its present-day size, which is likely a reasonable solution going back to 4.5 Gyr ago (Nesvorný et al., 2018). If we use the Main Belt asteroid population described in Bottke et al. (1994), which was benchmarked to have 682 Main Belt asteroids with $D > 50$ km, and we assume the intrinsic collision probability between Main Belt asteroids and Vesta is $P_i = 2.88 \times 10^{-18}$/km^2/yr (Bottke et al., 2020), we find that the probability that Vesta had 0, 1, 2, or 3 Rheasilvia- or Veneneia-size formation events over the last 4.5 Gyr is 50%, 35%, 12%, and 3%, respectively. This would suggest that Vesta beat the odds in a modest way (~12%) by having both Rheasilvia and Veneneia formed over its lifetime.

This calculation may have implications for the earliest history of Vesta. If we accept the crater retention ages discussed in Section 16.2 as reasonable, and one accepts the arguments made in Section 16.1, one must consider the possibility that Vesta experienced no Rheasilvia or Veneneia-style impact events when the Main Belt was undergoing its formative era. This sets limits on the nature of the early bombardment population on Vesta and the rest of the Main Belt.

For example, some have argued that shortly after planetesimal formation, the Main Belt was much larger than it is today, with most of this mass ejected out of the Main Belt zone by dynamical processes related to giant planet migration (e.g., Walsh et al., 2011; Morbidelli et al., 2015; Clement et al., 2018). We can test this using a simple back of the envelope calculation.

As a thought experiment, we will assume the Main Belt population was once 1,000-times larger. This value would mean that an Earth-mass of material was once in the Main Belt zone. While we choose this number to make our point in the next calculation, estimates of the Solar Nebula suggest it may approximate reality, though no one knows how much of this material was eventually turned into planetesimals (e.g., Morbidelli et al., 2009).

Next, we will assume this material was dynamically removed within $4.5\,\mathrm{Gyr}/1{,}000 = 4.5\,\mathrm{Myr}$ of the formation of the first solids. This would imply that the Main Belt experienced the equivalent of 4.5 Gyr of standard collision evolution within that short time. This calculation is overly simplistic, because ejected objects on excited orbits can continue to batter the Main Belt for an extended time (e.g., Bottke et al., 2005), but that can trade off against the objects being ejected earlier than 4.5 Myr. To any event, the probability that this lost population would produce an additional Rheasilvia/Veneneia on Vesta would be roughly 50%, the same as previously calculated.

In reality, to perform this calculation as accurately as possible, we would have to account for the precise timescale it takes for this material to be removed from Vesta-crossing orbits, the collision probability and impact velocities of this ejected material with Vesta over time, collisional evolution among the impacting population, and the physical nature of Vesta at this early stage. These modifications are unlikely to change the bottom line, though, namely that the lack of evidence for early Rheasilvia-type impacts may limit the size and nature of the impactor populations that have struck the primordial Main Belt.

16.6 LIMITS ON THE COLLISIONAL HISTORY OF VESTA FROM METEORITES

Impacts are potential sources of external heating for Vesta's surface materials. If the temperatures in the aftermath of a sizable cratering event are hot enough for a long enough period of time, it is possible to disturb or reset the radiometric ages of the rocks (e.g., Bogard, 1995, 2011). For many isotopic chronometers, the temperature thresholds required for resetting are too high to be easily met by many impact events, but $^{40}\mathrm{Ar}/^{39}\mathrm{Ar}$ ages appear to be an exception to this rule (e.g., Marchi et al., 2013a).

Past studies of the $^{40}\mathrm{Ar}/^{39}\mathrm{Ar}$ age distribution of the eucrites and howardites have shown three main features: a strong and fairly narrow spike at ~4.5 Gyr ago, which is mainly comprised of unbrecciated, cumulate, and monomict eucrites, relatively few ages between 4.1–4.4 Gyr ago, found mainly in howardites, brecciated, and polymict eucrites, and numerous ages between 3.4–4.1 Gyr ago, some of which were energetic enough to reset unbrecciated

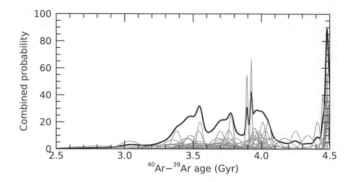

Figure 16.7 The $^{40}\mathrm{Ar}/^{39}\mathrm{Ar}$ shock degassing ages of eucrites and howardites as compiled by Marchi et al. (2013a). Each dated sample is reported with a Gaussian profile with center and width corresponding to the most probable age and 1-σ error. The black curve is the sum probability distribution obtained by the sum of all Gaussians divided by 3. Most eucrites that show few signs of impact alteration (unbrecciated, cumulate, monomict) have ages of about 4.45 Gyr ago. Note also that the thermal history of the samples may affect the height of the y-axis values, but not the ages displayed on the x-axis.

and cumulate eucrites at depth (e.g., Marchi et al., 2013a; Bottke et al., 2015b) (Figure 16.7).

More recent studies of brecciated eucrites show a similar profile. For example, Kennedy et al. (2019) report multiple reset ages between $4{,}534 \pm 56$ Ma to $4{,}491 \pm 16$ Ma, which might represent a single large impact event at ~4.5 Ga, and multiple events between $3{,}851 \pm 21$ Ma to $3{,}469 \pm 35$ Ma. As before, there appear to be few impact events represented between these time intervals. If one assumes that every sizable impact event on Vesta has an equal chance to produce a shock degassing event, this observation becomes quite puzzling; Vesta should have been hit frequently within the Main Belt during this era. Even after seeing Vesta in detail with Dawn, it is challenging to interpret these ages, unless there is some component we have missed.

One way to explain these ages comes from Marchi et al. (2013a), who suggested it might be indicative of high velocity impacts onto Main Belt bodies from an outside source. A surprising fact about impact-derived heating is that collision velocities >10 km/s can increase the net volume of material that reaches the $^{40}\mathrm{Ar}/^{39}\mathrm{Ar}$ reset temperature threshold by orders of magnitude over impacts at typical Main Belt velocities of 5 km/s. This behavior potentially explains why impact melt is so rare within Rheasilvia basin or elsewhere on Vesta (e.g., Le Corre et al., 2013). It also translates to dynamical behavior, in that only objects that orbit the Sun with high eccentricities and/or inclinations can strike Main Belt asteroids at >10 km/s. This implies that shock degassing ages are mainly telling us impactors that hit Main Belt targets were from orbits that were located outside of the Main Belt zone. Using this idea, we can speculate about the age relationship of shock degassing ages on Vesta, starting with the oldest ages.

16.6.1 Vesta Samples with 4.5 Gyr Old Ages and "Bogard's Rule"

Bogard and Garrison (2003) were the first to argue that the unbrecciated eucrites with ~4.5 Ga ages might represent relatively undamaged basaltic debris that was ejected off of Vesta by an impact event (see also Kennedy et al., 2019). This material would have potentially agglomerated into the form of small rubble-pile asteroids. While these bodies could be destroyed by further bombardment, the proportion of shock-heated rocks would remain lower in these bodies than those coming later from Vesta.

Scott and Bottke (2011) called this behavior "Bogard's Rule," named for Don Bogard, who suggested it in several papers on shocked meteorite samples (e.g., Bogard 1995, 2011). Bogard found that the abundance of meteorite samples with $^{40}Ar/^{39}Ar$ reset ages was much higher among lunar samples than eucrites (from ~500 km diameter Vesta) and was lowest among ordinary chondrites (with their parent bodies only a few hundred kilometers in diameter). In addition, this increase in the abundance of shock-reheated samples was proportional to parent body size; smaller bodies tend to lose their ejecta much more readily than larger bodies.

The timing of the ~4.5 Ga impact is close to the time of the giant impact that made the Moon (e.g., Barboni et al., 2017). Bottke et al. (2015b) argued that debris from this impact, the largest one to ever take place in the terrestrial planet region, could have potentially produced high velocity impactors, numerous crater events on Vesta and other Main Belt bodies, and shock degassing ages among many Main Belt bodies, including Vesta. As an aside, we note that Bottke et al. (2015b) argued for an age of ~4.45 Ga for this event, but they used $^{40}Ar/^{39}Ar$ ages derived from an older ^{40}K decay constant (Steiger & Jäger, 1977). The use of newer values favored by some researchers (Trieloff et al., 2003; Renne et al., 2010, 2011; Schwarz et al., 2011) would lead to a systematic shift of all of the $^{40}Ar/^{39}Ar$ ages used in their paper to older ages by 10–30 Myr. This would favor a giant impact event on the Earth near ~4.5 Ga.

16.6.2 Vesta Samples with <4.5 Gyr Old Ages

We consider the paucity of numerous shock ages between 4.5 Ga and 4 Ga and the onset of numerous ages again between ~3.85 Ga and ~3.47 Ga to be puzzling (Kennedy et al., 2019). Vesta should have been struck by numerous impactors over this entire time interval, so why do we only see ages in the latter interval?

A clue that might help us understand this issue comes from Swindle et al. (2013), who found a similar $^{40}Ar/^{39}Ar$ age distribution among the shocked H chondrite meteorites. Two prominent meteorite groups from two different worlds showing the same pattern is intriguing. It weakens the argument that the ages simply represent peculiar sampling among the eucrites. Moreover, if we apply Bogard's Rule, it would say that the largest Main Belt bodies, like Vesta and the H chondrite parent body, which may be larger than 275 km in diameter (e.g., Blackburn et al., 2017), are most likely to retain shocked material from large crater formation events.

The origin of this putative population of high velocity impactors that hit Vesta and the H chondrite parent body between ~3.4 and ~4 Ga has yet to be determined, but these ages do take place at a similar time to the putative Late Heavy Bombardment (LHB) on the Moon (e.g., see Bottke & Norman, 2017 for a recent review). At one time, it was argued that these collision events could represent the onset of late stage giant planet migration, which would liberate numerous small body populations from stable regions. New evidence, however, suggests that the last stage of giant planet migration occurred within the first 100 Myr of Solar System history (Nesvorný et al., 2018). If so, another source of external impactors may be needed to explain these shock degassing ages.

Kennedy et al. (2019) raised the possibility that the termination of impact ages at ~3.47 Ga could represent the formation time of a large impact basin. They favored the formation of the Rheasilvia basin at this time but, as previously discussed, superposed crater counts on Rheasilvia terrains indicate its age is ~1 Gyr. Instead, we suggest that Veneneia basin has crater constraints that provide a better match to this age, as discussed in Section 16.2.2.

Rheasilvia's formation age may be better represented by the $^{40}Ar/^{39}Ar$ ages of feldspar grains within the brecciated howardite Kapoeta (Lindsay et al., 2015). They were reset by a thermal event between 0.6 and 1.7 Ga ago.

The bottom line is that the formation ages of Rheasilvia and Veneneia may only be marginally sampled in the eucrite and howardite shock degassing ages. The easiest way to explain this is that both may have been formed from Main Belt projectiles striking Vesta at ~5 km/s, a low enough value that it does not provide sufficient heating to trigger a $^{40}Ar/^{39}Ar$ reset event. These conclusions also seem consistent with the inferred crater retention ages for these basins.

16.7 LIMITS ON THE COLLISIONAL HISTORY OF THE MAIN BELT FROM V-TYPE ASTEROIDS

Many models of planetesimal and planet formation assume the asteroid belt was once much larger than it is today (see Chapters 13 and 15). From a purely statistical standpoint, this scenario raises interesting issues concerning the existence and evolution of Vesta itself.

Consider a primordial asteroid belt that once had considerably more mass shortly after planetesimal formation. Modeling work suggests most of this material was eliminated at an early Solar System time by a dynamical depletion process such as sweeping resonances associated with planet migration, ejection of material by interaction with planetary embryos, Jupiter migrating across the asteroid belt in the Grand Tack model, or giant planet migration leading to excitation of the Main Belt population (e.g., Walsh et al., 2011; Morbidelli et al., 2015; Nesvorný et al., 2017; Clement et al., 2018). Given the uniqueness of Vesta in the Main Belt today, that could mean that Vesta was either a lucky survivor of this larger population or that the primordial Main Belt once had additional "sister" objects to Vesta that were ejected by dynamical loss mechanisms (e.g., Bottke, 2014).

The evidence for an extremely ancient component of the Vesta family in the orbital element distribution of V-type asteroids within the inner Main Belt seems to be limited. Ejecta from the 3.2–3.5 Ga Veneneia basin appears capable of explaining most V-types in the inner Main Belt located beyond the nominal limits of Vesta's family (Nesvorný et al., 2008).

This scenario, however, needs to be considered against the existence of a scattered population of small V-type asteroids in the central and outer Main Belt (semimajor axis $a > 2.5$ AU). This population, almost all which are $D_{ast} < 10$ km, have been investigated by several groups (Moskovitz et al., 2008; Solontoi et al., 2012; Licandro et al., 2017; Mansour et al., 2020). Their spectral signatures are similar to Vesta's basaltic crust, and they span the

Main Belt in semimajor axis, eccentricity, and inclination. Their numbers, though, are limited relative to the size of the Vesta family in the inner Main Belt. Their distributed dynamical distribution and small population raises interesting questions concerning how they got there.

Dynamical simulations indicate these V-types in the central and outer Main Belt are not part of the nominal Vesta family that are generally associated with the formation of the ~1 Gyr old Rheasilvia basin. That family is limited to the inner Main Belt, and no concentrated sub-population of V-types has yet been identified on the other side of the 3:1 mean motion resonance with Jupiter at ~2.5 AU (J3:1) (Binzel & Xu, 1993; Moskovitz et al., 2008; Solontoi et al., 2012; Licandro et al., 2017; Mansour et al., 2020). If ejecta from the ~1 Ga old Rheasilvia basin had made it across this gap, they would be expected to have similar inclinations to the nominal population.

Related simulations show it is difficult for V-types from the nominal Vesta family in the inner Main Belt to enter the J3:1 via the Yarkovsky thermal drift and then dynamically jump across it to reach the central Main Belt (Nesvorný et al., 2008). The fraction of V-types that make a successful jump is low, and even those that make it cannot then evolve to the orbits of the observed V-types in the central and outer Main Belt over the age of the Solar System.

An interesting variation on this theme was presented by Brasil et al. (2017). They assumed than an ancient Vesta family was created prior to the migration of Jupiter into its current orbit. Planetary perturbations then might allow V-types to get beyond 2.5 AU prior to the formation of the present-day J3:1. An advantage of this model is that dynamical resonances produced by late giant planet migration would push many V-types onto the same eccentricities and inclinations as the observed objects. With the exception of the D_{ast} ~17–29 km V-type asteroid (1459) Magnya (Delbo et al., 2006; Masiero et al., 2012), the semimajor axes of the central and outer Main Belt V-types could then be potentially reproduced through an extended phase of Yarkovsky drift.

A disadvantage of this scenario is that the majority of this putative early Vesta family would be emplaced in the inner Main Belt, where they would spread by the effects of giant planet migration. While many would be lost by collisional and dynamical evolution, it seems likely that many putative ancient V-types would still reside in the inner Main Belt today at many different eccentricities and inclinations. The issue of whether this scenario is consistent with observational constraints in the inner Main Belt has yet to be thoroughly tested. The scenario might also require a Veneneia-sized basin to form on Vesta to provide sufficient mass. This would imply that Veneneia is much older than ~3.5 Ga, which does not fit the existing data, or that a Veneneia-sized basin and its associated graben were erased by early viscous relaxation in a newly formed Vesta with a hot interior. The latter possibility sets up an interesting issue for geophysical modelers to investigate in the future.

At the least, though, the Brasil et al. (2017) scenario cannot easily explain the origin of (1459) Magnya, which is too large to undergo extensive Yarkovsky drift. Moving Magnya from Vesta to its current orbit by gravitational perturbations during planet migration processes may be possible, but it seems likely the same mechanism that does this would also produce far more mixing between large S-types and C-types than observed in the Main Belt (Morbidelli et al., 2015). This hints at the possibility that there is more to this problem than meets the eye.

The alternative scenarios to create Magnya and the V-types are that (i) they were scattered into those regions from outside the Main Belt by the same dynamical processes that shaped the Main Belt, or (ii) they are fragments left there by a population of lost Vesta-like bodies. Bottke (2014) considered this latter possibility – in part – using Boulder, a collisional code capable of simulating the dynamical depletion and collisional fragmentation of multiple small body populations using a statistical particle-in-the box approach (Morbidelli et al., 2009). An estimate of the primordial Main Belt size distribution, stretched across many semimajor axis zones, was input into the code, as was a pre-set number of Vesta-like objects. In these simulations, if the primordial Main Belt was assumed to be N times larger than the currently observed population, we would also run trials that assumed there were 1, 2, 3, ..., N Vestas in that population.

In their main scenario, Bottke (2014) tracked these extra Vestas and their fragments between 4 and 4.5 Ga. At 4 Ga, they assumed late giant planet migration took place, which led to a dynamical depletion event that removed the remainder of the Main Belt's excess mass. From that point, collisional grinding over the next 4 Gyr had to reproduce the current Main Belt population and match the constraints from the central and outer Main Belt V-types. Over all of this, Bottke (2014) assumed that impacts on Vesta-like objects would produce fragments that were distinct (in spectroscopic signatures, colors, and albedos) from background asteroids, as the V-types are today in the Main Belt. When the model reached the current time, a successful run was one that was able to reproduce the current Main Belt and the observed V-type size distributions in the inner and central/outer Main Belt within tolerance limits.

In this scenario, it was found that an excited primordial Main Belt with more than 3–4 Vestas after the first few Myr of its history tends to produce too many V-type fragments between 4 and 4.5 Ga compared to observations; collisional and dynamical evolution over 4.5 Gyr was unable to get rid of all of the evidence.

Different dynamical depletion scenarios, however, will lead to different outcomes. For example, consider the dynamical depletion of the primordial Main Belt from the effects of a giant planet instability and post-nebula giant planet migration, as simulated by Nesvorný et al. (2017). In their preferred model, they found they could reproduce the semimajor axis, eccentricity, and inclination distribution of diameter $D_{ast} > 10$ km bodies across the Main Belt with considerable accuracy, but their central and outer Main Belt populations only lose a factor of two-to-three in mass during this time. If a giant planet instability is the primary source of dynamical depletion from the primordial Main Belt, the number of putative Vesta sisters that existed there is limited to a small number that had to leave early; if N were large, the probability is high that one or more would never escape, violating observations.

Even then, low N values for Vesta's sisters still have challenges to meet constraints. For example, suppose a single Vesta sister was placed in the outer Main Belt. A large impact would be needed to eject a Magnya-sized body, but this event would also produce a large population of smaller V-types, possibly a population comparable to the present-day Vesta family. The dynamical loss of a factor of two-to-three from this population via giant planet migration would not strongly affect the size of the V-type population, and collision and dynamical processes would be unlikely to get rid of the excess number of bodies over 4.5 Ga. Accordingly, this model run would fail. More work on this topic is needed to confirm this prediction.

In the end, we would argue that the most plausible solution is that the V-types in the central and outer Main Belt are fragments of Vesta-like bodies that were originally located outside of the Main Belt zone. Objects in the terrestrial planet zone have higher collision probabilities and impact velocities with each other, so we can expect these Vesta-like bodies to produce numerous V-type fragments in response to impacts from planetesimals scattered by planet formation processes. A small fraction of all of these V-types might find a dynamical pathway into the primordial Main Belt as the terrestrial planets form and/or the giant planets migrate to their current orbits (e.g., Bottke et al., 2006b; Scott et al., 2015; Vokrouhlický et al., 2016; Raymond & Izidoro, 2017). A low capture efficiency in the Main Belt would naturally explain why the majority of V-types in the central and outer Main Belt have small sizes and a large spread of semimajor axis, eccentricity, and inclination values.

16.8 CONCLUSIONS

The success of the Dawn mission, combined with the on-going exploration of the HED meteorites, has helped us to not only better understand Vesta's history but that of the Main Belt itself. Here we summarize the main conclusions of this chapter, as well as a few mysteries that require future work.

Vesta's surface is dominated by two gigantic overlapping impact basins: the older ~400 km Veneneia basin and the younger ~500 km diameter Rheasilvia basin. Their age and nature, along with the ejecta they have produced in the form of so-called V-type asteroids, can help us probe Vesta's evolution.

The crater retention age of Rheasilvia basin, as determined using a Main Belt model of crater production, is between ~0.6 and ~1.5 Ga. This age range is consistent with age constraints from the eucrite meteorites and collisional modeling of the Vesta family. Rheasilvia also has crater spatial densities lower than those found on Gaspra, whose inferred surface age from craters, its family age, and samples from Itokawa all suggest it is ~1.3 Ga. We infer from this that Rheasilvia is plausibly ~1 Ga.

Veneneia's crater retention age basin is more difficult to determine because it is largely covered by the younger Rheasilvia basin. Crater spatial densities of troughs and ridges produced by this basin-forming event in the northern hemisphere of Vesta, collectively called Saturnalia Fossae, suggest that Veneneia basin is ~3.2–3.5 Ga.

Impact simulations using 3D Smooth Particle Hydrodynamics (SPH) impact codes indicate the observed topography on Vesta can be obtained using projectiles with diameters of $D = 64$ km for Veneneia basin and $D = 66$ km for Rheasilvia basin. In this circumstance, the size and nature of the Rheasilvia impact is affected by the damage done to Vesta by the formation of Veneneia basin. Coriolis forces acting during the formation and collapse of Rheasilvia are also likely responsible for the characteristic swirl-like pattern observed in its topography.

The formation of Rheasilvia and Veneneia basins should have dredged up material within Vesta from depths of >50 km. The central mound of Rheasilvia and various basin overlap regions may sample depths of up to ~60–100 km. The paucity of mantle samples among the HED meteorites, together with the non-detection of olivine-rich rock on the surface, may indicate alternative explanations are needed for Vesta' formation and composition. One intriguing possibility is that olivine may only exist in abundance within Vesta's deep mantle, making it difficult to sample via these basin events (see Chapter 4).

There is no obvious topographic or compositional evidence that additional Rheasilvia–Veneneia size basins events took place at very early times on Vesta. At this time, though, we cannot rule out the possibility that a basin this size formed and was viscously relaxed away when Vesta's interior was still hot. The probability of forming 0, 1, 2, or 3 Rheasilvia–Veneneia-size basins events over the last 4.5 Gyr, assuming the Main Belt had a size and excitation comparable to the current Main Belt, is 50%, 35%, 12%, and 3%, respectively. This result suggests Vesta slightly beat the odds to have both basins (~12%).

There is only modest evidence at best for the formation time of Veneneia and Rheasilvia in the eucrite and howardite meteorite record. The absence of an obvious spike of $^{40}Ar/^{39}Ar$ shock degassing ages at ~1 Ga and ~3.5 Ga may be a consequence of low Main Belt impact velocities (<5 km/s) that produce limited heating. Shock degassing ages from HED meteorites, which may be indicative of high velocity impactors on orbits outside the Main Belt zone striking Vesta, suggest the following impact pattern: numerous impacts near ~4.5 Ga, relatively few between ~4 and 4.4 Ga, and numerous impacts between 3.4 and 3.9 Ga. This age distribution is similar to the one seen among the H chondrites. The source of this putative population of high velocity impactors that hit Vesta and the H chondrite parent body between ~3.4 and ~4 Ga has yet to be determined.

V-type asteroids, with spectroscopic signatures similar to the surface of Vesta, are mostly in the Vesta family and were derived from the Rheasilvia and Veneneia formation events. A more limited number are spread in semimajor axes, eccentricities, and inclinations throughout the inner, central, and outer Main Belt.

The V-types in the inner Main Belt that are not in the nominal Vesta family are plausibly ejecta from the 3.2–3.5 Ga Veneneia basin-forming event. Those in the central and outer Main Belt are highly scattered and limited in number. Their origin is unknown, but possible sources include (i) ejecta from Vesta at a very early time in Solar System history, (ii) extra Vesta-like bodies in the primordial Main Belt that were lost early-on via dynamical processes, and (iii) Vesta-like bodies outside the Main Belt whose fragments were scattered into the Main Belt by dynamical events associated with terrestrial planet formation and/or giant planet migration. We modestly favor the latter scenario at this time.

REFERENCES

Ammannito, E., De Sanctis, M. E., Palomba, E., et al. (2013) Olivine in an unexpected location on Vesta's surface. *Nature*, 504, 122.

Asphaug, E. (1997) Impact origin of the Vesta family. *Meteoritics & Planetary Science*, 32, 965–980.

Asphaug, E., Collins, G., & Jutzi, M. (2015) Global scale impacts. In P. Michel, F. DeMeo, & W. F. Bottke (eds.), *Asteroids IV*. Tucson: University of Arizona Press, pp. 661–677.

Barboni, M., Boehnke, P., Keller, B., et al. (2017) Early formation of the Moon 4.51 billion years ago. *Science Advances*, 3, e1602365.

Binzel, R. P., & Xu, S. (1993) Chips off of asteroid 4 Vesta – Evidence for the parent body of basaltic achondrite meteorites. *Science*, 260, 186–191.

Blackburn, T., Alexander, C. M. O., Carlson, R., & Elkins-Tanton, L. T. (2017) The accretion and impact history of the ordinary chondrite parent bodies. *Geochimica et Cosmochimica Acta*, 200, 201.

Bogard, D. D. (1995) Impact ages of meteorites: A synthesis. *Meteoritics*, 30, 244–268.

Bogard, D. D. (2011) K–Ar ages of meteorites: Clues to parent body thermal histories. *Chemie der Erde*, 71, 207–226.

Bogard D. D., & Garrison, D. H. (2003) $^{39}Ar/^{40}Ar$ ages of eucrites and the thermal history of asteroid 4 Vesta. *Meteoritics & Planetary Science*, 38, 669–710.

Bottke, W. F. (2014) On the origin and evolution of Vesta and the V-type asteroids. *Vesta in the Light of Dawn: First Exploration of a Protoplanet in the Asteroid Belt*, February 3–4, Houston, TX, 2024.

Bottke, W. F., Brož, M., O'Brien, D. P., et al. (2015a) The collisional evolution of the asteroid belt. In P. Michel, F. DeMeo, & W. F. Bottke (eds.), *Asteroids IV*. Tucson: University of Arizona Press, pp. 701–724.

Bottke, W. F., Durda, D. D., Nesvorny, D., et al. (2005) Linking the collisional history of the main asteroid belt to its dynamical excitation and depletion. *Icarus*, 179, 63–94.

Bottke, W. F., Nesvorný, D., Grimm, R. E., Morbidelli, A., & O'Brien, D. P. (2006a) Iron meteorites as remnants of planetesimals formed in the terrestrial planet region. *Nature*, 439, 821–824.

Bottke, W. F., Nolan, M. C., Greenberg, R., & Kolvoord, R. A. (1994) Velocity distributions among colliding asteroids. *Icarus*, 107, 255–268.

Bottke, W. F., & Norman, M. (2017) The late heavy bombardment. *Annual Review of Earth and Planetary Science*, 45, 619–647.

Bottke, W. F., Vokrouhlický, D., Ballouz, R.-L., et al. (2020) Interpreting the cratering histories of Bennu, Ryugu, and other spacecraft-explored asteroids. *The Astronomical Journal*, 160, 14.

Bottke, W. F., Vokrouhlický, D., Marchi, S., et al. (2015b) Dating the Moon-forming impact event with asteroidal meteorites. *Science*, 348, 321–323.

Bottke, W. F., Vokrouhlický, D., Rubincam, D. P., & Nesvorný. D. (2006b) The Yarkovsky and YORP effects: Implications for asteroid dynamics. *Annual Review of Earth and Planetary Science*, 34, 157–191.

Bowling, T. J., Johnson, B. C., Melosh, H. J., et al. (2013) Antipodal terrains created by the Rheasilvia basin forming impact on asteroid 4 Vesta. *Journal of Geophysical Research: Planets*, 118, 1821–1834.

Brasil, P. I. O., Roig, F., Nesvorný, D., & Carruba, V. (2017) Scattering V-type asteroids during the giant planet instability: A step for Jupiter, a leap for basalt. *Monthly Notices of the Royal Astronomical Society*, 468, 1236.

Brož, M., Morbidelli, A., Bottke, W. F., et al. (2013) Constraining the cometary flux through the asteroid belt during the late heavy bombardment. *Astronomy & Astrophysics*, 551, A117.

Buczkowski, D. L., Wyrick, D. Y., Iyer, K. A., et al. (2012) Large-scale troughs on Vesta: A signature of planetary tectonics. *Geophysical Research Letters*, 39, L18205.

Clement, M. S., Kaib, N. A., Raymond, S. N., & Walsh, K. J. (2018) Mars' growth stunted by an early giant planet instability. *Icarus*, 311, 340–356.

Clenet, H., Jutzi, M., Barrat, J.-A., et al. (2014) A deep crust-mantle boundary in the asteroid 4 Vesta. *Nature*, 511, 303–306.

Consolmagno, G. J., Golabek, G. J., Turrini, D., et al. (2015) Is Vesta an intact and pristine protoplanet? *Icarus*, 254, 190–201.

De Sanctis, M. C., Ammannito, E., Capria, M. T., et al. (2012) Spectroscopic characterization of mineralogy and its diversity across Vesta. *Science*, 336, 697–700.

Delbo, M., Gai, M., Lattanzi, M. G., et al. (2006) MIDI observations of 1459 Magnya: First attempt of interferometric observations of asteroids with the VLTI. *Icarus*, 181, 618.

DellaGiustina, D. N., Kaplan, H. H., Simon, A. A., et al. (2021) Exogenic basalt on asteroid (101955) Bennu. *Nature Astronomy*, 5, 1–8.

Emsenhuber, A., Jutzi, M., & Benz, W. (2018) SPH calculations of planet-scale collisions: The role of the Equation of State, material rheologies, and numerical effects. *Icarus*, 301, 247–257.

Ermakov, A. I., Zuber, M. T., Smith, D. E., et al. (2014) Constraints on Vesta's interior structure using gravity and shape models from the Dawn mission. *Icarus*, 240, 146–160.

Holsapple, K. A., & Housen, K. R. (2007) A crater and its ejecta: An interpretation of deep impact. *Icarus*, 187, 345–356.

Hopkins, M. D., Mojzsis, S. J., Bottke, W. F., & Abramov, O. (2015) Micrometer-scale U–Pb age domains in eucrite zircons, impact resetting, and the thermal history of the HED parent body. *Icarus*, 245, 367–378.

Ivanov, B. A., & Melosh, H. J. (2013) Two-dimensional numerical modeling of the Rheasilvia impact formation. *Journal of Geophysical Research: Planets*, 118, 1545–1557.

Jaumann, R. J., Williams, D. A., Buczkowski, D. L., et al. (2012) Vesta's shape and morphology. *Science*, 336, 687–690.

Jourdan, F., Kennedy, T., Benedix, G. K., Eroglu, E., & Mayer, C. (2020) Timing of the magmatic activity and upper crustal cooling of differentiated asteroid 4 Vesta. *Geochimica et Cosmochimica Acta*, 273, 205.

Jutzi, M., & Asphaug, E. (2011) Mega-ejecta on asteroid Vesta. *Geophysical Research Letters*, 38, 1102.

Jutzi, M., Asphaug, E., Gillet, P., Barrat, J.-A.,n & Benz, W. (2013) The structure of the asteroid 4 Vesta as revealed by models of planet-scale collisions. *Nature*, 494, 207–210.

Jutzi, M., Michel, P., & Richardson, D. C. (2019) Fragment properties from large-scale asteroid collisions: I: Results from SPH/N-body simulations using porous parent bodies and improved material models. *Icarus*, 317, 215–228

Kennedy, T., Jourdan, F., Eroglu, E., & Mayers, C. (2019) Bombardment history of asteroid 4 Vesta recorded by brecciated eucrites: Large impact event clusters at 4.50 Ga and discreet bombardment until 3.47 Ga. *Geochimica et Cosmochimica Acta*, 260, 99.

Knežević, Z., & Milani, A. (2003) Proper element catalogs and asteroid families. *Astronomy & Astrophysics*, 403, 1165–1173.

Le Corre, L., Reddy, V., Schmedemann, N., et al. (2013) Olivine or impact melt: Nature of the "Orange" material on Vesta from Dawn. *Icarus*, 226, 1568.

Licandro, J., Popescu, M., Morate, D., & de Leon, J. (2017) V-type candidates and Vesta family asteroids in the Moving Objects VISTA (MOVIS) catalogue. *Astronomy & Astrophysics*, 600, A126.

Lindsay, F. N., Delaney, J. S., Herzog, G. F., et al. (2015) Rheasilvia provenance of the Kapoeta howardite inferred from 1 Ga $^{40}Ar/^{39}Ar$ feldspar ages. *Earth and Planetary Science Letters*, 413, 208.

Mandler, B. E., & Elkins-Tanton, L. T. (2013) The origin of eucrites, diogenites, and olivine diogenites: Magma ocean crystallization and shallow magma chamber processes on Vesta. *Meteoritics & Planetary Science*, 48, 2333–2349.

Mansour, J.-A., Popescu, M., de Leon, J., & Licandro, J. (2020) Distribution and spectrophotometric classification of basaltic asteroids. *Monthly Notices of the Royal Astronomical Society*, 491, 5966.

Marchi, S., Bottke, W. F., Cohen, B. A., et al. (2013a) High-velocity collisions from the lunar cataclysm recorded in asteroidal meteorites. *Nature Geosciences*, 6, 303–307.

Marchi, S., Bottke, W. F., O'Brien, D. P., et al. (2013b) Small crater populations on Vesta. *Planetary and Space Science*, 103, 96–103.

Marchi, S., Chapman, C. R., Barnouin, O. S., Richardson, J. E., & Vincent, J.-B. (2015) Cratering on asteroids. In P. Michel, F. DeMeo, & W. F. Bottke (eds.), *Asteroids IV*. Tucson: University of Arizona Press, pp. 725–744.

Marchi, S., Ermakov, A. I., Raymond, C. A., et al. (2016) The missing large impact craters on Ceres. *Nature Communications*, 7, eid. 12257.

Marchi, S., McSween, H. Y., O'Brien, D. P., et al. (2012) The violent collisional history of asteroid 4 Vesta. *Science*, 336, 690–693.

Masiero, J. R., DeMeo, F. E., Kasuga, T., & Parker, A. H. (2015) Asteroid family physical properties. In P. Michel, F. DeMeo, & W. F. Bottke (eds.), *Asteroids IV*. Tucson: University of Arizona Press, pp. 323–340.

Masiero, J. R., Mainzer, A. K., Grav, T., et al. (2012) Preliminary analysis of WISE/NEOWISE 3-band cryogenic and post-cryogenic observations of Main Belt asteroids. *The Astrophysical Journal Letters*, 759, 5.

McSween, H. J., Ammannito, E., Reddy, V., et al. (2013) Composition of the Rheasilvia basin, a window into Vesta's interior. *Journal of Geophysical Research*, 118, 335–346.

McSween, H. Y., Raymond, C. A., Stolper, E. M., et al. (2019) Differentiation and magmatic history of Vesta: Constraints from HED meteorites and Dawn spacecraft data. *Chemie der Erde – Geochemistry*, 79, 125526.

Michel, P., Benz, W., & Richardson, D. C. (2003) Disruption of fragmented parent bodies as the origin of asteroid families. *Nature*, 421, 608.

Michel, P., Benz, W. Tanga, P., & Richardson, D. C. (2001) Collisions and gravitational reaccumulation: Forming asteroid families and satellites. *Science*, 294, 1696.

Mizzon, H., Monnereau, M., Toplis, M. J., et al. (2015) A numerical model of the physical and chemical evolution of Vesta based on compaction equations and the olivine-anorthite-silica ternary diagram. *Lunar and Planetary Science Conference*, 46, 1832.

Morbidelli, A., Bottke, W. F., Nesvorný, D., & Levison, H. (2009). Asteroids were born big. *Icarus*, 204, 558–573.

Morbidelli, A., Walsh, K. J., O'Brien, D. P., Minton, D. A., & Bottke, W. F. (2015) The dynamical evolution of the asteroid belt. In P. Michel, F. DeMeo, & W. F. Bottke (eds.), *Asteroids IV*. Tucson: University of Arizona Press, pp. 493–508.

Moskovitz, N. A., Jedicke, R., Gaidos, E., et al. (2008) The distribution of basaltic asteroids in the Main Belt. *Icarus*, 198, 77.

Nesvorný, D., Brož, M., & Carruba, V. (2015) Identification and dynamical properties of asteroid families. In P. Michel, F. DeMeo, & W. F. Bottke (eds.), *Asteroids IV*. Tucson: University of Arizona Press, pp. 297–321.

Nesvorný, D., Roig, F., & Bottke, W. F. (2017) Modeling the historical flux of planetary impactors. *The Astronomical Journal*, 153, 103.

Nesvorný, D., Roig, F., Gladman, B., et al. (2008) Fugitives from the Vesta family. *Icarus*, 193, 85.

Nesvorný, D., Vokrouhlický, D., Bottke, W. F., & Levison, H. F. (2018) Evidence for very early migration of the Solar System planets from the Patroclus-Menoetius binary Jupiter Trojan. *Nature Astronomy*, 2, 878–882.

Neumann, W., Breuer, D., & Spohn, T. (2014) Differentiation of Vesta: Implications for a shallow magma ocean. *Earth and Planetary Science Letters*, 395, 267–280.

Otto, K. A., Jaumann, R., Krohn, K., et al. (2016) The Coriolis effect on mass wasting during the Rheasilvia impact on asteroid Vesta. *Geophysical Research Letters*, 43, 12340–12347.

Park, J., Turrin, B. D., Herzog, G. F., et al. (2015) ^{40}Ar/^{39}Ar age of material returned from asteroid 25143 Itokawa. *Meteoritics & Planetary Science*, 50, 2087–2098.

Parker, A. H., Ivezić, Ž., Jurić, M., et al. (2008) The size distributions of asteroid families in the SDSS Moving Object Catalog 4. *Icarus*, 198, 138–155.

Raymond, C. A., Russell, C. T., & McSween, Jr. H. Y. (2017) Dawn at Vesta: Paradigms and Paradoxes. In B. P. W. Linda, & T. Elkins-Tanton (eds.), *Planetisimals: Early Differentiation and Consequences for Planets*. Cambridge: Cambridge University Press, pp. 321–339.

Raymond, S. N., & Izidoro, A. (2017) The empty primordial asteroid belt. *Science Advances*, 3, e1701138.

Renne, P. R., Balco, G., Ludwig, K. R., Mundil, R., & Min, K. (2011) Response to the comment by W. H. Schwarz et al. on the "Joint determination of ^{40}K decay constants and ^{40}Ar*/^{40}K for the Fish Canyon sanidine standard, and improved accuracy for ^{40}Ar/^{39}Ar geochronology" by P. R. Renne et al. (2010). *Geochimica et Cosmochimica Acta*, 75, 5097–5100.

Renne, P. R., Mundil, R., Balco, G., et al. (2010) Joint determination of ^{40}K decay constants and ^{40}Ar*/^{40}K for the Fish Canyon sanidine standard, and improved accuracy for ^{40}Ar/^{39}Ar geochronology. *Geochimica et Cosmochimica Acta*, 74, 5349–5367.

Russell, C. T., Raymond, C. A., Coradini, A., et al. (2012) Dawn at Vesta: Testing the protoplanetary paradigm. *Science*, 336, 684–686.

Schenk, P., O'Brien, D. P., Marchi, S., et al. (2012) The geologically recent giant impact basins at Vestas South Pole. *Science*, 336, 694–697.

Schwarz, W. H., Kossert, K., Trieloff, M., & Hopp, J. (2011) Comment on the "Joint determination of ^{40}K decay constants and ^{40}Ar*/^{40}K for the Fish Canyon sanidine standard, and improved accuracy for ^{40}Ar/^{39}Ar geochronology" by Paul R. Renne et al. (2010). *Geochimica et Cosmochimica Acta*, 75, 5094–5096.

Scott, E. R. D., and Bottke, W. F. (2011) Impact histories of angrites, eucrites, and their parent bodies. *Meteoritics & Planetary Science*, 46, 1878–1887.

Scott, E. R. D., Keil, K., Goldstein, J. I., et al. (2015) Early impact history and dynamical origin of differentiated meteorites and asteroids. In P. Michel, F. DeMeo, & W. F. Bottke (eds.), *Asteroids IV*. Tucson: University of Arizona Press, pp. 573–596.

Solontoi, M. R., Hammergren, M., Gyuk, G., Puckett, A. 2012. AVAST survey 0.4–1.0 µm spectroscopy of igneous asteroids in the inner and middle Main Belt. *Icarus*, 220, 577.

Spoto, F., Milani, A., & Knežević, Z. (2015) Asteroid family ages. *Icarus*, 257, 275–289.

Steiger, R. H., & Jäger, E. (1977) Subcommission on geochronology: Convention on the use of decay constants in geo- and cosmochronology. *Earth and Planetary Science Letters*, 36, 359–362.

Stickle, A. M., Schultz, P. H., and Crawford, D. A. (2015) Subsurface failure in spherical bodies: A formation scenario for linear troughs on Vesta's surface. *Icarus*, 247, 18–34.

Swindle, T. D., Kring, D. A., & Wierich, J. R. (2013) ^{40}Ar/^{39}Ar ages of impacts involving ordinary chondrite meteorites. In Advances in 40Ar/39Ar Dating: from Archeaology to Planetary Sciences. *Geological Society of London, Special Publications*, 378, 333–347.

Trieloff, M., Jessberger, E. K., Herrwerth, I., et al. (2003) Structure and thermal history of the H-chondrite parent asteroid revealed by thermo-chronometry. *Nature*, 422, 502–506.

Vernazza, P., Jorda, L., Ševeček, P., et al. (2020) A basin-free spherical shape as an outcome of a giant impact on asteroid Hygiea. *Nature Astronomy*, 4, 136–141.

Vokrouhlický, D., Bottke, W. F., Chesley, S. R., Scheeres, D. J., & Statler, T. S. (2015) The Yarkovsky and YORP effects. In P. Michel, F. DeMeo, & W. F. Bottke (eds.), *Asteroids IV*. Tucson: University of Arizona Press, pp. 509–532.

Vokrouhlický, D., Bottke, W. F., & Nesvorný, D. (2016) Capture of trans-Neptunian planetesimals in the main asteroid belt. *The Astronomical Journal*, 152, 39.

Vokrouhlický, D., Bottke, W. F., & Nesvorný, D. (2017) Forming the Flora family: Implications for the near-Earth asteroid population and large terrestrial planet impactors. *The Astronomical Journal*, 153, 172.

Walsh, K. J., Morbidelli, A., Raymond, S. N., O'Brien, D. P., & Mandell, A. M. (2011) A low mass for Mars from Jupiter's early gas-driven migration. *Nature*, 475, 206–209.

Yingst, R. A., Mest, S. C., Berman, D. C., et al. (2014) Geologic mapping of Vesta. *Planetary and Space Science*, 103, 2.

17

Epilogue

The Renaissance of Main Belt Asteroid Science

SIMONE MARCHI, CAROL A. RAYMOND, AND CHRISTOPHER T. RUSSELL

17.1 INTRODUCTION

More than 200 years separate us from the scientific turmoil generated by the discovery of Ceres, the largest Main Belt asteroid, in a region of the Solar System that was expected to harbor a planet. Yet scientists are still wrestling with fundamental questions such as: How did asteroids form? What are they made of? What do they tell us about the evolution of our Solar System? These questions are at the core of modern planetary science and are drivers for the robotic exploration of the Solar System.

Asteroid science progressed at a slow pace in the decades after Ceres' first observations. As more asteroids were discovered in the nineteenth and twentieth centuries, their formation and physical properties remained largely unknown, but one thing became immediately clear: asteroids are tiny compared to planets. So, one may ask, are asteroids just a minor detail in the grand scheme of the Solar System?

In the 1970s, technology finally caught up and sophisticated telescopic observations became possible (e.g., McCord et al., 1970). Around the same time, it was established that asteroids are the likely source of meteorites found on Earth (Arnold, 1965; Wetherill, 1967), allowing a first quantitative understanding of their nature based on meteorite properties. The first serendipitous flyby of Main Belt asteroid Gaspra in 1991 by the NASA *Galileo* spacecraft enroute to Jupiter was a pivotal moment in planetary sciences (Belton et al., 1992), when asteroids became resolved astronomical objects like the Moon and Mars. As more data from telescopic observations, theoretical models, and space exploration started to accumulate, it became clear that the Main Belt was a fundamental part of the structure of our Solar System, a natural divide between the inner rocky terrestrial planets and the outer gaseous planets.

This book depicts a vivid and vibrant image of modern Main Belt asteroid science. In the last decade, thanks to the exploration by the NASA Dawn mission and advent of high-resolution Earth-bound observations (see Chapters 1 and 2), we have entered a renaissance of Main Belt asteroid science. Formation theories, dynamical models, meteorite geochemical data, remote and *in-situ* observations synergistically show asteroids are leftover building blocks of planetary formation and tracers of important evolutionary processes (e.g., collisions, orbital migration) that have shaped the evolution of the early Solar System. And, perhaps, asteroids will be exploited for their resources in the future.

17.2 MAIN BELT ASTEROID SCIENCE STATE-OF-THE-ART

Looking back at the progress of Main Belt asteroid science reveals the scientists' constant struggle to make sense of new information. An early theory proposed by Wilhelm Olbers in 1802 envisioned that asteroids were formed by the explosion of a larger planet (e.g., see Cunningham & Orchiston, 2013). This view quickly became obsolete when more asteroids were discovered to span a wide range of heliocentric distances and orbital inclinations. In addition, the cumulative mass of the Main Belt is very low compared to nearby Mars and Jupiter, implying a discontinuity in the distribution of material in the proto-solar nebula (Kuiper, 1956). The broad orbital distribution of asteroidal eccentricities and inclinations were explained via self-stirring by sizable planetary embryos embedded in the Main Belt, which would have subsequently been removed together with 99.9% of the original Main Belt mass (Wetherill, 1992). Or, perhaps, the Main Belt was never so massive, and the asteroid's orbital stirring was caused by external processes, such as dynamical perturbations from nearby Mars and Jupiter (Ward, 1981).

To make the picture even fuzzier, spectroscopic observations of asteroids and meteoritical analyses indicate a large compositional variability in the Main Belt (see Chapter 1). Yet, isotopic data from meteorites favor the existence of two non-contiguous regions of asteroid formation, arguably the terrestrial planet region and the Jupiter–Saturn region (see Chapter 14). Recent dynamical models predict that the Main Belt could have started off (nearly) empty, and was populated at a later time due to planetesimal–planet orbital scattering (see Chapter 15). In this view, the Main Belt is a receptor of planetesimals from a wide range of heliocentric distances.

To examine this idea more closely, Figure 17.1 shows the distribution of Main Belt asteroids larger than 100 km (diameter), and their spectral types. These objects are considered to be primordial, that is, planetesimals that are not fragments of larger parent bodies and therefore better represent their distribution in the Main Belt, while smaller asteroids are thought to be fragments of collisional evolution of larger parent bodies (Bottke et al., 2005; Morbidelli et al., 2009). The spectral types of asteroids are proxy for their compositions, as inferred by comparison with meteorites (see Chapter 1). The figure shows that pristine, volatile-rich asteroids represented by the B, C, and D types are preferentially found in the outer Main Belt, leading to the idea of a gradual increase of volatiles (and water) with heliocentric distance. At a closer look,

262

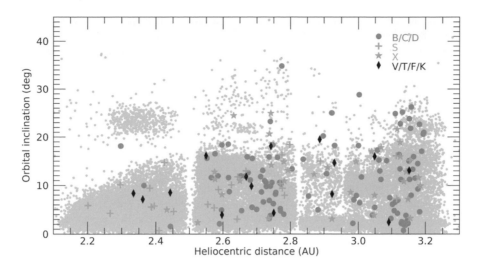

Figure 17.1 Distribution of inclination versus heliocentric distance (proper semi-major axis) for Main Belt asteroids (small light gray dots). Asteroids larger than 100 km in diameter are indicated by their spectral types: B/C/D represents volatile-rich objects, S indicates volatile-poor objects, X are not well characterized, and could range from metal-rich to more pristine assemblages; V/T/F/K are peculiar compositions typically interpreted as highly evolved (e.g., Vesta is a V-type).

however, several notable outliers are evident: a few large pristine asteroids (e.g., Nemausa, Fortuna, Zelinda) are found in the inner Main Belt amidst volatile-poor asteroids (e.g., Vesta), and vice versa in the outer Main Belt. Scattered throughout the Main Belt, elusive X-type asteroids have been linked to both highly evolved bodies (e.g., rich in metal such as Psyche) to primitive objects (similar to CM carbonaceous meteorites; Fornasier et al., 2011). It is difficult to imagine that asteroids with such wildly different compositions may have formed close to each other as reflected in their current orbital arrangement. These observations seem to indicate, instead, that the Main Belt is the result of widespread mixing. Dawn's detailed observations at Vesta and Ceres have contributed to further refinements of these arguments and at the same time have unveiled unexpected details.

Vesta is volatile-poor and the howardite–eucrite–diogenite (HED) meteorites that show evidence of being sourced from Vesta constrain its bulk water content to have been a few percent at most (see Chapters 3 and 4). On the contrary, Ceres has a bulk water content of up to 30–40% (see Chapters 11 and 12). Thus, the two most massive objects in the Main Belt comprising about 45% of the total mass have water content that differs by an order of magnitude, yet their semi-major axes are separated by only about 0.4 AU (indeed, Vesta and Ceres orbits intersect each other). Such organization would require the "fine tuning" of the location of the water condensation line in the protoplanetary disk to fall between Vesta ($\sim$2.36 AU) and Ceres ($\sim$2.76 AU). This fine tuning, however, is contrary to our understanding of how the protoplanetary disk evolved. For instance, the disk radial temperature profile is expected to dramatically migrate radially during the first Myrs of evolution due to viscous dissipation (e.g., Bitsch et al., 2015). Therefore, it is unclear how disk evolution could have resulted in a sharp compositional difference between Vesta and Ceres. In addition, ammonia may have been an important raw ingredient of the materials from which Ceres formed (see Chapter 9). If one assumes that the condensation of super volatile ammonia hydrate ($\sim$80 K) occurred at Ceres' present heliocentric distance, then the entire Main Belt would have been within the water condensation zone ($\sim$180 K), contrary to the presence of many volatile-poor asteroids.

The Dawn mission showed beyond any reasonable doubt that most (if not all) of the HED meteorites originate from Vesta based on its surface composition (see Chapter 6), and geomorphology (see Chapter 5), including the presence of two large impact basins near the south pole (see Chapter 16). The Vesta–HED connection is of particular importance for our understanding of the early Solar System evolution, as it allowed a full closure of modeling, remote observations, and sample analysis, which has been previously possible only for the Moon and Mars (e.g., Marchi et al., 2013).

Collectively, these arguments suggest that the majority of large Main Belt asteroids, including Vesta and Ceres, formed elsewhere in the Solar System. If so, where did they form?

Dynamical models (see Chapter 15) show that there are multiple pathways by which planetesimals from the terrestrial planet regions (1–2 AU) and the outer Solar System (>5 AU) could end up in the Main Belt. This double implantation could also explain isotopic evidence for two distinct reservoirs (inner versus outer Solar System), potentially due to a barrier resulting from the formation of Jupiter that would have temporarily limited exchange of materials between the two reservoirs (see Chapter 14). While this view of a divide in the formation zones of planetesimals and their subsequent radial mixing is rapidly gaining consensus, open issues remain. Consider the following examples.

Elemental and isotopic data from HEDs seem to require that Vesta accreted from material that was ¾ H chondrites and ¼ CM chondrites, within 1–2 Myr after Solar System formation (see Chapter 4). Interestingly, H and CM chondrites belong to the two distinct meteorite reservoirs based on isotopic data, implying that mixing of planetesimals between the two reservoirs occurred very early in the Solar System. However, CM meteorites are thought to have accreted between 2.5 and 3.5 Myr (Kruijer et al., 2020).

Ceres could have accreted beyond Jupiter to account for its ammonia and high carbon content (see Chapters 8 and 9). It is therefore possible that Ceres accreted in the same reservoir as other carbonaceous chondrites, but if so, it is unclear why these meteorites have lower concentrations of ammonia and carbon than Ceres. Another interesting aspect is how Ceres grew to its current size. Formation models envision planetesimals formed due to the collapse of pebble clouds with a final preferred size of ~100 km (see Chapter 13). In this scenario, subsequent growth via planetesimal collisions or pebble accretion is required. It could be that this second stage of accretion is responsible for Ceres's compositional differences from carbonaceous chondrites. Ceres could be one of many 1,000-km objects that formed in this way and ended up in the Main Belt, while its siblings remain in the outer Solar System.

There is also another fundamental consequence of the different bulk volatile content between Vesta and Ceres. The presence of water, in particular, has the capability to limit the rise of internal temperature powered primarily by ^{26}Al, ^{60}Fe radioactive decay. As a result, Vesta reached internal temperatures that were high enough for metal–silicate differentiation (see Chapter 4), while Ceres' internal temperatures remained low enough to stunt large-scale metal–silicate differentiation (see Chapter 11). As a result of this radically different evolution these objects evolved on divergent paths, as inferred by their internal structure (see Chapter 12), surface composition (see Chapters 6 and 7), and geomorphology (see Chapters 5 and 10). Still, there remain many uncertainties to precisely understand the chemical pathways that led to Ceres' complex surface composition, and its possible recent geological activity (Raymond et al., 2020).

17.3 LOOKING AHEAD

The twenty-first century renaissance of Main Belt asteroid science reflects the fact that asteroids hold crucial and unique clues about the formation and evolution of our Solar System. But as in an unfinished painting of Renaissance master Leonardo da Vinci, there is more left for us to imagine, explore, and eventually discover. Planned missions such as NASA's *Lucy* and *Psyche* (scheduled to launch in 2021 and 2022) will surely provide additional colorful strokes to our ever-evolving portrait of the Main Belt. *Psyche* will expand Main Belt asteroid exploration to the uncharted territory of metal-rich objects, while *Lucy* will study unexplored Trojan asteroids, a class of objects orbiting the Sun in Jupiter's stable Lagrangian points. Both missions will provide new detailed information about the formation of the Solar System, including asteroid migration and implantation processes that could have been important for the Main Belt.

The *Psyche* and *Lucy* missions will hopefully pave the way for a new generation of bold Main Belt-oriented missions. These missions should not be limited to remotely observing asteroids (flybys or rendezvous) but should also sample their surfaces for in-situ chemical and mineralogical analysis or bring those samples back to Earth. It is generally considered that the ensemble of known meteorites are samples of roughly 80–100 distinct parent bodies (Greenwood et al., 2020), a number that is 50% of primordial asteroids in the Main Belt. This suggests that our understanding of the building blocks of the Solar System is limited and potentially strongly biased by the few meteorites in our labs. Synergy between spacecraft exploration of large planetesimals, analysis of returned samples, and advances in measuring isotopic characteristics of individual components of meteorite samples promises to accelerate our understanding of the many clues held within the Main Belt asteroids that still await discovery.

As this book goes to press, the JAXA *Hayabusa-2* mission has returned to Earth samples from the small near-Earth asteroid Ryugu, and NASA's *OSIRIS-Rex* is heading back to Earth with its precious cargo of samples from Bennu. These small asteroids originate from the Main Belt, as likely fragments of larger parent bodies. We see these ambitious missions as precursors of similar efforts to land on and sample large, Main Belt asteroids. It is the authors' belief that only with these future investigations will the portrait of the Main Belt be, hopefully, completed.

REFERENCES

Arnold, J. R. (1965) The origin of meteorites as small bodies. III. General considerations. *Astrophysical Journal*, 141, 1548.

Belton, M. J. S., Veverka, J., Thomas, P., et al. (1992) Galileo encounter with 951 Gaspra: First pictures of an asteroid. *Science*, 257, 1647–1652.

Bitsch, B., Johansen, A., Lambrechts, M., & Morbidelli, A. (2015) The structure of protoplanetary discs around evolving young stars. *Astronomy & Astrophysics*, 575, A28.

Bottke, W. F., Durda, D. D., Nesvorný, D., et al. (2005) Linking the collisional history of the main asteroid belt to its dynamical excitation and depletion. *Icarus*, 179, 63–94.

Cunningham, C. J., & Orchiston, W. (2013) Olbers's planetary explosion hypothesis: Genesis and early nineteenth-century interpretations. *Journal for the History of Astronomy*, 44, 187–205.

Fornasier, S., Clark, B. E., & Dotto, E. (2011) Spectroscopic survey of X-type asteroids. *Icarus*, 214, 131–146.

Greenwood, R. C., Burbine, T. H., & Franchi, I. A. (2020) Linking asteroids and meteorites to the primordial planetesimal population. *Geochimica et Cosmochimica Acta*, 277, 377–406.

Kruijer, T. S., Kleine, T., & Borg, L. E. (2020) The great isotopic dichotomy of the early Solar System. *Nature Astronomy*, 4, 32–40.

Kuiper, G. P. (1956) The formation of the planets, part III. *Journal of the Royal Astronomical Society of Canada*, 50, 158.

Marchi, S., Bottke, W. F., Cohen, B. A., et al. (2013) High-velocity collisions from the lunar cataclysm recorded in asteroidal meteorites. *Nature Geoscience*, 6, 303–307.

McCord, T. B., Adams, J. B., & Johnson, T. V. (1970) Asteroid Vesta: Spectral reflectivity and compositional implications. *Science*, 168, 1445–1447.

Morbidelli, A., Bottke, W. F., Nesvorný, D., & Levison, H. F. (2009) Asteroids were born big. *Icarus*, 204, 558–573.

Raymond, C. A., Ermakov, A. I., Castillo-Rogez, J. C., et al. (2020) Impact-driven mobilization of deep crustal brines on dwarf planet Ceres. *Nature Astronomy*, 4, 741–747.

Ward, W. R. (1981) Solar nebula dispersal and the stability of the planetary system: I. Scanning secular resonance theory. *Icarus*, 47, 234–264.

Wetherill, G. W. (1967) Collisions in the asteroid belt. *Journal of Geophysical Reserch*, 72, 2429–2444.

Wetherill, G. W. (1992) An alternative model for the formation of the asteroids. *Icarus*, 100, 307–325.

INDEX